CONODONTS:
Investigative Techniques and Applications

BRITISH MICROPALAEONTOLOGICAL SOCIETY SERIES

This series, published by Ellis Horwood Limited for the British Micropalaeontological Society, aims to gather together knowledge of a particular faunal group for specialist and non-specialist geologists alike. The original series of Stratigraphic Atlas or Index volumes ultimately will cover all groups and will describe and illustrate the common elements of the microfauna through time (whether index or long-ranging species) thus enabling the reader to identify characteristic species in addition to those of restricted stratigraphic range. The series has now been enlarged to include the reports of conferences, organized by the Society, and collected essays on specialist themes.

The synthesis of knowledge presented in the series will reveal its strengths and prove its usefulness to the practising micropalaeontologist, and to those teaching and learning the subject. By identifying some of the gaps in this knowledge, the series will, it is believed, promote and stimulate further active research and investigation.

STRATIGRAPHICAL ATLAS OF FOSSIL FORAMINIFERA
Editors: D. G. JENKINS, The Open University, and J. W. MURRAY, Professor of Geology, University of Exeter
MICROFOSSILS FROM RECENT AND FOSSIL SHELF SEAS
Editors: J. W. NEALE, Professor of Micropalaeontology, University of Hull, and M. D. BRASIER, Lecturer in Geology, University of Hull
FOSSIL AND RECENT OSTRACODS
Editors: R. H. BATE, Stratigraphic Services International, Guildford, E. ROBINSON, Department of Geology, University College London, and L. SHEPPARD, Stratigraphic Services International, Guildford
A STRATIGRAPHICAL INDEX OF CALCAREOUS NANNOFOSSILS
Editor: A. R. LORD, Department of Geology, University College London
A STRATIGRAPHICAL INDEX OF CONODONTS
Editors: A. C. HIGGINS, Geological Survey of Canada, Calgary, and R. L. AUSTIN, Department of Geology, University of Southampton
CONODONTS: Investigative Techniques and Applications
Editor: R. L. AUSTIN, Department of Geology, University of Southampton
PALAEOBIOLOGY OF CONODONTS
Editor: R. J. ALDRIDGE, Department of Geology, University of Nottingham
MICROPALAEONTOLOGY OF CARBONATE ENVIRONMENTS
Editor: M. B. HART, Professor of Micropalaeontology and Head of Department of Geological Studies, Plymouth Polytechnic

ELLIS HORWOOD SERIES IN GEOLOGY

Editors: D. T. DONOVAN, Professor of Geology, University College London and J. W. MURRAY, Professor of Geology, University of Exeter

This series aims to build up a library of books on geology which will include student texts and also more advanced works of interest to professional geologists and to industry. The series will include translation of important books recently published in Europe, and also books specially commissioned.

A GUIDE TO CLASSIFICATION IN GEOLOGY
J. W. MURRAY, Professor of Geology, University of Exeter
THE CENOZOIC ERA: Tertiary and Quaternary
C. POMEROL, Professor, University of Paris VI
Translated by D. W. HUMPHRIES, Department of Geology, University of Sheffield, and E. E. HUMPHRIES
Edited by Professor D. CURRY and D. T. DONOVAN, University College London
INTRODUCTION TO PALAEOBIOLOGY: GENERAL PALAEONTOLOGY
B. ZIEGLER, Professor of Geology and Palaeontology, University of Stuttgart, and Director of the State Museum for Natural Science, Stuttgart
FAULT AND FOLD TECTONICS
W. JAROSZEWSKI, Faculty of Geology, University of Warsaw
RADIOACTIVITY IN GEOLOGY: Principles and Applications
E. M. DURRANCE, Department of Geology, University of Exeter

ELLIS HORWOOD SERIES IN APPLIED GEOLOGY

The books listed below are motivated by the up-to-date applications of geology to a wide range of industrial and environmental factors: they are practical, for use by the professional and practising geologist or engineer, for use in the field, for study, and for reference.

A GUIDE TO PUMPING TESTS
F. C. BRASSINGTON, Principal Hydrogeologist, North West Water Authority
QUATERNARY GEOLOGY: Processes and Products
JOHN A. CATT, Rothamsted Experimental Station, Harpenden, UK
PRACTICAL PEDOLOGY: Manual of Soil Formation, Description and Mapping
S. G. McRAE and C. P. BURNHAM, Department of Environmental Studies and Countryside Planning, Wye College (University of London)

CONODONTS:
Investigative Techniques and Applications

Editor:
RONALD L. AUSTIN, B.Sc., Ph.D.
Senior Lecturer
University of Southampton

Published by
ELLIS HORWOOD LIMITED
Publishers · Chichester

for
THE BRITISH MICROPALAEONTOLOGICAL SOCIETY

First published in 1987
ELLIS HORWOOD LIMITED
Market Cross House, Cooper Street,
Chichester, West Sussex, PO19 1EB, England
*The publisher's colophon is reproduced from James
Gillison's drawing of the ancient Market Cross, Chichester.*

Distributors:

Australia and New Zealand:
JACARANDA WILEY LIMITED
GPO Box 859, Brisbane, Queensland 4001,
Australia

Canada:
JOHN WILEY & SONS CANADA LIMITED
22 Worcester Road, Rexdale, Ontario, Canada

Europe and Africa:
JOHN WILEY & SONS LIMITED
Baffins Lane, Chichester, West Sussex, England

North and South America and the rest of the world:
Halsted Press: a division of
JOHN WILEY & SONS
605 Third Avenue, New York, NY 10158, USA

© **1987 The British Micropalaeontological Society/
Ellis Horwood Limited**

British Library Cataloguing in Publication Data
Conodonts: investigative techniques and applications. —
(British Micropalaeontological Society series)
1. Conodonts
I. Austin, R. L.   II. Series
562′.2   QE899

Library of Congress Card No. 86–20061

ISBN 0–85312–907–X (Ellis Horwood Limited)
ISBN 0–470–20697–7 (Halsted Press)

Phototypeset in Times by Ellis Horwood Limited
Printed in Great Britain by Butler & Tanner, Frome,
Somerset

Contents

Contributors

Péter Árkai,
Laboratory for Geochemical Research, Hungarian Academy of Sciences, H-1112 Budapest, Budaörsi u. 45, Hungary.

Howard A. Armstrong,
Department of Geology, The University, Newcastle upon Tyne NE1 7RU, England, UK.

Ronald L. Austin,
Department of Geology, The University, Southampton SO9 5NH, England, UK.

Matthew J. Avcin,
517 Ridge Avenue, Verona, PA 15147, USA.

Christopher R. Barnes,
Centre for Earth Resources Research, Department of Earth Sciences, Memorial University of Newfoundland, St. John's, Newfoundland A1B 3X5, Canada.

James E. Barrick,
Department of Geosciences, Texas Tech University, Box 4109, Lubbock, TX 79409, USA.

Zdzislaw Belka,
Institute of Geology, Warsaw University, Al Żwirkii Wigury 98, 02-089 Warsaw, Poland.

Stig M. Bergström,
Department of Geology and Mineralogy, The Ohio State University, Columbus, OH 43210, USA.

Richard D. Burnett,
Robertson Research International Ltd, Llandudno, Gwynedd LL30 1SA, Wales, UK.

Timothy R. Carr,
Exploration and Product Research, ARCO Oil and Gas Company, 2300 W. Plano Parkway, Plano, TX 75075, USA.

Kenneth J. Dorning,
Pallab Research, 58 Robertson Road, Sheffield S6 5DX, England, UK.

Samuel P. Ellison, Jr.,
Department of Geological Sciences, The University of Texas at Austin, Austin, TX 78713, USA.

Doris Fredholm,
Geological Institutionen, Lunds Universitet, Sölvegatan 13, S-223, 62 Lund, Sweden.

James E. Geitgey,
ARCO Exploration and Technology, 2300 W. Plano Parkway, Plano, TX 75023, USA.

William T. Holser,
Department of Geology, Arizona State University, Tempe, AZ 85287, USA.

Mario A. Hünicken,
Academia Nacional de Ciencias, C.C. 36, 5000 Córdoba, Argentina.

Noel P. James,
Centre for Earth Resources Research, Department of Earth Sciences, Memorial University of
    Newfoundland, St. John's, Newfoundland A1B 3X5, Canada.

Lennart Jeppsson,
Department of Geology, University of Lund, Sölvegatan 13, S-223, 62 Lund, Sweden.

Sandor Kovács,
Hungarian Geological Survey, H-1143 Budapest, Népstadion u.14, Hungary.

Maurits Lindström,
Geologiska Institute, Stockholms Universitet, S-10691, Stockholm, Sweden.

Norman MacLeod,
Programs in Geosciences, The University of Texas at Dallas, P.O. Box 830688, Richardson, TX
    75080, USA.

Hanna Matyja,
Instytut Nauk Geologicznycl and Polska Akademia Nauk, Zwirki i Wigbury 93, 02-089, Warsaw,
    Poland

Dagmar Merino,
Centro de Tecnología Petrolera, YPFB, Casilla 727, Santa Cruz, Bolivia.

Glen K. Merrill,
Department of Natural Sciences, University of Houston — Downtown, One Main Street,
    Houston, TX 77002, USA.

Karol Miaskiewicz,
Department of Chemistry, Warsaw University, ul. Pasteura 1, Warsaw, Poland.

James F. Miller,
Department of Geosciences, Southwest Missouri State University, Springfield, MO 65804, USA.

Michael A. Murphy,
Department of Earth Sciences, University of California, Riverside CA 92521, USA.

Rodney D. Norby,
Illinois State Geological Survey, 615 E. Peabody Drive, Champaign, IL 61820, USA.

Godfrey S. Nowlan,
Geological Survey of Canada, 601, Booth Street, Ottawa, Ontario, K1A 0E8, Canada.

Felicity H. C. O'Brien,
Centre for Earth Resources Research, Department of Earth Sciences, Memorial University of Newfoundland, St. John's, Newfoundland A1B 3X5, Canada.

Michael J. Orchard,
Geological Survey of Canada, 100 West Pender, Vancouver V6B 1R8, Canada.

Gladys C. Ortega,
Cátedra de Paleontología, Facultad de Ciencias Exactas, Físicas y Naturales, Universidad Nacional de Córdoba, C.Correo 395, 5000, Córdoba, Argentina.

Susanne L. Pohler,
Centre for Earth Resources Research, Department of Earth Sciences, Memorial University of Newfoundland, St. John's, Newfoundland A1B 3X5, Canada.

Frank H. T. Rhodes,
300, Day Hall, Cornell University, Ithaca, NY 14853, USA.

Mario Suárez Riglos,
Centro de Tecnología Petrolera, YPFB, Casilla 727, Santa Cruz, Bolivia.

C. Christopher Ryley,
Centre for Earth Resources Research, Department of Earth Sciences, Memorial University of Newfoundland, St. John's, Newfoundland A1B 3 X5, Canada.

Scott M. Ritter,
Department of Geology and Geophysics, University of Wisconsin, Lewis G. Weeks Hall, 1215, W. Dayton Street, Madison, WI 53706, USA.
now at: T. Boone Pickens, Jr. School of Geology, Oklahoma State University, Stillwater, OK 74078, USA.

Robin Saunders,
Department of Geology, The University, Southampton SO9 5NH, England, UK.

M. Paul Smith,
Department of Geology, The University of Nottingham, University Park, Nottingham NG7 2RD, England, UK.
now at: Cambridge Arctic Shelf Programme, West Building, Gravel Hill, Huntingdon Road, Cambridge CB3 0DJ, England, UK.

Jeremy Stone,
Department of Geology, The University, Southampton SO9 5NH, England, UK.

Rosalind Strens,
Department of Geology, The University, Newcastle upon Tyne 7RU, England, UK.

Andrew Swift,
Department of Geology, University of Nottingham, University Park, Nottingham NG7 2RD, England, UK.

Emilio N. Vaccari,
Departamento de Geología, Facultad de Ciencias Exactas, Físicas y Naturales, Universidad
    Nacional de Córdoba, C. Correo 395, 5000 Córdoba, Argentina.

W. John Varker,
Department of Earth Sciences, The University, Leeds LS2 9JT, England, UK.

Beatriz Waisfeld,
Departamento de Geología, Facultad de Ciencias Exactas, Físicas y Naturales, Universidad
    Nacional de Córdoba, C. Correo 395, 5000 Córdoba, Argentina.

Karsten Weddige,
Institut für Paleontologie und historische Geologie, Richard-Wagner-Strasse 10, D-8000 München
    2, FRG.

Judith Wright,
Department of Geology, University of Oregon, Eugene, OR 97403, USA.
now at: Department of Geology, Arizona State University, Tempe, AZ 85287, USA.

Zhang Youqiu,
Department of Geological Exploration, The East China Petroleum Institute, Dongying, Chan-
    dong, China.

Willi Ziegler,
Senckenberg-Museum, Senckenberganlage 25, D-6000 Frankfurt am Main, FRG.

Tomasz Zydorowicz,
Institute of Geology, Warsaw University, Al Zwirki i Wigury 93, 02-089 Warsaw, Poland.

Preface

When the original organizing committee (Drs. R. J. Aldridge, R. L. Austin and A. C. Higgins) for ECOS IV met to discuss the meeting at Nottingham it was agreed that two aspects of conodont studies were timely for debate. The first, on the palaeobiology of conodonts, obviously was attractive in view of the then recently discovered fossilized remains of orgnisms containing conodonts within the outline of the body. The second, on investigative techniques and their applications, was appropriate because of the increased awareness both of the usefulness of conodonts in applied micropalaeontology and of the hazards to human health associated with techniques for the extraction and concentration of conodonts. Subsequently there was a satisfactory response to our call for papers and over a hundred persons from 29 countries attended the meeting during the 25–29 July 1985 when 62 contributions were presented. A selection of these is included in this book with a further selection in the companion book *Palaeobiology of Conodonts,* edited by R. J. Aldridge. During a portion of the Symposium, attention focussed on techniques, which generated much discussion. Consequently a number of further contributions were offered and some are included in this book. Techniques and their applications,

as presented in this book, cover seven broad categories as follows.

(1) Extraction and concentrations of conodonts (Chapters 1–7).

(2) Scanning electron microscopy and contact microradiography of conodont elements (Chapters 8 and 9).

(3) Micromorphics and the analysis of shape in conodonts (Chapter 10).

(4) Organic metamorphism of conodonts (Chapters 11–13).

(5) Other geochemical and physical analyses (Chapters 14–16).

(6) Biostratigraphical investigations (Chapters 4–7 and 16–19).

(7) Palaeoecological studies (Chapters 20–23).

The early conodont workers experimented and it is only relatively recently that emphasis again has been placed on experimentation to test popular beliefs regarding the theories of chemical and physical reactions. A review of investigative techniques is provided by Stone in Chapter 1. Chapter 2 covers recent developments in techniques to recover conodonts from a variety of rock types. It is interesting to note developments relating to the handling of large samples. Chapter 3, dealing with developments in conodont concentration techniques, includes important contributions on non-toxic techniques for separation and concentration of conodonts. These methods are alternatives to the use of heavy liquids such as bromoform and tetrabromethane, which are narcotics easily absorbed, suspected carcinogens and potentially cumulative metabolic poisons. Other novel ideas covered in this chapter include the use of coffee filters for separations, as alternatives to standard filter papers, and the technique of electrostatic picking of conodonts. Chert as a rock type has long been avoided in the search for conodonts, with the notable exception of the sporadic occurrences of conodonts on chert bedding-plane surfaces. Ellison illustrates examples of this latter type of occurrence. Recently a new technique has involved the use of hydrofluoric acid to dissolve chert. Excellent biostratigraphical information can be provided, as indicated by both Orchard and Barrick, who also detail procedures for processing samples. Shales represent another rock type which, because of the difficulty of disintegration, have largely been avoided by researchers. This has possibly led to some bias in the information available. The contribution by Hünicken and his co-workers provides information based in part on a study of shale sequences.

The application of scanning electron microscopy has opened up a whole new dimension for the study of conodonts. Burnett reviews this application and outlines a technique for the study of etched and polished surfaces of conodonts, whilst Smith outlines the merits of back-scattered imaging. Norby and Avcin discuss the use of contact microradiography for the study of conodont assemblages.

Macleod and Carr discuss at length techniques of morphometrics and the analysis of shape in conodonts.

Nowlan and Barnes review the many applications of colour alteration index (CAI) data with examples of recent investigations for hydrocarbon and mineral exploration, for regional tectonic interpretations and for the recognition of hot spots. Armstrong and Strens document colour changes in conodonts away from a basaltic dyke and correlate these with theoretically derived country-rock temperatures. Kovacs and Árkai relate conodont alteration in metamorphosed rocks from the Carpathians of northern Hungary, to carbonate texture, illite crystallinity and vitrinite reflectance. Belka and his colleagues describe the electron spin resonance (ESR) technique, a technique applicable for the dating of the completion of the last heating of conodont-bearing rocks. The technique, based on results from conodonts of Ordovician to Triassic age, suggests a correlation between the conodont CAI and the structure of the ESR spectra.

Conodonts contain a host of trace elements. Geitgey and Carr categorize geochemical studies of conodonts as follows.

(1) Strontium concentration and isotopic analysis, with variations in the $^{87}Sr/^{86}Sr$ ratio through time being useful for correlation.

(2) Radiometric dating using fission track dating, or U–Th–Pb dating.

(3) Neodymium isotope studies for palaeocirculation and palaeogeographical reconstructions.

(4) Rare-earth element analysis.

(5) Neutron activation analysis.

Their own contribution to this book relates to oxygen-isotope analysis of conodonts for determination of palaeotemperatures in the Ordovician–Pennsylvanian interval of time. Wright and her colleagues report a cerium anomaly in rare-earth patterns. They note that sections at the Cambrian–Ordovician boundary display correlative variations in the trace-element contents of apatite. This research, referred to as chemostratigraphy, provides a new method of stratigraphic correlation, which has interesting potential in poorly fossiliferous strata.

Murphy discusses the possibility of a Lower

Devonian equal-increment time scale based on lineages in Lower Devonian conodonts. To use conodonts as aids for the dating of rocks, it is essential to know the sequence of conodont faunas through each of the periods of geological time in which they are represented. The standard for reference is the stratotype section. In a selected example, Bergström and his colleagues present details of the conodont succession for the Llanvirn–Llandeilo and Llandeilo––Caradoc Series. The contribution of Riglos and his co-workers has been selected as an example to illustrate the application of conodonts to date rocks of previously unknown age from a part of the world where the conodont sequence has not as yet been investigated in detail.

Weddige and Ziegler present some ratios of conodont sample data as indicators of facies. Pohler and her colleagues demonstrate the use of conodonts to reconstruct a lost faunal realm. Matyja interprets the conodont record from Polish borehole material to recognize and relate sedimentary environments to conodont biofacies. Understanding of biofacies permits a more refined biostratigraphy, as illustrated by Ritter in his analysis of Early Permian conodont faunas from the central and western parts of the USA.

Manuscripts submitted for inclusion in this book were refereed. The editor thanks the referees and also acknowledges the assistance given by Mr. T. Clayton and Drs. I. W. Croudace, R. P. Foster, A. P. Gize and J. A. E. Marshall, Department of Geology, University of Southampton. The editor also wishes to thank all the contributors to this book. Drs. G. K. Merrill and L. Jeppsson have given considerable additional information and advice as to the content of Chapters 1–3. Dr. D. Moore has generously given his time to assist with editing. Mrs. Dawn Trenchard assisted with the indexing. Technical and clerical support has been provided by members of the Department of Geology, University of Southampton. The University of Nottingham kindly hosted the Symposium. Members of the Department of Geology, University of Nottingham, in particular R. J. Aldridge, M. P. Smith, M. Dean, A. Swift and S. J. Tull, are thanked for their assistance with the organization and smooth operation of the thematic section. The Symposium initiated by the Pander Society was generously supported by the Universities of Nottingham and Southampton, the British Council, the British Micropalaeontological Society, British Petroleum PLC, Britoil PLC, E. K. Hull Microslide Company, Ellis Horwood Limited, ERICO, the Midland Bank PLC and Stratigraphic Services International.

Last, but by no means least, the editor sincerely thanks James Gillison and the staff of Ellis Horwood for their constant advice during the production of this book.

Ronald L. Austin

1

Review of investigative techniques used in the study of conodonts

J. Stone

A review is made of the techniques currently employed by conodont workers with respect to the following: sampling; processing of limestones, shales, black shales, marls, ironstones and cherts; sieving; separation and concentration including electromagnetic, heavy liquid and interfacial techniques; picking; mounting; photography.

DISCLAIMER

Whilst every effort has been made to include safety advice where necessary, the author, editor and publisher accept no liability for injury or damage resulting from use of any of the techniques outlined. It is suggested that, before using any new technique or piece of equipment, the safety aspects should be thoroughly considered.

1.1 SAMPLING

On rare occasions, conodonts may be visible in the field and can be studied with a hand lens. For ease of study it may be advantageous to make a cast of the elements (Matthews, 1969; Ellison, this volume; Hünicken *et al.*, this volume). Any cast has to be taken from a mould, and there is a variety of materials available for making such moulds (Rixon, 1976, p. 193). Visible conodonts usually occur on

bedding planes where they may appear either as distinct apparatuses (Schmidt, 1934) or as discrete elements (Matthews, 1969). The majority of conodonts, however, are collected 'blind' in the field, and samples of the host-rock are returned to the laboratory for breakdown, which has the advantage over bedding-plane assemblages of allowing a three-dimensional study. Conodont biostratigraphy and palaeoecology have advanced to a stage where meaningful studies require exact sampling at small intervals, taking into account any change in lithofacies. When sampling in order to establish a biostratigraphy, care must be taken that exposures sampled are fully representative and also well correlated in order to provide adequate geographical and stratigraphical coverage (Collinson, 1965, pp. 97, 98). The nature of the sampling procedure is determined by the lithology, the purpose of the study, and the time and facilities available. Collinson (1965) outlines a three-phase operation: initial reconnaissance comprises continuous channel (i.e. bed-by-bed) sampling of the best exposed representative units, followed by resampling of zones of high abundance to investigate diversity of the fauna, and finally bulk resampling may occur where faunas are sparse but significant. Time efficiency must be weighed against representati-

It is important that sample sites are exactly located and accurately recorded. In some cases it may be advisable to mark sample sites with suitable weatherproof paint. Sample blocks or sample bags should be clearly marked in permanent ink as each sample is taken. Jones (1969, p. 9) suggests leaving the clearly labelled samples at their appropriate sites and photographing the whole locality. Some workers use a Polaroid camera to record each sample site and annotate the photograph in the field. At each site, notes should be made on the lithology and macrofauna, etc. (e.g. whether the microfauna is likely to be autochthonous or allochthonous).

Collinson (1965) points out that, because of the durability of conodonts, extra care must be taken to guard against contamination. Sample bags should never be re-used and should be well sealed and checked for rips or punctures. Tools should be cleaned before each sample is taken.

In the majority of cases, sample size should be about 2 kg, and because of sample size, a tempered steel chisel saves much time. It may be worthwhile to take a crowbar into the field. Saftey goggles are essential when collecting well-indurated rocks. After sampling, the site should be left clean and tidy without any litter or loose rock chippings.

1.2   PROCESSING OF ROCK TO YIELD CONODONTS

Conodonts may be found in almost any marine rocks (Collinson, 1965), from Cambrian to Triassic in age e.g. limestones, shales, black shales, ironstones, cherts and some marls and sandstones. Depending on the lithology, a variety of techniques may be used to extract the conodonts (Fig. 1.1).

(a) Limestones and carbonate-cemented rocks
Since Graves and Ellison (1941) illustrated the effectiveness of the acid-digest technique, limestones have come to be amongst the most commonly studied conodont-bearing lithologies (Lindström, 1964, p. 132; Collinson, 1965, p. 95). Collinson (1965) suggests collecting

samples of approximately 2 kg, although this can vary from 0.1 kg (Lindström, 1964, p. 131) to over 75 kg (Jeppsson, 1983). Depending on the purpose of the study and the lithology involved, it is often advisable to keep some of the sample back for reference, e.g. a thin section, or for probe work. In order to speed up the digestion process, samples are often crushed into 3 cm chips, although the mechanical stress may damage the elements, and some workers feel there is no long-term gain from crushing (Jeppsson and Fredholm, this volume), particularly with sparse or metamorphosed faunas (Dégardin, 1975).

The sample is dissolved in dilute acid in polyethylene buckets (Collinson, 1963); the three most commonly used acids have been acetic, monochloracetic and formic, and these must be diluted as they will not react strongly unless ionized by water. When using acetic acid at concentrations greater than 10–15%, a proportion of the acid is wasted (Rixon, 1976, p. 84). Formic acid is commonly used in concentrations as high as 45% (Merrill, pers. commun., 1985).

Müller (1962) suggests that, when dealing with muddy limestones, samples should be supported in the bucket in a plastic sieve, so as to prevent the insoluble mud from settling on the limestone and choking the reaction. Jeppsson (pers. commun., 1985) recommends supporting samples halfway up the bucket in a plastic bowl which has a number of 1–2 cm holes drilled in it. This will speed up the reaction (possibly because of improved circulation and the decrease in hydrostatic pressure which allows CO_2 to escape more easily) and, as most conodont elements drop through the holes with the mud–clay fraction, the elements are protected from mechanical damage.

The amount of rock per bucket may range from 0.5 kg (von Bitter, pers. commun., 1985) to 2.5 kg depending on the situation, although 1–2 kg is about the optimum (Merrill, pers. commun., 1985). It is good practice to use covers on the buckets as the effervescence due

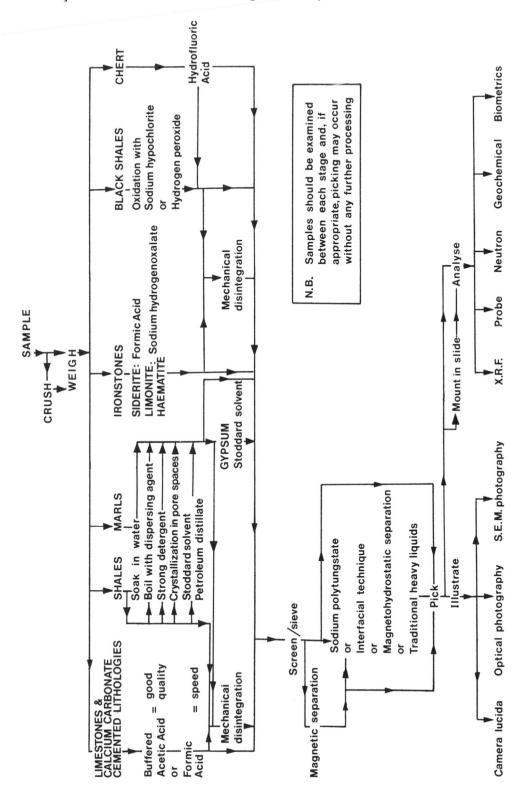

Fig. 1.1 — Flow diagram to illustrate the range of techniques used in processing samples for conodont studies: SEM, scanning electron microscopy; XRF, X-ray fluorescence. (After Merrill, pers. commun.).

to expelled CO_2 can be quite violent, particularly in rocks rich in carbonate.

The residue resulting from the digestion process is usually wet sieved into different size fractions (see Rixon, 1976). Yields in conodont-bearing limestones are commonly about 10–15 per kg (Collinson, 1965, p. 96) but may range from 0 (Nicoll, 1976) to 20 000 (Jeppsson, pers. commun., 1985).

A number of acids have been utilized for conodont studies (see Jeppsson, this volume). Acetic acid (ethanoic acid, CH_3COOH) was suggested by St. Clair (1935) as a means of obtaining insoluble residues from limestones and was used for conodont studies at least as early as 1941 (Graves and Ellison, 1941). Beckmann (1952) suggested the use of monochloracetic acid ($ClCH_2COOH$) as a faster and more efficient alternative when working with dolomites and muddy limestones. More recently, many workers have substituted formic acid (methanoic acid, $HCOOH$) which works more quickly than acetic acid and probably dissolves argillaceous limestones more effectively than either of the above (Ziegler et al., 1971). However, the physiological hazards associated with the use of formic acid are greater than with acetic acid, as formic acid liquid can cause bad skin burns, and the vapour can cause eye damage (Rixon, 1976, p. 111). Although unbuffered hydrochloric acid (HCl) is corrosive to conodonts, Dorning (this volume) outlines an integrated conodont and palynomorph extraction technique for carbonate rocks using HCl. Jeppsson et al. (1985) also report a controlled experiment involving dissolution of limestone using 37% (concentrated) HCl. Finally, although developed for extraction from siliceous rocks (see p. 24), the hydrofluoric acid (HF) method, in which the apatite component of the element becomes fluoridized, is potentially applicable to any rock type (Orchard, this volume). Unfortunately, the quality of the elements recovered is often rather poor.

A number of workers have subsequently studied the damage caused to elements by the acid used in the digestion. Ziegler et al. (1971)

highlight the dangers of leaving samples in monochloracetic acid for long periods of time and report a 95% increased yield for the same sample when sieved every 48 h as opposed to being left for 2 weeks. Even then, some elements showed signs of corrosion. Ziegler et al. also suggest that samples in formic acid should be sieved every 24 h to avoid damage to elements, although no quantitative data are presented.

Dégardin (1975) reports that, even with acetic acid, metamorphosed conodonts may be damaged if left for over 6 h. Jeppsson et al. (1985) illustrate that, at least in some cases, $\frac{1}{2}$ h in acetic acid is enough to cause damage. They suggest that it may be the production of calcium acetate and the change in pH, in the reaction between $CaCO_3$ and acetic acid, which prevents damage to the conodonts. Hence, until enough calcium acetate has been formed to buffer the solution, conodonts (particularly the smaller elements) will be attacked. If so, elements in rocks with comparatively low $CaCO_3$ content (e.g. calcareous shales or carbonate-cemented sandstones) are particularly at risk. To overcome this problem, Jeppsson et al. suggest buffering all samples by using an acid solution consisting of 7% concentrated acetic acid with 63% water and 30% of the filtered solution left after the digestion of previous samples.

HCl is known to dissolve conodont elements (von Bitter and Millar-Campbell, 1984; Jeppsson et al., 1985), and von Bitter and Millar-Campbell suggest that the free chloride present in some grades of acetic acid could ionize to form dilute HCl and thus could explain the destructive effects of acetic acid. However, Jeppssson et al. (1985) and Dorning (this volume) show that, when buffered (as above for acetic acid), HCl is non-destructive and can be used to extract conodonts. Mixtures of acetic acid and formic acid are highly corrosive to conodonts; therefore great care must be taken to wash samples thoroughly if they are being changed from one acid to another (Rixon, 1976, p. 111).

The damaging effects of acids on faunas are

variable. Since the work of Ziegler *et al.* (1971), use of monochloracetic acid has largely been abandoned (Jeppsson *et al.*, 1985). With acetic acid, faunas particularly at risk appear to be those which have undergone metamorphism (Dégardin, 1975) or those occurring in rocks comparatively poor in $CaCO_3$. The total time spent in acid before the solution is buffered by calcium acetate seems to be more critical than the concentration of acid used, and it is suggested that all samples should be buffered as proposed by Jeppsson *et al.*, (1985). Concentrations of acetic acid commonly used range from 15% to 8%. Although less work seems to have been done on the damaging effects of formic acid, Merrill (pers. commun., 1985) uses it at a concentration of 45% with no apparent signs of etching if the sample is sieved every 24 h.

Both Ziegler *et al.* (1971) and Jeppsson *et al.* (1985) suggest that some of the unbalanced, sparse or barren faunas reported in the literature may be reflections of the destructive powers of some acid digests.

When dealing with large volumes of rock, it would seem advantageous to automate the acidizing process (see Barnes, this volume; Jeppsson, this volume, and Zhang, this volume). With particularly difficult limestones or carbonate-cemented rocks, the speed and efficiency of the digest may be increased by the use of a Soxhlet extractor (Mann and Saunders, 1975, p. 37), although risk of acid damage to conodonts may be increased (Merrill, pers. commun., 1986). This is commercially available apparatus (see Fig. 1.2) which circulates continuously purified hot solvent (aqueous acetic acid) through the sample (limestone) and continuously removes all dissolved material which then crystallizes as calcium acetate in a separate flask. Using a standard 2 l capacity system, 500 g of $CaCO_3$ may be dissolved using 600 g of glacial acetic acid. The flask should never be heated by a direct flame or hot-plate; rather an electric heating mantle is recommended.

(b) Shales, black shales and marls

The breakdown potential of these rocks varies greatly, depending on the degree of metamorphism and induration which they have undergone and on their detailed lithology. For this reason, no standard method can be recommended for processing such rocks. Shales comparatively rich in organic matter may be X-rayed to establish the presence of conodonts prior to breakdown (Peters *et al.*, 1970; Norby and Avcin, this volume).

With the softer marls or calcareous shales, claystones and semiconsolidated materials, simply soaking in water for 1–2 h may cause sufficient disintegration for the residue to be poured through a sieve stack (see section 1.3). Gentle kneading with the fingers and use of a water jet with a 'rose' may also be required during sieving.

Hard clays and clay-cemented rocks which do not disintegrate on soaking, may respond to slow boiling for 1–2 h with a dispersing agent mixed in water. The boiling process should be slow to avoid unnecessary agitation and, in order to increase surface area avilable for reaction, the sample may first be crushed into 2–3 cm fragments. Following disintegration (e.g. after 2 h) the sample is poured through a sieve stack (see section 1.3) and may be washed with tap water followed by distilled water (Rixon, 1976, p. 118). A variety of dispersing agents have been suggested: sodium carbonate (washing soda, Na_2CO_3) is most commonly used, although it is highly corrosive to aluminium containers. Only small quantities are needed — a spoonful per reaction. Alternatives to sodium carbonate are sodium hydroxide (caustic soda, NaOH) and sodium hexametaphosphate (Calgon or Graham's salt $(NaPO_3)_6$). Sodium hydroxide may be used as a 20% solution (Pokorny, 1963, p. 10), and sodium hexametaphosphate should be added to water until the solution feels slippery when rubbed between the fingers. (Note that the partly dissolved material is dangerously sharp, and so care must be taken during stirring (see

Rixon, 1976, p. 116)). Any loss through evaporation can be replaced during boiling.

Zingula (1968) describes the use of Quaternary 'O' for disaggregating hard shales and argillaceous sandstones or limestones. This is a tertiary amine compound marketed in the USA by Ciba Geigy as a detergent with good wetting properties. A 20% solution is made with warm water and added to boiling water containing the sample. Using a cloth-based heating element or hot-plate, the mixture is kept boiling until the sample breaks down. Stirring may be necessary during the boiling to stop chippings from sticking to the bottom of the beaker and to combat froth build-up. To overcome the problem of smell, samples may be processed in a fume cupboard.

Jones (1969, p. 10) suggests using powerful detergents such as trisodium phosphate in a slightly different way. The dry crushed sample is heated in a bowl until quite hot and then one teaspoonful per pint of water detergent solution is suddenly added; this technique has the advantage of being quite rapid.

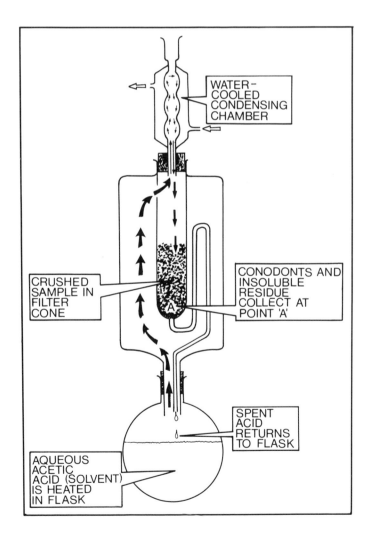

WATER-
COOLED
CONDENSING
CHAMBER

CRUSHED
SAMPLE IN
FILTER
CONE

CONODONTS AND
INSOLUBLE
RESIDUE
COLLECT AT
POINT 'A'

SPENT
ACID
RETURNS
TO FLASK

AQUEOUS
ACETIC
ACID (SOLVENT)
IS HEATED
IN FLASK

Fig. 1.2 — Stylized diagram of a Soxhlet extraction system.

Crystallization of a solution which has entered the pores and cracks in a rock can result in mechanical disintegration. Rixon (1976, p. 116) and Jones (1969, p. 11) suggest boiling the crushed sample in a concentrated solution of sodium thiosulphate (photographic hypo, $Na_2S_2O_3 \cdot 5H_2O$) until the rock fragments are saturated. The saturated chippings are then poured into a shallow dish and left to dry. As the solution evaporates, crystals of sodium thiosulphate grow in the rock, which disintegrates. Pokorny (1963, p. 9) suggests sodium sulphate (glauberite, $Na_2SO_4 \cdot 10H_2O$) as an alternative and notes the potential for automation when hundreds of samples are to be processed in a day (see Wicher, 1942). An even cheaper alternative is to use ice, by letting water fill the pore spaces and cracks in the rock and then freezing it. Pokorny (1963, p. 100) suggests covering the sample in water and applying a vacuum until no more bubbles escape, in order to expel all air from the pore spaces. The sample is then returned to atmospheric pressure and frozen. Alternatively, if no vacuum is available, the sample can be boiled and then frozen at $-30\,°C$. Jones (1969, p. 11) has also successfully disintegrated shales through rapid heating and cooling using a blow-torch, and through boiling under pressure in an autoclave. It seems feasible to split a shale by 'cooking' it in a microwave oven (Athersuch, Jeppsson and Merrill, pers. commun., 1985), but no particular method seems to be established. Perhaps some experimentation with regard to soaking, freezing, etc., prior to 'cooking' is indicated. Jeppsson (pers. commun., 1985) suggests a 'popcorn' method involving sealing of the pores with oil after soaking in water.

For indurated rocks not responding to the above expansion-contraction methods, a number of options are available. Calcareous shales may be treated with acetic or formic acid as described in section 1.2(a) for limestones, although digestion will probably be much slower.

Black shales often respond to oxidation of the organic matter they contain. Pokorny (1963) suggests crushing the sample and oxidizing it by soaking in a 10% to 15% solution of hydrogen peroxide (H_2O_2) in water. Boiling may occur in a matter of seconds or may take 10–15 min. and disintegration is often total. A drop of diluted ammonium hydroxide (NH_4OH) can increase the reaction dramatically; however, heat is produced and the reaction may be violent. Higgins (in Lindström 1964, p. 132) and Higgins and Spinner (1969) suggest a 15% solution of sodium hypochlorite (NaClO) in water as an alternative oxidizing medium. Sodium hypochlorite is contained in household bleach (4%–6%). Duffield and Warshauer (1979) suggest a technique involving oxidation followed by deflocculation of the remaining clastic particles with Quaternary O. A 1 Kg sample is crushed, placed in a plastic bucket and soaked in 3–4 l of household bleach for 2 weeks. 100 g of sodium hydroxide beads (NaOH) are added to facilitate the reaction. After breakdown the sample is sieved and the material not disintegrated is returned for further breakdown. After the final sieving the material not disintegrated is boiled in Quaternary O (see above) and sieved while still warm. Then and Dougherty (1983) describe a speedier adaptation of the above in which 150 g of the crushed sample is first dry sieved through a 10 mesh (2000 μm) sieve, then placed in a glass beaker and barely covered with 30% hydrogen peroxide. Effervescence may occur in a matter of seconds and water may be added to dampen violent reactions. After 1 h, or when bubbling has stopped, the sample is soaked in 300 ml of a 50% solution of water and household bleach, on an oscillating hot plate, for 1–2 h. To increase the reaction a 0.3 M base solution (10 g NaOH pellets to 750 ml of water) is substituted for the water. The sample is then wet sieved and soaked in a beaker of 20 ml stock Quarternary O (5% solution) and 300 ml water for a few hours. Wet sieving is then followed by oscillation of 1–2 h. Resistant samples are returned for further breakdown. (Note: hydrogen peroxide and sodium hydroxide can cause severe skin burns and rubber gloves, a lab coat and a face mask should be worn. Processing should be carried out in a fume cupboard as noxious fumes are produced when treating sulphides with bleach (Merrill 1980).

Lighter-coloured shales not responding to any of the above techniques may react to treatment with an organic solvent, e.g. dry-cleaning fluid (Stoddard solvent), petrol (gasoline in the USA; Layne, 1950), paraffin (kerosene in the USA; Varker, this volume), petroleum ether (Swift, this volume) or white spirit. Collinson (1963) describes a technique using Stoddard solvent (or any low-volatility low-flammability dry-cleaning fluid). The sample is thoroughly dried, e.g. in an oven at 150–200 °C (Mapes and Mapes, 1982), and is then allowed to soak in the fluid for approximately 2 h. The fluid is drained off and filtered for re-use. Hot water with detergent is added to the sample, which, after soaking, should disintegrate to a sludge which can be sieved. Any shale not broken down can be dried and treated again. If, after sieving, the residue consists of a large percentage of gypsum, Mapes and Mapes (1982) suggest a technique for its removal. The residue is totally redried, e.g. 24 h at 150–200 °C, and then resoaked in Stoddard solvent for about 1 h. The solvent is then decanted, and the residue soaked in water for no longer than 30 min, followed immediately by washing through appropriate sieves. Note that these researchers point out that the actual chemical function of the Stoddard solvent is unknown, although Solakius (1983) suggests that in the case of petrol (gasoline) the hot water causes the organic solvent in the pore spaces to expand, causing mechanical breakdown. Haynes (1981, p. 13) notes that the technique was originally discovered by one of the French oil companies drilling in the Sahara, where it was noticed that dark shales outcropping near the drilling platform disintegrated after being soaked in waste petroleum.

Jones, (1969, p. 11) suggests that black fissile shales may be examined for conodonts by breaking the sample into 8–10 cm blocks and then splitting each block into thin laminae using single-edged razor blades. These surfaces are then examined using a binocular microscope.

If all else fails, including use of HF (see section 1.2(d)), some recognizable conodont fragments may be recovered by simply crushing the shale quite finely and then using a separation method (see section 1.4).

(c) Ironstones or rocks cemented by iron minerals

In siderite-rich samples, formic acid may be used, as outlined on p. 18 for limestones (Merrill, pers. commun., 1985). For goethite (limonite)-rich samples, Freeman (1982) and Merrill (this volume) suggest use of sodium hydrogen oxalate solution, which is made up by mixing 100 parts of anhydrous oxalic acid with 93 parts of sodium bicarbonate in distilled (i.e. calcium-free) water. Samples are processed in quantities of approximately 10 g per each 150 ml of solution, by boiling for 1 h. Thioglycollic acid in 5% aqueous solution (Howie, 1974) may be used to digest goethite and haematite but it can dissolve conodonts and so should be buffered with 9 g of calcium phosphate (e.g. bone meal) per litre for pristine specimens. The major drawback to the use of thioglycollic acid is that of cost, and for goethite Freeman (1982) suggests that sodium hydrogen oxalate (see above) is superior. Pyrite may be oxidized to goethite (Merrill, 1980) by being immersed in a $7\frac{1}{2}\%$ aqueous solution of sodium hypochlorite and NaOH for 1–3 h (Merrill, pers. commun., 1985). Alternatively, pyrite may be heated until it can be removed magnetically, or conodonts may be 'floated off' from pyrite using methylene iodide in a density separation (Hamar, 1966) as outlined in section 1.4(b). Note that sodium hypochlorite is toxic and corrosive and releases chlorine when mixed with acids.

(d) Cherts (and siliceous rocks)

When processing siliceous rocks, a smaller sample (approximately 200 g) than for limestones or shales is usually taken. All carbonate should first be removed by one of the standard techniques outlined in section 1.2(a). This avoids the formation of insoluble fluorides during the HF treatment. Passagno and Newport (1972) outline the general procedure by which samples should be crushed, using a rock crusher or steel mortar and pestle, into 1–5 cm fragments. These fragments are placed in a plastic beaker of dilute HF in a fume cupboard. The beaker should never be more than half full to allow for substantial frothing during digestion. Relative concentrations of acid used vary from 5% or 6% (Orchard, this volume; Sevastopulo and Keegan, 1980) to 10% (Barrick, this volume), but it may be necessary to experiment in order to achieve the best results. For some rocks, stronger solutions and shorter digestion times (e.g. 50% HF for from 5 min to 1 h) may be more effective, although effervescence is more vigorous and may damage the conodonts (Sevastopulo and Keegan, 1980). Care must be taken when calculating the percentage of acid as it is usually supplied at a specified concentration, e.g. 60%. After 24 h the HF is decanted off and the sample is washed in dilute HCl, to drive off any remaining HF and to alter any remaining fluorides into soluble chlorides. The HCl is decanted off and replaced with water several times until the liquid remaining is neutralized. The sample can then be sieved, and any undigested material returned for further breakdown. Orchard (this volume) reports that no buffers are necessary during processing but that the samples should be sieved every 24 h. After each batch has been processed, the acid may be recovered and re-used, although it is safer and more straightforward if the acid is discarded. (Note that HF should be neutralized before disposal and the system used should be carefully considered with regard to local safety requirements.). Barrick (this volume) reports that abundances recovered are highest on days 2, 3 and 4 of the process and then fall off rapidly. Total abundances are often as high as thousands of elements per kilogram.

When treated with HF, the apatite component of the conodont element is fluoridized, resulting in a volume increase and consequent cracking and distortion of the element.

Although few good-quality elements are recovered, the fragments are often large enough to be identified.

If dissolution with HF alone proves slow or inefficient, or if speed is the overwhelming consideration, Magné and Dufaure (1964) describe a combined HF and hydrogen peroxide (H_2O_2) technique. 100 g of the crushed sample is placed in a 1 l plastic beaker, and 0.2 l of 43% H_2O_2 and 0.3 l of 'pure' HF (i.e. the strongest concentration available) is poured on. After 20–30 min the reaction goes strongly exothermic for a few minutes and then dies down. The liquid is decanted off and water is gently added until the liquid is neutralized. The sample is washed gently with water and sieved without rubbing. Merrill (pers. commun., 1985) has experience with the method and suggests sitting beakers in a polyethylene washing-up bowl, in the fume cupboard, to deal with any overflow. Fresh 50% H_2O_2 was diluted to the correct volume just prior to use and 52% HF was found to be the maximum concentration available. Results are varied, although in general the technique works well, and in some cases produces beautifully preserved conodonts. It appears that most of the dissolution actually takes place within the few tens of seconds when the reaction is at its most violent. Drawbacks to the technique are that, with some rocks, dissolution is only partial or slow and that, when the technique is carried out in small quantities, the cost is prohibitive, particularly with respect to providing 'fresh' supplies of H_2O_2. Nevertheless the technique is remarkably rapid, such that a sample may be received from the field, processed and a conodont age determination given, in as little as 1 h.

It is emphasised that HF is extremely dangerous and should only be used with a properly designed fume cupboard, protective clothing, rubber gloves and a face mask.

1.3 SIEVING

After the conodont-bearing rock has been broken down (as described in section 1.2), the conodont-bearing size fraction of the insoluble residue must be separated out. This is most commonly done by pouring the entire sample through a sieve stack (see Rixon, 1976, p. 126). Lids and bottom containers, supplied with the commercially available sieve stacks, are not used. A suitable range of mesh sizes is 200 (75 μm), 120 (125 μm), 60 (250 μm) and 25 (710 μm). Much of the fraction retained in the 25 (710 μm) mesh consists of undigested rock, which may be returned for further breakdown. The above size range is only a general guide and, when studying some of the larger elements (e.g. some of the Devonian pectiniform conodonts), a coarser sieve should be substituted for the 25 (710 μm) mesh. Water should be gently sprayed through the loaded sieve stack until no further fine or clay material is being washed out. Care must be taken not to damage the plumbing system by washing down the clay fraction. Use of a plastic 'rose' to break up the spray helps to overcome the problem of mechanical breakage of the conodonts.

If a sample is particularly clay-rich, the entire stack may be fully immersed and sieved prior to washing, or the sample may be subdivided, or the entire sample may be washed through a fine mesh of commercially produced cotton fabric (i.e. a mesh between 40 μm and 75 μm) prior to sieving, to eliminate the clay fraction.

After sieving, the residue from each sieve is washed into a labelled filter paper, allowed to drain and dried in a cool drying oven. Splitting of the conodont-bearing size range into two or more fractions makes it easier to achieve efficient separations (see section 1.4). Note that between each sample the sieves should be thoroughly cleaned to prevent any possibility of contamination. Methylene blue may be used to check cleanliness of the sieves.

1.4 SEPARATION AND CONCENTRATION (see also Fig. 1.1)

It is common for the dried residue from shale and limestone samples to consist of over 99% unwanted 'gangue', the nature of which will vary depending on the original rock. To facilitate accurate and untaxing picking of the conodont elements, they should be concentrated in some way; a number of methods are used. Chert residues are usually small enough for no concentration to be necessary.

(a) Electromagnetic separation (Dow, 1960, 1965)

As the magnetic susceptibility of conodonts (diamagnetic), quartz (diamagnetic) and dolomite (paramagnetic) is different, a two-phase separation is possible. The first phase separates out the ferromagnetic component (e.g. goethite, haematite) and the second phase separates the remaining less magnetic components.

Using either a Franz- or Cook-type electromagnetic separator which consists of a powerful adjustable electromagnet and a vibrating chute divided into two channels, an inward side slope of between 5° (Merrill, pers. commun., 1985) and 10° (Dow, 1960) is set, and the forward slope is set between 20° (Dow, 1960) and 30° (Merrill, pers. commun., 1985). The forward slope controls the speed of the particles, and experimentation may be necessary as a steeper slope is required when dealing with finer particles (Jeppsson, pers. commun., 1985). The washed and dried residue is fed into the hopper and allowed to run down the chute as a continuous single-file stream of particles. The vibrator should be on and initial amperage set between 0.6 (Merrill, pers. commun., 1985) and 1.0 (Dow, 1960). The 'non-magnetic' portion (inner channel) should contain all conodonts and the quartz–dolomite component, whereas the outer-channel portion should comprise the ferromagnetic component. If this is not so, the amperage should be adjusted empirically until the required separation is achieved (settings may vary from one machine to another). Note that clogging and false results may occur if the initial amperage, the speed of the particles down the chute or the feed rate of the sample is too high. At every stage of the separation, two or more runs may be necessary.

Settings for the second phase of the operation depend on the nature of the 'non-magnetic' portion. If a large proportion consists of conodonts, further separation may not be necessary. If dolomite makes up a large proportion, further digestion with buffered acetic acid (in extreme cases) or a second run with side slope $-2°$ and forward slope of $10°$ (Dow, 1960) should be made. Amperage should be set at a maximum and the vibrator should be on. At these settings, the dolomite (which is paramagnetic as it contains small amounts of iron) goes into the magnetic channel whereas conodonts (and quartz) go into the 'non-magnetic' channel. Separation of iron-free quartz from conodonts using the isodynamic separator is considered by many workers to be too delicate and intricate to be worthwhile. The same settings are used as for dolomite, with an amperage of approximately 0.9. As conodonts are slightly less diamagnetic than quartz, they go into the 'magnetic' channel.

Depending on the particular conditions involved with each different type of sample, slight variations on the above procedure may prove profitable. Varker (this volume) suggests a number of modifications for use with shale samples of low conodont concentration. A recently available addition to the standard magnetic separators (from GBL Electronics Inc., St. Bruno, Quebec) is an attachable 'vibrating track–feeder' which allows increased control over the flow rate.

(b) Heavy-liquid separation
The relative density of conodonts (specific gravity, 2.84–3.10) (Ellison, 1944) is greater than that of common gangue minerals, e.g. quartz (specific gravity, 2.65) and calcite (specific gravity, 2.715); hence one of the most commonly used separation techniques is density separation in a heavy liquid. The two liquids most commonly used are bromoform (tribromomethane, $CHBr_3$; specific gravity, 2.89) and tetrabromoethane (acetylene tetrabromide, $(CHBr_2)_2$; specific gravity, 2.96) and, in the less usual case of needing to separate conodonts from a heavier gangue, methylene iodide (diiodomethane,

CH_2I_2; specific gravity, 3.325) (Swift, this volume). Recently there has been an increased awareness of the dangers to health involved with the use of such heavy liquids (Brem et al., 1974; Sax, 1979 pp. 338, 434, 820; Hauff and Airey, 1980, pp. 2, 5, 8). Sodium polytungstate ($3Na_2WO_4 \cdot 9WO_3 \cdot H_2O$), a non-toxic alternative with a range in relative density from a specific gravity of 1.0 to 3.1 (manufacturer's values), is being introduced.

Great care should be taken when using bromoform or tetrabromoethane (Collinson, 1963, 1965), as both are easily absorbed systemic poisons and tetrabromoethane is a known carcinogen. The entire process should be carried out in an efficient fume cupboard using rubber gloves, protective clothing and a face mask. Firstly the specific gravity should be adjusted to 2.8 by diluting with ethyl alcohol or acetone, so that a ramiform conodont element will sink and calcite will float. The sample is added to the liquid in a separating funnel with an outlet large enough to prevent clogging and is stirred vigorously. The quantity of the sample should be small enough to allow free movement of all particles. The sample should be stirred at $\frac{1}{2}$ h intervals, or a continuous stirring mechanism may be used (Charlton, 1969; Davis and Webster, 1985). After at least 3 h, when no further settling-out is visible, the 'heavy' fraction containing the conodonts may be run off into a labelled filter paper in a funnel over a large recovery bottle. Because of the expense, as much heavy liquid as possible is recovered. Standard grade 4 filter paper is commonly used, although Barnes and O'Brien (this volume) suggest using coffee filters. After being allowed to drain, the filter paper and separated sample are thoroughly washed, either by using a squirter bottle of acetone or similar solvent (e.g. petrol–gasoline or alcohol) or by full immersion in a series of beakers of solvent (Sevastopulo, pers. commun., 1985). When using a squirter bottle, it is important to start washing at the edge of the filter paper.

The remaining 'light' fraction and heavy liquid is run off as above; any clogging may be overcome by forcing air through (Ryley, this volume) or by inverting the funnel and washing the residue out of the wider end using a wash bottle of solvent. The separating flask should be thoroughly washed out with solvent between each sample to avoid contamination. The washed filter papers may be left in a fume cupboard to dry. If any trace of the smell of the heavy liquid remains, the washing process must be repeated or the sample may be cleaned in a Soxhlet extractor (Fig. 1.2) using acetone as the solvent. (See also Freeman, 1975). Care must be taken throughout the above process because of the high degree of inflammability of the commonly used solvents.

Merrill and Ziegler (pers. commun., 1985) use a basic magnetic separation (see p. 26) as a matter of course before using heavy liquids — to increase efficiency of separation. Ziegler also suggests boiling samples in NaOH for 10–15 min before using heavy liquids as this cleans the samples and increases percentage recovery (as a result of decreased rafting of conodonts in the 'light' fraction).

A number of methods are used to reclaim the heavy liquid from the acetone–heavy-liquid washings. As outlined by Krumbein and Pettijohn (1938), Turner (1966), Benjamin (1971) and Hauff and Airey (1980), the solution may be mixed with water in excess, and then the water and its contained solvent can be continually removed (i.e. poured off) and replaced with fresh water until the proper density (specific gravity, 2.8) is achieved. This method involves loss of heavy liquid from 2% to 44%. More simply, the solvent can be allowed to evaporate from an evaporating dish placed in a fume-cupboard, until the correct density is reached. Hauff and Airey (1980, p. 16) outline a 'flash evaporation' method of recovering both the heavy liquid and the solvent. Mileson and Jeppsson (1983) suggest a more sophisticated development of this in which the solution is not heated. The mixture is placed in a flask connected to another flask twice its volume, which in turn is connected to a vacuum pump. The pressure is lowered until the mixture starts to boil, whereupon the system is sealed, the vacuum pump is disconnected and the larger flask is placed in a freezer at approximately $-25\,°C$. The solvent evaporates from the small flask (at room temperature) and condenses in the large flask. This method allows more complete recovery of the heavy liquid than 'excess-water' methods and does less damage to the heavy liquid.

Halogenated hydrocarbons (e.g. bromoform, tetrabromoethane) break down and darken in colour over time, particularly if exposed to heat, water and ultraviolet radiation (Hauff and Airey, 1980, p. 17). With repeated use the liquid may darken so much as to make it difficult to observe the effect of separation. von Bitter et al. (1978) compare several methods for decolourizing tetrabromoethane, with respect to effectiveness, cost and speed. They recommend that 250 ml of the discoloured liquid should be passed through a 5 cm × 60 cm glass column filled with 250 g of 30–60 Fuller's earth (attapulgus clay). In view of the recommendation made by Hanna (1927) that Fuller's earth be used for decolorizing bromoform, von Bitter et al. assume their results to be equally applicable to bromoform.

Sodium polytungstate is a recently introduced commerical compound marketed by the Metawo Company of Berlin, and separations are carried out using the same principle as with the traditional heavy liquids (see Merrill, this volume). Procedures differ slightly as the liquid is both non-toxic and water-soluble. Early reports are varied, some workers experiencing difficulties with evaporation, crystallization and viscosity and others finding it most satisfactory; Jeppsson (pers. commun., 1985) suggests that, in view of the health risks outlined above, both bromoform and tetrabromoethane should be abandoned in favour of sodium polytungstate.

(c) Interfacial methods of separation

The standard technique (Freeman, 1982; Merrill, 1985, this volume) relies on the relative hydrophobic qualities of calcium phosphate such that, unlike most gangue minerals, when immersed in a mixture of water and kerosene––paraffin (or any insoluble organic liquid), the calcium phosphate is preferentially wetted by the kerosene which forms a film enveloping the particle (e.g. conodont element). The particles and their 'envelopes' can then be removed by being made to adhere to a substrate which the kerosene softens (e.g. paraffin wax) so that the particle becomes attached, allowing substrate and particle to be removed together.

Freeman (1982) describes two variations of the above method using the same principles. One method substitutes an unfoamed polystyrene food container (yoghurt carton) as the substrate, and tetrachloroethylene as the insoluble organic liquid. The other method traps calcium phosphate in the interface between the water and the kerosene–paraffin.

The procedure for the standard technique was demonstrated at the ECOS IV meeting by Merrill and Freeman. A 1% detergent solution is made up (Fairy liquid in the UK or Dawn in the USA is suitable, but the choice of detergent is not critical), the detergent ensuring that the kerosene–paraffin disperses to form a suspension. The washed sample is poured into a screw-topped beaker and thoroughly wetted with kerosene, detergent solution is added to a depth of 2–3 cm and the beaker is briefly shaken. Paraffin wax flakes (1–2 cm long and 1–2 mm thick) are then added and the beaker is agitated by hand for 5 min. For ease of pouring off, the beaker is completely filled with detergent solution and tapped lightly to dislodge any gangue particles 'rafted' on the wax. The wax flakes are poured into a filter paper of the type designed for cleaning oil in deep-fat dryers and placed (in a funnel over a beaker) in an oven just hot enough to melt the wax, until all the wax has filtered through (this is then remelted and new wax flakes are made). The procedure is repeated another four times with the remainder of the sample. Any residual traces of paraffin wax remaining in the sample concentrate can be removed either by using a Soxhlet extractor (Merrill, pers. commun., 1985) with 60–80 petroleum ether as the solvent (see Fig. 1.2) or

by repeated washing with a dispersion of kerosene in hot aqueous detergent solution, followed by washing in the hot aqueous detergent solution, before air drying (Freeman, pers, commun., 1985). (Note that kerosene–paraffin is highly inflammable and should be used with care.)

The advantage of the above method is that it provides an alternative to the poisonous and carcinogenic heavy liquids. It is also cheap and requires no expensive equipment. There is much scope for experimentation and adaptation. However, although recovery can be nearly total (Merrill, pers. commun., 1985), many workers find that yields are often rather low.

(d) Magnetohydrostatic separation

Stone and Saunders (this volume) outline an application of work by Andres (1976) and Parsonage (1977, 1978) in which a high-density liquid is created by placing a paramagnetic solution in an inhomogeneous magnetic field. This method shows great potential for providing a safe, cheap, efficient and elegant density separation.

(e) High-frequency dielectrical separation

Following the criteria outlined by Wang Yu-Xian (1981) it should be possible to separate apatite (dielectric constant, 7.41–9.5) from quartz (dielectric constant, 4.27–4.34), although the dielectric constants of dolomite and calcite appear to be too close to that of apatite. An application of this method appears to offer much scope for experimentation, particularly when trying to separate materials of similar density and magnetic properties. Note that, for testing the efficiency of any extraction and concentration system, attention is drawn to the work of von Bitter and Millar-Campbell (1984) who introduce a known number of distinctive apatite grains (specific gravity, 3.1–3.2) to the sample prior to processing and then check the percentage recovery after processing.

1.5 PICKING AND MOUNTING

Even after judicious use of the processing and concentration techniques outlined above, the sample residue is likely to consist of a large proportion of gangue. A number of methods are used to pick out the conodonts. Most commonly, the residue is spread out in a home-made or commercial shallow picking tray, the bottom of which is marked in numbered squares. At no point should the sample layer be more than one particle thick. Using a binocular microscope, each square is examined in turn and all conodont fragments are carefully lifted out using a slightly wetted fine paint-brush, a bristle which is either dampened or slightly rubbed with wax (or see Barnes *et al.*, this volume, for an electrostatic method), or a mounted needle, the tip of which is finely covered with Plasticine (or similar proprietary modelling clay). Alternatively, Stinemeyer (1965) describes construction of a vacuum-needle picking device. The elements are transferred to a cavity slide (see below) and may at this stage be loosley grouped into pectiniforms, ramiforms or 'others' to await more detailed examination and classification. There is always a danger of dropping or damaging specimens *en route* between picking tray and storage slide, as this is usually done without the microscope; hence a little dexterity and concentration are required. In order for the whole process to occur under the microscope a 'holed' tray may be used in which the holes are positioned directly above a secured storage slide, and each element is 'posted' through. The latter technique is refined by the use of electrostatic picking.

If a sample is particularly poor in conodonts and pristine condition is not a necessity, a method outlined by Schopf and Simpson (1970) may be used to speed up the picking. A piece of filter paper is soaked in an ammonium molybdate solution (200 ml ammonium molybdate solution plus 20 ml of 10% HCl plus a few small crystals of stannous chloride) for 15 min. When it turns blue, the filter paper is removed and, whilst still damp, the sample residue is sprinkled over it and allowed to dry under a heat lamp for about 5 min. A white spot should occur surrounding every particle of phosphate in the sample. Formation of the spot depends on the formation of ammonium molybdophosphate, i.e. phosphate is removed from the conodont elements in the reaction; consequently, there is some etching and minor damage in the process.

Once the sample residue has been picked clean of all its conodonts, it can be bagged, labelled and put aside. The conodonts can be more thoroughly examined using the binocular microscope and identified. Proprietary cavity slides are commonly provided with 48, 60 or 100 divisions and may contain an auxiliary cavity for preliminary picking. Before use, the slides should be coated with a semipermanent glue which allows some repositioning when hydrated, such as gum arabicum which consists of 30 g of Arabic gum in 60 g of water plus 2 ml of glycerine and 5 ml of carbolic acid (Jeppsson, 1974, p. 5) or gum tragacanth. (Note the latter is carcinogenic and care must be taken not to lick the paint-brush during its use.) Single- or multiple-hole slides, for mounting type specimens for example, are also provided. The built-in sliding cover-slip of clear plastic which is often provided with proprietary slides may be replaced by a glass cover-slip if specimens cling to the plastic beacuse of static electricity (Jones, 1969, p. 16). Cullison (1934) describes construction of a slide in which a mirror makes up half the base, thus allowing simultaneous examination of both top and bottom sides of the conodont under the binocular microscope.

Some careful cleaning of the conodont elements is usually possible using a very fine paintbrush and much water, but the elements are very brittle, particularly those of high colour alteration index (CAI) and extreme care must be taken. Ultrasonic cleaning tends to fragment the conodonts (van den Boogard, pers. commun., 1985). Finally all the elements and their slides can be catalogued. An invaluable aid to identification is *Catalogue of Conodonts* edited by Ziegler (1973, 1975, 1977, 1981) and the

Revised treatise of conodonts (Robison, 1981).

1.6 PHOTOGRAPHY

Early workers and many current researchers produce illustrations of conodonts using light photography (see Ellison, this volume). However, transmission electron microscopy and more particularly scanning electron microscopy (SEM) have now come into routine use and have provided magnifications large enough to spark off a whole new area of conodont research, i.e. detailed studies of conodont ultrastructure (see Burnett and Smith, this volume). G. Jones (pers. commun., 1985) suggests taking stereo-pair photographs of a natural or artificial conodont mould and then displaying them reversing the stereopair; this will produce pseudoscopy and show the mould as a positive feature.

(a) Light photography

Photographs may be taken using a low-power photomicroscope, either as stereo-pairs for three-dimensional viewing (Evitt, 1949) or as single plates. Using a very small amount of gum arabicum (Jeppsson, 1974) or a similar glue, specimens may be mounted either directly on a dark background slide or on a small (1–3 cm^3) piece of glass and fixed to a universal stage (Jeppsson, pers. commun., 1985). The obvious advantage of the latter is the ease of adjustment of orientation of the specimens without risk of damage. Barnett (1970) describes how to construct a cheap Plexiglas 'orienting stage' which is less cumbersome than the universal stage and simple enough to be constructed in the workshop of most institutions.

Specimens being photographed to illustrate CAI (Epstein *et al.*, 1977) or distribution of white matter, etc., should be uncoated (Jeppsson, pers. commun., 1985). Black, transparent or particularly shiny specimens are usually lightly coated either with magnesium oxide dust (or 'smoke') or with ammonium chloride. For the former the specimen, mounted on its slide, is held above burning magnesium ribbon. In order to direct the coating an upturned (heat-proof!) funnel may be supported between the ribbon and the specimen (Merrill, pers. commun., 1985). The magnesium oxide 'dust' washes off with water. Dusting with ammonium chloride is perhaps more common but, since it is hygroscopic, the process should be carried out immediately prior to photographing in conditions of low humidity. Either it can be applied as a fine spray in liquid form, or a quantity of ammonium chloride crystals may be placed in a 'puffer' (Cooper, 1935) and heated until the smoke from the 'puffer' is blue, when the slide is held about 20 cm away and lightly coated. It is very difficult to get a fine even coating on complex surfaces, and any excess, which may result in frost-like spikes, should be cleaned off with alcohol, which does not affect the water-soluble glue (Sparling, pers. commun., 1985). Ammonium chloride should be carefully removed after photographing as it will eventually produce HCl which can damage the conodont (Jeppsson *et al.*, 1985).

For high-quality photographs, the specimen should be lit from two sides at 45° to the stage and the intensity of the lights balanced until a satisfactory image is obtained (Marsh, pers. commun., 1985). To avoid reflections from an uncoated specimen, either it may be immersed in a very small cup of alcohol mounted on the stage (Jeppsson, 1974), or diffuse lighting could be used. To produce the latter, either a 'ring lamp' or very bright normal lighting with a cylinder of white paper between the lights and the specimen can be used (Jeppsson, pers. commun., 1985).

When using a high magnification, a major limitation with the photomicroscope is the restricted depth of field obtainable. By closing down the aperture depth of field is increased but resolution is lost. Normally it is impossible to increase depth of field above about $\frac{1}{10}$ mm at 100 × magnification (Hay and Sandberg, 1967) Although light photography must be used for CAI illustrations, the combination of small overall size and high relief, particularly in the more ornate genera, causes many workers to feel that optical methods are not satisfactory for routine photography. Magnifications are inadequate for detailed studies of ultrastructure.

(b) Scanning electron microscope

Since the late 1960s (Hay and Sandberg, 1967; Honjo and Berggren, 1967), SEM photography has developed as the most accurate, efficient and frequently used means of illustrating microfossils for publication. The great advantage over light photography is the much greater depth of field available with high resolution, over a range of magnifications from 15× to 500000× (Hay and Sandberg, 1967).

Specimens to be photographed should be mounted on an SEM stub (usually an aluminium disc). A number of methods are employed for holding the specimen in place. To provide a smooth background, the surface of the stub may be covered with a thin layer of melted wax W. Either a portion of the wax is made tacky with a hot needle and the specimen is positioned, or the wax is left to harden and a small amount of water-soluble glue is used (e.g. gum tragacanth, gum arabicum) or a 'glue-stick', such as a UHU stick may be used (Uyeno, pers. commun., 1985). Alternatively the stub can be covered with a thin film of melted Lakeside 70 cement, which is soluble in alcohol, and is treated in the same way as wax W to affix the specimen (Hansen, 1968). With a smooth stub, specimens can be affixed using a thin coating of acrylic polymer varnish (e.g. shellac) in which the specimen is positioned while the surface is still just tacky (Britten, pers. commun., 1985). Merrill (pers. commun., 1985) suggests the use of double-sided sticky tape although this must be of a type which does not shrink during irradiation! Honjo and Berggren (1967) use Noeplen glue for mounting, and the Cambridge Scanning Company use Durofix glue (Hansen, 1968), but these glues are not easy to remove from the specimens.

Using the standard mode of operation, with an accelerating voltage of 20 kV, the surface of the specimen must be electrically conductive to avoid build up of charge. Hence the specimen is coated with an ultrathin layer of noble metal (e.g. gold, platinum, or a mixture of these), using either vacuum evaporation or sputtering techniques. An advantage of using only gold is that it can be removed afterwards (see below). Other substances may be used for coating, e.g. carbon (see below), silicon monoxide, aluminium and certain organic polymers (Merrill, pers, commun., 1985). Many non-conductive surfaces can be examined without coating if the accelerating voltage is greatly reduced, usually to between 1 kV and 3 kV (Hay and Sandberg, 1967), although some resolution is lost. Following the specific operating instructions, the stub is locked in the stage and is ready for viewing. The stage is capable of rotation and tilt around two axes as well as linear motion along x, y and z axes to enable orientation of the specimen. Stereoscopic pairs of photographs for true three-dimensional representation can be taken easily by tilting the specimen about one of the rotational axes (6° is standard for printed viewing).

A major disadvantage of the standard method outlined above, particularly with type specimens, is that, once covered in noble metal, the conodont element can no longer be examined 'as new' and discrimination between white matter and hyaline matter is impossible. To overcome this problem, Repetski and Brown (1982) outline a technique in which the specimen is coated with carbon using a vacuum chamber, and in which the gun potential of the scanning electron microscope is reduced from 20 kV to 10 kV. Although extremely high (greater than 4000×) magnifications are not possible, the details of the specimen are not obscured. Alternatively, Hansen (1968) describes a method of removing gold from plated specimens, in which the specimen is removed from the stub and placed in a small flask of an aqueous solution of 0.1–1.0% NaCN with one or two drops of 1 molar NaOH solution. Air is bubbled through the flask for about 15 min, in which time the gold should have been fully removed. Note that this entire process should be carried out in a fume hood as the CN compounds are highly poisonous.

The final prints may be trimmed, and the edges darkened using a felt pen and then set out on a suitable background (e.g. black card, or exposed photographic paper) with any numbers and labels added as required. Using standard light photography, 'plates' for publication can then be made.

1.7 ACKNOWLEDGEMENTS

I offer my sincere thanks to Dr. R. Austin, Dr. J. Hudec, Dr. L. Jeppsson and Dr. G. Merrill for their numerous suggestions, ideas and improvements. I am most grateful to Dr. D. Moore for editing the manuscript, and to everyone at ECOS IV with whom I discussed the various techniques. I also thank Mrs. A. Dunkley who drafted the figures.

1.8 REFERENCES

Andres, U. 1976, *Magnetohydrodynamic and Magnetohydrostatic Methods of Mineral Separation*, John Wiley, New York, 1–224.

Barnett, S. G. 1970. A new stage for orienting microfossils, *Journal of Paleontology*, **44**, 1133.

Beckmann, H. 1952. Zur Anwendung von Essigsäure in der Mikropaläontologie. *Paläontologische Zeitschrift*, **26**, 138.

Benjamin, R. E. K. 1971. Recovery of heavy liquids from dilute solutions. *American Mineralogist*, **56**, 613–619.

Brem, H., Stein, A. B. and Rosenkrantz, H. S. 1974. The mutagenicity and DNA-modifying effect of haloalkanes. *Cancer Research*, **34**, 2576–2579.

Charlton, D. S. 1969. An improved technique for heavy liquid separation of conodonts. *Journal of Paleontology*, **43**, 590–592.

Collinson, C. W. 1963. Collection and preparation of conodonts through mass production techniques. *Illinois State Geological Survey Circular*, **343**, 1–16.

Collinson, C. W. 1965. Conodonts. In B. Kummel and D. Raup (Eds.), *Handbook of Paleontological Techniques*, Freeman, San Francisco, California, 94–102.

Cooper, C. L. 1935. Ammonium chloride sublimate apparatus. *Journal of Paleontology*, **9**, 357–359.

Cullison, J. S. 1934. A suitable tray for comparative examination of minute opaque objects under the binocular microscope. *Journal of Paleontology*, **8**, 247.

Davis, L. and Webster, G. D. 1985. A modified funnel for heavy mineral separations. *Journal of Paleontology*, **59**, 1505–1506.

Dégardin, J. M. 1975. Method of extraction of Silurian conodonts from slightly metamorphosed limestones from the Eseva valley. *Geologica et Palaeontologica*, **9**, 61–63.

Dow, V. E. 1960. Magnetic separation of conodonts. *Journal of Paleontology*, **34**, 738–743.

Dow, V. E. 1965. Magnetic separation of conodonts. In B. Kummel and D. Raup (Eds.), *Handbook of Paleontological Techniques*, Freeman, San Francisco, California, 263–267.

Duffield, S. and Warshauer, S. M. 1979. Two step process for extraction of microfossils from indurated organic shales. *Journal of Paleontology*, **53**, 746–747.

Ellison, S. 1944. The composition of conodonts. *Journal of Paleontology*, **18**, 133–140.

Epstein, A. G., Epstein, J. B. and Harris, L. D. 1977. Conodont color alteration — an index to organic metamorphism. *U.S. Geological Survey Professional Paper*, **995**, 1–27.

Evitt, W. R. 1949. Stereophotography as a tool of the paleontologist. *Journal of Paleontology*, **23**, 566–570.

Freeman, E. F. 1975. The isolation and ecological implications of the microvertebrate fauna of a Lower Cretaceous lignite bed. *Proceedings of the Geological Association*, **86**, 307–312.

Freeman, E. F. 1982. Fossil bone recovery from sediment residues by the 'interfacial method'. *Palaeontology*, **25**, 471–484.

Graves, R. W. and Ellison, S. 1941. Ordovician conodonts of the Marathon Basin, Texas, *Missouri University, School of Mines and Metallurgy*, *Bulletin Technical Series*, **14** (2), 1–26, 3 plates.

Hamar, G. 1966. The Middle Ordovician of the Oslo Region, Norway, 22. Preliminary report on conodonts from the Oslo-Asker and Ringerike districts. *Norsk Geologisk Tidsskrift*, **46**, 27–83.

Hanna, M. A. 1927. Clarification of oil-discolored bromoform. *Journal of Paleontology*, **1**, 145.

Hansen, H. J. 1968. A technique for removing gold from plated calcareous microfossils. *Micropaleontology*, **14**, 499–500.

Hauff, P. L. and Airey J. 1980. The handling, hazards and maintenance of heavy liquids in the geologic laboratory. *US Geological Survey Circular*, **827**, 1–24.

Haynes, J. 1981. *Foraminifera*, Macmillan, London, 1–433, 15 plates.

Higgins, A. C. and Spinner, E. G. 1969. Techniques for the extraction of selected microfossils. *Geology*, 12–28.

Honjo, S. and Berggren, W. A. 1967. Scanning electron microscope studies of planktonic foraminifera. *Micropaleontology*, **13**, 393–406.

Howie, F. M. P. 1974. Introduction of thioglycollic acid in preparation of vertebrate fossils. *Curator*, **17**, 159–166.

Jeppsson, L. 1974. Aspects of Late Silurian conodonts. *Fossils and Strata*, **6**, 1–54, 12 plates.

Jeppsson, L. 1983. Silurian conodont faunas from Gotland. *Fossils and Strata*, **15**, 121–144.

Jeppsson, L., Fredholm, D. and Mattiasson, B. 1985. Acetic acid and phosphatic fossils — a warning. *Journal of Paleontology*, **59**, 952–956.

Jones, D. J. 1969. *Introduction to Microfossils*, Hafner, New York, 1–406.

Krumbein, W. C. and Pettijohn, F. J. 1938. *Manual of*

Sedimentary Petrography, Appleton–Century–Crofts, New York, 1–549.

Layne, N. M. 1950. A procedure for shale disintegration. *Micropaleontologist*, **4**, 21.

Lindström, M. 1964. *Conodonts*, Elsevier, Amsterdam, 1–196.

Magné, J. and Dufaure, G. 1964. Une méthode nouvelle pour l'extraction rapide des microfossiles. *Revue de Micropaléontologie*, **7**, 77–79.

Mann, F. G. and Saunders, B. C. 1975. *Practical Organic Chemistry* (4th edition), Longmans, London, 1–585.

Mapes, R. H. and Mapes, G. 1982. Removal of gypsum from microfossiliferous shales. *Micropaleontology*, **28**, 218–219.

Matthews, S. C. 1969. Two conodont faunas from the Lower Carboniferous of Chudleigh, S. Devon. *Palaeontology*, **12**, 276–280, 1 plate.

Merrill, G. K. 1980. Removal of pyrite from microfossil samples by means of sodium hypochlorite. *Journal of Paleontology*, **54**, 633–634.

Merrill, G. K. 1985. Interfacial alternatives to the use of dangerous heavy liquids in micropaleontology. *Journal of Paleontology*, **59**, 479–481.

Mileson, P. and Jeppsson, L. 1983. Heavy liquid and solvent recovery — an economical method. *Journal of Sedimentary Petrology*, **53**, 673–674.

Müller, K. J. 1962. Ein einfacher Behelf für die Lösungstechnik. *Paläontologische Zeitschrift*, **36**, 265–267.

Nicoll, R. S., 1976. The effect of Late Carboniferous–Early Permian glaciation on the distribution of conodonts in Australia. In C. R. Barnes (Ed.), *Conodont Paleoecology, Geological Association of Canada Special Paper*, **15**, 273–278.

Parsonage, P. 1977. Small-scale separations of minerals by use of paramagnetic liquids. *Transactions of the Institution of Mining and Metallurgy, Section B*, **86**, B43–B46.

Parsonage, P. 1978. Design and testing of paramagnetic liquid separation systems. *Report, Warren Spring Laboratory, Department of Industry, Gunnels Wood Road, Stevenage, Hertfordshire*, 1–38.

Passagno, E. A. and Newport, R. L. 1972. A technique for extracting radiolaria from radiolarian cherts. *Micropaleontology*, **18**, 231–234.

Peters, W. G., Wallace, R. J., Odom, I. E. and Frost, S. 1970. Application of radiography to the study of microfossils in black shales and other sediments. *Geological Society of America Abstracts with Programs*, **2** (6), 400–401.

Pokorny, V. 1963. In E. Ingerson (Ed.), *Principles of Zoological Micropalaeontology, International Series of Monographs on Earth Sciences*, Pergamon Press, Oxford, 1–652.

Repetski, J. E. and Brown, W. R. 1982. On illustrating conodont type-specimens using the scanning electron microscope — new techniques and a recommendation. *Journal of Paleontology*, **56**, 908–911.

Rixon, A. E. 1976. *Fossil Animal Remains, their Preparation and Conservation*, Athlone Press of the University of London, London, 1–304.

Robison, R. A. (Ed.) 1981. *Treatise on Invertebrate Paleontology, Part W, Miscellanea, Supplement 2, Conodonta*, Geological Society of America, Boulder, Colorado, and University of Kansas Press, Lawrence, Kansas, 1–202.

Sax, N. I. 1979. *Dangerous Properties of Industrial Materials* (5th edition), Van Nostrand Reinhold, New York, 1–1118.

Schmidt, H. 1934. Conodonten-Funde in Ursprunglichen Zusammenhang. *Paläontologische Zeitschrift*, **16**, 76–85, 6 plates.

Schopf, T. J. M. and Simpson, D. R. 1970. A method for finding conodonts in large nearly-barren samples. *Journal of Paleontology*, **44**, 164–165.

Sevastopulo, G. D. and Keegan, J. B. 1980. A technique for revealing the stereom structure of fossil crinoids. *Palaeontology*, **23**, 749–756.

Solakius, N. 1983. Use of the 'petrol method' for extracting foraminifera from various types of sedimentary rocks. In L. I. Costa (Ed.), *Palynology Micropalaeontology: Laboratories, Equipment and Methods*, Norwegian Petroleum Directorate Bulletin, **2**, 93–95.

St. Clair, D. W. 1935. The use of acetic acid to obtain insoluble residues. *Journal of Sedimentary Petrology*, **5**, 146–149.

Stinemayer, E. H. 1965. Microfossil vacuum-needle segregating pick. In B. Kummel and D. Raup (Eds.), *Handbook of Paleontological Techniques*, Freeman, San Francisco, California, 276–283.

Then, D. R. and Dougherty, B. J. 1983. A new procedure for extracting foraminifera from indurated organic shale. In *Current Research, Part B, Geological Survey of Canada*, Paper 83–1B, 413–414.

Turner, W. M. 1966. An improved method for recovery of bromoform used in mineral separation. *U.S. Geological Survey Professional Paper*, **550C**, 224–227.

von Bitter, P. H., Plint Geberl, H. A. and Miller, H. A. L. 1978. Decolorization of heavy Liquids — an economic assessment. *Canadian Journal of Earth Sciences*, **15**, 1872–1875.

von Bitter, P. H. and Millar-Campbell, C. 1984. The use of spikes in monitoring the effectiveness of phosphatic microssil recovery. *Journal of Paleontology*, **58**, 1193–1195.

Wang Yu Xian. 1981. High frequency dielectric separation of microfossils. *Lethaia*, **14**, 261–268.

Wicher, C. A. 1942. *Praktikum der augewandten Mikropaläontologie*, Bomtrager, Berlin.

Zeigler, W. (Ed.) 1973. *Catalogue of Conodonts, I*, Schweizerbartsche, Stuttgart, 1–504, 27 plates.

Ziegler, W. (Ed.) 1973. *Catalogue of Conodonts, II*, Schweizerbartsche, Stuttgart, 1–404, 25 plates.

Ziegler, W. (Ed.) 1977. *Catalogue of Conodonts, III*, Schweizerbartsche, Stuttgart, 1–574, 39 plates.

Ziegler, W. (Ed.) 1981. *Catalogue of Conodonts, IV*, Schweizerbartsche, Stuttgart, 1–445, 40 plates.

Zeigler, W., Lindström, M. and McTavish, R. 1971. Monochloracetic acid and conodonts — a warning. *Nature*, **230**, 584–585.

Zingula, R. P. 1968. A new breakthrough in sample washing. *Journal of Paleontology*, **42**, 1092.

2

Recent developments in rock disintegration techniques for the extraction of conodonts

with contributions from **C. R. Barnes; Zhang Youqiu; L. Jeppsson** and **D. Fredholm; W. J. Varker; A. Swift; G. K. Merill; L. Jeppsson;** and **K. J. Dorning**

Recommendations are proposed concerning the design of racks for buckets containing acid and the use of a peristaltic pump for dispensing concentrated acid. We describe a new tower method which allows rocks to be dissolved more quickly, with the production of more numerous and better-preserved conodonts. Data are presented to indicate the relationship between temperature and limestone dissolution. The rate of dissolution can be doubled if the temperature is raised by 20 °C and if the solution recommended by Jeppsson *et al.* (1985) is used. Crushing a limestone sample has a very limited effect on the dissolution time. Methods for extracting conodonts from non-calcareous shale using paraffin (kerosene) and petroleum ether for the disaggregation of clastic sediments are outlined, as also is the use of sodium hydrogen oxalate to remove limonite from 'limonite'-cemented sandstones. Only a part of the total time spent on extraction of conodonts from a sample is related to the size of the sample, but the relative importance of that part increases with the size of the sample. To keep it within reason, other methods are needed in large-scale work. Buckets with dilute acid should, in future, be supplemented by larger vessels in wheeled supports placed in a 'fume-room' and filled and emptied using a semiautomatic system. Similarly, hand washing should be supplemented by a an automatic washing process, such as may be useful even when handling small but strongly argillaceous samples. An integrated conodont and palynomorph preparation of small samples for biostratigraphical and palaeotemperature analysis is described.

DISCLAIMER

Whilst every effort has been made to include safety advice where necessary, the authors, editor and publisher accept no liability for any injury or damage resulting from use of any of the techniques outlined. It is suggested that, before using any new technique or piece of equipment, the safety aspects should be thoroughly considered.

2.1 IMPROVED LABORATORY TECHNIQUES FOR PROCESSING CONODONT SAMPLES
(by C. R. Barnes)

The bulk processing of large samples using a variety of weak organic acids (e.g. acetic acid or formic acid) is a standard technique in most conodont studies. The basic procedures have been outlined by Collinson (1963, 1965) and Lindström (1964). Specific laboratory techniques are commonly constrained by the laboratory design, size and venting, and by the degree and efficiency of technical assistance. Two specific recommendations are proposed herein to aid efficiency and to improve safety in acid digestion of conodont samples. These concern the design of racks for buckets containing acid and the use of a peristaltic pump for dispensing concentrated acid; both have been used for over a decade in my own work and have proved to be of great value. They are used in certain other laboratories with similar success.

In some laboratories, acid in buckets is placed in fume-hoods but this severely limits the number of samples that can be processed concurrently. If the laboratory itself has good overall ventilation, buckets can be located within the laboratory on specially constructed mobile racks. The actual dimensions should suit the laboratory and the operator. Buckets with rectangular cross-sections (as opposed to circular) are more efficient in terms of space and the area of each rack shelf can be designed to take a specific number of such buckets. I have found that three mobile rack systems, approximately 2 m high by 1.1 m wide by 0.8 m deep with four or five shelves provide space for about 80 buckets and some shelf space for drying sieved samples. The racks are readily constructed with an outer framework of angle iron, with solid plywood shelves to prevent any spills from contaminating lower buckets. The racks have a heavy-duty castor (wheel) on each corner to allow them to be wheeled to the sink where acid can be decanted. This enables buckets to be decanted with minimal disturbance of the insoluble residue on the bottom and the easy supply of new water to the buckets via a long hose from the sink. Acid changes are thus rapid, safe and less tiring for the operator. If buckets are not purchased with lids, evaporation and excessive fumes can be reduced by placing sheets of Plexiglas (e.g. 1.1 m×0.4 m) over each row of (four) adjacent buckets. Buckets can have plastic grids or perforated containers inside to keep

the undissolved sample clear of the fine insoluble residue and to enhance continued acid digestion of the sample.

After the used acidic solution in the buckets has been decanted and replaced with fresh water, new concentrated acid must be added to give the 10–15% final concentration. It is cheaper to purchase acid in large volume and as commercial rather than reagent grade. An efficient size is 200 l (approximately 45 gal or 55 US gal) (e.g. 98% glacial acetic acid). A large barrel can be placed on a low barrel dolly (with wheels for easy mobility in the laboratory). The principal problems in dispensing acid to the buckets are speed and safety. Using a tap or hand pump on the barrel is time-consuming and strong fumes may be hazardous or unpleasant. A rapid, easy and safe way of dispensing acid is to use an electric peristaltic pump. This is available from most large chemical and laboratory supply companies. A model found to be especially good is model 410-G High VAC variable-speed pump, $\frac{1}{3}$ hp, 115 V, 1725 rev min^{-1} totally enclosed fan-cooled motor, with 2 m of Tygon R-3603 tubing (especially acid-resistant tubing) (available from TAT Engineering Corporation, 300 Shaw Avenue, North Branford, Connecticut 06471, USA). The acid flow can be controlled and can be started and stopped immediately through an electrical switch that should be incorporated and placed on top of the barrel near the pump. Each bucket receives sufficient acid in only a few seconds. The tubing end can be placed close to the water in the bucket to minimize strong acid fumes. Acid can be added to all the buckets on the shelves rapidly and sequentially; racks can be rotated on their wheels if necessary to allow easier access to all buckets.

These changes in rack design and the use of a peristaltic pump have been found most effective in processing hundreds of samples annually. Although they require an initial cost outlay, this is quickly offset by the increased efficiency and in developing a safe and less odorous working environment in the laboratory. In permitting bulk processing of samples, the unit cost (acid cost and employee's time) for processing is also reduced.

2.2 THE TOWER METHOD: AN IMPROVED METHOD FOR CONODONT RECOVERY FROM CALCAREOUS ROCKS

(by Zhang Youqiu)

(a) Introduction

For several decades, acetic acid or other organic acids have been used to dissolve limestones and other calcareous rocks for the recovery of conodont elements and other microfossils. The standard laboratory procedure (Fig. 2.1) has involved immersing the rock in 10% acetic acid in buckets or beakers, followed by wet sieving and heavy-liquid and/or magnetic concentration (Collinson, 1963).

In my experience, conodont elements retrieved from the buckets are commonly broken, and the speed of disaggregation of the rock is rather slow. In order to increase the rate of reaction and to minimize damage to the specimens, I have designed and constructed a tower apparatus (Fig. 2.2). The principle is to circulate the acid over the sample and to protect the freed specimens from damage that may be caused by contact with undissolved rock.

(b) Technique

Fresh acid solution is placed in container A and is allowed to flow through a tube into the lower part of the tower. Within the tower is a nylon sieve (150 mesh, or finer) on which the sample, crushed to fragments of about 1 cm, is supported. As the tower fills, acid flows out of the top through tubes to container B, from which it can be pumped back into container A for recycling through the apparatus. A rate of flow of about 10 l h^{-1} is sufficient, and the reaction can be further enhanced by the insertion of a stirrer into the upper part of the tower. When the sample has been dissolved, the acid may be removed from the tower through a tap (K1) into

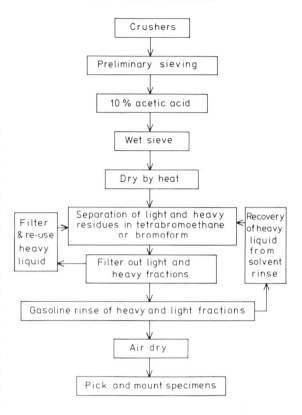

Fig. 2.1 — Sequence of laboratory procedures. (Modified after Collinson, 1963.)

container C. A circular door with a lid in the upper part of the tower permits removal of the nylon sieve, and the residue can be washed gently with water before concentration and examination.

As an example of the relative efficiency of this method, I have processed two 500 g splits of the same sample by the traditional 'soak' method and by using the tower apparatus. Cold acetic acid was used in both tests and the results are shown in Table 2.1. The soaked sample had only partly dissolved in 5 days and very few good quality specimens were recovered. In contrast, the sample in the tower was dissolved in 2 days and eight times as many well-preserved elements were found.

With my apparatus, 0.5–1 kg samples may be used, but a larger tower could cope with larger amounts of rock. The thickness of the rock layer on the sieve should not exceed 3 cm. The procedure may be expanded for mass production by the employment of a series of towers, through which flow could be generated by a single pump.

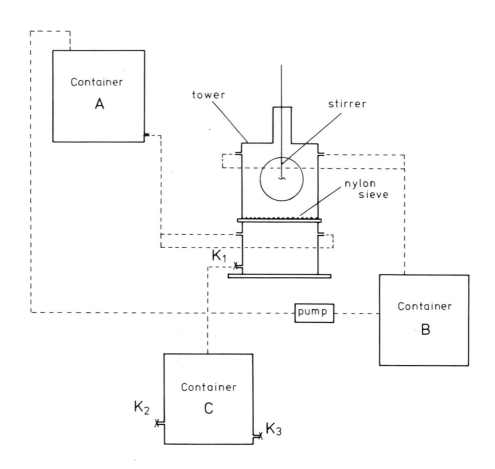

Fig. 2.2 — Diagram of the apparatus utilized in the tower method.

Table 2.1 — Test runs of the soak and tower methods using 500 g splits of the same sample.

Sample number	Sample mass (g)	Method	Number and quality of specimens				Number of days
			Good	Medium	Poor	total	
Cm3f–9	500	Soak	5	75	28	108	5
Cm3f–9	500	Tower	40	129	88	257	2

This work is only preliminary. According to an old Chinese saying, 'you must throw out a brick in order to bring forth a jade'. My work is only a 'brick'; I hope it will enable all conodont wokers to gather more 'jades' in the future.

2.3 TEMPERATURE DEPENDENCE OF LIMESTONE DISSOLUTION IN CONODONT EXTRACTION
(by L. Jeppsson and D. Fredholm)

(a) Introduction

When a sample is placed in a solution of acetic acid, it is a general experience that if warm water is used in the mixing of the solution instead of cold water, the reaction becomes more vigorous. Theoretical calculations (courtesy of Lennart Sjöberg, Department of Geology, Stockholm, based on Sjöberg and Rickard, 1984a, 1984b) indicate that an approximate 50% increase in the dissolution rate could be achieved by raising the temperature about 15 °C above room temperature (using only cool tap water might instead lower the temperature below room temperature and thus lower the rate of dissolution initially). The study of Jeppsson et al. (1985) showed (1) that 5–15% acetic acid solution destroys conodont elements in less than 1 day, (2) why (with luck) conodont elements can be extracted despite the fact that it usually takes many days to dissolve an average sample, (3) that the destructive effect decreases with the amount of carbonate dissolved and (4) how a safe solution should be mixed, so that none of the conodont elements present in the sample is lost because of the acid. These studies were done at room temperature (or more correctly at a slightly lower temperature due to evaporation from the vessel). It was to be expected that a warm solution would be even more aggressive towards the phosphatic fossils. Recently, it has been possible to borrow the equipment necessary for the running of two further experiments.

(b) Experiments

In the first experiment, one large lump, a small lump and some chips, together totalling 13.3 kg of crinoid limestone (sample G85-14LJ from the Hemse Beds at Gannes 3, (Laufeld, 1974), 1.70–1.80 m above the reference level, the rippled surface described by Sundquist (1982)) were placed just below the surface of the liquid in a 165 l vessel. The solution used was that recommended by Jeppsson et al. (1985), except that warm tap water was used instead of cool. Using two old aquarium-type immersion heaters totalling 250 W, placed in the liquid, the temperature was kept at about 33–36 °C for the first few days. On the fourth day, a thermocline had developed. During that and the following days, the temperature was between 32 °C and 33 °C after mixing, whereas the surface temperature was between 37° C and 41 °C.

Jeppsson et al. (1985) showed that the initial solution was most destructive; thus only that had to be tested. To do so, some pieces of conodont elements were placed in a small vial, some of the initial solution added, and the vial was corked and incubated in the vessel. The fragments were found to be destroyed after 4 days. Further incubation experiments, using freshly mixed solution were undertaken, including one at about 60 °C run for over 10 h and another at about 37 °C for approximately 30 days. No damage at all could be found on any of the elements.

There are two possible explanations for the destruction of the elements, which are not mutually exclusive. There is always a certain variation in the composition of the acetate solution ('acetate soup'). For example, a surplus of solution may have been used when it was produced, or the solution may have been exchanged too early. In both cases it will contain less calcium acetate and more remaining acetic acid than usual. Especially when strongly argillaceous samples are dissolved, the composition of the 'acetate soup' may easily be outside the normal limits. If such a solution is used, then less acid than usual needs to be added, and correspondingly more of the 'acetate soup' should be added. It is possible that such an 'acid-rich' solution happened to be used (the experiments were done when the laboratory was closed for rebuilding and the equipment and chemicals were evacuated). To avoid damaging some conodont elements because of the normal variation in composition of the 'acetate soup', Jeppsson *et al.* (1985) recommended that 30% of it was used. This was considered to give a reasonable margin to these variations. If, as expected, the solution becomes more reactive when the temperature rises, then this safety margin may have disappeared with the 20 °C rise in temperature.

The data of Jeppsson *et al.* (1985, Fig. 2) indicate that the 'acetate soup', if used in a cool solution, must be diluted to a tenth of its normal concentration before the solution will destroy conodont elements in 4 days. If the rate of destruction of phosphate fossils also doubles in the warm solution, then we expect that correspondingly an 'acetate soup' diluted to a quarter of its normal concentration could allow the solution to cause damage. The rate of dissolution of phosphate might well increase more rapidly with temperature. It is unlikely that the 'acetate soup' was less than about two-thirds of its normal strength. Thus we assume that the destruction was caused by a combination of a higher rate of dissolution and the fact that the solution becomes more reactive with a rise in temperature.

After it had been established that the conodont elements were not damaged by a correctly mixed warm solution, an experiment was designed to study the possible reduction in processing time which could be achieved. To do this it was necessary to have two samples identical for example in weight, surface area and type of limestone. Therefore a sample of about 20 kg of limestone, closely similar to that used in the first experiment, was selected (sample G84-54LJ from the Hemse Beds at Rudvier 1, 0.7 m below the top of the uppermost crinoid limestone unit). The sample was crushed into pieces up to about 70 mm across. Starting with the largest, the pieces were hand sorted into two portions,

taking care that the number of pieces, their combined weight (8793 g and 8798 g respectively) and surface area should be as close to equal as possible. The weight was checked several times during sorting to keep the distribution of weight equal over the size range. Two 150 l vessels of the same shape were each equipped with three plastic basins (diameter, 200 mm; height, 90 mm; each with about 20 holes, 8 mm in diameter) hanging just below the surface of the liquid to hold the samples. The solution was mixed using the 'acetate soup' from the first experiment (thus, independent of its exact composition, it would be equal in both vessels). The liquid in one of the vessels was mixed using warm tap water and the other using cool tap water. On the second day the former was partly insulated with foamed plastic; thereby the temperature could be raised to about 37–38 °C from the initial 35 °C. This temperature could be maintained for most of the time. The temperature in the other vessel started at 16 °C, but from the second day onwards it stayed at about 17–18 °C, about 3 °C below room temperature. The plastic basins were lifted out and drained and their weight determined every day during the first week and then most working days for 31 days, after which time the experiment had to be interrupted. The calculated loss in weight (Table 2.2) refers to the amount of limestone broken down. However, the weight of the clay (about 30 g was in the warm vessel) and the fraction above 63 μm (3.3 g in the warm vessel) was less than 0.5%; thus the amounts of carbonate dissolved were about the same.

Further, small pieces of limestone may have slipped through the holes; notably many small calcium carbonate crystals were moving around in the vessels during the first few days (kept in suspension by gas bubbles), some clay remained in the basins and the draining of the basins may have varied slightly from day to day. Control weight determinations showed that the latter affected the weight by a few grams. These errors partly counteract each other and, further, are closely similar for the two vessels; they do not affect the results. Of course the surface area available for etching decreased during the experiment, after the initial rise. However, the experiment was designed to develop the method of conodont-element extraction; thus no efforts to counteract this reduction were made.

The pH was measured three times: on the seventh day it was 5.00 in the cool vessel and 5.13 in the warm, on the seventeenth day it was 5.26 and 5.33 respectively and after the experiment was interrupted it was 5.44 and 5.60 respectively.

(c) Results

As is evident from Table 2.2, half of the total amount dissolved was achieved in less than 22 h in the warm vessel and in 2 days in the cool vessel. This difference in the rate of dissolution (it took twice as long to dissove a given quantity in the cool vessel as in the warm one) was kept until about 91% had been dissolved (counted on the total amount dissolved in the warm vessel). This amount (91%) was dissolved after 16 days and 8 days respectively. When about 80–90%

had been dissolved, the clay had settled and the only activity seen was a few bubbles when the basins were gently tapped. Thus, in routine work, the acid would probably have been exchanged at that time.

It is the changes in pH and in calcium acetate concentration that slow down the reaction. Therefore, comparisons between the rate of dissolution should not be related to the day but to the amount of limestone already dissolved. Such a comparison shows that the rate of dissolution is highest in the warm vessel, over the whole range studied. In a semilogarithmic diagram (logarithmic time) plots of the amounts dissolved gave a straight line for each sample, up until about 87% had been dissolved. When extrapolated, the lines intersected at about 46 days and 7.65 kg. Whether this is the theoretical amount that it is possible to dissolve or whether this lacks significance is not known.

A comparison between the rate of dissolution of a crushed and an uncrushed sample is also of interest. The large lump used in the first experiment was probably as far away from a crushed sample as any ordinary sample is likely to be; a slab would give a larger surface area per

Table 2.2 — Data from selected days in experiment 2.

Day	Time (h)	Amount dissolved (g)		Amount dissolved (%)	
		Cold	Warm	Cold	Warm
0	22.3	2582	3901	35	52
1	23	3872	5201	52	70
2	23	4552	5781	61	78
3	23	5022	6119	68	82
6	23	5832	6669	78	90
7	22	6062	6759	82	91
10	21.5	6361	6870	86	92
13	23.5	6608	7017	89	94
16	0.8	6729	7090	91	95
20	23.5	6964	7237	94	97
24	23.5	7109	7318	96	98
29	23.0	7265	7419	—	—
30	22	7293	7432	98	100

unit of weight. In contrast the second sample was not crushed as much as is generally recommended (the size of a walnut (Lindström, 1964)); the largest volume was probably in the size range 40–60 mm. After 7 days, the activity was recorded as very low and the pH was 4.94 when the uncrushed sample was dissolved, but the solution was not exchanged until the eleventh day, when the second experiment was started. The temperature had been some degrees Celsius lower than in the warm vessel where the crushed sample was dissolved. Thus the dissolution rate should also be lower. After compensating for that, the remaining effect of the crushing is found to be about 1 day or even less. The reason for this very low saving of processing time (for a considerable increase in the working time) is probably that the dissolution rate depends on the amount dissolved. Thus the gradual decrease in the dissolution rate will be delayed. (The only reason the limestone was crushed in the second experiment was that it was the only practical way to obtain two equal samples; a theoretical alternative had been to use a stone saw to obtain two identical pieces.)

(d) Recommendations

(1) Warm water can be used when starting to dissolve a sample or when the solution is exchanged. The gain in processing time depends on how quickly the liquid cools down; had the temperature stayed at about 35 °C for 22 h, then 1 day would be gained compared with a sample at 17 °C.

(2) Further gain is possible if the vessel is isolated and heated with an immersion heater or incubated in some other way. Half the processing time would be saved, if the temperature was raised about 20 °C and still more with further increase in temperature (the stability of the conodont elements in liquids warmer than 60 °C has not been tested). A lid to decrease evaporation would probably raise the temperature a few

degrees Celsius and thus help to raise the dissolution rate somewhat.

(3) If speed in the dissolution means more than economy, then a large surplus of liquid should be used. Without waste of acid, it is possible to divide the samples into a few which must be dissolved as fast as possible and those that can be allowed to take their time (at least 70% of the total weight). The former are dissolved in a large surplus of liquid (either in a large vessel or by exchange of the solution every day) which, when a calculated amount of acid and water has been added, gives a solution identical in composition with that recommended by Jeppsson *et al.* (1985).

(4) The most rapid method would be to combine an elevated temperature and a surplus of liquid.

(5) If the amount of calcium ions could be maintained at the initial level (corresponding to about 2.5% calcium acetate), e.g. in the way suggested by Jeppsson (this volume), then the speed of dissolution could be kept at its maximum throughout the processing for all samples dissolved.

(6) Crushing of a limestone sample means that part of the sample is wasted (the fraction below 1 mm), that a proportion of the conodont elements is broken and that a separate room is required to avoid the risk of contamination because of splinters. All three factors add considerably to the amount of working time spent on each sample (independent of whether it is field-work time or laboratory technician time) but saves only about a day or less (much less if the sample consists of one or more slabs) of processing time when a pure limestone is dissolved. Vessels large enough to dissolve the sample without exchange of the acid have a much larger effect on the processing time.

2.4 A METHOD OF EXTRACTING CONODONTS FROM NON-CALCAREOUS SHALE SAMPLES USING PARAFFIN (KEROSENE)
(by W. J. Varker)

(a) Introduction

The breakdown of calcareous shales is not difficult, provided that the sample is sufficiently calcareous to allow the application of the same acid-digestion techniques which would be applied to limestone samples. In such cases the problem is one of handling the large volume of insoluble residue which usually results, rather than the actual breakdown of the rock. However, non-calcareous shales do present much greater difficulty and a number of disintegration techniques are described by Stone in this volume, none of which is universally successful for all shales. The technique here described has the advantages of being very cheap, simple to apply, relatively safe (although care must of course be taken to ensure that the fume-cupboard is clearly labelled as containing flammable materials) and, if it is going to work, very rapid. The disadvantage of the technique is that, whilst breakdown of some samples is spectacular, others are totally unaffected and must then be washed in preparation for other methods of breakdown. Unfortunately it is difficult to predict which shales will succumb to this method and it seems mostly to depend upon cement type or the relative degree of induration of the sample. The method is more frequently successful with friable shales.

(b) Method

The sample should first be broken into approximately 3 cm fragments. This is not essential if large and/or rich samples are being treated and there is to be no numerical check on the concentration of specimens. It is necessary, however, if the original sample consists of large fragments and total breakdown is required. The sample should then be thoroughly dried. This is a most important stage of the process; surface drying at room temperature is not usually sufficient. Normally the rock fragments require to be dried in a relatively cool oven (75–100 °C) for approximately 24 h, preferably spread as a thin layer over a shelf. After drying, use a polyethylene bucket to immerse the rock fragments in sufficient paraffin (kerosene) to give about 1 cm coverage above the sample and leave to soak in a fume-cupboard for approximately 2 days. A greater depth of paraffin than this is neither necessary nor advisable, bearing in mind the fire risk. Shallower immersion may cause some rock fragments to become exposed after the reduction in level which is bound to occur because of evaporation and penetration of the paraffin into the rock. At this stage there is no visible change in the sample. Decant the paraffin, which may be cleaned and re-used, and then sluice the sample in the bucket using a strong jet of tap water. Almost immediately, those samples for which this method is going to be successful will break down into a soft disaggregated sludge which can be washed and sieved in the normal way.

(c) Conclusions

Experience with Namurian shales indicates that approximately 25–50% of shale samples will respond to this method. In some samples the breakdown is complete and immediate; in others the sample has been softened sufficiently to enable the fragments to be helped through the sieve by kneading them gently between the fingers whilst at the same time sluicing with tap water.

Since this method results in the purely physical disaggregation of the sample, there is no volume-reduction in the production of the sludge. This creates some problems since the very large amounts of sludge to be washed result in the rapid choking of the sieves. This problem is enhanced because paraffin is released by the sample during the washing process. This paraffin must be washed through the sieves together with the finer residue and on no account should even gentle overflow from the sieve be permitted since the paraffin slick may carry numerous conodonts (see also Merrill's interfacial method C). Large volumes of water are therefore required during the initial washing of the residue, the most useful method being to sluice the residue in the bucket whilst at the same time allowing the bucket to overflow into the sieves. In this way some or much of the finer material can be removed in suspension, enabling the volume of the residue to be reduced sufficiently to allow the remaining residue itself to be poured into the sieve. Sustained washing is advisable since it is often possible to reduce considerably the volume of even the 'conodont fraction' of the residue under the tap.

2.5 THE PETROLEUM ETHER METHOD FOR THE DISAGGREGATION OF CLASTIC SEDIMENTS
(by A. Swift)

This is recommended for the controlled recovery of a microfossil assemblage in undamaged condition; petroleum ether of boiling point 100–120 °C is most suitable for this process. Indurated and well-cemented samples do not respond to this treatment, which works best on soft unmetamorphosed mudstones and shales. The method of processing is to pre-dry the samples, to soak in petroleum ether for at least 1 h, to pour off the ether and finally to inundate with either cold or preferably hot (not boiling) water. The process involves the exploitation of interlayer and/or interclast spaces within expandable clay minerals, with subsequent voltalization and expansion of the ether when hot water is applied. The resulting pressure causes weaknesses in the bonding structure of the rock, resulting in breakdown to a muddy slurry.

2.6 USE OF SODIUM HYDROGEN OXALATE TO REMOVE 'LIMONITE' FROM 'LIMONITE'-CEMENTED SANDSTONES
(by G. K. Merrill)

As Freeman (1982, p. 474) has pointed out, 'limonite' (assumed to be mixtures of various

hydrous iron oxides, principally geothite FeO(OH)), is soluble in hot sodium hydrogen oxalate. Sodium hydrogen oxalate is prepared by adding 100 weight units of anhydrous oxalic acid to 93 weight units of sodium bicarbonate in a volume of demineralized water consistent with the volume of the container. The acid is thereby buffered to protect phosphate from dissolution. Considerable success has been obtained in disaggregating 'limonite'-cemented sandstones by boiling up to a kilogram of fragments of the crushed rock with 1.0 kg of oxalic acid and 0.93 kg of sodium bicarbonate in a large stainless steel container holding about 6 l of distilled water. Boiling is continued for a couple of hours until the liquid turns a bright green from the reduced iron or until all the yellow colour of the 'limonite' has disappeared. The reagent solution is replaced as necessary. By this method, significant numbers of unetched conodonts have been recovered.

There are three cautions in using this procedure. First, oxalic acid is a dangerous poison and a powerful skin irritant. Considerable care should be used in handling this substance. Second, oxalic acid is corrosive and has the potential to destroy conodonts if not properly buffered. Although pyrite has been successfully removed on an experimental basis from samples by alternating sodium hypochlorite–sodium hydroxide treatment with boiling in sodium hydrogen oxalate, a safer and recommended procedure would be to remove the oxidized pyrite with the magnetic separator after the hypochlorite–hydroxide treatment (Merrill, 1980). Finally, some conodonts treated with this two-step process are very much paler in colour (ivory rather than amber) than they were initially, indicating alteration of their colour alteration index (CAI) values. As van den Boogaard (1983, p. 22) has pointed out, conodonts may be made more reddish in colour by the sodium hypochlorite–sodium hydroxide treatment, probably by altering minute pyrite crystals in the interlamellar spaces to 'limonite'. Removal of this material in conjunction with destruction of the organic matter by the hypochlorite can lighten the colour of the specimens considerably. If CAIs are to be determined, they should be done prior to either treatment.

2.7 SOME THOUGHTS ABOUT FUTURE IMPROVEMENTS IN CONODONT EXTRACTION METHODS
(by L. Jeppsson)

(a) Introduction
Ideally the size of a conodont sample should be governed only by our requirements regarding the size of the extracted fauna. However, often the sample size is limited to a few kilograms or even less than 1 kg. Often the reason for this is not a field-work transport cost consideration but laboratory related.

My own experience is based mainly on handling unmetamorphosed samples, lacking any detectable thermal effect. Most samples consist of marl, argillaceous limestone or rather pure limestone. Few of them contain large quantities of dolomite or coarse terrigenous material. In most cases only 0.5 kg has been dissolved, but when possible several kilograms have been treated, with approximately 100 samples between 5 kg and 50 kg and some tens of samples larger than 50 kg. However, some of the following comments may also be relevant when working with other kinds of samples or other sizes of samples.

(b) An analysis of some problems with large samples
Here I shall consider some of the effects if the sample size is increased by ten or a hundred times. As a measure of the gain possible through large-scale methods, the working time needed to process such a large sample may be compared with that needed to process 10 or 100 separate samples of the original size respectively.

As a starting point we may take a 0.5 kg sample dissolved in the most labour-efficient way. Thus, a single slab of limestone from which any overgrowth of lichens and the like was trimmed away in the field is treated as follows.

(1) Clean with a wire-brush and, if necessary, brief immersion in hydrochloric acid (HCl) (but do not crush).

(2) Determine the weight.

(3) Support (e.g. in a plastic sieve or colander) high up in a 15 l plastic waste-paper bin (waste-paper bins are higher sided than buckets and thus require less fume-hood surface, the volume is great enough to overcome the need to exchange the acid, and the sample is dissolved in about 5–10 days).

(4) Dissolve in buffered acetic acid (Jeppsson *et al.*, 1985).

(5) Hand wash on a 63 μm (or finer) screen when all the carbonate is dissolved and, when needed, treat with one or more of the methods (6)–(9).

(6) Petroleum ether.

(7) Heavy liquid (sodium polytungstate).

(8) Magnetic separation, in the case of much pyrite this is preceded by treatment with sodium hypochlorite (Merrill, 1980).

(9) Hand pick electrostatically (Barnes *et al.*, this volume).

The working time needed for each of these processes can be divided into two parts, one of which is independent of the size of the sample and a second part which depends on the size of the sample. The former includes such matters as book keeping, the fetching of equipment, chemicals, sample or residue, cleaning everything afterwards and putting it back into storage, and similar handling processes.

For a small sample the sample-size-dependent time is nil (e.g. it does not take longer time to treat 2 cm^3 with petroleum ether than it takes to treat 1 cm^3) but, as size increases, this time becomes significant. If the sample is large enough, this time becomes longer than the size-independent working time, and for very much larger samples the total working time becomes approximately proportional to the size of the sample (e.g. it takes about ten times as long to hand wash 10 l of sticky clay as it takes to wash 1 l). Before this point has been reached, there is a stage at which an increase in sample size gives an increase in the quantity processed without a corresponding increase in time spent. The critical point differs for different methods; eventually the importance of this increased yield with regard to time spent gradually levels off. Another consequence is that, when the size of the sample is increased the relative amount of time spent on each method changes markedly. For example, if the sample contains a typical amount of clay, the washing time and picking time will quickly become the most time-consuming parts of the process. However, during the first step (from 0.5 kg to 5 kg) the increase in picking time can usually be kept down by utilizing all the available residue-reduction methods. This is because the residue after dissolution of 0.5 kg is often so small that only one or two of the available methods are used. Further, methods such as the interfacial method (Freeman, 1982; Merrill, 1985; Freeman and Merrill, 1985), may be needed to keep the picking time within reasonable limits in the next power of ten scaling-up step; some increase in volume processed per unit time is also to be expected. However, the relative effect of this gain will be only marginal, considering the fact that the majority of the time is spent in using methods for which no further increase in volume processed per unit time is possible. Thus, to achieve any substantial progress, these methods need to be adapted for large-scale work. In the following sections I shall discuss briefly some possible improvements regarding the dissolution of large samples and the subsequent washing to remove the clay and silt.

(c) Dissolution of large samples

When scaling up my own conodont extraction process, I have worked with the requirement that the amount of manual labour per sample should be kept as close as possible to that required for dissolving a 0.5 kg sample. Preferably also the time for which the sample occupies space in the laboratory should be kept to a minimum. Easy handling of large vessels requires that they are placed in frames equipped with wheels. Large vessels of the necessary quality with fitted frames and wheels are widely used in commerce, e.g. for meat handling, and are thus readily available. To empty the vessel easily, it must be possible to tilt it obliquely so that one corner is used as a spout. To permit this, the upper part of the frame must be pivoted at two points about an axis perpendicular to the bisector of the corner used as a spout, and slightly towards the spout side of the centre of gravity of the vessel (see Fig. 2.3). In this way the vessel can be tilted with one hand, whilst the other hand is free to hold the small vessel (e.g. bucket) into which the residue is washed after the clay has been removed (see below). The extra cost to have the frame built in this way is relatively small, if several frames are ordered directly from the factory and at the same time. It should be possible to handle vessels of at least 1 m^3 capacity in this way.

Filling and emptying such large vessels (and also smaller ones) requires an electric pump, for ease, rapidity and efficiency. The handling of water and acid causes few problems, but a 10% solution of calcium acetate (equals the spent acid) is likely to crystallize in the valves and to clog the system. Therefore the pipes and valves must be flushed with water after spent acid has been pumped. To get the correct volumes of acetic acid and acetate solution a flow-meter is necessary. Flow-meters and electric pumps which can cope with acid, particles and air bubbles and which can be reversed (to flush the pipes and valves) are rather expensive. Thus, their number should be kept down. It is possible to construct a system in which only one (non-reversible) pump and one flow-meter are needed (Fig. 2.4). In principle the pipes form a figure 8, with the pump in the centre and all outlets to the vessels on one of the loops and the outlets for the storage tanks and a drain and tap-water connection on the other loop. Several valves have to be open and shut for each phase. However, if solenoid valves are used, then all the switches can be assembled on a control panel, each with a single control lamp. Then a quick glance is enough to check that no mistakes are made. To prevent particles from being

retained in the valves, sanitary solenoid valves should be used (these are widely used in food-processing plants to eliminate the possibility of bacterial contamination). Another drawback with this system is that only one phase can be handled at a time (e.g. when vessel 2 is being emptied it is impossible to use the system at the same time to fill vessel 1). Thus the capacity of the pump and pipes should be fairly high (as a guideline, I have suggested $20 \, l \, min^{-1}$ when planning a laboratory with 165 l vessels). To obtain the correct volumes with such a high capacity, it is convenient to have a flow-meter which gives a signal when the ordered volume has passed (that volume will vary depending on type of liquid and the size of the actual vessel).

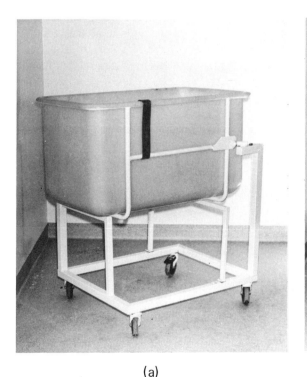

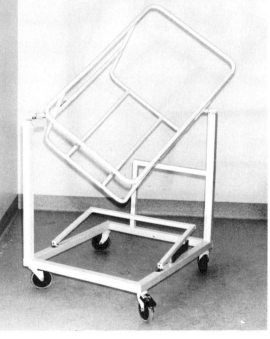

(a)   (b)

Fig. 2.3 — The prototype for a moveable support for the vessels: (a) the vessel in the frame, seen from the distal left corner (there is a ventilation outlet low on the wall in the laboratory, therefore the lower part of the support is 150 mm shorter than the vessel); (b) the support seen from the front right corner. The locking device for the frame is partly folded down and the frame is tilted halfway (the distal support for the moveable frame is placed far enough back for the frame to be rotated upside down, if necessary for cleaning). The frame and vessel were bought from Sinter-Plast, Bredaryd, Sweden.

To avoid overfilling and consequent spillage, a level transmitter should be placed in the vessel and the signal from it should be used to stop the pump automatically when a suitable amount of liquid has been added. A simple microprocessor can also be added to shut off the pump and to close the valves on the signal from the flow-meter and the level transmitter, and to inform the person in charge that the automatic system is ready for the next phase. A better but more expensive system would permit programming (computer software is produced by Peter Melander). Thus, it would be possible to order not only that the three liquids (two when the acid is exchanged) are added in prescribed order and in correct volumes but also that a number of vessels (of various sizes) are filled in order. To do this, there must be one outlet valve and one level transmitter for each large vessel; otherwise, two or more vessels may share one such valve.

When the acid is spent, a hose with a filter on that end which is placed in the vessel is connected to the valve, and the pump is used to suck out the acetate solution. The filter is necessary to prevent any conodont element or other fossil from entering the pipes. I have used a plastic funnel (diameter, about 75 mm) with a sieve cloth glued over the wide end and a hose on the other to siphon off the spent acid from the large vessels. As long as no clay has been suspended, the filter has not clogged but it has stopped many light particles from entering the hose. The filter should stop much smaller particles than those finally collected, so that any contamination consists of particles so small that they are washed away with the clay of the contaminated sample. The acetate solution is normally stored in a tank, which is protected from overflowing by an automatic switch which redirects any surplus solution to the drain, where it is discarded.

The acid tank can be placed outdoors, if 60% acetic acid instead of glacial acetic acid is used because the former solution freezes at -25 °C, whereas glacial acid can only withstand ca-16.6 °C.

The three liquids should be added to the

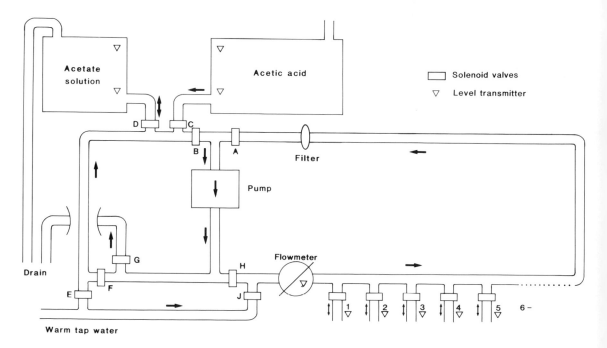

Fig. 2.4 — A routing diagram to illustrate pipe connections in the liquid-handling system.

vessel in the following order: acetic acid, acetate solution and water. In this way the pipes and valves are flushed automatically. Further, the flow-meter measures only the amount of liquid that has passed. Thus, when filling a rather small vessel, it may be that all the acid is still in the pipes when the flow-meter signals that the acetic acid-filling phase is finished. However, the acetate solution, which follows (except in the case of exchange) will force the acetic acid on towards the outlet, if the pipe from the acetate tank is connected as shown in Fig. 2.4. Similarly, if connected as shown, and if added through valve E in Fig. 2.4, the water will flush out both the acetic acid and the acetate solution. (Valve J is used only when washing the outlet loop). The smallest vessel that can be filled thus depends on the volume of liquid in the pipes between valve D and the outlet valve (i.e. a smaller vessel can be filled through valve 1 than through valve 5). This volume can be kept down by using pipes of the smallest convenient diameter and by arranging it so that the length of the pipes between valve D and the first outlet valve is as short as possible. Since this volume can be kept down to a few litres, it should be no problem to fill ordinary buckets but, if smaller vessels are used, it might be useful to fill them from a small tank to which the liquids are pumped and where they are allowed to mix.

There are many technical considerations regarding installation of the equipment which should be considered by a specialist. To take one example, all valves should be placed as close to the loop as possible so that liquids passing a closed valve have as little liquid to mix with as possible. When our laboratory is ready, Peter Melander will know most about such details.

The system suggested here is still only on the drawing-board, and thus it is not tested. However, because it only includes components that are widely used, it is anticipated that only minor modifications may be needed to get a functional system. The cost for the equipment depends on whether the parts are bought new or whether they are found on the second-hand market. I managed to get an offer on the equipment for our laboratory totalling about £4000 plus sales taxes. It includes tanks, pump, flow-meter, valves, level transmitters, filter, pipes and a control panel with a computer. The offer was calculated partly on second hand parts. The time and money saved will be considerable, considering the many samples that will be handled. In other countries the savings will of course not only depend on the number of samples but also on factors such as the cost per hour for a laboratory technician or a student, whether or not the installation and maintenance of the system has to be paid from grants and whether the university or institute has its own engineering and technical staff. Independent of whether the costs break even after 50, 100 or more large samples have been handled, the improvements in the working conditions are such that they alone may justify installing such a system.

(d) Removal of clay

The time needed for hand washing a sample is a matter of major concern when deciding how large a sample can be handled. I do not know of any mechanical washing equipment that is used by conodont students. I assume that the two major reasons for this are related to the fact that the amount of equipment involved is large; thus, the time that it takes to put it in use and to clean it afterwards counteracts much of the gain in time. Second, even with careful cleaning, the risk of contamination increases considerably.

I have tested a prototype for new washing equipment which keeps both the contamination risk and the manual work time close to that of hand washing a small sample. The basic principle is that the screen is placed in the vessel where the sample was dissolved. By adding water in instalments, the clay is brought automatically into suspension and the clay–water mixture is siphoned off through the screen. The construction permits the use of any desired screen mesh and prevents clogging. Thus, no time is needed to empty a large vessel containing many kilograms of wet sticky clay and to transfer it into any washing machine. Other than the screen which anyhow would have been used, very few parts need to be cleaned. To keep a high clay content in the discharge water, the prototype has needed 1 min of supervision every 10–30 min or so during the final washing time. Possibly this minor problem will be solved before the equipment is put on the market. A few of my samples have contained large quantities of very fine sand (I have used a 63 μm mesh screen) and this sand has tended to cover the unwashed part of the residue so that more supervision has been needed and a certain quantity of clay has remained in the sample. Apart from that, the prototype has been a complete success and even samples which otherwise would have required more than a full day of manual washing contain so little clay after automatic washing that it only takes a few minutes of hand-washing to remove it. During the test of the prototype we have moved it from vessel to vessel, which has taken some time; when the laboratory is rebuilt, a space is to be set by where we plan to have two units permanently mounted. Thus, after most of the acetate solution has been sucked out of the vessel, we plan to transfer the vessel in its frame out of the 'fume-room' to this place for washing and then on to a nearby station where it is emptied into a smaller vessel and finally cleaned.

(e) Choice of acid

If we are successful in reducing the amount of working time per kilogram of sample and if we use this saving to work with larger samples (we tend to use available resources whether in the form of grants or technician-hours available; larger samples permit a higher accuracy in our conclusions, and the increased quality of our work may give us the larger resources needed to cope with the law of diminishing return), we shall come to a point where the cost of acid and other chemicals are of major concern. It has been shown that both monochloracetic acid and ordinary acetic acid (Ziegler *et al.*, 1971; Jepps-

son, in Sweet, 1982; Jeppsson *et al.*, 1985) destroy phosphatic fossils. However, the discovery that this can be avoided by buffering the acid with Ca^{2+} and keeping the pH high enough (Jeppsson *et al.*, 1985), indicate that also other acids may be used. Already it has been established that it is possible to use HCl (Jeppsson *et al.*, 1985; Dorning, 1985) and there seems to be no reason against testing other acids along the same lines. Some acids, such as monochloracetic acid, can probably be used in the same way as acetic acid, by buffering with spent acid and may also be usable in the same way as HCl by adding calcium acetate. Some acids cannot be used directly; for example, sulphuric acid (H_2SO_4) would cause gypsum to be precipitated. A three-step process may perhaps overcome that problem.

(1) Dissolve the limestone with buffered acetic acid according to Jeppsson *et al.* (1985) and recover the residue.
(2) To recover the buffered acetic acid, add a calculated amount of H_2SO_4 and let gypsum precipitate.
(3) To prevent the small amount of sulphate ions which remain in the acid from forming further gypsum when the recovered buffered acetic acid is re-used, it may be necessary to remove those ions by, for example, adding a small amount of barium chloride solution to precipitate barium sulphate before re-use.

With this or a similar process, it is H_2SO_4 that is used (however indirectly) while the acetic acid is recovered and used over and over again. However much experimental work remains, it seems reasonable to conclude that it may be possible to use the cheapest (calculated on capacity and not cost per litre) locally available acid.

(f) Postscript

Anders Martinsson, Uppsala, once told me of the need for a sample ten times as large. (Collect a sample and describe its fauna. However, there is not enough material of a new species. You need a sample that is ten times as large. Collect that sample. Now you have enough material of that new species, but there is also some material of another undescribed species, but not enough. You need a sample that is ten times as large … .).

2.8 INTEGRATED CONODONT AND PALYNOMORPH PREPARATION OF SMALL SAMPLES FOR BIOSTRATIGRAPHICAL AND PALAEOTEMPERATURE ANALYSIS
(by K. J. Dorning)

(a) Introduction
The integrated conodont and palynomorph method, applicable to marine carbonates, is particularly suitable where the rock available from any horizon is limited, as in core samples. The mass of rock prepared for conodonts is normally greater than that ordinarily used for palynological preparations. However, the total number of palynomorphs per gram of rock is often low in limestones, and so the use of a larger rock sample is beneficial to provide adequate palynomorphs for quantitative analysis. A small sample of about 500 g of Palaeozoic rocks will regularly provide conodont data which may be directly compared with acritarch, chitinozoa, scolecodont and spore data.

(b) Procedure
Palynological analyses will readily detect very small quantities of acid-insoluble organic materials. In order to minimize contaminants, all apparatus must be kept absolutely clean and chemicals must lack organic particles. Analytical-grade chemicals are normally suitable. The rock sample must also be clear from contamination such as drilling mud, drilling additives, algae and lichens by scrubbing and etching in dilute HCl as required. For quantitative preparations, the weight of rock to be processed is recorded, preferably leaving at least a small rock sample for lithological reappraisal.

Quantities are given for a preparation of a 500 g rock sample. Place the cleaned rock sample in a clean 10 l plastic bucket. Impure limestones may process faster if crushed to lumps of about 20 mm diameter. Place the plastic bucket in a fume cupboard and add 500 ml of 10% acetic acid. The pH of the solution may be monitored during dissolution of the calcium carbonate by using a permanent indicator strip. After several hours the rate of carbon dioxide production slows and the pH will rise from 3.5 to 5.0. For dolomitic limestones the addition of 5 g of calcium carbonate may be required to form sufficient calcium acetate buffer solution. The acetic acid may be partially replenished by repeated additions of 10 ml of 50% HCl at intervals of 1–2 h, until about 500 ml of 50% HCl has been added. For pure limestones, about 1 g of heavy kaolin may be added to reduce corrosion of phosphatic material in the residue. After the last addition of HCl, leave for at least 8 h until all reaction ceases. It is essential that some limestone remains at the end of dissolution. The weight of this residual limestone is required for quantitative calculations.

Decant the top liquid fraction into a separate clean 10 l plastic bucket and add water to the rock and residue. The liquid fraction may be sieved with a 5 μm nylon bolting cloth sieve to catch any floating palynomorphs. Repeated additions of water to the residue, followed by decanting the top liquid through the 5 μm mesh, is repeated until the calcium chloride is removed.

The residue is wet sieved to retain a fraction between 2000 μm and 75 μm. Fragments larger than 2000 μm are returned to the fine residue. The 75–2000 μm fraction is swirled in water on a large clock glass to float the lighter palynomorphs away from the heavier minerals and microfossils. The palynomorphs are added to the fine residue. After examination to confirm the absence of palynomorphs, the heavy residue remaining on the clock glass is ready for routine separation of the heavy conodonts from the lighter minerals and siliceous microfossils.

The fine residue together with the larger rock fragments and palynomorphs is usually sufficiently small to be placed in a 1000 ml polypropylene beaker. This fraction may be treated as a routine quantitative palynological preparation. This residue may be stored for many months if a small quantity of 1% phenol is added to discourage fungal growth and the sample is stored in a cool dark location.

The palynomorphs may be separated from the residue using routine HCl, then hydrofluoric acid (HF) and finally HCl treatment. After swirling or heavy liquid separation of the lighter organic fraction from the residual mineral, and sieving into size fractions as required, the residues may be strew mounted on microslides for visual observation. If aliquots of the residue are spread over a recorded number of microslides, quantitative data may be calculated.

The preparation method provides a convenient technique for routine analyses for biostratigraphy and palaeotemperature interpretation using both conodont and palynological data. Qualitative conodont and palynomorph preparations may also be prepared using the method, although the additional quantitative data are often invaluable in geological interpretation.

(c) Warning

The techniques described should only be undertaken in a laboratory with fume-cupboard facilities equipped for handling and disposal of strong mineral acids and HF. Protective safety clothing must be worn when handling equipment in the fume-cupboard. All waste acids for disposal, including dilute washings, must first be neutralized with an excess calcium hydroxide–water slurry prior to disposal. Sodium hydroxide or sodium carbonate are not suitable for neutralizing HF.

2.9 ACKNOWLEDGEMENTS

Zangh Youqiu thanks Dr. R. J. Aldridge and all at the Department of Geology, University of Nottingham, for their assistance in the preparation of his paper.

A question from Claes Bergman initiated the contribution from Lennart Jeppsson and Doris Fredholm and the manuscript was read critically by E. Lennart Sjöberg with Ann-Sofi Jeppsson and Jeremy Stone making the necessary linguistic changes in it. Grants from the Swedish Natural Science Research Council financed the research of Lennart Jeppsson.

Lennart Jeppsson acknowledges the stimulating discussions with Peter Mileson which produced many ideas regarding how problems could be handled, at least in theory, including the idea to automate the solution-handling system. Doris Fredholm as an undergraduate did much of the laboratory work when equipment and ideas were tested. If the first laboratory following the recommendations of the model is built in Lund, then it is thanks to the friendly and efficient cooperation of Edgar Persson, Byggnadsstyrelsen, in charge of the rebuilding of the laboratory, Lennart Larsson, Sinterplast, Bredaryd, and Peter Melander, Företagskonsult, Hövadsvägen 53, Veberöd, Sweden, and Plate Heat Exchanges Services Ltd., Unit 4, Commerce Street, Carrs Industrial Estate, Rossendale, Lancashire, England, UK. The original manuscript [Section 2.7] was improved by the many comments received from Jeremy Stone, University of Southampton, and Ann-Sofi Jeppsson. Sven Stridsberg took the photographs and the line drawing was drafted by Christine Andreasson. A grant from Kungliga Fysiografiska Sällskapet i Lund offset part of the costs for the prototype of the clay-removing equipment. Grants from The Swedish Natural Science Research Council paid Jeppsson's salary and research costs, including the remaining costs for the prototype.

Ken Dorning thanks R. J. Aldridge and A. Swift, University of Nottingham, and H. A. Armstrong, University of Newcastle upon Tyne, for advice and for separating trial conodont residues.

2.10 REFERENCES

Collinson, C. 1963. Collection and preparation of conodonts through mass production techniques. *Illinois State Geological Survey Circular,* **343,** 1–16.
Collinson, C. 1965. Conodonts. In B. Kummel and D. M.

Raup (Eds.), *Handbook of Paleontological Techniques,* Freeman, San Francisco, California, 94–102.

Dorning, K. J. 1985. Integrated conodont and palynomorph preparation for biostratigraphical and palaeotemperature analysis. In R. J. Aldridge, R. L. Austin and M. P. Smith (Eds.), *Fourth European Conodont Symposium (ECOS IV), Abstracts,* Private publication, University of Southampton, 9.

Freeman, E. F. 1982. Fossil bone recovery from sediment residues by the 'interfacial method'. *Palaeontology,* **25,** 471–484.

Freeman, E. F. and Merrill, G. K. 1985. Safe, cheap, and efficient concentration techniques for conodonts. In R. J. Aldridge, R. L. Austin and M. P. Smith (Eds.), *Fourth European Conodont Symposium (ECOS IV), Abstracts.* Private publication, University of Southampton, 11.

Jeppsson, L. Fredholm, D. and Mattiasson, B. 1985. Acetic acid and phosphatic fossils — a warning. *Journal of Paleontology,* **59,** 952–956.

Laufeld, S. 1974. Reference localities for palaeontology and geology in the Silurian of Gotland. *Sveriges Geologiska Undersökning C,* **705,** 1–172.

Lindström, M. 1964. *Conodonts,* Elsevier, Amsterdam, 1–196.

Merrill, G. K. 1980. Removal of pyrite from microfossil samples by means of sodium hypochlorite. *Journal of Paleontology,* **54,** 633–634.

Merrill, G. K. 1985. Interfacial alternatives to the use of dangerous heavy liquids in micropaleontology. *Journal of Paleontology,* **59,** 479–481.

Sjöberg, E. L. and Rickard, D. T. 1984a. Calcite dissolution kinetics: surface speciation and the origin of the variable pH dependence. *Chemical Geology,* **42,** 119–136.

Sjöberg, E. L. and Rickard, D. T. 1984b. Temperature dependence of calcite dissolution kinetics between 1 and 62 °C at pH 2.7 to 8.4 in aqueous solutions. *Geochimica et Cosmochimica Acta,* **48,** 485–493.

Sundquist, B. 1982. Palaeobathymetric interpretation of wave ripple-marks in a Ludlovian grainstone of Gotland. *Geologiska Föreningens i Stockholm Förhandlingar,* **104,** 157–166.

Sweet, W. C. (Ed.), 1982. *Pander Society Newsletter,* **14,** 1–40.

van den Boogaard, M. 1983. On some occurrences of *Diplognathodus* in Carboniferous strata of western Europe and North Africa. *Scripta Geologica,* **69,** 19–28.

Ziegler, W. Lindström, M. and McTavish, R. 1971. Monochloroacetic acid and conodonts—a warning. *Nature,* **230,** 584–585.

3

Recent developments in conodont concentration techniques

with contributions from **G. K. Merrill; A. Swift; C. C. Ryley; C. R. Barnes and F. H. C. O'Brien; W. J. Varker; J. Stone and R. Saunders; and C. R. Barnes, D. Fredholm and L. Jeppsson**

Interfacial methods form a series of techniques in which the lipophilic properties of phosphatic particles are exploited. Carbonates, clays, lithic fragments and pyrite should be removed from residues prior to treatment by interfacial methods to reduce the sample insofar as possible to silicious components and phosphates. A step-by-step technique (modified procedure B) is described, which is simple, rapid and effective at separating conodonts. The materials required for procedure B are detergent, wax substrate, a solvent and a filtration medium such as a funnel and filter paper, or a fine stainless steel mesh, or fine cloth. For the technique to be effective it is essential that every phosphate particle has an opportunity to become coated with water-insoluble oil and it is critical also that phosphate particles have the maximum probability of encountering and contacting a suitable substrate. A new high-density liquid, sodium polytungstate, has a relatively high viscosity which allows particles to settle slowly with a corresponding increase in separation time. It is reported that coniform and ramiform elements are underrepresented in concentrates. A method to improve the effectiveness of sodium polytungstate separations is outlined, which involves use of a centrifuge and a Dewar flask with liquid nitrogen. Pyrite, limonite and barite, which are present in heavy fractions after bromoform separation, sink in another toxic haloalkane heavy liquid, diiodomethane. Air pressure, using a polyethylene or rubber glove, will clear a clogged separating funnel. Coffee filters may be used as an alternative to standard filter papers in sample separation. A technique involving use of the Frantz isodynamic magnetic separator is a valuable method to concentrate conodonts from residues of low yield from ferruginous Namurian shales. Use of a hand magnet prior to separation is advantageous. A magnetohydrostatic separation technique is outlined. This is a density separation in which the heavy liquid is replaced by a paramagnetic solution or ferromagnetic suspension placed in an inhomogenous magnetic field. The construction of the separator from Perspex (lucite) for use with the standard Cook or Frantz-type isodynamic separator and its operation are described. Electrostatic picking is appropriate when conodonts are numerous in residues.

DISCLAIMER

Whilst every effort has been made to include safety advice where necessary, the authors, editor and publisher accept no liability for any injury or damage resulting from use of any of the techniques outlined. It is suggested that, before using any new technique or piece of equipment, the safety aspects should be thoroughly considered.

3.1 IMPROVEMENTS IN INTERFACIAL METHODS FOR CONODONT SEPARATIONS

(by G. K. Merrill)

(a) Introduction

Interfacial methods form a series of techniques new to micropalaeontology in which the lipophilic properties of phosphatic particles, such as conodonts, are exploited (Freeman, 1982; Merrill, 1985). Particles become coated with oil or other water-insoluble compounds and then are made to adhere to a substrate soluble in that compound (procedures A and B) or simply made to float because of their oil coating (procedure C). In exploiting the surface chemical properties of the particles, the interfacial methods differ significantly from most other microfossil concentration procedures (density separations and magnetic separations) that rely on mass properties of the particles.

(b) Characteristics of interfacial extractions

Because they rely on entirely different principles than more conventional concentration procedures, interfacial methods have unique advantages and capabilities as well as some limitations. The exact procedure that will be described below is one of many possible variations. Although techniques should be adapted for particular variables such as sample size and mineral composition, this procedure has worked well for samples ranging from Cambrian to Carboniferous in age, most of which are in the range of tens of grams in size, and whose conodonts have colour alteration index (CAI) values ranging from 1 to 5.

(i) Scales

Most processed conodont samples ready for interfacial methods are in the range of a few tens of grams and samples in this size range will be used as examples for the description of improved techniques that follows.

Samples of up to 1 kg have been processed manually in closed containers that have volumes of a couple of litres. Larger samples weighing up to 4.5 kg have been concentrated in open plastic buckets, the slurry being stirred with a paint stirrer attached to an electric drill. Freeman has used a small portable concrete mixer to concentrate samples weighing up to about 20 kg and even larger samples could be treated. With proper and appropriate equipment and materials, there are essentially no limits on sample size.

(ii) *Materials removed by interfacial methods and sample preparation*

Biogenic apatite (conodonts, ichthyoliths, phosphatic brachiopods and faecal pellets) and inorganic phosphates are readily separated by interfacial methods. Sulphides, at least pyrite, are also lipophilic. Some iron oxides in the form of 'limonite' separate also, possibly because they largely represent oxidized pyrite in which some sulphide remains. Pyrite can be oxidized with sodium hypochlorite and sodium hydroxide (Merrill, 1980) and then dissolved with sodium hydrogen oxalate (Merrill, this volume). Repeated treatments will completely oxidize pyrite to 'limonite' and then dissolve it. A more bothersome set of materials that ends up in the concentrate are the various phyllosilicates, especially deflocculated clays and micas. Clay is a particular problem because it tends to block filters and to make filtration slow and difficult. Previous disaggregation of lithic fragments with petroleum distillate (kerosene, Stoddard's solvent and Varsol) and water, followed by thorough washing, should greatly reduce or eliminate problems with clays. Some ultrastable minerals (part of the 'heavy-mineral suite') are also lipophilic and end up in the concentrate. Similarly, some flaky calcite fragments (ostracods, splits of brachiopod shells, etc.) are also likely to adhere to the substrate because of their large surface areas.

Interfacial methods, like conventional heavy liquids, can separate phosphate from a wide variety of materials. Both techniques are most effective at separating phosphate from quartz. However, the more non-quartz non-chert particles that can be removed prior to concentration by interfacial methods the better are the results obtained (this is also true with heavy liquids). In particular, acidization to remove carbonates, petroleum distillate–water treatment to disaggregate clays, washing, drying, magnetic separation to remove lithic fragments, glauconite, 'limonite', etc., and possible sodium hypochlorite–sodium hydroxide and sodium hydrogen oxalate to remove pyrite are advisable prior to treatment by interfacial methods to reduce the sample as nearly as possible to quartz (plus chert) and phosphate only.

(iii) *Effectiveness*
Microfossil sample-concentration techniques (sieving, magnetic separation, and now oil-froth flotation which are interfacial methods) are all scaled-down versions of operations used in the mineral-processing industry. Consequently, ore-concentration terminology is appropriate: the conodonts are ore, all other material is gangue, the material removed by the interfacial method is the concentrate and the material that is left behind constitutes the tailings. In any microfossil (or ore) concentration procedure, two related phenomena occur simultaneously: the first is *recovery*, the removal of ore from tailings; the second is *enrichment*, the elimination of gangue from the concentrate. Ideally, the treatment should remove all ore from the tailings, achieving 100% recovery and the concentrate would contain no gangue. Neither condition is ever achieved with any technique, of course, and some compromise between them is always necessary. Preliminary quantification and comparison of the interfacial method with tetrabromoethane separations suggests that recovery is better for the interfacial method at the expense of some enrichment, in contrast with the tetra-

bromoethane method that has poorer recovery but somewhat better enrichment (but also see experiments 23 and 24 in the paper by Freeman (1982)).

(c) **Material selection and management**
(i) *Detergent*
The functions of the detergent solution are to disperse the solvent into droplets, to act as a surfactant, to 'lubricate' the movement of the particles and to act as a 'stand-off' between sediment residue and substrate. There are, no doubt, a large number of chemicals that, when mixed with water, would perform these tasks. The ready availability and effectiveness of commerical dish-washing detergents have discouraged wide experimentation. A wide range of off-the-shelf dish-washing detergents works well, but there are some undesirable characteristics found in some of them. Some concentrated detergents are highly viscous and do not completely drain from the measuring device. Others when mixed with water produce more foam ('suds'), and foam is disadvantageous because it can raft gangue particles to the top of the separation together with the floating substrate and solvent. Under any circumstances, detergent solution should be allowed to stand for several hours after mixing to reduce the foam. Finally, some detergents have objectionable odours that become tiresome.

All dish-washing detergents that have been tried work well at concentrations of 1 part of detergent to 99 parts of water by volume. I have also discovered, quite accidentally, that some of them will work at much lower concentrations of detergent. During the reclamation phase of the detergent recovery cycle, several individual brands have been mixed with no loss in effectiveness.

After each extraction the detergent solution is filtered to recover wax flakes and other particulate matter. The filtration funnel is put in the neck of a large container (2.5 l 'Winchester'-type bottles or similar) in which the filtered detergent is collected. These bottles are initially slightly overfilled; so some floating oil overflows into the sink and is lost. After several hours, additional oil, degraded wax and some components of the detergent will form a 'scum' on the top of the detergent solution. This is poured off into the drain before re-using the detergent. Recycled detergent becomes cloudy and has less foam and odour than when fresh. All evidence indicates that recycled detergent is more effective in interfacial method separation than it was when fresh. Some solution is lost during separation by spillage, evaporation, deliberate overflow and the like; so fresh solution is used in small amounts to top up the containers and to avoid the problems (foam, etc.) probable with a completely new solution. With these small additions, there seems to be no limit to the lifespan of the solution.

(ii) *Wax substrates*
For interfacial method procedure B, the most convenient substrate used is paraffin wax. The most common and readily available source in the USA is probably in the form of small packages (about 0.5 kg) generally intended for home canning. 0.5 kg of wax, properly recycled, will last a long time. Larger amounts of paraffin wax are available in 10–20 kg blocks intended for candle making and, of course, candles could be melted down as a limited source of wax. Certain candles (so-called 'Plumber's candles') are likely to contain stearic acid that perhaps might adversely affect interfacial method separations.

Flakes of wax are prepared by pouring melted wax onto a polyethylene sheet as described by Merrill (1985). After use, the flakes are melted so that the molten paraffin passes through a filter into a beaker. It is then poured immediately and directly back onto the polyethylene sheet to make new flakes. Some detergent solution may also pass through the filter into the beaker and form a harmless layer beneath the molten wax. After cooling, the wax is broken into flakes that are 1 cm or so across. Smaller pieces are kept and used, however, because the small sizes of conodonts permits them to be carried by small pieces of wax substrate.

The paraffin wax also absorbs some oil or solvent and is thereby softened when recycled into new flakes. Some softening of the wax is advantageous because it promotes particle adhesion and so, like the detergent, old wax is better than new wax. Nevertheless, wax will become excessively soft with frequent recycling. Additions of some new wax to the beaker of melted wax not only will replenish that which is lost during filtration and cleaning but also will maintain optimum consistency. The wax can also be hardened simply by letting solvent evaporate for several days from the thin layer poured onto the polyethylene sheet.

(iii) *Solvents*
A variety of petroleum distillates has been used successfully for interfacial method procedure B. Other oils would probably work as well, but the ready availability of such solvents as kerosene (British 'paraffin') make them logical choices. These solvents are used in small amounts in the interfacial method procedures and are the only component of the system that is not recycled. The same kind of solvent used to extract the phosphate can also be used to flush and clean the concentrate.

(iv) *Filtration media*
The phosphate-bearing wax flakes can simply be melted in water and the floating molten wax releases most of the phosphate particles. Some of the smaller particles remain trapped between the detergent solution and the molten wax (called 'rafting' in Merrill (1985)). To overcome this, the simplest method is to melt the wax flakes in a funnel lined with filter paper. Laboratory filter papers, even the 'faster' grades, are intended to filter aqueous solutions and are not very effective at transmitting oils and melted wax. Better suited are a variety of papers marketed for filtering deep-fat-fryer oil. Several American-made filter papers have been tried, but they have been lint producing, very slow and commonly quite weak, developing holes at inconvenient times. Two papers marketed in

the UK by the Moulinex and Tefal companies have been used quite successfully. Both are strong, fast, essentially lint free and have sufficiently fine retention for conodont work. Each is designed to be used in a particular filtering device so that it has holes (Moulinex) or slits (Tefal) for attachment that reduce the effective diameter and depth when used in a funnel. In all other respects, these papers are excellent. Other papers are at present under investigation and this remains the one area of the current procedure B extractions that needs further improvement.

Alternatively, it might be possible to use fine stainless steel mesh or even fine cloth. (I have heard a rumour that someone is manufacturing sample bags with a mesh so fine that clay–shale samples are actually washed in the bags in which they were collected.)

Large (25 cm or larger) circles of filter paper can be folded either into quarters, and used singly or doubly, or into a 'Union Jack' (folds in the same direction at 0–180°, 90–270°, 45–235° and 135–315°), i.e. into octants, and then the paper is *reversed* and folded in the opposite direction at 22.5–202.5°, 112.5–292.5°, 67.5–247.5° and 157.5–337.5°. The result is a fluted filter paper that presents far more surface area than quartered circles and results in much faster filtration. I have obtained some large plastic funnels with internal spiral ridges to speed filtration and capable of tolerating temperatures in excess of 200 °C that are ideal for this work.

(d) Sample small-scale interfacial method procedure B extraction

The following is a step-by-step description of the concentration of a small size (tens of grams) conodont sample manually by means of interfacial method procedure B.

(i) *Preparation of materials* (Fig. 3.1)

(1) Sample which is cleaned of extraneous interfering materials, especially clay, and sieved (the narrower the size ranges of

particles in the sample, the more probable is efficient separation, although the usual range of conodont sizes is not excessive for extraction together; alternatively, some larger samples have been split into fractions of greater and less than 0.42 mm).

(2) Detergent solution, at least 2.5 l.

(3) Paraffin wax flakes (see materials).

(4) Solvent.

(5) Filter paper and funnel (the neck of the funnel inserted into a large (about 2.5 l) bottle or flask to receive the recycled detergent solution).

(6) A suitable container in which to perform the extraction (for small-scale manual or semiautomated extractions, wide-mouthed (5 cm) screw-capped 500 ml polyethylene bottles are ideal).

(ii) *Separation*

(7) The sample is poured into the extraction container. If it has a weight of the order of tens of grams, it will cover the bottom to a depth of a few centimetres.

(8) Sufficient detergent solution is poured into the container to cover the sample to a depth of no more than 1 cm.

(9) For samples of the size described (depth of about 2 cm), an amount of solvent in the 5.0–10.0 ml range is sufficient. Freeman (pers. commun., 1985) has suggested a ratio of $X/10$ ml of solvent for each X g of sample. This is appropriate for smaller samples, but I initially use 5.0 ml for small samples, which is added to the container.

(10) The container is agitated for a few seconds (usually swirled with the cap off), to disperse the solvent.

(11) Wax flakes are now added to the container. There is little point in using much more than sufficient wax to cover the surface of the solution.

(12) The container is now capped and agitated to bring the particles into contact with the wax substrate. If this is done manually, the agitation should be vigorous and largely

with a fore-and-aft motion in the long axis of the container. Vigorous agitation of this type for 5 min is usually sufficient to maximize particle adhesion.

(13) At the end of the agitation period the container is uncapped and gently filled with detergent solution.

(14) Now filled with detergent solution, the container is tapped and gently shaken to dislodge particles buoyed and carried by

Fig. 3.1 — Materials needed for a small-scale manual interfacial procedure B extraction: A, extraction container with sample; B, detergent solution; C, solvent (kerosene); D, measuring device for solvent; E, paraffin wax flakes; F, filter paper (quartered Moulinex) in funnel inserted into a receptacle.

the flakes but not attached to them. A surprising amount of material will thus be sent back to the bottom of the container.

(15) The flakes with their attached particles are now poured off the top of the container into the filter. It will be noticed that there is an oil film between the wax flakes and that this film carries some particles. Many of these are phosphatic (every procedure B separation also inevitably yields a procedure C separation) and these are also carried into the filter. The solution is poured off until the approximate 1 cm level of step (8) is restored.

(16) Before proceeding, it is prudent to examine the flakes in the filter. They should show adhering conodonts and other phosphatic particles. If the proportion of these seems to be low, it may be advisable to increase the amount of solvent in the next extraction, usually from 5.0 ml to 10.0 ml.

(17) Another 5.0 ml or 10.0 ml of solvent is added and steps (9)–(15) are repeated until five extractions have been performed. Under normal circumstances, it is possible to check the flakes in the filter after extraction for the amount of adhering particles. Some solvent remains in the container each time; so it is commonly possible to revert to 5.0 ml of solvent after having increased it to 10.0 ml. The flakes from subsequent extractions usually show decreasing numbers of particles, indicating that increasingly fewer remain in the container, as is intended. However, if many particles come out on the fifth extraction, it is advisable to continue extraction. Experience has shown that five extractions usually remove 90+% of the conodonts originally present in the sample.

(iii) *Post-extraction actions*

(18) After draining, the filter and funnel with flakes from the extractions are put in a beaker and the beaker is placed in a drying

oven at about 70 °C. Higher temperatures may degrade the wax and should be avoided.

(19) After a couple of hours the wax will have melted and most of it passed through the filter and funnel into the beaker ready for recasting into new flakes.

(20) The filter paper remains wet with molten wax and, together with the concentrate, must be cleaned. The filter is removed from the funnel and inserted into the thimble of a Soxhlet extractor. It can be pushed down so that its volume fits within the thimble (usually about 12 cm high). Loose samples can also be cleaned in the thimble when required. However, the Soxhlet extractor does an excellent job of cleaning the concentrate *and* the filter paper containing it simultaneously.

(21) The thimble is placed in the Soxhlet extractor, the extractor mounted below the condenser, and a flask of solvent (petroleum distillate) is mounted with a heat source below the extractor (see Fig. 3.2 with explanation).

(22) Water for the condenser and heat for the extractor are turned on and the apparatus is attended until proper and safe functioning is confirmed.

(23) The Soxhlet extractor is allowed to clean and filter continuously for up to several hours if desired. Because the solvent used is always clean when passing through the filter and sample, this time can be as little as a couple of hours. There is no harm and a good deal of added security in allowing extra time for the Soxhlet extractor to run and periods of 6–8 h are common when no other sample is ready to be cleaned.

(24) After cleaning is complete, the thimble, filter and sample are put in a beaker in the 70 °C oven for a few hours to evaporate the solvent. Further cleaning with other solvents, including water, is now greatly facilitated if required; otherwise the sample is ready to resieve and pick.

(25) The detergent solution is ready for re-use

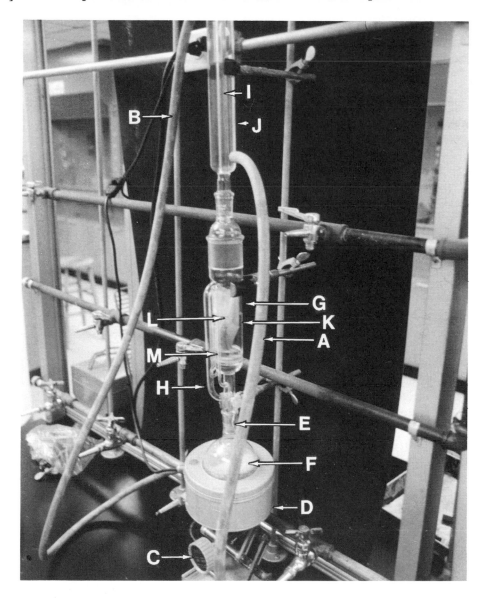

Fig. 3.2 — Soxhlet extractor in operation mounted on a distillation rack: A, water inlet; B, water outlet; C, height adjustment (laboratory jack); D, isomantle with temperature control (not shown); E, 500 ml round-bottomed Florence flask; F, solvent (Varsol, etc.); G, Soxhlet extractor body; H, vapour tube; I, condenser tube; J, condenser water jacket; K, thimble; L, sample in filter paper; M, overflow tube.

Functioning is as follows

(1) (Fill phase) solvent boils and vapour passes up vapour tube.
(2) Vapour condenses in Soxhlet body and/or condenser.
(3) As the condensed liquid level rises in the Soxhlet extractor body, it also rises in the overflow tube.
(4) (Flush phase) fluid spills over the top bend in the overflow tube and the resulting partial vacuum drains the overflow tube and the Soxhlet extractor body.
(5) Some solvent that had accumulated in the thimble is drawn through the sintered glass filter in the base of the thimble.
(6) (Fill phase) the fill phase is repeated.

after pouring off some accumulated scum as discussed previously.

(26) If nothing has been spilled or accidentally overflowed, the only cleaning required to avoid contamination is of the extraction container, which is washed in a laboratory dishwasher at about 100 °C. As already mentioned (Merrill, 1985), other purging of the wax and containers with hydrochloric acid (HCl) is possible, but the acid must be carefully removed or neutralized. Under normal circumstances, acid purging should not be necessary.

(iv) *Analysis of failures*

To evaluate the causes of unsatisfactory results with any technique, it is necessary to understand the underlying principles on which it operates to determine what may not have been satisfied.

Interfacial methods work because phosphate particles are lipophilic and selectively become coated with oils and other water-insoluble organic compounds in contrast with hydrophilic gangue materials, such as quartz, that remain wetted with water instead. If an oil-coated particle comes in contact with a 'substrate' material that is soluble in the coating material (solvent), the particle may be made to adhere to that substrate. From this it follows that, for the technique to be effective, it is necessary to maximize two variables. First, it is essential that every phosphate particle has an opportunity to become coated with water-insoluble oil. Second, it is critical that phosphate particles have the maximum probability of encountering and contacting a suitable substrate. The first condition is realized if an adequate volume of a suitable solvent is properly dispersed in such a manner that the phosphate particles have the maximum opportunity to become coated. Once coated, the grains must encounter the substrate. They will have a reduced probability of doing so if the volume of detergent solution is excessive. The least amount of detergent solution is employed that is

consistent with making the sediment mobile and lubricated. Additionally, agitation is important to bring particles into contact with the substrate. Insufficient agitation time or motion reduces the probability of adhesion dramatically. Some persons have expressed concern that the vigorous agitation might break some conodonts. This is unlikely to be significant in an oil- and water-buffered milieu, but it is possible to use less vigour on a tumble polisher or sample shaker, presumably for a longer time. Furthermore, breakage seems less likely under these conditions than stirring heavy liquids in a glass funnel with a glass stirring rod as is commonly done.

Most of the described failures seem to be attributable to one or more of these causes: improper amount of solvent, excessive detergent and/or insufficient agitation.

(e) Conclusions

Not only is the interfacial method procedure B as detailed herein extremely effective at separating conodonts and other phosphate from quartz and other gangue, but also the techniques described herein are actually less labour intensive than conventional heavy liquids using tetrabromoethane. It is clear that interfacial methods form a valuable set of concentration techniques for conodont samples which, when used in conjunction with other methods such as magnetic separation and/or density separations, offer an excellent method for concentrating conodonts and other phosphatic microfossils. The modified procedure B technique described here is both simple and rapid. A manual extraction such as that described here has a set-up time of a few minutes and five extractions of 5 min; when these are added to pour and fill times, they give a time of about 40 min; the clean-up afterwards involves a few additional minutes of actual work, most of the effort and time being expended by the oven and Soxhlet extractor.

Although procedure B extractions seem the best suited to the normal recovery of conodonts from most samples (especially quartzose ones), both procedure A and procedure C offer possibilities under certain circumstances. Procedure C might permit conodont extraction from some especially recalcitrant samples. Preliminary experiments suggest a possible use for procedure A. Well cuttings, still wet, are wet sieved to remove clay, drained and put in a plastic container and then detergent solution is added. Thereafter, 2 ml of trichloroethane ($CH_3 CCl_3$) is added and swirled to disperse it. The sample and liquid are poured out into another container and the walls of the first are examined for adhering conodonts. No method for quick preliminary determination could be faster. One thing is quite predictable about the broad spectrum of interfacial methods; people will use them with success and newer, better, easier and more effective ways to use the principles will be devised.

3.2 CRYOGENIC DENSITY SEPARATION OF CONODONTS

(by G. K. Merrill)

(a) Introduction

Because of the high density of apatite, conodonts have been separated from residues by means of high density liquids for more than half a century (Branson and Mehl, 1933, pp. 12–14). The liquids most commonly used for such separations have been bromoform ($CHBr_3$) (density, 2.89 g cm^{-3}) and tetrabromoethane ($CHBr_2CHBr_2$) (density, 2.94 g cm^{-3}). Both of these liquids are toxic with wide-ranging effects from immediate unconsciousness to short-term liver poisoning and long-term carcinogenesis. To an increasingly large number of conodont workers the use of these liquids is medically unacceptable and morally unjustifiable, either for their technicians or for themselves. In the hopes of bypassing these dangerous materials, there has been considerable interest in a 'new' high-density liquid: sodium polytungstate (or metatungstate ($3Na_2WO_4.9WO_3.H_2O$) furnished at a density of 3.00 g cm^{-3}) manufactured by Metawo, Falkenried 4, D-1000 Berlin

33, FRG. In spite of the promise of sodium polytungstate over the more commonly used tetrabromoethane which it is designed to replace, there have been a number of problems with its use. This note is intended to describe procedures for using sodium polytungstate that offer opportunities for avoiding many of these problems.

(b) Theory and problems

Heavy-liquid separations involve two important theoretical aspects. First the density of the liquid must be less (tetrabromoethane) or greater (methylene iodide (CH_2I_2) (density, 3.33 g cm^{-3})) than the particles it is wished to concentrate (apatite with a density range of approximately 2.84–3.10 g cm^{-3}). As a result, the particles should float in methylene iodide and sink in tetrabromothane or sodium polytungstate. The other important aspect influencing density separations where the sought-after particles are intended to sink involves their terminal velocities in a particular liquid. The terminal velocity is dependent upon several factors, but the chief are surface-area/mass ratios of the particles, the viscosity of the fluid through which they settle and possibly the buoyant turbulent flow within the vessel (separatory funnel in most cases).

The terminal velocities of most conodonts in tetrabromoethane are fairly high because of the low viscosity of that liquid. Although some particles remain suspended for considerable time and can be seen against a light-coloured background, relatively few of these are conodonts that characteristically settle quite rapidly in tetrabromoethane. Using tetrabromoethane, most workers stir the funnel contents to separate particles so that their cohesive forces are broken and each particle can behave independently, sinking or rising according to its own density and terminal velocity. Furthermore, the normal practice is to allow sufficient time for turbulence to cease and (nearly) all particles to settle, commonly several hours. Thus an open separatory funnel (under a fume-hood to reduce exposure risks) would be intermittently stirred and allowed to settle over a span of several hours. Evaporation of tetrabromoethane is slow and without significant change in the density of the remaining fluid.

There are several problems with sodium polytungstate when used in this manner. The most difficult is the significantly higher viscosity of sodium polytungstate so that the terminal velocities of phosphate particles are greatly reduced. This attenuation of terminal velocities makes all particles settle more slowly and separation times employing separatory funnels may be increased drastically as a result. Still worse, the terminal velocities are governed in large part by the shapes of the particles, some being attenuated so badly that they may not be collected at all. Coniform elements and elongate ramiform elements are reported to be particularly susceptible and could be significantly underrepresented in the concentrate. In addition to this possible taxonomic bias, the long wait for settling is counterproductive because the aqueous solution increases in density as the water evaporates. As the density and viscosity increase, the terminal velocities and particles removed decrease, with the additional problem of crystallization of the sodium polytungstate in the funnel edges and surroundings. Particles are commonly cemented together or to the container by this crystallization. It is also more difficult to recover the sodium polytungstate from both the concentrate and the tailings than is tetrabromoethane. The volume of concentrate should be small enough that the loss is relatively small, but that remaining in the tailings could quickly become a major expense. Additionally, the hydrophilic properties of minerals such as quartz cause a considerable amount of sodium polytungstate to adhere to them and to become tied up in the tailings (float). After all the sodium polytungstate that can be removed by normal washing with a wash (squeeze) bottle and distilled water has been recovered, the filter papers with both tailings (float) and concentrate (sink) are put into the thimble of a Soxhlet extractor and extracted for

a brief period with distilled water. This not only is highly effective at recovering the sodium polytungstate from the sample for later reclaiming but also removes all traces of sodium polytungstate crystals that may have grown on the specimens and were not removed by normal washing.

(c) Materials required (Fig. 3.3)
The method that I have used to increase the effectiveness of sodium polytungstate separations uses the following materials and equipment.

(1) Samples to be separated.
(2) Sodium polytungstate (density adjusted to 2.88 g cm^{-3}).
(3) Distilled water in a polyethylene wash (squeeze) bottle.
(4) Centrifuge.
(5) Centrifuge tubes or better, bottles with screw caps (these should be either polyethylene or polycarbonate).
(6) Dewar flask (preferably wide mouthed) with liquid nitrogen.
(7) Funnels, filter papers, test-tube rack, beakers, etc. (Fig. 3.3).

(d) Procedure
The procedure to separate apatite from gangue with sodium polytungstate is as follows.

(1) A centrifuge tube or bottle is partially filled with sodium polytungstate and the sample is added. I have found it helpful to top up the centrifuge tube with heavy liquid at this stage.
(2) The sample is 'spun down' in the centrifuge. Although the factors are not especially critical, I have found 10 min at 2000 rev min^{-1} quite effective. Only a small amount of material should remain in suspension in the middle part of the tube.
(3) The tubes are removed from the centrifuge and lowered into the liquid nitrogen and are held at a level so that in a few seconds the bottom of the column of liquid, including

the 'sink', is frozen, the upper part of the column with the 'float' remaining liquid.

(4) The liquid and float are poured off into a filter paper.

(5) The heavy-liquid 'ice' is allowed to melt with the centrifuge tube resting in a test-tube rack.

(6) After melting is complete, the remaining heavy liquid and 'sink' are poured into another filter.

(7) After it has drained through the filters, the sodium polytunstate is recovered.

(8) The tube and both filter papers with their contents are thoroughly washed with distilled water, the washings being saved for later sodium polytungstate recovery.

(e) Discussion

This technique was originally done with tetrabromoethane, and an assistant and I successfully separated over 400 samples in 1½ work days. Although 'spin-down' time may be a little longer with sodium polytungstate than with tetrabromoethane because of the higher viscosity, the use of the centrifuge makes this the fastest possible way to separate mineral grains by their densities. Not only is the terminal velocity increased by using the centrifuge, but evaporation of the sodium polytungstate is almost eliminated because of the short time involved. No increase in breakage has been noted from the freezing and thawing or the high-speed rotation. This is an improvement,

Fig. 3.3 — Cryogenic separation in progress. The centrifuge bottle is having the lower part of its liquid column frozen in a wide-mouthed Dewar flask of liquid nitrogen.

and not a panacea. Viscosity continues to create problems, not only with terminal velocity, but also simply with pouring cold, dense, viscous but unfrozen sodium polytungstate from a narrow-diameter centrifuge tube. Enrichment is superb, but recovery is still less complete than with competing techniques (tetrabromoethane or interfacial methods).

3.3 CONODONT CONCENTRATION USING DIIODOMETHANE

(by A. Swift)

In spite of problems of toxicity and stability, the high-density haloalkane bromoform (tribromo-methane $CHBr_3$), has been adopted by micro-palaeontology laboratories worldwide as the 'standard' separating medium for the concentration of conodonts (and other phosphates) from residues remaining after initial sample breakdown. Its specific density of 2.894 at 20 °C plus its great fluidity make it ideal for this purpose, and most residues will yield a relatively small and easily picked heavy fraction (i.e. that which sinks and contains conodonts) on separation in bromoform. However, there are a number of minerals of specific gravity greater than 2.894, particularly pyrite (4.8–5.1), limonite (3.6–4) and barite (4.5), which are commonly present in residues and will form part of a heavy fraction after bromoform processing. At times the quantities involved make hand picking a laborious and time-consuming task. The use of another haloalkane heavy liquid, diiodo-methane (CH_2I_2), as a further concentrating agent on a bromoform heavy fraction can prove very effective in countering this problem. A specific density of 3.325 at 20 °C means that conodonts and other phosphates, which are chiefly apatite and have a specific gravity range 3.17–3.23, will float in this medium. Most of the other minerals from a bromoform heavy fraction, including the troublesome pyrite, limonite and barite, will sink in diiodomethane. As with all halogen compounds of this type, diiodo-methane is toxic and needs careful handling in an efficient fume-cupboard; it also possesses irritant properties.

3.4 CLEARING CLOGGED SEPARATING FUNNELS

(by C. C. Ryley)

(a) Introduction

Clearing a clogged funnel can be very time-consuming, as well as frustrating. Often, a small-diameter rod is inserted from the top in an attempt to push the clogged light fraction through the funnel. In many cases, this serves only to jam the funnel more firmly. It is often done in conjunction with washing with water or another solvent, thus diluting the heavy liquid, whose reclamation is time-consuming. Outlined below is a method which generally clears clogged funnels without the addition of a solvent.

(b) Method

The first step for avoiding a clogged funnel is, once the heavy fraction has been removed, to stir the light fraction into the remaining heavy liquid. This disperses the light fraction and in most cases prevents clogging. If it fails and a clogged funnel results, air pressure through the top of the funnel will push the clogged portion through. I create this air pressure using a polyethylene or rubber glove. I blow into the glove, inflating it, and then place the inverted glove over the neck of the funnel. Squeezing the glove creates a positive pressure in the funnel, forcing the clogged material through. This process may need to be repeated several times. If this fails to clear the light fraction, it will generally force the heavy liquid through. It is not advisable to use the same glove more than once if the heavy liquid being used is tetra-bromoethane.

The advantages of using a glove are that the pressure is controlled and portable and requires little equipment which might otherwise clutter up a separating table or fume-hood. Of prime importance is the avoidance of diluting the heavy liquid, which is particularly critical in the use of the viscous sodium polytungstate.

3.5 COFFEE AND CONODONTS: CHEAP FILTRATION

(by C. R. Barnes and F. H. C. O'Brien)

Whatever heavy liquid is used to extract conodonts from the insoluble residue of dissolved samples, filter papers are required to catch the heavy and light fractions of the separated sample. Standard filter papers may be used; grade 4 papers allow more rapid flow of liquid through the paper. However, these are not inexpensive and alternative papers have been tried.

Sheets of standard paper towels can be used but tend to let through more clays. A more effective and cheap filter is close at hand in most laboratories next to the coffee machine. Coffee filters (e.g. Bunn Filter brand, no. 42, 25 cm in diameter) are ideal filters. They must be the flat basket type and not the seamed (Melita) cone filters since the glue holding the seams of the latter is soluble in nearly all solvents.

They are about one-tenth of the price of filter papers from chemical supply companies; they retain clays and yet allow a fairly rapid flow of heavy liquid. However, they do not pass most oils. They are available in various sizes but can be easily cut to the size required for the funnels. They are available in bulk from coffee supply companies and are probably already available in most departments, organizations or cafeterias, or from a local coffee addict.

3.6 ELECTROMAGNETIC SEPARATION OF CONODONTS FROM RESIDUES

(by W. J. Varker)

(a) Introduction

Merrill (1985; this volume) details the dangers associated with heavy liquids such as tetrabromoethane and bromoform, commonly used in the concentration of conodonts by specific gra-

vity, and presents a strong case that they should no longer be used. Merrill (1985) also considered magnetic separation to be at best a supplement to be used in conjunction with other techniques and not as a total replacement for density separation. There are many circumstances, however, when the electromagnetic separation of conodonts is a valuable method, with a number of distinct advantages, such as the fact that it is cheap (assuming the availability of separators), clean, dry (no further preparation of the residue is required after concentration), semiautomatic, in selected samples very efficient and, of course, almost totally safe. The technique was first described by Dow (1960, p. 738) and the reader should consult this reference for details of equipment and methodology. The general principle behind the method is that 'apatite' is among the least susceptible minerals in a magnetic field. Dow provided considerable detail concerning amperage settings, etc., for different types of residue. Merrill (pers. commun., 1985) has for many years also achieved considerable success with standard settings (i.e. a forward slope of 30°, a side slope of 5°, and two passes at each of his magnetic field settings of 0.6 A and full field strength), different from those described below. Standardized procedures can certainly save a great deal of time. It is equally clear, however, that there is also considerable scope for experimentation and the Frantz instrument, which was used in all these instances, gives considerable scope for success.

The technique outlined below is that used in the separation of conodonts from large Namurian shale residues of low concentration from the north of England. Many of these residues were strongly ferruginous, with significant concentrations of mainly 'limonite' and pyrite and thus produced large 'heavy' concentrates by the standard heavy-liquid techniques, which for residues of this size were expensive and time-consuming, as well as dangerous. Experience with the Namurian residues and several different separators led to the conclusion that it can be more effective to make four or five runs at progressively increasing field strength rather than to adhere to standard settings, since variations between instruments, even those of similar design, can produce effects equal to those produced by small changes in settings.

(b) Method

Residues containing large concentrations of strongly magnetic material such as 'limonitc', which might choke the separators even at their lowest settings, are first tested with a hand magnet. Wrap the magnet in a tissue or filter paper and gently stroke the surface of the residue. Strongly magnetic particles will be attracted to the tissue, from which they can be easily removed. If 2 min are spent on this process, it may save the frustration of choked slides later on.

The residue is then passed through the Frantz isodynamic magnetic separator. The positive side slope and the forward slope (Dow, 1960, p. 739) are both set at 10°, although these settings, particularly the latter, are not too critical. With the amperage first set at zero, begin to pass the residue through the instrument and gradually increase the magnetic field until 30–50% of the residue is being removed as a 'magnetic' fraction. Having determined the required setting empirically (this process takes no more than two or three minutes), return all the residue to the feed hopper to begin the first complete run. When the run has been completed, the 'magnetic' fraction is removed, labelled and stored, and the 'non-magnetic' fraction is returned to the hopper for the second run. Once again fix the strength of the new magnetic field empirically at a point where a further 30–50% of the residue is being removed into the 'magnetic' side, and restart for a complete second run. This process of re-running the 'non-magnetic' fraction at progressively higher settings is repeated twice or three times and during later runs it is safe to allow a higher proportion of the residue to be removed to the 'magnetic' side. By the end of four or five runs the least magnetic fraction will consist of material such as quartz, pyrite, perhaps muscovite, black bituminous material, fish remains, scolecodonts and other chitinous fragments and, hopefully, conodonts.

Provided that the feed and catch hoppers are large enough, and the particle size is suitable for the instrument, it should require no attention during a run. The time required for a run obviously depends upon the size of the residue and the speed at which it is fed through the instrument. Ideally, the aim should be to pass the particles through the chute in a single row, i.e. in single file, and certainly not to fill the chute 'wall to wall', thereby avoiding the danger that 'magnetic' particles may push 'non-magnetic' particles up the side slope onto the 'magnetic' side. With this in mind the first run of perhaps 1 l of residue may take 2–4 h. The time required for later runs is of course greatly reduced.

There are a number of ways in which the process may be safely accelerated. If the original residue is unduly large for the Frantz instrument to handle, it can be reduced by first using a Carpco electromagnetic separator, which is rapid and designed for more bulky samples than the Frantz instrument. Whilst its design is different (it uses magnetic rollers), the principle behind it is the same and it is the 'non-magnetic' fraction from the Carpco instrument which would be passed to the Frantz for the repeated runs outlined above. If more than one Frantz instrument is available, the whole process need take no more time than is required for the first run. A group of three Frantz instruments is ideal. As soon as the 'non-magnetic' fraction becomes available from the first instrument, it is passed to the second, for the second run to begin. Similarly, as soon as this instrument starts to yield a 'non-magnetic' fraction it is passed to the third instrument. In this way, three instruments, producing four different magnetic fractions, can be running simultaneously.

If the final 'non-magnetic' fraction is still unduly large for conodont picking, even after the procedures outlined above, its volume can be further reduced by decreasing the positive side slope of one of the instruments to perhaps 5° or even less if necessary and carrying out a further run.

By now the overall residue will have been separated into perhaps six fractions, each being smaller and progressively less responsive to a magnetic field than the one before. The conodonts can now be picked under the microscope in the normal way. If separation has been effective, all the conodonts will be in the smallest least magnetic fraction. After this fraction has been picked, the next fraction in line should be checked. If this is barren, separation has been complete and it can be safely assumed that there are no further condonts in any of the other fractions. If the second fraction to be checked contains conodonts, which is most likely when a very small (less than 5°) positive side slope has been employed, then the third in line must also be checked. In all cases the latter has proved to be barren.

3.7 MAGNETOHYDROSTATIC SEPARATION OF CONODONTS FROM RESIDUES

(by J. Stone and R. Saunders)

(a) Explanation of technique

Magnetohydrostatic separation, introduced by Andres *et al.* (1965), was so called to distinguish it from the older established 'magnetohydrodynamic separation' in which crossed electric and magnetic fields, producing vortexes, are used (Andres, 1975). In simple terms, magnetohydrostatic separation is a density separation in which the heavy liquid is replaced by a paramagnetic solution or ferromagnetic suspension placed in an inhomogeneous magnetic field. As shown in Parsonage (1977), when a paramagnetic solution (water plus a paramagnetic salt), e.g. manganous chloride solution, is placed in an inhomogeneous magnetic field, it takes on an apparent density directly proportional to its magnetic susceptibility and the strength of the magnetic field at any given point (see Fig. 3.4). Hence a density gradient is established, which is variable at the flick of a switch and, using manganous chloride, has a potential specific gravity range from 1.4 to greater than 9 (Alminas *et al.*, 1984). The behaviour of non-magnetic minerals is controlled purely by their density. Partially magnetic minerals behave as if they are slightly more dense than they actually are, and ferromagnetic minerals collect in the part of the system where the magnetic field is at its strongest.

(b) Construction of magnetohydrostatic separator for conodont studies

Parsonage (1978, Fig. 5) illustrates a simple laboratory-scale magnetohydrostatic separator using a standard Frantz or Cook-type isodynamic magnetic separator as the source for the magnetic field. Following his work, a system was constructed as shown schematically in Fig. 3.5, using a Cook separator. The separating chamber was constructed from two sheets of Perspex (lucite) 5 mm thick from which the channel was machined out before they were glued together (see Fig. 3.6 for dimensions). The connecting tubes were made from standard Perspex tubing with a bore of 7 mm diameter, which was curved gently in a tray of boiling water, where necessary, prior to assembly. The bungs were also machined out of Perspex. The components were glued together using standard Perspex glue. It is important to make the through channel as smooth as possible, and to ensure there are no 'backwaters' where particles may become trapped. To ensure that particles do not collect in the splitting chamber (see Fig. 3.6), the outflow tubes to the separating flasks should begin right at the end of the zone of magnetic influence (see Fig. 3.5). As the magnetic field ends slightly before the end of the pole pieces, it has proved necessary to remove the end 3.5 cm from the base of the air gap of our Cook separator to recess the outflow pipes. The position of the apparatus within the pole pieces of the magnetic separator is important, and some experimentation may be necessary. Fixing the channel slightly above the pole pieces gives the best results and also allows the operator to see what is happening in the splitting chamber.

It was found that the system operates most smoothly with a forward slope of 10°, and the vertical inflow and outflow pipes were constructed, bearing this in mind. The inflow pipe is connected to a large transparent funnel as shown in Fig. 3.5. To allow a flow through the system, a smaller hole was drilled through the Perspex bungs to accommodate an 8 cm length of tubing, which in turn was joined by flexible tubing to a Y-shaped joint to give a single tube (see Fig. 3.5). The single tube leads to a tap or valve to allow control of the flow rate and eventually into a large reservoir beaker. To prevent any particles from being sucked out of the separating flasks up these smaller pipes, the

end of the pipe in the flask is covered with a fine mesh (e.g. 35 μm nylon mesh).

(c) Operation of magnetohydrostatic separator
Parsonage (1978) suggests a number of paramagnetic liquids which could potentially be used in the system, e.g. aqueous solutions of the salts of manganese, iron, nickel, cobalt and the rare-earth elements. Although the latter have the highest magnetic susceptibilities, they are expensive and, for reasons of cost, chemical stability, pH 8, ease of regeneration, and lack of toxicity, a solution of manganous chloride is considered to be the most useful. A useful working solution strength is 1 Kg of $MnCl_2.4H_2O$ in 1.5 l of water.

Filling the apparatus may take 5–10 min. In order to eradicate all air bubbles, judicious squeezing of the flexible tubing may be

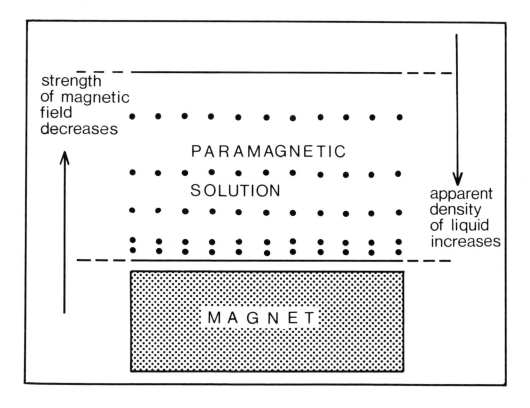

Fig. 3.4 — The relative factors involved in magnetohydrostatic separation.

required. It is advisable to cover the workings of the magnetic separator with polyethylene sheeting prior to the filling process.

Once set up (see Fig. 3.7), the system may be calibrated by using samples of minerals of known density, e.g. quartz, calcite, etc., and some selected pectiniform and ramiform conodont elements and by recording at which amperage the sample just enters the heavy, or lower, outlet and at which amperage the sample just enters the light, or upper, outlet. It is advantageous to make the piece of Perspex forming the splitter as thin a wedge as possible.

Given the 10° forward tilt, a relatively slow flow rate is required, e.g. 30 ml min^{-1}, to provide the best results. The sample of insoluble residue should be sieved prior to separation and split into size fractions (Stone, this volume). All clay, fine material and ferromagnetics should be removed prior to separation. The lower size limit of particle for which this

magnetohydrostatic separation is effective is approximately 75 μm. To overcome the problem that air bubbles raft particles higher than they should be, the sample is first gently shaken in a 100 ml screw-top beaker of manganous chloride solution and is added to the system by submerging the beaker totally in the funnel and then stirring the funnel. The particles become entrained in single file as they flow through the system and separate out according to density (see Fig. 3.5). At any time the flow rate of the solution can be reduced to zero and the content of the reservoir beaker poured back into the funnel. Alternatively a degree of automation could be added by introducing a pump to maintain a continuous flow of the solution.

Once the entire sample has been run (any lodged particles should be visible), the heavy and light fractions may be tapped off into funnels lined with 35 μm nylon mesh (acting like a folded filter paper). The manganous chloride

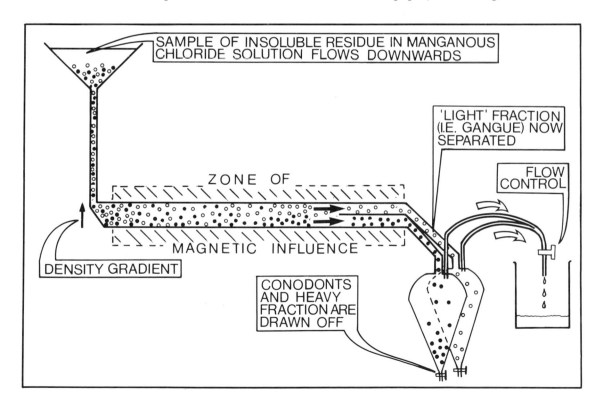

Fig. 3.5 — Schematic diagram to illustrate the method involved in a magnetohydrostatic separation.

solution is then recycled and the particle fractions can be washed into labelled filter papers for further washing (with water) and drying.

To avoid the possibility of contamination, it is advisable to scan the entire system with a hand lens between each sample. On evaporation the manganous chloride crystallizes out to leave a tacky residue. It is therefore advisable to flush the system through with distilled water after usc and to keep the system water filled when not in use.

(d) Results

Once the system is set up, each run takes approximately 20 min. Although some experimentation still remains to be done, early results

are very promising. A spike test (von Bitter and Millar-Campbell, 1984) in which 100 conodont elements were added to a known barren sample of assorted heavy and light minerals, produced a 98% recovery on a single run, and the heavy fraction comprised less than one-quarter of the original sample.

The concentration of the solution does not seem to be too critical and, although some manganous chloride does crystallize, the total volume of solution stays reasonably constant through use and does not seem to lose its effectiveness. If an appreciable amount of manganous chloride has crystallized, the system can simply be checked, using a few conodont fragments, to see whether a slight change in amperage is required.

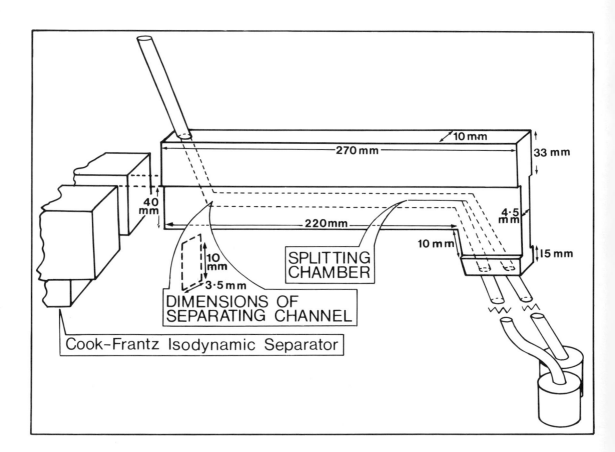

Fig. 3.6 — Diagram showing the dimensions of the apparatus used with a Cook isodynamic separator (not to scale).

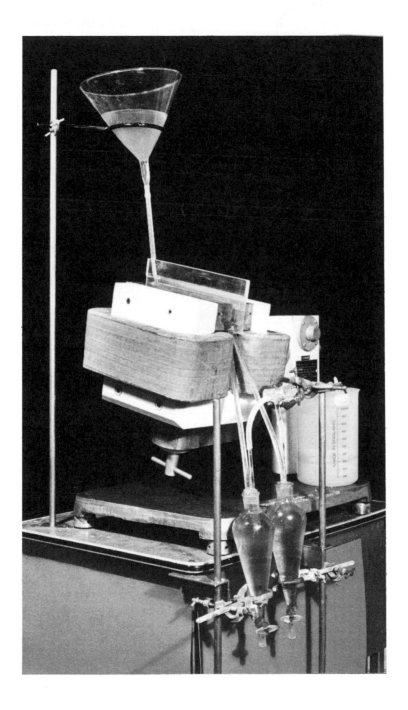

Fig. 3.7 — Magnetohydrostatic separator in operation with a Cook isodynamic separator as the source of the magnetic field.

(e) Advantages of magnetohydrostatic separations

The major advantage over the conventional heavy liquids is that manganous chloride is non-toxic and non-carcinogenic. There is also an increase in efficiency as one effect of the magnetic field is to entrain the particles to flow in single file, thus overcoming problems of rafting. As the density of the liquid is variable, conditions can be tailored to each situation, and the conodonts may be floated off as either a light or heavy fraction, even on consecutive runs. One separation takes 20 min whereas a conventional heavy-liquid separation takes a whole morning. After separation, the samples are ready for picking almost immediately, in contrast with heavy-liquid separations. Once operational, the system is extremely cheap to maintain, and savings are also made because water is used as the solvent rather than acetone.

A major disadvantage is that, if a department or institution does not already have an isodynamic magnetic separator, it may be expensive to acquire one (or an alternative magnetic source). The system is slightly more labour intensive than conventional heavy-liquid separations as it is useful to have an operator in attendance at all times. Manganous chloride solution is a slight irritant, particularly on broken skin; hence the use of rubber gloves is recommended when handling the solution.

(f) Future work

It is the intention to automate the system to a greater extent in the future with the addition of a pump and a removable trap at the start of the Perspex chamber for ferromagnetic particles. More experimental work is also planned to investigate problems which may arise from a variety of sample types, and to quantify the relative efficiency of magnetohydrostatic separation, interfacial and conventional heavy-liquid techniques in dealing with these samples.

3.8 IMPROVED TECHNIQUES FOR PICKING OF MICROFOSSILS
(by C. R. Barnes, D. Fredholm and L. Jeppsson)

(a) Introduction

After available concentration methods have been used, the final extraction of conodonts from the other fossils and insoluble residue remaining is typically undertaken by wet picking. A suitable part of the sample is spread evenly on a small tray, and the conodonts are extracted with a wet brush (saliva or distilled water) and placed on the slide. This requires either that the tray is replaced with the slide under the microscope or that the conodont is transported to the slide without the aid of the microscope. The latter may be safely achieved with medium-sized and large conodonts, but it is nearly impossible to see the removal of a small specimen from the brush with the naked eye. After the elements have all been picked, they are sorted in the storage slide. Independent of which method is used, this picking technique is very time-consuming especially when large faunas are extracted. Electrostatic picking is much more suited for handling large faunas.

(b) Technique (Fig. 3.8)

The storage slide (lightly pre-coated with a water-soluble glue (preferably gum arabicum, as recommended by Jeppsson 1975, p. 5) is placed directly under the microscope and a special picking tray, as described by Triebel (1947) but made of metal, is placed on top of the slide. A 'brush' with a single hair is charged electrostatically by rubbing it once or twice on the operator's hair or face, a piece of cloth or similar material. The operator's hair (if straight) is most convenient.Cat hair becomes stronger charged and is better than human hair. When a conodont element is touched with the tip of the 'brush' hair, the electrostatic forces are sufficient to lift it. The special picking tray has a number of small holes on top of small 'craters' or 'lips'. The conodont element usually drops from the hair when touched against the wall of a crater, falling through the hole into the slide. In low-humidity environments, the charge of the hair is enough to extract many conodonts in sequence, and it can be recharged without taking one's eyes from the microscope. The holes are placed in rows forming a grid and, at a low magnification, four holes appear in the field of view. Thus, when a good sample is picked, the result will be four piles of conodont elements in the slide. Partial sorting is thus possible during picking.

The slide must be fixed so that no part of the field of view is outside it. This is easily done using two pieces of corrugated cardboard. The first is cut to fit in the frame of the microscope. In the second a hole is cut in which the slide fits (the long sides should be widened at one end of the hole, so that it is easy to remove the slide). This second piece is glued to the first in such a way that, for example, the right end of the slide is in the field of view. In this way most of the slide is available to sort the elements.

The single hair may not charge enough to lift a very heavy conodont element or a large fish scale. In such cases a narrow strip of plastic is useful (the best plastic is the polyvinylchloride–poleyeten, of the type found in, for example, small packs with a clear plastic cap to show the

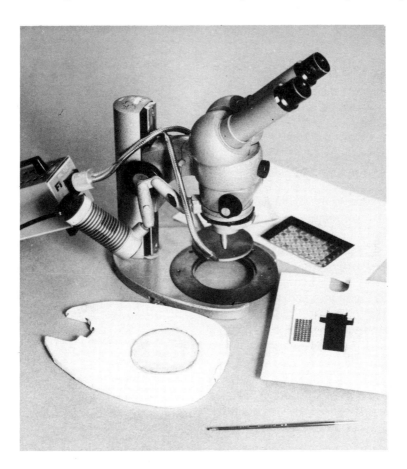

Fig. 3.8 — Counterclockwise around the microscope beginning at 8 o'clock. The slide-holder, upside down, a brush (with the single hair in the upper end), another slide holder (for another microscope model) showing the hole for the slide, a slide (with a conodont pile), a picking tray with a black bottom and, partly behind the microscope, another picking tray with a white background, and the round metal piece which is removed so that the slide holder fits in a fixed position.

contents). The strip can be fixed onto the brush, parallel to the hair, so that a slight twist of the brush handle is enough to select the proper tool when picking a fauna with both large and small fossils. Detergents with antistatic ingredients spoil charging of the hair.

A piece of paper should be placed below the tray when the sample is spread on it, since some material passes through the holes. The tray should also be tapped slightly above the paper in case any particles are balancing on the edges of the craters. Alternatively, a tray without holes may be purchased to nest below the perforated tray when the sample residue is being spread.

Metal picking trays are available with either a black or a white background (e.g. Fema-Salzgitter BRD sell them under the designation 'Hand-picking scales made of brass, Punched'; they are available from Fa. Rudolf Stratmann, 3327 Salzgitter, Friedrich Ebert Strasse 53, FRG).

(c) Advantages of the methods
The advantages include the following.

(1) Hundreds of conodont elements can be picked without removing one's eyes from the microscope.
(2) The risk of dropping a specimen (usually the most valuable one) somewhere between the tray and the slide is avoided.
(3) The conodont elements need not be washed free from the collecting slide; they are already within the sorting slide and a slight touch with a wet brush is enough to remove them from the pile.
(4) The risk of creating a mess on the picking tray if touched with a wet picking brush is avoided; thus, it is very easy to train students and assistants in the picking technique.
(5) Sorting can be interrupted or postponed as necessary.

Considerable time is thus saved when handling faunas with hundreds or more conodont elements.

3.9 ACKNOWLEDGEMENTS
Glen Merrill thanks Eric Freeman for helpful discussions of interfacial methods and Barbara Rodgers for suggesting the deep-fat-fryer filter paper.

Jeremy Stone and Robin Saunders thank Dr. G. Sevastopulo for originally suggesting magnetohydrostatic separation as a possible alternative to heavy-liquid separations, and they are grateful to Mr. C. Balcombe for constructing the apparatus, Mr. B. Marsh for photography and Mrs. A. Dunkley for her patience in drafting the figures.

3.10 REFERENCES
Alminas, H. V., Marceau, T. L., Hoffman, J. D. and Bigelow, R. C. 1984. A laboratory scale magnetohydrostatic separator and its application to mineralogic problems. *US Geological Survey Bulletin,* **1541**.

Andres, U. 1975. Magnetohydrodynamic and magnetohydrostatic separation — a new prospect for mineral separation in the magnetic field. *Minerals Science and Engineering,* **7**, 99–109.

Andres, U., Dorfman, Y. C. *et al.* 1965. Method of preparation of dense liquids for mineral separation according to specific gravity. *Russian Patent,* **177365**.

Branson, E. B. and Mehl, M. G. 1933. Conodont studies. *University of Missouri Studies,* **8**, 1–343.

Dow, V. E. 1960. Magnetic separation of conodonts. *Journal of Paleontology,* **34** (4) 738–743.

Freeman, E. F. 1982. Fossil bone recovery from sediment residues by the 'interfacial method'. *Palaeontology,* **25**, 471–484.

Jeppsson, L. 1975. Aspects of Late Silurian conodonts. *Fossils and Strata,* **6**, 1–79.

Merrill, G. K. 1980. Removal of pyrite from microfossil samples by means of sodium hypochlorite. *Journal of Paleontology,* **45**, 633–634.

Merrill, G. K. 1985. Interfacial alternatives to the use of dangerous heavy liquids in micropaleontology. *Journal of Paleontology,* **59**, 479–481.

Parsonage, P. 1977. Small scale separation of minerals by use of paramagnetic liquids. *Transactions of the Institution of Mining and Metallurgy, Section B,* **86**, B43–B46.

Parsonage, P. 1978. Design and testing of paramagnetic liquid separation systems. *Report,* Warren Spring Laboratory, Department of Industry, Gunnels Wood Road, Stevenage, Hertfordshire.

Triebel, E. 1947. Methodische und technische Fragen der Mikropaläontologie. *Senckenberg-Buch 19,* Kramer, Frankfurt am Main, 1–47.

von Bitter, P. H. and Millar-Campbell, C. 1984. The use of spikes in monitoring the effectiveness of phosphatic microfossil recovery. *Journal of Paleontology,* **58**, 1193–1195.

4

Examples of Devonian and Mississippian conodont lag concentrates from Texas

S. P. Ellison, Jr.

Conodont zones within Devonian and Mississippian rocks of Texas are interpreted as lag concentrates. Thin layers of mostly pectiniform (platform) conodont elements in siliceous shales and siltstones of the Canutillo Formation (Upper Devonian age) of the Hueco Mountains and the lower part of the Tesnus Formation (Upper Mississippian age) of the Marathon Basin have such conodont concentrations and may represent density current deposition. The hard siliceous rocks require that the conodonts be studied in place or from latex casts or moulds.

The Ives Breccia member of the Houy Formation and adjacent Chappel Limestone (both with Lower Mississippian conodonts) in central Texas are concentrates in which the specimens found in the carbonate portions can be freed by using organic acids. These conodonts are also mostly pectiniforn elements and many of the occurrences have Ordovician and Devonian conodonts in the faunas, indicating reworking. These beds are related to a major unconformity at the base of the Mississippian rock column.

Lag concentrates are made of heavier and harder mineral or fossil particles. In the case of the thin layers, the concentrates are coarser grained than the layers of the enclosing strata. Conodonts, with their greater specific gravity, greater hardness and greater resistance to weathering, assume the role of heavy minerals both in turbidites and in basal beds of unconformities. There is a geological history of particle sorting and distribution in the processes between the death of the conodont organism and the final resting place in sedimentary rocks.

4.1 INTRODUCTION

Four localities (Fig. 4.1) in Texas have concentrations of conodonts in which the conodont elements are particularly abundant. Thin beds of siliceous shales and siltstones in the Devonian and Mississippian rocks of west Texas have such concentrations, both as phosphatic conodont material as well as weathered moulds of conodonts. The central Texas localities have concentrations of phosphatic conodont elements in chert breccias and in limestones near the base of the Mississippian rock sequence.

Distribution of the conodonts in each of

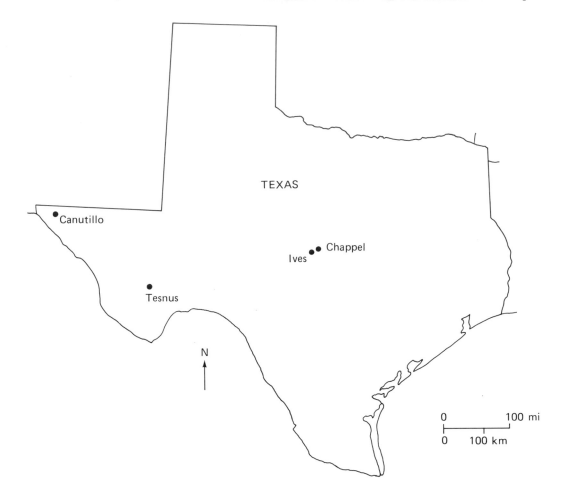

Fig. 4.1 — Conodont lag concentration localities of Texas.

these occurrences suggests that conodont elements react as heavy minerals, are resistant to weathering and wear and are distributed and deposited by water. Specimen counts have shown that 40–80% of the conodont elements in these concentrations are pectiniform (platform) elements and thus the use of statistical evidence to reconstruct multielement conodont apparatuses is futile when applied to these particular rocks. In the west Texas siliceous shales and siltstones, some of the concentrates may be dominantly made of pectiniform elements and others may be made only of ramiform elements. The elements needed to reconstruct multiele-

ment units are present but the ratios of the elements do not fit those of the multielement patterns.

The lag concentrates of conodonts in west Texas are in thin beds (1 mm or less in thickness) and are found in layers that are slightly coarser than the beds above or below. The chert breccias and the limestones that contain concentrates of conodonts in central Texas are in beds less than 30 cm thick.

The main reason for this chapter is to call attention to the unusual abundance of conodonts in lag cencentrates and to point out field techniques for finding these concentrates. In every case the lag concentrates are the result of sedimentary depositional processes.

Cloud and Barnes (1984) referred to the Ives Breccia member of the Houy Formation in central Texas as a lag breccia. Hence, the writer has chosen to refer to the abundance of conodonts in the localities cited here as examples of lag concentrates.

4.2 TECHNIQUES

Conodonts in the siliceous shales and siltstones of west Texas and the chert breccias of central Texas cannot be freed from the matrices of the rocks by laboratory methods of boiling, decanting, washing or use of organic acids. The use of hydrofluoric acid (HF) resulted in granulation of the conodont specimens. The phosphatic conodont specimens and the weathered moulds of conodonts must be studied in place in the rock samples or from latex peels. Organic acids work well to free the conodonts from the limestones in central Texas.

The field collecting techniques for the silicified strata and chert breccias require hand-lens examination of each bedding plane because interval sampling or channel sampling could easily miss the conodont concentrations. Limestone samples usually required two samplings, first to ascertain the conodont distribution and second to resample the abundant zones.

Latex peels for securing moulds and casts of conodonts were made on the siliceous material from west Texas. Applying the latex so that it was free of air bubbles was difficult and white latex required staining before photography the conodont replicas.

Black-and-white photographs were made using Panatomic-X Kodak 35 mm film with an ASA 32°. The pictures were taken with an f16 stop for 10 s. Two lenses were employed depending upon the magnification needed. These were Photar Leitz Wetzlar 1:2.5/25 and Photar Leitz Wetzlar 1:4/25 lenses. Colour slides were made using Ektachrome Kodak film with an ASA 200°. The use of a Pentax camera with automatic exposure control was helpful.

All identifications were made as discrete conodont elements and no attempt was made to recognize multielement conodont forms.

4.3 CANUTILLO FORMATION, HUECO MOUNTAINS, TEXAS

Following field collecting and section measuring in the Hueco Mountains, Texas, in 1959 and 1963, the writter recognized two thin conodont concentrations in the lower part of the shale portion of the Canutillo Formation at exposures 1.6 km southeast of Helms Peak, El Paso County, Texas (Fig. 4.2). These concentrations are in a silicified siltstone immediately above the chert portion of the formation. Conodonts occur in small numbers both above and below the concentration beds. The chert portion of the Canutillo Formation weathers into a topographic bench along the western side of the Hueco Mountains and the lag concentrates are exposed in a continuous mat of mostly pectiniform conodonts on the upper surface of this bench. Examples of these mats are shown in Figs. 4.3 and 4.6. As seen in the thin section (Figs. 4.4 and 4.5) the lag concentrate zones are 1–2 mm in thickness and the conodont elements are embedded in the siltstone matrix.

Moyaud Shafiq (pers. commun., 1969–1976, in an unfinished and unpublished Ph.D. Dissertation) studied these lag concentrates and recognized the Upper Devonian age of the conodonts. Similarly, the conodonts listed below (using discrete conodont-element nomenclature) are interpreted to be of Upper Devonian age. Platform (pectiniform) elements comprise (60–80% of the fauna)

Platform (pectiniform elements)
Palmatolepis glabra Ulrich and Bassler, 1926.
Palmatolepis rugosa Branson and Mehl, 1934.
Palmatolepis sp.
Polylophodonta sp.
Ancyrodella sp.
Ancyrognathus sp.
Polygnathus nodocostatus Branson and Mehl, 1934.

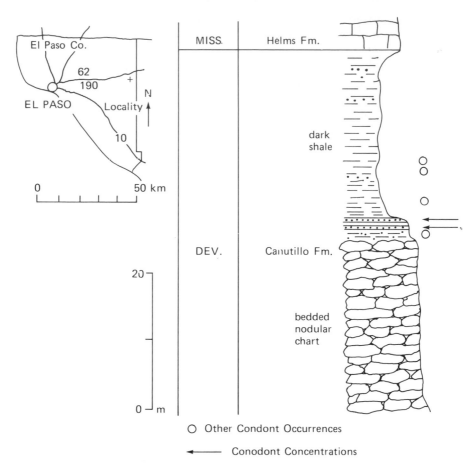

Fig. 4.2 — Canutillo conodonts, Hueco Mountains, Texas. 1.6 km southeast of Helms Peak, El Paso County, Texas. Same as loc. D of King and Knight (1945).

Polygnathus sp.

Blade-bar (ramiform) elements
Bryantodus sp.
Spathognathodus sp.
Hindeodella sp.
Ligonodina sp.
Lonchodina sp.
Synprioniodina sp.

4.4 TESNUS FORMATION, MARATHON BASIN, TEXAS

Conodonts from the Tesnus Formation in west Texas were first discovered by J. Dan Powell (pers. commun., 1961) and were reported in abstract form by Ellison (1962) but the complete fauna remains unpublished.

Six lag concentrate zones occur in the lower part of the Tesnus Formation ranging from 4 m to 120 m above the base of the formation (Fig. 4.7). These six conodont concentrations, like the Hueco Mountain Devonian occurrences, are in siliceous shales and siltstones that are thin bedded and suggestive of turbidites. Collections include both the weathered external moulds of conodonts (Fig. 4.8 and the original phosphatic conodonts embedded in the rock (Fig. 4.9). The use of latex moulding and casting compound was needed for some identifications of the weathered moulds (Fig. 4.10).

0   1
|_____|
mm

Fig. 4.3 — Unweathered conodonts in the Canutillo mats.

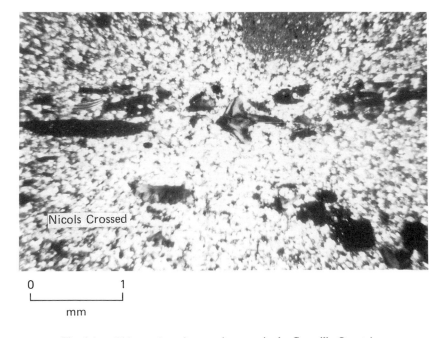

Nicols Crossed

0   1
|_____|
mm

Fig. 4.4 — Thin section of a conodont mat in the Canutillo Quartzite.

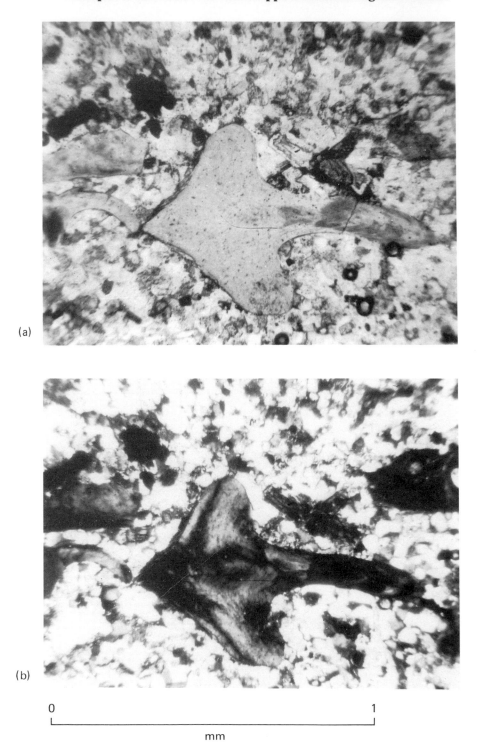

(a)

(b)

0   1
└──┘
                              mm

Fig. 4.5 — Canutillo conodonts, in a thin section: (a) Nicol prisms not crossed; (b) Nicol prisms crossed.

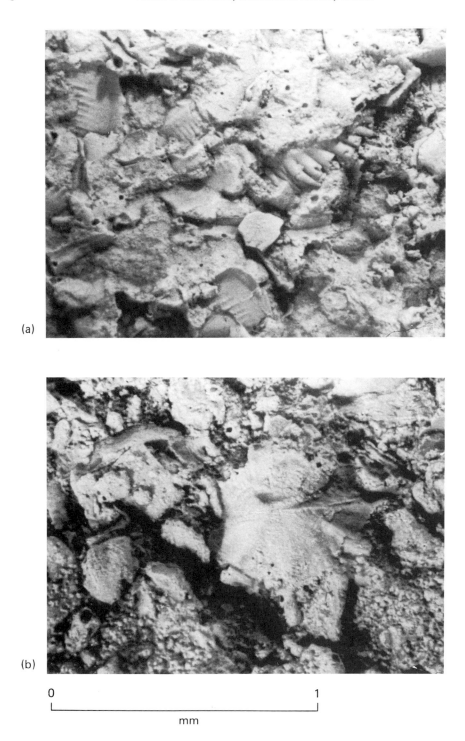

Fig. 4.6 — Canutillo conodonts, stained plaster casts.

The best lag concentrates were found on the east side of East Bourland Anticline and on the Combs' Ranch klippe. Some of the concentrates were all pectiniform elements and some were all ramiform elements. There were also examples of mixtures. Multiple sampling along a single bed often yielded different conodont concentrations. It is suggested that the lag concentrates are the result of differential sorting.

The conodonts listed below (using discrete conodont-element nomenclature) are interpreted to be Upper Mississippian in age. The platform (pectiniform) elements comprise 40% of the fauna.

Platform (pectiniform elements)

Gnathodus bilineatus (Roundy) Hass, 1953.
Gnathodus inornatus Hass, 1953.
Gnathodus sp.
Geniculatus claviger (Roundy) Hass, 1953.
Geniculatus sp.

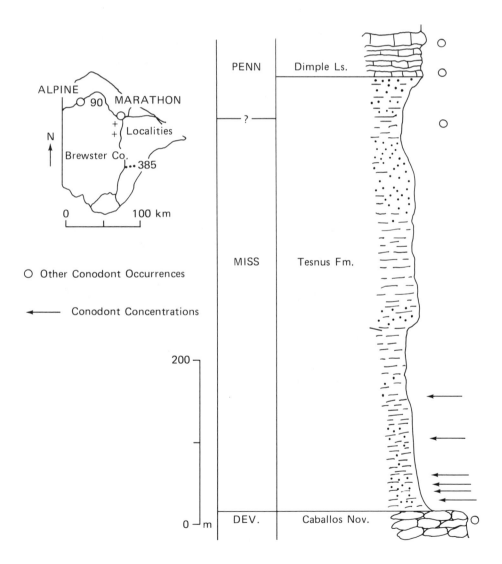

Fig. 4.7 — Tesnus conodonts, Marathon Basin, Texas. 9.6 km and 24 km south of Marathon, Brewster County, Texas. Exposures on East and West Bourland Mountain and on Comb's Ranch klippe.

(a)

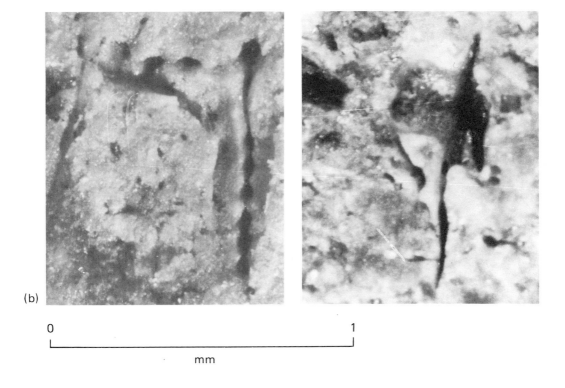

(b)

0   1

mm

Fig. 4.8 — Tesnus conodonts, weathered exterior moulds in siliceous shale.

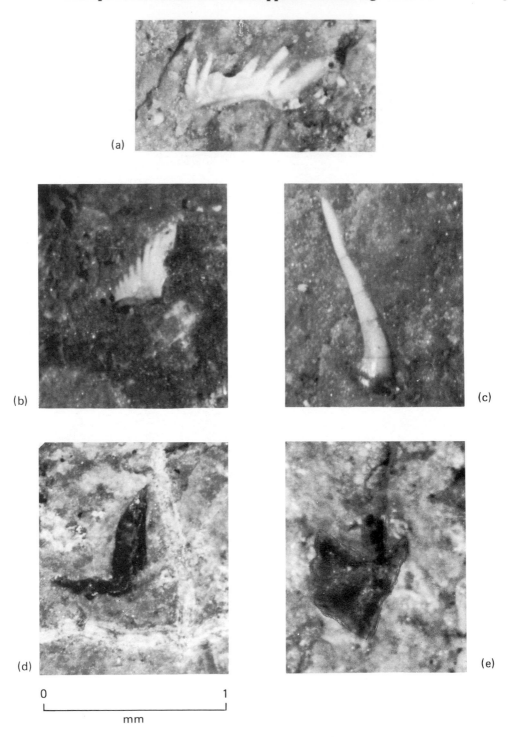

(a)

(b)

(c)

(d)

(e)

0   1
mm

Fig. 4.9 — Tesnus conodonts in siliceous shale.

(a)

(b)

Fig. 4.10 — Tesnus conodonts, stained plaster casts.

Blade-bar (ramiform) elements
  Ozarkodina sp.
  Hindeodella sp.
  Metalonchodina sp.
  Lonchodina sp.
  Ligonodina sp.
  Roundya sp.
  Neoprioniodus singularis (Hass) Stanley, 1958.
  Neoprioniodus ligo (Hass) Elias, 1959.
  Prioniodina sp.
  Prioniodus inclinatus Hass, 1953.
  Spathognathodus sp.

4.5 IVES BRECCIA, CENTRAL TEXAS

The concentration of conodonts in the Ives Breccia of the Houy Formation is in the matrix of a very coarse chert breccia (Figs. 4.11–4.13).

These occurrences were noted by Cloud and Barnes (1948) and were interpreted to be conodont specimens of Grassy Creek (Upper Devonian) age. Later, Seddon (1970) found Lower Mississippian conodonts mixed with Upper Devonian forms at King Springs, 19 km south-west of San Saba, San Saba County, Texas. This is not the type of locality of the Ives Breccia but the exposures are excellent. The faunal list of conodont elements prepared by Seddon (1970) from this locality includes conodonts that are interpreted as Ordovician, Devonian and Lower Mississippian in age. However, the majority of specimens are Devonian in age and a full 70–80% of the conodont specimens are pectiniform elements. The presence of the conodont elements *Pinacognathus*, *Siphonodella*, *Pseudopolygnathus*, *Elictognathus* and *Gnathodus* is ample evi-

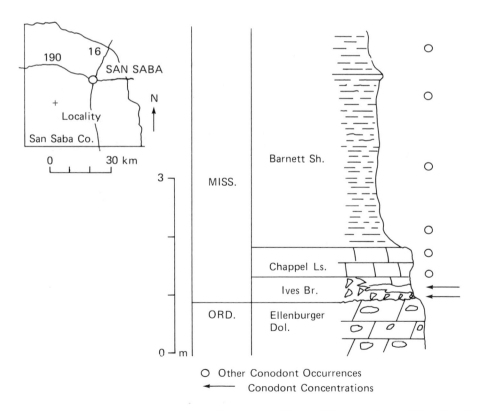

O Other Conodont Occurrences
← Conodont Concentrations

Fig. 4.11 — Ives conodonts, King Springs, San Saba County, Texas. 19.2 km southwest of San Saba, Texas. Same as loc. C-21 of Hass (1959).

0.5

m

0

Fig. 4.12 — Coarse Ives breccia. (After Cloud *et al.*, 1967).

0   3

mm

Fig. 4.13 — Abundant Ives breccia conodonts, freed from matrix.

dence to imply a Lower Mississippian age. This
does not mean that breccias in other places on
the Llano Uplift must also be of Lower Mississippian age. The abundance of conodont specimens in relationship to the breccia is difficult to
ascertain but some of the matrix contains fully
50% conodonts.

Admixtures require that rocks of an older
age must be weathered and the conodonts
freed; the specimens are then deposited
together with the conodonts of the younger
strata. Conodonts seldom show signs of weathering or wear and here in the Ives Breccia the
Ordovician, Devonian and Lower Mississippian
conodonts all appear with the same amber
brown colour and no signs of weathering. The

extensive faunal list given by Seddon (1970)
agrees closely with the list of specimens collected by the writer. These lists are not repeated
here.

4.6 TYPE CHAPPEL LIMESTONE, CENTRAL TEXAS

Conodonts are known throughout the Chappel
Limestone at the type locality (Sellards, 1932)
which is 4 km southeast of San Saba, San Saba
County, Texas (Figs. 4.14 and 4.15). The entire
formation here is 0.8 m thick and the lower
0.2 m of crinoidal grey limestone has very abundant conodonts obtainable by dissolving the
limestone with organic acid.

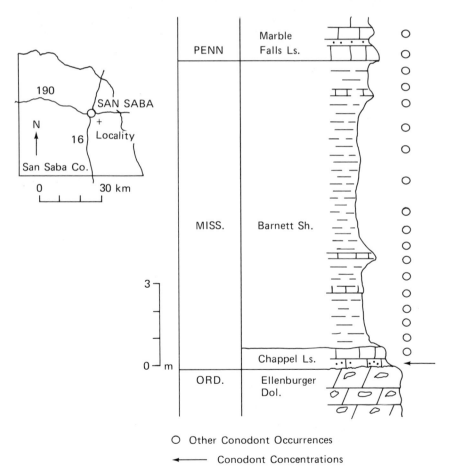

Fig. 4.14 — Type Chappel Limestone conodonts, San Saba County, Texas. 4 km southeast of San Saba,
Texas. Same as loc. C-1 of Hass (1953, 1959).

Fig. 4.15 — Type Chappel Limestone outcrop, San Saba County, Texas.

Girty (1926) and Roundy (1926) described the fossils from these strata but interpreted their collections to be from Mississippian rocks of Boone age. Hass (1959) made a thorough study of the conodonts here and recognized three conodont zones, the lowermost of which was the *Siphonodella cooperi* Zone of Kinderhookian age. He suggested that the lower 0.12–0.31 m of the crinoidal limestone, typically Chappel, be referred to the Houy Formation. Lithologic separation of this zone from the Chappel Limestone is impossible. It could only be done on the conodont content. At this type locality the limestone rests directly on cherty light-grey Ellenburger Dolomite (of Lower Ordovician age).

The faunal list of the Chappel Formation basal unit published by Hass (1959) does not contain as many Devonian forms compared with the Ives Breccia faunal lists. Further the percentage of pectiniform elements is less and the limestone may contain up to 4000 specimens of conodont elements per kilogram of original rock (Fig. 4.16). With Devonian and Lower Mississippian conodont elements, this fauna,

like the Ives fauna, is an admixture. The age determination of the fauna is based on the presence of the well-known Lower Mississippian forms *Siphonodella cooperi* Hass, 1959, *Elictognathus lacerata* (Branson and Mehl) Hass, 1951, *Gnathodus delicatus* Branson and Mehl, 1934, and various species of *Pseudopolygnathus*. The Devonian forms in the fauna must have been freed by weathering and then reworked into Lower Mississippian strata. The faunal list of conodont elements of the Lower Chappel Limestone published by Hass (1959) agrees closely with the list prepared by the writer. These lists are not repeated here.

4.7 CONCLUSIONS

The abundance of conodont elements in the four examples cited above are interpreted to be lag concentrates in which conodonts are prevalent because of their heavy-mineral characteristic and because the conodonts are harder and more resistant to weathering. The two examples in west Texas appear to be thin bedded and the result of deposition by density currents. Those

0   1
└──────────────────────────────┘
                    mm

Fig. 4.16 — Abundant Lower Chappel conodonts, freed from matrix.

of central Texas are related to concentration and admixing along the basal Mississippian unconformity. Close field examination of each bedding plane with a hand lens in the field is necessary to find the conodonts that cannot be extracted by laboratory methods. In all examples there is a geological history of particle sorting and distribution between the death of the conodont organism and the final burial in the sedimentary rock.

4.8 ACKNOWLEDGEMENTS

Appreciation and thanks go to the many students and colleagues who have aided these studies. Thanks are due to Rosemary Brant for typing the manuscript and to David Stephens for photography of a portion of the figures. Acknowledgement is made to the aid provided by the Dorothy Ogden Carsey Memorial Fund of the Geology Foundation of the University of Texas at Austin.

4.9 REFERENCES

Branson, E. B. and Mehl, M. G. 1934 (1933). Conodonts from the Grassy Creek Shale of Missouri. *University of Missouri Studies*, **8**, 171–264.
Cloud, P. E., Jr. and Barnes, V. E. 1948. The Ellenberger Group of Central Texas. *University of Texas Publication*, **4621**, 42–49, 192–198.
Cloud, P. E., Jr., Barnes, V. E. and Hass, W. H. 1967. Devonian–Mississippian transition in Central Texas. *Geological Society of America Bulletin*, **68**, 807–816.
Elias, M. K. 1959. Some Mississippian conodonts from the Ouachita Mountains. In *Geology of the Ouachita Mountains Symposium, Dallas and Ardmore Geological Societies Guidebook*, 141–165.
Ellison, S. P., Jr. 1962. Conodonts from the Trans-Pecos Paleozoic of Texas (Abstract). *American Association of Petroleum Geologists Bulletin*, **46**, 266.
Girty, G. H. 1926. Geologic age and correlation, Part 3. The macro-fauna of the limestone of Boone age. In P. V. Roundy, G. H. Girty and M. I. Goldman, *Mississippian formations of San Saba County, Texas, U.S. Geological Survey Professional Paper*, **146**, 3–43.

Hass, W. H. 1951. Age of the Arkansas novaculite. *American Association of Petroleum Geologists Bulletin*, **35**, 2526–2541.

Hass, W. H. 1953. Conodonts of the Barnett Formation of Texas. *U.S. Geological Survey Professional Paper*, **243-F**, 69–94, 3 plates.

Hass, W. H. 1959. Conodonts from the Chappel Limestone of Texas. *U.S. Geological Survey Professional Paper*, **293-J**, 5 plates.

King, P. B. and Knight, J. B. 1945. Geology of the Heuco Mountains, El Paso and Hudspeth County, Texas. *U.S. Geological Survey, Oil and Gas Investigation Preliminary Map*, **36**, 2 sheets.

Roundy, P. V. 1926. The micro-fauna, Part 2. In P. V. Roundy, G. H. Girty and M. I. Goldman, *Misissippian formations of San Saba County, Texas, U.S. Geological Survey Professional Paper*, **146**, 5–23, 4 plates.

Seddon, G. 1970. Pre-Chappel conodonts of the Llano region. Texas. *University of Texas Bureau of Economic Geology Report of Investigations*, **68**, 1–130.

Sellards, E. H. 1932. The pre-Paleozoic and Paleozoic systems in Texas. *University of Texas Bulletin*, **3232**, 15–230.

Stanley, E. A. 1958. Some Mississippian conodonts from the high resistivity shale of the Nancy Watson No. 1 well in northeastern Mississippi. *Journal of Paleontology*, **32**, 459–476, 6 plates.

Ulrich, E. O. and Bassler, R. S. 1926. A classification of the tooth-like fossils, conodonts, with descriptions of American Devonian and Mississippian species. *US National Museum Proceedings*, **68**, Article 12, 1–63, 11 plates.

5

Conodonts from western Canadian chert: their nature, distribution and stratigraphic application

M. J. Orchard

The Canadian Western Cordillera includes large areas underlain by chert, much of it regarded as of relatively deep-water origin. The chert occurs in a variety of stratigraphic situations: interlayered with pillow lavas, in the matrix of mélange deposits, associated with important mineral deposits and as clasts in younger sediments. The age of the chert has long been problematic because of the absence of macrofossils. Similarly, stratigraphic relations are invariably obscure because the structure is commonly complex. One hundred samples of chert from western Canada processed with dilute hydrofluoric acid have yielded conodonts. The conodonts provide a means of dating the cherts in spite of frequent breakage. Those of Carboniferous to Triassic age are differentiated into a minimum of eight faunas. This is a practical first step in a chert-based conodont biostratigraphy for the Canadian Western Cordillera. The role of the conodonts in structural studies throughout the region is discussed. In the Yukon Territories, areas of older chert within the Selwyn Basin are also dated. The taxonomic make-up and palaeo-environmental significance of these conodont faunas from chert is stressed.

5.1 INTRODUCTION

The Canadian Cordillera includes large areas underlain by chert, much of it regarded as of relatively deep-water 'oceanic' origin. Some of the chert is interlayered with volcanics, some constitutes the matrix of mélange deposits, and much of it occurs in sedimentary sequences lacking carbonate rock types. The cherts are usually devoid of macrofossils and, because their stratigraphic relationships are invariably obscure as a result of structural complexity, their age has long been problematic.

Routine processing of chert with dilute hydrofluoric acid (HF) has yielded numerous conodonts that are generally determinable to at least generic level in spite of frequent breakage. The age dating of these cherts is providing considerable insight into Cordilleran geological problems. The faunas are of added interest because they reveal the character of conodont remains from a poorly explored non-carbonate regime.

In North America, relatively few conodonts from chert have previously been described. Triassic conodonts recovered from some Cordilleran localities were illustrated by Wardlaw and Jones (1981), and others have formed parts of studies by Orchard (1984a, 1985). This paper summarizes the data from western Canada. This is based on samples collected by many geologists, who are acknowledged in the locality registry (see Appendix).

5.2 TECHNIQUES

The extraction of microfossils from siliceous rocks by dissolution in HF is a relatively simple technique that is employed routinely by radiolarian workers (Pessagno and Newport, 1972). Its application in conodont research began with the work of Hayashi (1968, 1969), who recovered a rich Triassic fauna from chert of the Adoyama Formation, central Japan. With the impetus generated by the growth of radiolarian biostratigraphy, and its initial dependence on associated conodonts for zonation in Palaeozoic and Triassic rocks, an increasing number of workers, notably in Japan (Isosaki and Matsuda, 1980, 1982, 1983; Ishida, 1981; Igo and Koike, 1983), are extracting conodonts by this method.

The technique employed is adapted from that used by radiolarian workers. Cherts are broken to walnut-sized pieces and about 500 g placed in each of two 1 l polypropylene beakers in a stainless steel fume-hood. The chert is then covered in a 5% solution of HF and left for 24 h. Sodium carbonate is then slowly added to the solution until the unspent acid is fully neutralized. The sample can then be safely screened using Canadian Standard Sieve Series 12 (1.70 mm), 80 (180 μm) and 200 (75 μm) mesh sieves, washed thoroughly but gently and dried.

This procoedure results in the reduction of the original sample by only about 20–40 g but results in a quantity of +200 residue that can be comfortably picked. This fraction will contain most of the conodonts although larger ones may be found in the +80 residue. However, frag-

mentation and fragility is a common characteristic of conodonts recovered by the HF method, and large intact specimens are rare. Reduction of the residue is not possible by the use of heavy liquids because the conodont apatite is chemically altered and its specific gravity is reduced. In two cases this was determined to be about 2.6, which compares with a specific gravity of 2.65 for silica, the principal component of the residue. Occasionally, some reduction by electromagnetic separation is worthwhile. Repeated rerunning of the sample is often necessary to obtain a diagnostic fauna.

Some variations on this standard procedure have been carried out. Immersion times of 48 and 72 h on several samples did not result in an appreciably different yield, possibly because conodonts released from the matrix early in the process may become increasingly etched or fragmented, whilst further etching of 'fresh' conodonts continues. Isozaki and Matsuda (1980) described conodonts from cherts etched for 2–3 h in 20% HF.

Breaking the chert samples into smaller pieces prior to acid treatment substantially increases the surface area and results in a larger residue, but not necessarily in an increased yield of conodonts. This may be because conodonts are concentrated in parts of the chert sample rather than evenly distributed throughout. Inspection of etched chert fragments will sometimes reveal conodonts, especially when there is a colour contrast, and so subsequent processing can concentrate on these. Similarly, conodontbearing chert may be identified by careful observation in the field.

5.3 CONODONT BIOSTRATIGRAPHY

Details of 100 condont-bearing chert locations in western Canada are provided in this paper (Fig. 5.1 and Appendix). They range from Ordovician to Triassic in age, but the majority are of Late Palaeozoic and Triassic age. Of these, the age of a faunule can generally be determined in terms of one of the eight intervals (1–8) delineated in Fig. 5.2. These are as follows.

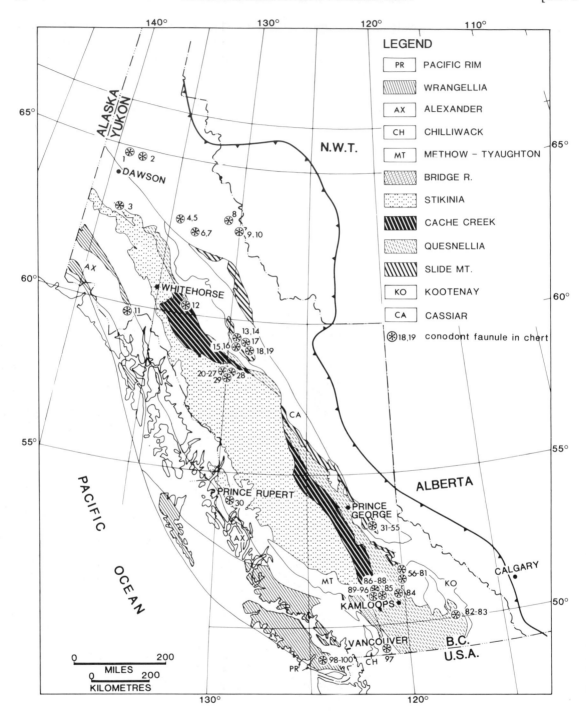

Fig. 5.1 — Conodont-bearing chert localities in western Canada. The base map delineates tectonostratigraphic terranes (after Monger and Berg, 1984): NWT, North West Territories; BC, British Columbia. The solid line in the east defines the eastern limit of deformation.

Interval 1: *Siphonodella–Pseudopolygnathus*
            fauna.
Interval 2: *Gnathodus* fauna.
Interval 3: *Idiognathoides–Idiognathodus*
            fauna.
Interval 4: *Neogondolella±Sweetognathus,*
            Neostreptognathodus fauna.
Interval 5: *Neogondolella serrata* group fauna.
Interval 6: *Neospathodus* fauna.
Interval 7: *Neogondolella* fauna.
Interval 8: *Epigondolella* fauna.

Many faunules are determinable more precisely

in terms of conodont zonal schemes established in carbonate regimes elsewhere. Where appropriate, this is given below in the description of the faunas.

(a) *Siphonodella–Pseudopolygnathus* fauna (Antler Formation, Fennel Formation, Sicker Group)

Siphonodella, which ranges chiefly within the Kinderhookian 'Series' of the Mississippian 'Subsystem', is known from the sediment-sill unit of the Sicker Group (nos. 98, 99) and from the Antler Formation of the Slide Mountain

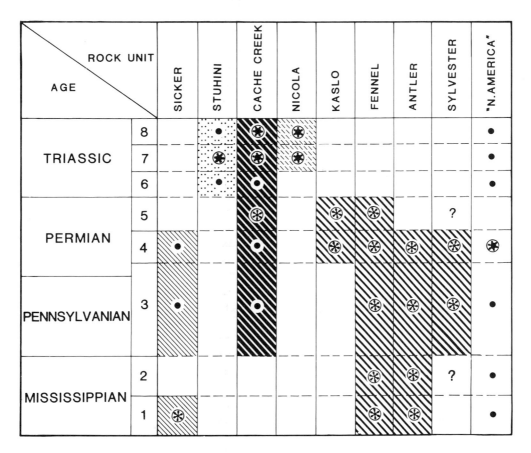

Fig. 5.2 — The age and correlation of chert-bearing rock units of western Canada based on conodonts. The ages 1–8 are explained in the text under conodont biostratigraphy. The circled asterisks denote faunule(s) from chert, and the solid dots denote faunule(s) from carbonate; the two are sometimes combined. The shading of the time stratigraphic intervals follows the patterns in Fig. 5.1 and indicates the tectonostratigraphic terrane in which the rock unit occurs. 'North America' includes those areas regarded as autochonous with respect to the craton.

Terrane (nos. 31–54; Struik and Orchard, 1985). Therein, it is accompanied by *Pseudopolygnathus*, which ranges from latest Devonian through middle Osage (late Tournaisian) in age. Only the latter genus is known at present from the Fennel Formation (no. 63). Both genera are characterized by relatively smooth-plated species, which may be due to their early stage of growth. Cavusgnathoids and coarsely ornate pseudopolygnathids have not been found in chert.

(b) *Gnathodus* fauna ('North America', Sylvester Group?, Antler Formation, Fennel Formation)

Several gnathodid species are known from chert, all of them regarded as characteristic of Late Mississippian (Viséan–early Namurian) time. These are *G. bilineatus, G. texanus, G. girtyi* and *Lochriea* ex gr. *commutata*. The fauna is well represented in the Slide Mountain Terrane, particularly within the Fennel Formation (nos. 57, 60, 76, 80), and the Antler Formation at Slide Mountain (Struik and Orchard, 1985). In one faunule from the Fennel Formation (no. 72), *G. bilineatus* is associated with Pennsylvanian indices. A single faunule is known from the Yukon (no. 9). *Cavusgnathus* and *Kladognathus*, typical of Late Mississippian near-shore (shallow-water) environments, are absent.

(c) *Idiognathoides–Idiognathodus* fauna (Sylvester Group, Antler Formation, Fennel Formation)

Idiognathoides is the most common Upper Carboniferous (Pennsylvanian) genus in chert faunas from the Slide Mountain Terrane. In one sample from the Antler Formation, the genus is prolific and includes rare *Rhachistognathus*. The presence of *Idiognathoides* permits the recognition of Early or Middle Pennsylvanian strata. Representatives of the *Idiognathodus-Streptognathodus* plexus are common in chert faunules too but they are less diagnostic: their presence defines a long (Pennsylvanian and earliest Permian) interval. Gondolelloids,

including both costate and smooth-plated *Gondolella* and the broad-plated '*Neogondolella*', occur sporadically in this interval. A possible specimen of *Declinognathodus* and a few specimens of *Neognathodus* are known from the Fennel Formation (no. 62). Faunules containing several of the above genera may be assigned to early, middle or late parts of fauna 3. For example, the appearance of *Streptognathodus expansus, Gondolella* ex gr. *magnus* and *Streptognathodus* ex gr. *elongatus* are thought to represent successively younger strata in the Antler Formation (Struik and Orchard, 1985), as they do elsewhere.

(d) *Neogondolella±Sweetognathus, Neostreptognathodus* fauna ('North America', Sylvester Group, Antler Formation, Fennell Formation, Kaslo Group)

Typical Permian species of *Neogondolella* occur alone, and commonly with ramiform elements, in four units of the Slide Mountain Terrane, i.e. the Sylvester and Kaslo Groups and the Antler and Fennel Formations. The genus is also widespread in the Yukon autochthon (nos. 4, 6, 7, 10), where *Neogondolella bisselli* occurs with simple sweetognathids (no. 6), here called *Sweetognathus* aff. *S. whitei* (see Orchard, 1984b). The same fauna occurs in chert of the Fennell Formation (no. 70). The association is regarded as Early Permian (Wolfcampian) in age. Similarly, in one chert from the Yukon autochthon (no. 10) and in a second from the Fennell Formation (no. 58), the slightly younger *Neostreptognathodus pequopensis,* of early Leonardian age, occurs. Both *Neostreptognathodus* and *Sweetognathus* are rare in the chert faunas. Most neogondolellids from the Slide Mountain Terrane are composed of specimens which correspond to a broad concept of *Neogondolella idahoensis* and are regarded as Leonardian in age.

(e) *Neogondolella* ex gr. *serrata* fauna (Fennell Formation, Kaslo Group, Cache Creek Group)

The appearance of serrations on the anterior platform of *Neogondolella* is indicative of the Late Permian, *N. serrata* complex, which is known from the Fennell Formation (nos. 74, 77), the Kaslo Group (no. 82) and the Cache Creek Mélange unit (nos. 86, 87). These faunules are devoid of species of *Hindeodus, Diplognathodus* and *Sweetognathus* which characterize Upper Permian limestones of the Marble Canyon Formation of the Cache Creek Group (Orchard, 1984a).

(f) *Neospathodus* fauna (Stuhini Group, Cache Creek Group, Hope)

Neospathodus accompanied by *Neogondolella* characterizes Early Triassic rocks in dark limestones associated with chert both in the Cache Creek Group (Orchard, 1981) and in the Stuhini Group. Clasts in the Nicola Group, possibly derived from the Cache Creek Group, are of this age too. *Neospathodus* was also found in a chert clast associated with volcanics in the Hope map area (no. 97).

(g) *Neogondolella* fauna (Stuhini Group, Cache Creek Group, Nicola Group)

Several Middle Triassic species of *Neogondolella* are identified on the basis of their high blade-like carina (*N. regale–N.* sp. *A–N. excelsa*). Both chert and limestone from the Cache Creek Group (no., 94) and the Stuhini Group (nos. 20, 21, 23–26, 29, 30) contain species of this complex. Elsewhere, interpillow carbonate of the Nicola Group collected by Smith (1979) near Kamloops includes similar Middle Triassic conodonts, as does the eastern sedimentary equivalents of the Nicola Group, the black phyllite unit.

(h) *Epigondolella* fauna (Nicola Group, Cache Creek Group, 'Windy Craggy')

'*Paragondolella*' ex gr. *polygnathiformis*, '*Metapolygnathus*' *nodosus* and *Epigondolella* species identify the Carnian, Carnian–Norian boundary and Norian strata respectively. Of the eight zones recognized within the Norian Stage by Orchard (1983), three are identifed in chert. Carnian chert occurs in the Cache Creek Group (nos. 91, 92), and chert dating from the Carnian–Norian boundary interval (*primitia* Zone)

occurs at Cache Creek (no. 90) and at Teslin (no. 12). Early Norian siliceous sediments are identified close to the Windy Craggy deposit (no. 11) and in the eastern part of the Nicola Group (no. 84).

5.4 THE ROLE OF CHERT-DERIVED CONODONTS IN STRUCTURAL, STRATIGRAPHIC AND TECTONIC STUDIES IN WESTERN CANADA

The following account describes the geological context of the conodonts derived from chert and summarizes their significance. The sample numbers in the headings for sections 5.4(a)–5.4(m) correspond to the sample numbers in the appendix, and in Fig. 5.1, which also depicts currently recognized tectonostratigraphic terranes in western Canada (Monger and Berg, 1984). NTS 1:250 000 map areas are shown in parentheses.

(a) Road River Formation (Dawson, Sheldon Lake) (samples 1, 2, 8)

Three samples of 'Road River' chert from the Yukon have yielded Ordovician conodonts. The most diagnostic is *Pygodus serra* of late Llanvirn (Middle Ordovician) age. *Periodon* and a variety of coniform elements also occur. In the Dawson area, Green (1972) described the Road River Formation (Jackson and Lenz, 1962) as an invariably highly faulted and folded unit consisting of black to drab shale with a varying amount of thinly bedded dark chert, chert-pebble conglomerate and impure gritty quartzite. Calcareous rock types are uncommon. Fossil collections, mainly graptolites, demonstrate that the unit ranges from Middle Ordovician to Late Silurian (Green, 1972) in age. The cherts of the Road River Formation provide a virtually unexplored source for conodonts in this region.

(b) Clast in the 'Indian River Formation' (Stewart River) (sample 3)

The 'Indian River Formation' was named informally by Lowey (1984). It consists of Lower Cretaceous sandstone, shale, conglomerate and minor coal deposited in a marginal marine basin by a southward-prograding fan delta complex (Lowey, 1984). A section of core through this formation included a conglomerate in which pyritized radiolarians were observed. Acid etching produced well-preserved specimens of *Capnodoce* and typical Triassic ramiform elements. This deposit is thought to have accumulated prior to 450 km strike–slip movement on the Tintina Fault, and therefore the source rocks for the chert pebble probably lay far to the south. At present, radiolarian and conodont elements identical with those in the conglomerate are known in Whitehorse Trough in southern Yukon, and in several units in British Columbia.

(c) Upper Palaeozoic chert, Yukon Territory (Glenlyon, Tay River, Nahanni) (samples 4–7, 9, 10)

In the Nahanni map area, Gordey (1981) has mapped a unit (Pt) about 160 m thick which includes orange to gray weathering, pale green to blue-gray chert and minor pale green and splintery shale. Conodonts recovered from this unit are Early Permian or, in one case, Late Mississippian in age. This correlates in part with the Fantasque Formation of the southwestern district of MacKenzie (Gordey *et al.*, 1982a). Further to the west, chert from the Glenlyon and Tay River map areas has produced closely comparable Permian conodont faunules. Apparently, Late Palaeozoic chert is widely distributed in the Yukon.

(d) The Windy Craggy strata-bound massive sulphide deposit (Tatshenshini River) (sample 11)

In the Alsek Ranges of the St. Elias Mountains in northwest British Columbia lies a massive sulphide stratiform deposit known as the Windy Craggy which takes its name from the high (2200 m) rugged mountainous terrain. The deposit, which contains significant amounts of copper, cobalt and gold (MacIntyre, 1984), lies in an area underlain by complexly deformed Palaeozoic clastic and carbonate rocks (Camp-

bell and Dodds, 1983). The rocks in the immediate vicinity of the deposit are intermediate to mafic submarine volcanic units with variable amounts of interbedded limy argillaceous sedimentary rocks. About 25 collections of Late Triassic conodonts have now been recovered from the sediments associated with this important deposit, and consistently date the host rocks as of Late Triassic rather than of Palaeozoic age. The first diagnostic collections, from drill-core, were recovered in siliceous rocks processed in HF.

(e) Cache Creek Group, Teslin (Teslin) (sample 12)

At the northwest end of the French Range Subterrane of the Cache Creek Terrane (Monger and Berg, 1984) in northern British Columbia, previously undated radiolarian chert interbedded with greywacke has yielded Late Triassic conodonts (Carnian–Norian boundary). According to Monger and Berg (1984, loc. 7, Table 1), the greywacke is lithologically similar to coeval rocks in Stikinia to the west, onto which the Cache Creek Terrane is thrust, and the dating of the chert supports the suggestion that in the latest Triassic time the two terranes were amalgamated.

(f) Sylvester Allochthon (McDame, Cry Lake) (samples 13–19)

The stratigraphy of the Sylvester Group was originally described by Gabrielse (1963), who included what is now recognized to be both autochthonous and allochthonous strata, the latter representing an obducted oceanic asemblage lying above North American miogeoclinal rocks. In common with other components of the Slide Mountain Terrane (Antler Formation, Fennell Formation, Kaslo Group; see below), the Sylvester Allochthon comprises chert, greenstone, clastic and ultramafic rock; carbonates appear to be more common than in the southern parts of the terrane. Conodonts from both chert and limestone were first reported by Gordey *et al.* (1982b), who showed that the group constituted an allochthon in which at least three discrete, mildly deformed fault-bounded assemblages were present. In particular, thin-bedded black chert from the lower thrust sheet yielded Early Permian conodonts, and a volcanic unit within the upper thrust sheet was bracketed within the Early–Middle Pennsylvanian by conodonts. Harms (1984, 1985), working to the southeast, has recently recovered additional faunules that support the proposed structural style.

(g) The Stuhini Group (Dease Lake, Cry Lake, Spatzizi, ?Terrace) (samples 20–29, ?30)

The Stuhini Group (Kerr, 1948; Souther, 1971) is based on an extremely variable succession of eugeosynclinal sedimentary and volcanic rocks which crop out in the northern part of the Stikinia Terrane. In the type area of the Stuhini Group, a basal conglomerate uncomformably overlies Lower Permian rocks and is succeeded by a thick sequence of andesitic flows interlayered with pyroclastic rocks. The upper part of the group, consisting primarily of poorly sorted clastic sedimentary rocks with minor lenses of impure limestone (Souther, 1971), has been differentiated as the King Salmon Formation in the Tulsequah Map Area; it contains Late Triassic, principally Carnian molluscs (Tozer, in Souther, 1971, pp. 78–79). A slightly older Middle Triassic, Ladinian shale (map unit 4) was recognized in the adjacent Telegraph Creek Map Area (Souther, 1972).

To the north and east, Stuhini Group strata of the Stikine Canyon consists of a basal unit of grey argillite, siltstone and chert with intercalated augite porphyry breccia and tuff, overlain by a thick sequence of augite porphyry volcanic rocks with thin sedimentary intercalations (Read, 1983). Conodonts extracted from both limestones and cherts in these units demonstrate that sedimentation began there in Early Triassic time and persisted through the Middle Triassic. The colour alteration index (CAI) characteristics of the conodonts, as well as structural aspects of these rocks, show that

there is a significant difference between Triassic and older rocks, possibly as a result of the Sonoma Orogeny at the Palaeozoic–Mesozoic boundary (Read et al., 1983).

Read (1984) extended the study of the Stuhini Group eastwards and further conodont collections confirmed the existence of an extensive pre-Upper Triassic stratigraphy in this part of Stikinia, for which the informal 'Tsaybahe Group' was introduced. A sample of chert collected far to the south in the Terrace map area occupies a similar stratigraphic position between Permian carbonates and Upper Triassic volcaniclastics (G. J. Woodsworth, pers. commun. 1985). This chert is also Middle Triassic in age and provides an important link between distant parts of Stikinia.

(h) Antler Formation, Slide Mountain (McBride) (samples 31–54)

The Antler Formation at Slide Mountain, which gives its name to the terrane, is dominated by basic volcanic rocks, chert, argillite and locally some coarser sediments. Because the formation extends over a large area, early estimates of its thickness ranged upwards from 1100 m (3600 ft in Campbell et al. (1963)). In the absence of fossil data from the Antler Formation, age constraints were provided by the apparently underlying Early Mississippian Greenberry Limestone and the overlying Triassic 'black phyllite'. We now know from the studies of Struik (1980) and the conodont dating documented by Orchard and Struik (1985), that the underlying strata is as young as Early Permian and that the whole of the Antler Formation is thrust onto it. About 25 collections of conodonts have been recovered from the cherts of the Antler Formation and range from Early Mississippian to Early Permian in age. Furthermore, the conodonts demonstrate that the formation consists internally of several thrust sheets that repeat part of or all the unit (Struik and Orchard, 1985). In contrast with original estimates, 300 m of strata represent the total age-range of the formation.

(i) Fennell Formation (Bonaparte Lake, Seymour Arm) (samples 55–81)

The Fennell Formation (Uglow, 1922) lies on the western flank of the Shuswap Metamorphic Complex in south–central British Columbia. Campbell and Tipper (1971) recognized that the formation was correlative with the Antler Formation and referred it to the Slide Mountain Group. Preto and Schiarizza (1982) mapped the formation in the Adams Plateau to Clearwater area and have outlined the geology. The formation consists largely of massive and pillowed basalt interbedded with chert and phyllite. Schiarizza (1981, 1982) has distinguished two units: an eastern ?lower part is characterized by well-bedded, predominantly grey and green chert to cherty mudstone, separated by thinner argillaceous partings, and a western ?upper unit in which chert is generally restricted to small discontinuous pods within greenstone. The rocks have undergone three phases of folding and are intensely sheared in places. 26 conodont faunules ranging in age from early Mississippian to Late Permian are the only fossils known from the formation and facilitate the recognition of several thrust sheets within it (Preto and Schiarizza, 1982).

(j) Kaslo Group (Lardeau) (samples 82, 83)

The Kaslo Group lies in the central part of the Kootenay Arc, a major north-trending arcuate structural zone in southeastern British Columbia. The group is dominated by volcanic rocks and no fossils were previously known from the relatively minor siliceous sediments. This resulted in various interpretations of the structure in this complex area. The discovery of Permian conodonts in the Kaslo Group was therefore of considerable significance and has enabled a rational stratigraphy to be assembled (Klepacki and Wheeler, 1985). The conodonts were extracted from a channel sample of chert assigned to the lower plate units of Klepacki and Wheeler (1985), which consists of tholeiitic pyroxene-porphyry pillow lava, flows and tuffaceous greenstone interbedded with green and white cherty tuff. Orchard (1985) has described the composite conodont fauna, which is now known to include Late Permian species also.

(k) Nicola Group (Ashcroft) (samples 84, 85)

Within Quesnellia, the Upper Triassic and Lower Jurassic sequence is represented by mainly marine calc-alkaline to alkaline volcanic rocks, comagmatic intrusions and intercalated argillite, volcanic sandstone and local limestone. According to Monger and McMillan (1984), the volcanic facies are, in general, older (late Carnian–Early Norian) and acidic in the west and younger (Late Norian) and basic in the east, where they interdigitate with and in part overlie sedimentary rocks.

One sample of chert (no. 84) from unit uTrN5 (breccia, tuff and argillite) within the eastern outcrop yielded Late Triassic–Early Norian conodonts. This lies in an area formerly mapped as being of Palaeozoic age by Cockfield (1948) and supplements Carnian conodont ages reported by Smith (1979) from nearby carbonates.

A second chert faunule was recovered from a clast within unit uTrN3 (breccia, sandstone and local shale) on the western edge of Quesnellia in Nicola Group strata adjacent to the eastern (mélange) belt, or Bonaparte Subterrane of the Cache Creek Terrane. Clasts of both limestone and chert within a conglomerate, which also includes greenstone and ultramafic clasts (Travers, 1978; Monger, 1981), have yielded conodonts of probable Early and ?Middle Triassic age. Because these clasts are older than any sediment known from the Nicola Group, and because blocks of Nicola-like volcanic clastics occur in the mélange unit of the Cache Creek Group, Monger and Berg (1984) suggest that the Bonaparte subterrane was associated with Quesnellia by Upper Triassic time (Monger and Berg, 1984, loc. 13, Table 1).

(l) Cache Creek Group, Cache Creek (Ashcroft) (samples 86–96)

The Cache Creek Terrane consists of structurally complex radiolarian chert, argillite and basalt, extensive bodies of shallow-water carbonate rocks, alpine-type ultramafic rocks, mélange containing these components and locally high-grade blueschist and eclogite (Monger and Berg, 1984). Corals and fusulinids from the limestone are similar to those in Japan, China, Indonesia and the Himalayan region (e.g. Ross and Ross, 1983) and have led to the suggestion that the terrane is far travelled. Several workers (e.g. Monger, 1981) have regarded the Cache Creek Group as an early Mesozoic subduction complex and the Nicola Group as its corresponding volcanic arc.

At Cache Creek, the terrane has been divided into three units: the eastern belt consists predominantly of chert, argillite and basic volcanics with limestone, chert and ultramafic bodies of variable size that are interpreted as olistoliths within the so-called mélange unit. The central belt, or Marble Canyon Formation (Duffell and McTaggart, 1952), consists primarily of bedded and massive carbonates which yield the 'Tethyan' verbeekinid fusulinaceans. Minor amounts of chert, argillite and volcanics also occur in this belt, particularly in the south. The western belt consists mainly of chert and argillite, with some carbonate and volcanic rock.

Prior to the discovery of conodonts in the Cache Creek Group (Orchard, 1981), most of the fossil dates were based on foraminiferids from the carbonate rocks (see Trettin, 1980, for a summary). This led to the assumption that the group was wholly Palaeozoic in age. Radiolarians reported by Travers (1978) and conodonts described by Orchard (1984a; this volume) show that the matrix of the 'mélange unit' is Late Permian–Late Triassic in age. Conodont data from the Marble Canyon Formation, including Early and Middle Triassic faunules associated with cherts and tuffs, demonstrate that a considerable, perhaps major, part of the formation is Triassic in age. Furthermore, conodonts from the carbonate olistoliths within the mélange are older than both the matrix of the mélange unit and carbonate of the Marble Canyon Formation, from which derivation might logically have been assumed. On the

assumption of continuity of chert sedimentation in the Mélange unit, the eastern and central belts of the Cache Creek Group appear to have coexisted during much of Late Permian and Triassic time. At present, there are no compelling palaeontological data for a common origin for the two belts.

(m) Clast in chert breccia (Hope) (sample 97)
An unnamed volcanic greenstone with sparse interpillow chert breccia forms a basement to the Pasayton Trough, a northerly extension of the Methow Basin in Washington State. Between the greenstone and the unconformably overlying Ladner Group there is a chert breccia 1 m thick from which conodonts of Early Triassic age have been recovered. G. Ray (pers. commun. 1985) regards this deposit as having formed during pre-Ladner weathering and suggests that the age of the conodonts is that of the underlying formation, analysis of which indicates an oceanic ridge origin. The conodonts, which represent the only fossil control, imply that the time gap beneath the Lower to Middle Jurassic Ladner Group is considerable (Ray, in press).

(n) Sicker Group, Vancouver Island (Alberni) (samples 98–100)
The Sicker Group consists of pillow lavas and pyroclastics in the lower part, overlain by cherty tuffs and argillites, siliceous sediments and finally the carbonate of the Buttle Lake Formation. Muller (1980) has proposed a four-fold division of the group, including an informal sediment–sill unit beneath the Pennsylvanian and Early Permian Buttle Lake Formation. The unit, which comprises thinly bedded turbidite-like massive argillite and siltstone cut by diabase sills, was thought by Muller (1980) to be either coeval with, or older than, the Buttle Lake Limestone. Since virtually all the fossils from the Sicker Group come from the upper limestone, the earlier history of sedimentation has not been well constrained. Recently, Pessagno (in Muller, 1980, addendum) reports a radiolarian faunule of middle Kinderhookian through

early Meramecian age within the sediment–sill unit near Cowichan Lake. Several collections of chert made by M. Brandon from the same area have now yielded conodonts of Early Mississippian (Kinderhookian) age too, thus confirming a long history of Carboniferous sedimentation in the region.

5.5 ACKNOWLEDGEMENTS

D. Jones and B. Murchey (US Geological Survey, Menlo Park) kindly advised on chert-processing techniques during the initial phase of this study. This chapter documents conodonts faunules collected by many Cordilleran geologists (listed in the appendix) who have freely contributed information on the stratigraphic context of their samples; their contribution is gratefully acknowledged. Over several years, a series of technicians was involved in the processing of numerous cherts, including many barren samples not discussed here; these included Anne Walton, Carlo Lagatolla and Peter Krauss. P. Krauss was also responsible for much of the conodont photography. T. Oliveric drafted the figures.

5.6 APPENDIX: LOCALITY REGISTRY OF CONODONT-BEARING CHERTS

Each sample number corresponds to a number in Fig. 5.1. The following information is provided: Geological Survey of Canada locality number. Collector of the sample, year; field number. National Topographic System number and map name: latitude; longitude. Geographic location. Stratigraphic context. Conodont taxa. Colour alteration index (CAI). Age. (Published reference.)

The conodonts are deposited in the Geological Survey of Canada collections.

Sample 1.   C-101574. R. I. Thompson, 1982; 82-TW-265. 116B/10, Dawson: 64° 41′ 25″; 138° 44′ 30″. Seela Pass. Road River Formation. *Dapsilodus?* sp., *Periodon* sp., oistodiform and ramiform elements. CAI, 5. Ordovician in age.

Sample 2.   C-103896. R. I. Thompson, 1983; 83-TW-77. 116B/9, Dawson: 64° 44′; 138°

14′. Rein Barite. Road River Formation, chert from above graptolitic shale. *Periodon* sp., *Pygodus serra*, ramiform elements. CAI, 5. Middle Ordovician in age.

Sample 3.   C-102374. G. W. Lowey, 1981; 81-TO-137-2-3. 1050/N Stewart River: 63° 43′; 139° 08′. Haystack Mountain. Indian River Formation, radiolarian chert pebble in Cretaceous conglomerate. *Capnodoce* sp., ramiform element. CAI, 2–3. Late Triassic in age according to B. Murchey.

Sample 4.   C-103768. S. P. Gordey, 1982; 84-GGA-82-55B2. 105L/15, Glenlyon: 62° 46′ 50″; 134° 40′ 05″. Unnamed chert. *Neogondolella* sp., ramiform elements. CIA, 5. Probably Permian in age.

Sample 5.   C-103769. S. P. Gordey, 1982; 84-GGA-82-59A3. 105L/19, Glenlyon: 62° 47′ 05″; 134° 36′ 40″. Unnamed chert. Gnathodid?, ramiform elements. CAI, 6. Permo-Carboniferous in age.

Sample 6.   C-103779. S. P. Gordey, 1983; 84-GGA-83-2D1. 105K/12, Tay River: 62° 44′ 30″; 133° 40′ 00″. Unnamed chert, *Neogondolella* sp., *Sweetognathus* cf. *S. whitei*, ramiform elements. CAI, 5–6. Early Permian in age.

Sample 7.   C-103813. S. P. Gordey, 1983; 83-GGA-2B2. 105K/12, Tay River: 62° 41′ 30″; 133° 39′ 10″. Unnamed chert. *Neogondolella* cf. *N. gracilis* ramiform elements, CAI, 6. Early Permian (Leonardian) in age.

Sample 8.   C-101905. S. P. Gordey, 1982; 82-GGA-1B2. 105J/14, Sheldon Lake: 62° 57.3′; 131° 13.5′. Road River Formation. *Periodon* sp., ramiform element, coniform elements. CAI, 5. Ordovician in age.

Sample 9.   C-102656. S. P. Gordey, 1982; 82-GGA-40.354. 105I/13, Nahanni: 62° 52.5′; 129° 49.7′. Unnamed chert. *Gnathodus* cf. *G. bilineatus*, *Paragnathodus?* sp., ramiform elements. CAI, 4. Probably Late Mississippian in age.

Sample 10.   C-087592. S. P. Gordey, 1982; GGA-80-6B-2. 105I/13, Nahanni: 62° 48.7′; 129° 42.0′. Unnamed chert. *Neogondolella* cf. *N. bisselli*, *Neostreptognathus pequopensis*, ramiform elements. CAI, 5. Early Permian (early Artinskian) in age.

Sample 11.   C-101387. M. J. Orchard, 1982; 82-MJO-Falconbridge #28536. 114P, Tatshenshini River: 59° 43′; 137° 43′. Windy Craggy DDH 5b, 106–107 ft. *Epigondolella* sp., ramiform element. CAI, 5?. Late Triassic (Norian, probably Early) in age.

Sample 12.   C-101892. J. W. H. Monger, 1980; 80-

MV-T3. 105C, Teslin: 60° 05′ 50″; 133° 25′ 40″. Ridge top, 6 miles southsoutheast of Snafu Lake. Cache Creek Group, Unit 4a of Mulligan, 1963. *Epigondolella primitia–E. abneptis* subsp. A of Orchard, ramiform element. CAI, 5. Late Triassic–Early Norian in age.

Sample 13. C-103665. J. F. Psutka, 1983; 83-MJO-P83-62F. 104P, McDame: 59° 16′ 03″; 129° 40′ 20″. Cassiar at 3550 ft, 750 m due west of Snowy Creek and 3.35 km from peak 6644′ at 240°. Sylvester Group. *Idiognathoides?* sp., ramiform elements. CAI, 5–6. Early–Middle Pennsylvanian in age.

Sample 14. C-103666. J. F. Psutka, 1983; 83-MJO-P83-63F. 104P, McDame: 59° 15′ 55″; 129° 39′ 55″. Cassiar, 3480 ft, 450 m due east of Snowy Creek and 3.55 km from peak 6644′ at 222°. Sylvester Group. *Idiognathoides* sp., ramiform elements. CAI, 6–7. Early–Middle Pennsylvanian in age.

Sample 15. C-087682. S. P. Gordey, 1981; 81-GGAS-5B-1. 104P/3, McDame: 59° 12′ 12″; 129° 24′ 12″. 3.3 km southwest of Mt. Pendleton. Sylvester Group, lower shale–chert unit. *Neogondolella* sp., ramiform elements. CAI, 5. Permian (probably Early) in age. (Gordey, *et al.*, 1982b, no. 4)

Sample 16. C-087679. S. P. Gordey, 1981; 81-GGAS-13A-1. 104P, McDame: 59° 08.1′; 129° 28.2′. 4.5 km south of Juniper Mountain. Sylvester Group. *Neogondolella* sp., ramiform element, CAI, 6–7. Permian in age. (Gordey *et al.*, 1982b, no. 5.)

Sample 17. C-087683. H. Gabrielse, 1981; 81-GA-red. 104P, McDame: 59° 02′; 128° 47′. Sylvester Group. *Neogondolella* sp., gnathodid?. CAI, 6–7. Permian in age. (Gordey *et al.*, 1982b, no. 8.)

Sample 18. C-116040. T. Harms, 1984; 84-GAH-322-R. 104I, Cry Lake: 58° 59′ 22″; 128° 40′ 07″. Sylvester Group. *Neogondolella* sp., ramiform elements. CAI, 7?. Permian in age.

Sample 19. C-116428. T. Harms, 1984; 84-GAH-331-R. 104I, Cry Lake: 58° 55′ 45″; 128° 31′ 00″. Sylvester Group. *Idiognathoides* sp., *Idiognathodus?* sp., *Gondolella* ex gr. *laevis*, ramiform elements. CAI, 6?. Pennsylvanian (Early–Middle) in age.

Sample 20. C-102901. P. B. Read, 1983; 83-MJO-R83-25F. 104J, Dease Lake: 58° 12′ 49″; 130° 13′ 11″. Stikine River, 4500 ft elevation, 11.7 km at 267° fron north end of Tsenaglode Lake. 'Tsaybahe' Group.

Neogondolella sp., CAI, 6?. Triassic (probably Middle) in age. (Read, 1984, no. F8.)

Sample 21. C-102924. J. F. Psutka, 1983; 83-MJO-P83-45F. 104J, Dease Lake: 58° 10′ 59″; 130° 18′ 06″. Stikine River, 3490 ft elevation east bank of Pallen Creek. Stuhini Group. *Neogondolella* sp. A, ramiform elements, CAI, 5. Middle Triassic in age.

Sample 22. C-102920. J. F. Psutka, 1983; 83-MJO-P83-41F. 104J, Dease Lake: 58° 08′ 36″; 130° 23′ 12″. Stikine River, 1700 ft elevation north bank of Stikine, 2.5 km downstream from mouth of Pallen Creek. Stuhini Group. *Metapolygnathus nodosus*, ramiform elements. CAI, 4?. Late? Triassic (Carnian?) in age.

Sample 23. C-102764. J. F. Psutka, 1982; 82-MJO-P30F. 104J/1W, Dease Lake: 58° 07′ 10″; 130° 16′ 14″. Stikine, Site Z drill-core. Stuhini Group, grey massive chert in dark grey siliceous argillite. *Neogondolella* cf. *N. regale*, *N.* sp. A, ramiform elements. CAI, 5. Middle Triassic (Anisian) in age. (Read, 1983, no. F12b.)

Sample 24. C-102765. J. F. Psutka, 1982; 82-MJO-P31F. 104J/1W, Dease Lake: 58° 07′ 10″; 130° 16′ 14″. Stikine, site Z. Stuhini Group, grey massive to bedded chert in argillite. *Neogondolella* sp., ramiform elements. CAI, 6?. Triassic (probably Middle) in age. (Read, 1983, no. F12c.)

Sample 25. C-102768. J. F. Psutka, 1982; 82-MJO-P35F. 104J/1W, Dease Lake: 58° 07′10″; 130° 16′ 44″. Stikine, site Z. Stuhini Group, grey massive chert. *Cratognathus* sp., *Neogondolella* cf. *N. constricta*, ramiform elements. CAI, 5. Middle Triassic (probably late Anisian–early Ladinian) in age. (Read, 1983, no. F12d.)

Sample 26. C-102767. J. F. Psutka, 1982; 82-MJO-P34F. 104J/1W, Dease Lake: 58° 07′ 07″; 130° 16′ 32″. Stikine, site Z. Stuhini Group, grey chert bed in black siliceous argillite. *Neogondolella* sp., ramiform elements. CAI, 5. Triassic (probably Middle) in age. (Read, 1983, no. 12F.)

Sample 27. C-102771. P. B. Read, 1982; 82-MJO-R12F. 104J, Dease Lake: 58° 06′ 33″; 130° 15′ 37″. Stikine River, site Z drill-core. 'Tsaybahe' Group, light-grey to white chert in graphitic siltstone. *Neogondolella?* sp., CAI, 6?. Triassic (probably Middle or Late) in age. (Read, 1983, no. F15p; Read, 1984, no. F78.)

Sample 28. C-087706. P. B. Read, 1981; 81MJO-

H81-1F. 104I/4W, Cry Lake: 58° 10′ 53″; 129° 59′ 50″. At 5225 ft elevation, 0.5 km at 232° from Thenatlodi Mountain. Stuhini Group, light- to dark-grey chert. *Neogondolella* sp., ramiform elements. CAI, 6. Triassic in age.

Sample 29. C-087707. P. B. Read, 1981; 81MJO-H81-10F. 104/13, Spatsizi: 57° 58′ 31″; 129° 49′ 47″. 4925 ft elevation, 2.8 km at 336° from Tsaybahe Mountain peak. Stuhini Group, medium-grey tuffaceous argillite with local chert layers. *Neogondolella* cf. *N. constricta*, *N.* cf. *N. regale*, *N.* sp. A, ramiform elements. CAI, 5. Middle Triassic (probably Anisian) in age. (Read, 1984, no. F124.)

Sample 30. C-101424. G. J. Woodsworth, 1984; 84-WV-19c-1. 103I, Terrace: 54° 31′ 00″; 128° 22′ 30″. North side of Copper River. From black-and-white thin-bedded chert stratigraphically above Permian carbonate and below Jurassic basal Telkwa polymictic breccia. *Neogondolella* sp. A, ramiform elements. CAI, 5. Middle Triassic in age.

Samples 31–54. C-102658–C-102685 (also C-86277–C-86281). M. J. Orchard and L. C. Struik, 1981; 81-MJOSCB-33A through AB. 93H, McBride: 53° 09′; 121° 29′. Sliding Mountain, near Barkerville. Antler Formation, sequence of chert samples collected from near peak of mountain (33A) and from southern and southwestern flank at successively lower structural levels (33B–33AB). CAI, 5–7. Late Devonian?, Early Mississippian–Early Permian in age. (Struik and Orchard, 1985.)

Sample 55. C-102844. L. C. Struik, 1984; 84-SCB-Bowron. 93H, McBride: 53° 09′ 05″; 121° 30′ 02″. Bowron Lake Road. Antler Formation. *Neogondolella* sp., ramiform elements. CAI, 6–7. Permian in age.

Sample 56. C-102470. P. Schiarizza, 1981; PS81-121. 92P/9, Bonaparte Lake: 51° 36′ 57″; 120° 03′ 13″. 3200 ft elevation, south of Clearwater between Russell Creek and North Thompson River. Fennell Formation. *Gondolella* cf. *G. laevis*, *Idiognathoides* sp., ramiform elements. CAI, 5–6. Early––Middle Pennsylvanian in age.

Sample 57. C-102472. P. Schiarizza, 1981; PS81-123. 92P/9, Bonaparte Lake: 51° 36′ 33″; 120° 03′ 48″. 3350 ft elevation, south of Clearwater between Russell Creek and North Thompson River. Fennel Formation. *Gnathodus girtyi, Lochriea* ex gr. *commu-*

tata, ramiform elements. CAI, 5–6. Late Mississippian in age.

Sample 58. C-102458. P. Schiarizza, 1981; PS81-50. 92P/9, Bonaparte Lake: 51° 36′ 02″; 120° 00′ 02″. 4000 ft elevation, slopes west of head of Hascherk Creek. Fennell Formation. *Neogondolella* sp., *Neostreptognathodus* cf. *N. pequopensis*, ramiform elements. CAI, 7. Early Permian (Leonardian) in age.

Sample 59. C-102459. P. Schiarizza, 1981; PS81-51. 92P/9, Bonaparte Lake: 51° 36′ 02″; 120° 00′ 02″. 4000 ft elevation west of Eagle Bay contact, slopes west of head of Hascheak Creek. Fennell Formation. Ramiform elements. CAI, 6–7. Ordovician––Triassic in age.

Sample 60. C-102461. P. Schiarizza, 1981; PS81-66. 92P/9, Bonaparte Lake: 51° 32′ 53″; 120° 04′ 20″. 4600 ft elevation, west of Mount McCarthy microwave station. Fennell Formation. *Gnathodus* cf. *G. bilineatus*, *G. girtyi*, ramiform elements. CAI, 6–7. Late Mississippian in age.

Sample 61. C-102473. P. Schiarizza, 1981; PS81-124. 92P/9, Bonaparte Lake: 51° 31′ 35″; 120° 03′ 40″. 4300 ft elevation, south of Mount McCarthy, between Axel Creek and Joseph Creek. Fennell Formation. *Gondolella* cf. *G. laevis*, *Idiognathoides* sp., *Idiognathodus* spp., *Streptognathodus expansus*. CAI, 7. Early–Middle Pennsylvanian in age.

Sample 62. C-102442. P. Schiarizza, 1981; PS81-1. 92P, Bonaparte Lake: 51° 31′ 15″; 120° 07′ 38″. 2300 ft elevation, small roadside outcrop north of Hallamore Lake. Fennell Formation, western belt. *Declinognathodus?* sp. CAI, 5–6. Early? Pennsylvanian in age.

Sample 63. C-102465. P. Schiarizza, 1981; PS81-81. 92P/9, Bonaparte Lake: 5° 31′ 02″; 120° 04′ 20″. 4600 ft elevation on south-trending ridge north of Joseph Creek and south of Mount McCarthy microwave station. Fennell Formation. *Pseudopolygnathus* cf. *P. nudus*, ramiform elements. CAI, 5–6. Early Mississippian in age.

Sample 64. C-102478. P. Schiarizza, 1981; PS81-135. 92P/9, Bonaparte Lake: 5° 30′ 56″; 120° 00′ 04″. 5600 ft elevation north slopes of Joseph Creek. Fennell Formation. *Neogondolella* sp., ramiform elements. CAI, 6–7. Permian in age.

Sample 66. C-102483. P. Schiarizza, 1981; PS81-225. 92P/8, Bonaparte Lake: 51° 26′ 10″; 120°

05′ 20″. 4950 ft elevation, east of Dunn Lake. Fennell Formation. Ramiform elements. CAI, 7. Ordovician Triassic in age.

Sample 67. C-102436. V. A. Preto, 1980; V2. 92P/81, Bonaparte Lake: 51° 22′ 47″; 120° 02′ 55″. 0.5 miles east of Chu Chua Deposit. Fennell Formation, ribbon chert within massive and pillowed basalts. *Neogondolella* sp., ramiform elements. CAI, 6–7. Permian in age.

Sample 68. C-102439. V. A. Preto, 1980; V83. 92P/8h, Bonaparte Lake: 51° 22′ 13″; 120° 00′ 32″. Fennell Formation, eastern belt. *Gondolella* cf. *G. laevis*, *G.* cf. *G. magna*. CAI, 7. Middle? Pennsylvanian in age.

Sample 69. C-102431. V. A. Preto, 1980; S140. 92P/8a, Bonaparte Lake: 51° 21′ 14″; 120° 00′ 35″. Fennell Formation, eastern belt, well-bedded chert with massive and pillowed basalt. *Neogondolella* sp., ramiform elements. CAI, 6–7. Permian in age.

Sample 70. C-102441. V. A. Preto, 1980; V90A. 82M, Seymour Arm: 51° 20′ 59″; 119° 59′ 52″. Fennell Formation. *Neogondolella* sp., *Sweetognathus* aff. *S. whitei*, ramiform elements. CAI, 6. Early Permian in age.

Sample 71. C-102440. V. A. Preto, 1980; V90. 82M/5d, Seymour Arm: 51° 20′ 57″; 119° 59′ 45″. Fennell Formation, eastern belt, ribbon chert very close to limestone of Kinderhookian age. *Idiognathodus* sp., *Idiognathoides* sp., *Idioprioniodus?*sp., ramiform elements. CAI, 6. Early–Middle Pennsylvanian in age.

Sample 72. C-102423. V. A. Preto, 1980; S124. 92P, Bonaparte Lake: 51° 19′ 51″; 120° 00′ 13″. Fennell Formation. *Gondolella* sp., *Gnathodus bilineatus*, *Idiognathodus* sp., ramiform elements. CAI, 5–6. Early? Pennsylvanian in age.

Sample 73. C-102416. V. A. Preto, 1980; S9. 82M, Seymour Arm: 51° 19′ 48″; 119° 59′ 12″. Fennell Formation. *Neogondolella* sp., ramiform elements. CAI, 6–7. Permian in age.

Sample 74. C-102417. V. A. Preto, 1980; S13. 82M, Seymour Arm: 51° 17′ 47″; 119° 58′ 33″. Fennell Formation, vertical beds of chert–argillite–siltstone. *Neogondolella serrata* group, ramiform elements, reworked protoconodonts?. CAI, 5. Late Permian (Guadalupian) in age.

Sample 75. C-102408. V. A. Preto, 1980; D68. 82M, Bonaparte Lake: 51° 18′ 01″; 120° 00′ 48″.

Fennell Formation. '*Spathognathodus?*' sp. CAI, 6. Devonian–Mississippian? in age.

Sample 76. C-102468. P. Schiarizza, 1981; PS81-108. 92P/8, Bonaparte Lake: 51° 15′ 02″; 120° 01′ 09″. 3200 ft elevation, east slopes of Bothel Creek. Fennell Formation. *Gnathodus* cf. *G. bilineatus*, *G. girtyi*, *Lochriea* ex gr. *L. commutata*, ramiform elements. CAI, 5–6. Late Mississippian in age.

Sample 77. C-102415. V. A. Preto, 1980; MC11. 92P, Bonaparte Lake: 51° 14′ 34″; 120° 04′ 57″. Fennell Formation, associated with massive and pillow basalts. *Neogondolella serrata* group. CAI, 7. Late Permian (Guadalupian) in age.

Sample 78. C-102409. V. A. Preto, 1980; M18. 92P/1h, Bonaparte Lake: 51° 14′ 22″; 120° 03′ 02″. Fennell Formation, bedded, buff and greenish chert within pillow basalt. *Neogondolella* sp. CAI, 7. Late? Permian in age.

Sample 79. C-102412. V. A. Preto, 1980; MC4. 92P, Bonaparte Lake: 51° 13′ 21″; 120° 04′ 12″. Fennell Formation. *Neogondolella* sp., ramiform elements. CAI, 6–7. Early? Permian in age.

Sample 80. C-102413. V. A. Preto, 1980; MC7. 92P, Bonaparte Lake: 51° 13′ 56″; 120° 04′ 55″. Fennell Formation. *Gnathodus* cf. *G. texanus*, *Rhachistognathus?* sp. CAI, 5–6. Late Mississippian in age.

Sample 81. C-102414. V. A. Preto, 1980; MC9. 92P, Bonaparte Lake: 51° 12′ 28″; 120° 05′ 12″. From Barriere, east to East Barriere Lake by roadside. Fennell Formation, pillow basalts and cherts. *Gondolella* cf. *G. laevis*, *Idiognathodus* spp., *Idiognathoides* sp., *Idioprioniodus* sp., *Neognathodus* sp., *Streptognathodus?* sp., ramiform elements. CAI, 7. Middle? Pennsylvanian in age.

Sample 82. C-103451. D. W. Klepacki, 1983; 83-WBDK-53-e1. 82K/3, Lardeau: 50° 08′ 00″; 117° 16′ 45″. Headwaters of Keen Creek. Kaslo Group, lower pale chert. *Neogondolella* sp., *N. serrata* group, ramiform elements. CAI, 5–7. Permian in age (Orchard, 1985, no. 19.)

Sample 83. C-103452. D. W. Klepacki, 1983; 83-WBDK-53-e1. 82K/3, Lardeau: 50° 08′ 00″; 117° 16′ 45″. Headwaters of Keen Creek. Kaslo Group, lower pale chert. *Neogondolella* sp. CAI, 5–7. Permian in age (Orchard, 1985, no. 19.)

Sample 84. C-081795. J. W. H. Monger, 1981; 81-

MV-4b. 92I, Ashcroft: 50° 52' 40"; 120° 15' 35". North Thompson Highway, approximately 25 km north of Kamloops. Nicola Group, outcrop of cherty argillite. *Epigondolella abneptis*. CAI, 5. Late Triassic, Early Norian in age. (Monger and McMillan, 1984, no. N31.)

Sample 85. C-087333. J. W. H. Monger, 1981; 81-MV-77-2. 92I/14, Ashcroft: 50° 47' 10"; 121° 17' 07". Above east bank of Bonaparte River, 2.7 km southsoutheast Cache Creek. Nicola Group, chert clasts. *Neogondolella* sp., *Cratognathus?* sp., ramiform elements. CAI, 3–4. Middle? Triassic in age. (Monger and McMillan, 1984, no. N1.)

Sample 86. C-102500. M. J. Orchard, 1980; 80-MJOS-LOON3. 92I/14, Ashcroft: 50° 58' 25"; 121° 27' 08". 0.6 km northnortheast of junction of turn-off to Loon Lake and Highway 97. Cache Creek Group, well-bedded ribbon chert in mélange unit. *Neogondolella postserrata* s.l., ramiform elements. CAI, 5–6. Late Permian (Guadalupian) in age. (Orchard, 1984a, no. 4.)

Sample 87. C-102552. K. R. Shannon, 1980; 80-MVS-191. 92I/13E, Ashcroft: 50° 54' 55"; 121° 36' 02". 7.5 km northwest of junction of Robertson Creek and Hat Creek Highway. Marble Canyon Formation, at base, interbedded cherts, phyllites and greenstones. *Neogondolella* sp., ramiform elements. CAI, 6. Permian (post-Wolfcampian) in age. (Orchard, 1984a, no. 7.)

Sample 88. C-087650. K. R. Shannon, 1980; 80-MVS-182. 92I/14, Ashcroft: 50° 53' 13"; 121° 27' 10". 1.45 km west of Carquile, large knob above Highway 12. Cache Creek Group, bedded ribbon chert in mélange unit. *Neogondolella* sp., ramiform elements. CAI, 5–6. Permian in age. (Orchard, 1984a, no. 8.)

Sample 89. C-116191. W. R. Danner, 1984; 85-OF-DAN. 92I, Ashcroft: 50° 49'; 121° 20'. Cache Creek, road cut behind Post Office. Cache Creek Group, mélange unit, grey chert. Ramiform elements. CAI, 5. Triassic in age.

Sample 90. C-116179. M. J. Orchard, 1985; 85-OF-CC11B. 92I, Ashcroft: 50° 48'; 121° 18'. Cache Creek, east of Bonaparte River. Cache Creek Group?, black chert about 1 m above base of section. *Epigondolella primitia*. CAI, 5–6. Late Triassic (Carnian–Norian boundary) in age.

Sample 91. C-116180. M. J. Orchard, 1985; 85-OF-

CC11C. 92I, Ashcroft: 50° 48'; 121° 18'. Cache Creek, east of Bonaparte River. Cache Creek Group?, black chert near top of section. *'Paragondolella'?* sp. CAI, 5–6. Late Triassic (Carnian?) in age.

Sample 92. C-102551. K. R. Shannon, 1980; 80-MVS-66a. 92I, Ashcroft: 50° 44' 40"; 121° 20' 10". On small knob 2.5 km northnorthwest of Ashcroft main junction. Cache Creek Group, grey ribbon chert in mélange unit. *'Paragondolella'* ex gr. *polygnathiformis*, ramiform elements. CAI, 5?. Late Triassic (Carnian) in age. (Orchard, 1984a, no. 18.)

Sample 93. C-087649. K. R. Shannon, 1980; 80-MVS-65C. 92I, Ashcroft: 50° 44' 40"; 121° 20' 10". On small knob 2.6 km northnorthwest of Ashcroft Manor junction on Highway 1. Cache Creek Group, ribbon chert in the mélange unit. *Neogondolella* sp., ramiform elements. CAI, 5. Permian (post-Wolfcampian) in age.

Sample 94. C-118486. M. J. Orchard, 1984; 84-MJO-CH14B. 92I, Ashcroft: 50° 40'; 121° 28'. Cornwall Hill, on track 2.3 km from look-out. Cache Creek Group, radiolarian chert in Marble Canyon Formation. *Neogondolella* cf. *N. excelsa*, ramiform elements. CAI, 6–7. Middle Triassic in age.

Sample 95. C-087651. K. R. Shannon, 1980; 80-MVS-296g. 92I/12E, Ashcroft: 50° 42' 04"; 121° 32' 42". 3.75 km northwest of north end of Bedard Lake, on logging road cut. Cache Creek Group, chert interbedded with fusulinid limestone, greenstone and argillite in Marble Canyon Formation. *Neogondolella* sp., ramiform elements. CAI, 5. Permian (post-Wolfcampian) in age. (Orchard, 1984a, no. 23).

Sample 96. C-087652. K. R. Shannon, 1980; 80-MVS-322. 92I/12E, Ashcroft: 50° 42' 04"; 121° 32' 42". Collected adjacent to no. 95. Ramiform elements. CAI, 5. Permian? in age. (Orchard, 1984a, no. 24.)

Sample 97. C-102363. G. E. Ray, 1981; 81-MJO-GER-87. 92H/6, Hope: 49° 26' 40"; 121° 15' 30". 10 km southsoutheast of Carolin Mine. 'Spider Peak' Formation interpillow breccias at base below Ladner Group. *Neospathodus* sp., ramiform elements. CAI, 5. Early Triassic in age.

Sample 98. C-127519. M. T. Brandon, 1984; 84-YB-8498-1A. 92F, Alberni: 49° 00' 27"; 124° 26' 17". Top of main fork of Shaw Creek, Cowichan Lake. Sicker Group, green ribbon chert with visible radiolaria, part of a large recumbent fold. *Pseudopolygnathus?* sp., *Siphonodella* sp. CAI, 6. Early Mississippian in age.

Sample 99. C-127520. M. T. Brandon, 1984; 84-YB-8498-1B. 92F, Alberni: 49° 00′ 27″; 124° 26′ 17″. Collected within several metres of no. 98. *Pseudopolygnathus* sp., *Siphono-*

della sp., ramiform elements. CAI, 5–6. Early Mississippian in age.

Sample 100. C-127521. M. T. Brandon, 1984; 84-YB-8498-1C. 92F, Alberni: 49° 00′ 27″; 124° 26′ 17″. Collected within several metres of no. 98. *Gnathodus* sp., ramiform elements. CAI, 5–6. Mississippian in age.

PLATE 5.1

All are Pa elements unless stated otherwise. GSC denotes Geological Survey of Canada.

Peridon sp.
Plate 5.1, Figs. 1, 3. Fig. 1, lateral view, bipennate S element, specimen GSC 69093, ×100. Road River Formation (no. 1).
Fig. 3, lateral view, geniculate coniform M element, specimen GSC 69094, ×100. Road River Formation (no.1).

Pygodus serra (Hadding)
Plate 5.1, Fig. 2, upper view, sinistral stelliscaphate element, specimen GSC 69095, ×100. Road River Formation (no. 2).

Oistodiform element, undetermined
Plate 5.1, Fig. 4, lateral view, specimen GSC 69096, ×100. Road River Formation (no. 1).

Polygnathus communis Branson and Mehl
Plate 5.1, Fig. 5. Upper view, specimen GSC 69097, ×100. Antler Formation (nos. 31–54) (GSC loc. C-86280).

Siphonodella spp.
Plate 5.1, Figs. 6, 10, 14. Fig. 6, upper view, specimen GSC 65822, ×125, Antler Formation (nos. 31–54) (GSC loc. C-102679). Fig. 10, upper view specimen GSC 69098, ×150. Sicker Group (no. 99). Fig. 14, upper view, specimen GSC 69099, ×100. Sicker Group (no. 98).

Pseudopolygnathus spp.
Plate 5.1, Figs. 7, 8, 9, 12. Fig. 7, lower view, specimen GSC 69100, ×100, Antler Formation (nos. 31–54) (GSC loc. C-102679). Fig. 8, upper view, specimen resembles *P. nudus* Pierce and Langenheim, specimen GSC 69101, ×100, Fennell Formation (no. 63). Fig. 9, upper view, specimen resembles *P.* cf. *P. micropunctatus* Bischoff and Ziegler, sensu Sandberg and Ziegler (1979), specimen GSC 69102, ×150, Antler Formation (nos. 31–54) (GSC loc. C-102658). Fig. 12, upper view, specimen GSC 69103, ×80, Antler Formation (nos. 31–54) (GSC loc. C-102679).

Rhachistognathus? sp.
Plate 5.1, Fig. 11. Upper view, specimen GSC 69104, ×100. Fennell Formation (no. 80).

Gnathodus cf. *G. texanus* Roundy
Plate 5.1, Fig. 13. Upper view, specimen GSC 69105, ×80. Fennell Formation (no. 80).

Lochriea ex gr. *commutata* (Branson and Mehl)
Plate 5.1, Fig. 15. Upper view, specimen GSC 65883, ×100. Antler Formation (nos. 31–54) (GSC loc. C-102678).

Gnathodus girtyi Hass
Plate 5.1, Fig. 16. Upper view, specimen GSC 69106, ×100. Fennell Formation (no. 57).

Gnathodus bilineatus Roundy
Plate 5.1, Fig. 17. Upper view, specimen GSC 69107, ×60. Fennell Formation (no. 72).

Pl. 5.1]   **Conodonts from western Canadian chert**   111

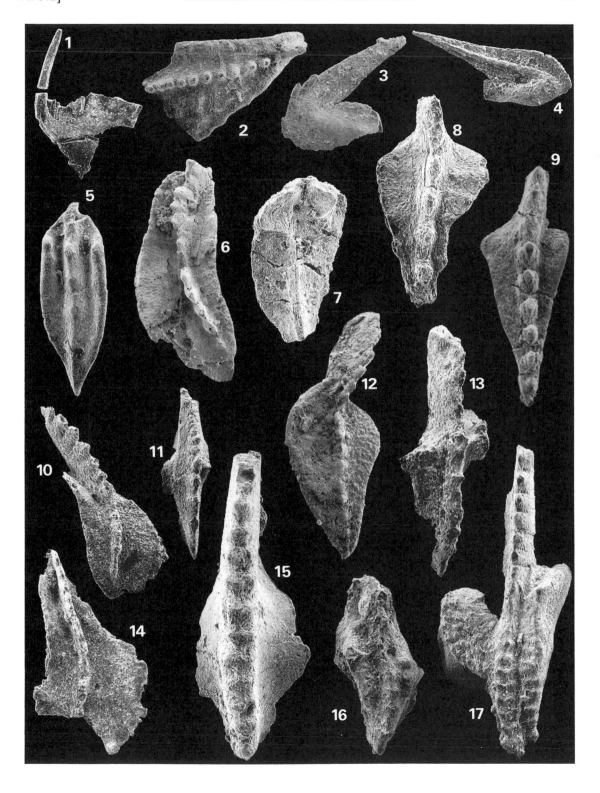

PLATE 5.2

All are Pa elements unless stated otherwise. GSC denotes Geological Survey of Canada.

Declinognathodus? sp.
Plate 5.2, Fig. 1. Upper view, specimen GSC 69108, ×100. Fennell Formation (no. 62).

Rhachistognathus prolixus Baesemann and Lane.
Plate 5.2, Figs. 2, 3. Fig. 2, upper view, specimen GSC 69109, ×80. Antler Formation (nos. 31–54) (GSC loc. C-86280). Fig. 3, lower view, specimen GSC 69109, ×80. Antler Formation (nos. 31–54) (GSC loc. C-86280).

Idiognathoides spp.
Plate 5.2, Figs. 4–7. Fig. 4, upper view, specimen resembles *I. convexus* (Ellison and Graves), specimen GSC 69110, ×80). Antler Formation (nos. 31–54) (GSC loc. C-86820). Fig. 5, upper view, specimen resembles *I. convexus* (Ellison and Graves), specimen GSC 69111, ×100. Antler Formation (nos. 31–54) (GSC loc. C-86820). Fig. 6, upper view, early growth stage, specimen GSC 69112, ×120. Sylvester Group (no. 14). Fig. 7, upper view, specimen resembles *I. sinuatus* Harris and Hollingsworth, specimen GSC 69113, ×100. Fennell Formation (no. 71).

Diplodelliform element, undetermined Sa element
Plate 5.2, Fig. 8, posterior view, specimen GSC 69114, ×80. Antler Formation (nos. 31–54) (GSC loc. C-86277).

Ozarkodiniform element, undetermined Pb element
Plate 5.2, Fig. 9, lateral view, specimen GSC 69115, ×100. Antler Formation (nos. 31–54) (GSC loc. C-86820).

'Streptognathodus' expansus Igo and Koike
Plate 5.2, Fig. 10. Upper view, specimen GSC 65884, ×80. Antler Formation (nos. 31–54) (GSC loc. C-86820).

Idiognathodus–Streptognathodus plexus
Plate 5.2, Figs. 11, 12, 18, 19. Fig. 11, upper view, specimen GSC 69116, ×100. Antler Formation (nos. 31–54) (GSC loc. 102682). Fig. 12, upper view, early growth stage, specimen GSC 69117, ×100. Fennell Formation (no. 72). Fig. 18, upper view, specimen GSC 69118, ×70. Fennell Formation (no. 81). Fig. 19, upper view, specimen GSC 69119, ×100. Fennell Formation (no. 81).

Neognathodus sp.
Plate 5.2, Fig. 13. Upper view, specimen GSC 69120, ×80. Fennell Formation (no. 81).

Gondolella sp.
Plate 5.2, Fig. 14. Lateral view, early growth stage, specimen GSC 69121, ×120. Fennell Formation (no. 72).

Gondolella ex gr. laevis Kosenko and Kozitskaya
Plate 5.2, Fig. 15. Upper view, specimen GSC 69122, ×100. Antler Formation (nos. 31–54) (GSC loc. C-102667).

Gondolella ex gr. *magnus* Stauffer and Plummer
Plate 5.2, Fig. 16. Upper view, posterior fragment, specimen GSC 69123, ×80. Antler Formation (nos. 31–54) (GSC loc. C-102667).

Streptognathodus elongatus Gunnell
Plate 5.2, Fig. 17. Upper view, specimen GSC 65886, ×100. Antler Formation (no. 31–54) (GSC loc. C-102664).

Pl. 5.2]   **Conodonts from western Canadian chert**   113

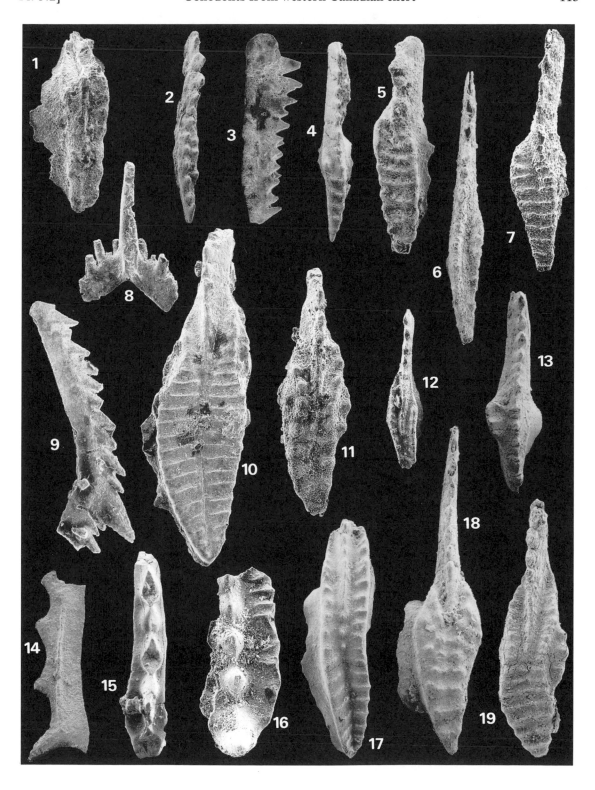

PLATE 5.3

All are Pa elements unless otherwise stated. GSC denotes Geological Survey of Canada.

Neogondolella cf. *N. bisselli* Clark and Behnken
Plate 5.3, Fig. 1. Upper view, specimen GSC 69124, ×100. Unnamed chert (no. 10).

Neogondolella cf. *N. gracilis* (Clark and Ethingrton)
Plate 5.3, Fig. 2. Upper view, specimen GSC 69125, ×100. Unnamed chert (no. 7).

Sweetognathus aff. *S. whitei* (Rhodes)
Plate 5.3, Figs. 3, 4. Fig. 3, upper view, specimen GSC 69126, ×100. Fennell Formation (no. 70). Fig. 4, upper view, specimen GSC 69127, ×150. Unnamed chert (no. 6).

Neogondolella serrata (Clark and Ethington) group
Plate 5.3, Figs. 5?, 9, 11, 12, 14, 15, 19–21. Fig. 5?, upper view, specimen GSC 69128, ×100. Fennell Formation (no. 78). Fig. 9, upper view, specimen GSC 69129, ×100. Fennell Formation (no. 77). Fig. 11, lateral view, specimen GSC 69130, ×80. Fennell Formation (no. 77). Fig. 12, upper view, specimen GSC 69131, ×90. Fennell Formation (no. 77). Fig. 14, upper view, specimen GSC 69132, ×150. Fennell Formation (no. 77). Fig. 15, upper view, anterior fragment, specimen GSC 69133, ×80. Kaslo Group (no. 82). Fig. 19, lateral view, specimen GSC 69134, ×80. Fennell Formation (no. 74). Fig. 20, upper view, specimen GSC 69135, ×100. Fennell Formation (no. 74). Fig. 21, upper view, specimen GSC 69136, ×100. Fennell Formation (no. 74).

Cypridodelliform element, undetermined
Plate 5.3, Fig. 6, posterior view, specimen 69137, ×100. Marble Canyon Formation (no. 87).

Neostreptognathodus pequopensis Behnken
Plate 5.3, Fig. 7. Upper view, specimen GSC 69138, ×200. Unnamed chert (no. 10).

Neogondolella spp.
Plate 5.3, Figs. 8, 10, 13, 16–18. Fig. 8, upper view, specimen GSC 69139, ×60. Sylvester Group (no. 17). Fig. 10, upper view, specimen GSC 69140, ×90. Fennell Formation (no. 79). Fig. 13, upper view, specimen GSC 69096, ×150. Kaslo Group (no. 83). Fig. 16, lateral view, specimen 69094, ×80. Kaslo Group (no. 82). Fig. 17, upper view, specimen 69094, ×80. Kaslo Group (no. 82). Fig. 18, lower view, specimen 69093, ×80. Kaslo Group (no. 82).

Pl. 5.3]   **Conodonts from western Canadian chert**   115

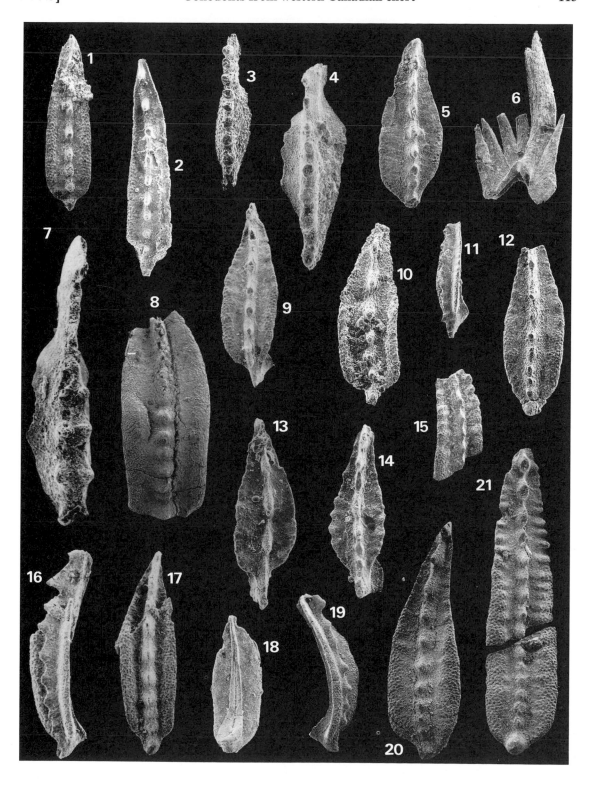

PLATE 5.4

All are Pa elements. GSC denotes Geological Survey of Canada.

Cratognathus? sp.
Plate 5.4, Fig. 1. Lateral view, specimen GSC 69141, ×100. Nicola Group (no. 85).

Neogondolella sp. A
Plate 5.4, Figs. 2, 6, 7. Fig. 2, upper view, specimen GSC 69142, ×80. Stuhini Group (no. 21). Fig. 6, upper view, specimen GSC 69143, ×100. Stuhini Group? (no. 30). Fig. 7, lateral view, specimen GSC 69143, ×100. Stuhini Group? (no. 30).

Neogondolella excelsa Mosher?
Plate 5.4, Figs. 3, 4, 9. Fig. 3, upper view, specimen GSC 66051, ×100, Cache Creek Group, Marble Canyon Formation (no. 94). Fig. 4, lateral view, specimen GSC 66052, ×100. Cache Creek Group, Marble Canyon Formation (no. 94). Fig. 9, upper view, specimen GSC 66053, ×100. Cache Creek Group, Marble Canyon Formation (no. 94).

Neogondolella cf. *N. regale* Mosher
Plate 5.4, Figs. 5, 10. Fig. 5, lateral view, specimen GSC 69144, ×80. Stuhini Group (no. 29). Fig. 10, upper view, specimen GSC 69144, ×100. Stuhini Group (no. 29).

Neogondolella constricta (Mosher and Clark)
Plate 5.4, Figs. 8, 11–13. Fig. 8, upper view, specimen GSC 66054, ×80. Stuhini Group (no. 29), Fig. 11, upper view, specimen GSC 66055, ×80. Stuhini Group (no. 29), Fig. 12, upper view, specimen GSC 66056, ×80. Stuhini Group (no. 29), Fig. 13, lateral view, specimen GSC 66057, ×80. Stuhini Group (no. 29).

'*Paragondolella*' ex gr. *polygnathiformis* (Budurov and Stefanov)
Plate 5.4, Figs. 14–16. Fig. 14, lower view, specimen resembles *N. tadpole* (Hayashi) sensu Kovacs (1983), specimen GSC 66058, ×100, Cache Creek Group (no. 92). Fig. 15, lateral view, specimen resembles *N. tadpole* (Hayashi) sensu Kovacs (1983). specimen GSC 66059, ×100. Cache Creek Group (no. 92). Fig. 16, upper view, specimen resembles *N. tadpole* (Hayashi) sensu Kovacs (1983), specimen GSC 66059, ×100, Cache Creek Group (no. 92).

Metapolygnathus nodosus (Hayashi)
Plate 5.4, Figs. 17, 18. Fig. 17, upper view, specimen GSC 66060, ×100. Stuhini Group (no. 22). Fig. 18, upper view, specimen GSC 66061, ×100. Stuhini Group (no. 22).

Epigondolella primitia Mosher
Plate 5.4, Fig. 19. Upper view, specimen GSC 66062, ×150. Cache Creek Group (no. 90).

Epigondolella abneptis (Huckriede)
Plate 5.4, Figs. 20–22. Fig. 20, upper view, specimen corresponds to *E. abneptis* subsp. A of Orchard (1983), specimen GSC 66063, ×100. Cache Creek Group (no. 12). Fig. 21, upper view, early growth stage, specimen resembles *E. bidentata*, specimen GSC 66064, ×150. Windy Craggy (no. 11). Fig. 22, upper view, specimen corresponds to an early form of *E. abneptis* subsp. B of Orchard (1983), specimen GSC 66065, ×100. Nicola Group (no. 84).

Pl. 5.4] **Conodonts from western Canadian chert** 117

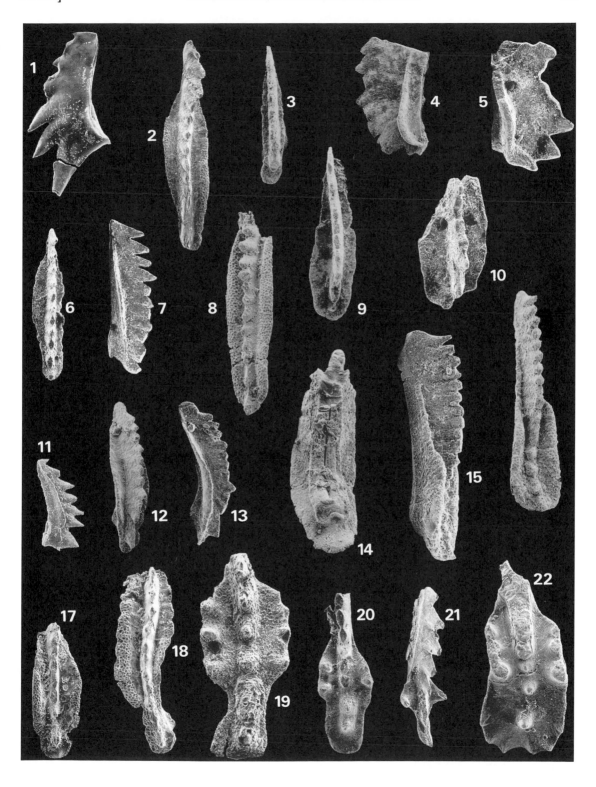

5.7 REFERENCES

Campbell, R. B. and Dodds, C. J. 1983. Geology, Tatshenshini map area (114P). *Geological Survey of Canada Open File Map*, **926**.

Campbell, R. B., Mountjoy, E. W. and Young, F. G. 1973. Geology of McBride map-area, British Columbia. *Geological Survey of Canada Paper*, **72–35**, 1–104.

Campbell, R. B. and Tipper, H. W. 1971. Geology of Bonaparte Lake map-area, British Columbia. *Geological Survey of Canada Memoir*, **363**, 1–96.

Cockfield, W. W. 1948. Geology and mineral deposits of Nicola map-area, British Columbia. *Geological Survey Memoir*, **249**, 1–164.

Duffell, S. and McTaggart, K. C. 1952. Ashcroft map-area, British Columbia. *Geological Survey of Canada Memoir*, **262**, 1–122.

Gabrielse, H. 1963. McDame map-area, British Columbia. *Geological Survey of Canada Memoir*, **319**, 1–138.

Gordey, S. P. 1981. Geology of Nahanni Map-Area (105I), Yukon Territory and District of Mackenzie. *Geological Survey of Canada Open File Map*, **780**.

Gordey, S. P., Abbott, J. G. and Orchard, M. J. 1982a. Devono–Mississippian (Earn Group) and younger strata in east–central Yukon. *Geological Survey of Canada Paper*, **82–1B**, 93–100.

Gordey, S. P., Gabrielse, H. and Orchard, M. J. 1982b. Stratigraphy and structure of Sylvester Allochthon, southwest McDame map area, northern British Columbia. *Geological Survey of Canada Paper*, **82–1B**, 101–106.

Green, L. H. 1972. Geology of Nash Creek, Larsen Creek, and Dawson map-areas, Yukon Territory. *Geological Survey of Canada Memoir*, **364**, 1–157.

Harms, T. 1984. Structural style of the Sylvester Allochthon, northeastern Cry Lake map area, British Columbia. *Geological Survey of Canada Paper*, **84–1A**, 109–112.

Harms, T. 1985. Pre-emplacement thrust faulting in the Sylvester Allochthon, northeast Cry Lake map area, British Columbia. *Geological Survey of Canada Paper*, **85–1A**, 301–304.

Hayashi, S. 1968. The Permian Conodonts in Chert of the Adoyama Formation, Ashio Mountains, central Japan. *Earth Science*, **22**, 63–77, 4 plates.

Hayashi, S. 1969. Extraction of conodonts through HF method. *Fossil Club Bulletin*, **2**, 1–9.

Igo, H. and Koike, T. 1983. Conodont biostratigraphy of cherts in the Japanese islands. In A. Iijima, J. R. Hein and R. Siever (Eds.), *Siliceous Deposits in the Pacific Region*, Elsevier, Amsterdam, 65–78.

Ishida, K. 1981, Fine stratigraphy and conodont biostratigraphy of a bedded chert member of the Nakagawa Group. *Journal of Science of College of General Education, University of Tokushima*, **14**, 107–137, 7 plates.

Isozaki, Y. and Matsuda, T. 1980. Age of the Tamba Group along the Hozugawa 'Anticline', Western Hills of Kyoto, southwest Japan. *Journal of Geosciences, Osaka City University*, **23**, Article 3, 115–134, 1 plate.

Isozaki, Y. and Matsuda, T. 1982. Middle and Late Triassic conodonts from bedded chert sequence in the Nimo-Tamba Belt, southwest Japan, Part 1: *Epigondolella. Journal of Geosciences, Osaka City University*, **25**, Article 7, 103–137, 6 plates.

Isozaki, Y. and Matsuda, T. 1983. Middle and Late Triassic conodonts from bedded chert sequence in the Nimo-Tamba belt, southwest Japan, Part 2: *Misikella* and *Parvigondolella. Journal of Geosciences, Osaka City University*, **26**, Article 3, 65–86, 14 plates.

Jackson, D. E. and Lenz, A. C. 1962. Zonation of Ordovician and Silurian graptolites of Northern Yukon, Canada. *Association of Petroleum Geologists Bulletin*, **46**, 30–45.

Kerr, F. A. 1948. Lower Stikine and western Iskut River Areas, British Columbia. *Mines and Geology Branch, Geological Survey Memoir*, **246**, 1–94.

Klepacki, D. W. and Wheeler, J. O. 1985. Stratigraphic and structural relations of the Milford, Kaslo and Slocan Groups, Goat Range, Lardeau and Nelson map areas, British Columbia. *Geological Survey of Canada Paper*, **85–1A**, 277–286.

Kovacs, S. 1983. On the evolution of the *excelsa*-stock in the Upper Ladinian–Carnian (Conodonta, genus *Gondolella*, Triassic). *Schriftenreihe der Erdwissenschaftlichen Kommissionen*, **5**, 107–120, 6 plates.

Lowey, G. W. 1984. The stratigraphy and sedimentology of siliciclastic rocks, west–central Yukon, and their tectonic implications. Unpublished Ph.D. Thesis, University of Calgary.

MacIntyre, D. G. 1984. Geology of the Alsek–Tatshenshini Rivers area (114P). In *Geological Fieldwork 1983: A Summary of Field Activities, British Columbia Department of Mines Paper*, **1984-1**, 173–184.

Monger, J. W. H. 1981. Geology of parts of western Ashcroft map area, southwestern British Columbia. *Geological Survey of Canada Paper*, **81–1A**, 185–189.

Monger, J. W. H. and Berg, H. C. 1984. Lithotectonic terrane map of western Canada and southeastern Alaska. In N. J. Silberling and D. L. Jones (Eds), *Lithotectonic Terrane Maps of the North American Cordillera, US Geological Survey Open-File Report*, **84–523**, B1–B31.

Monger, J. W. H. and McMillan, W. 1984. Bedrock geology of Ashcroft (921) map area. *Geological Survey of Canada Open-File Map*, **980**.

Muller, J. E. 1980. The Paleozoic Sicker Group of Vancouver Island, British Columbia. *Geological Survey of Canada Paper*, **79-30**, 1–23.

Mulligan, R. 1963. Geology of Teslin map-area, Yukon Territory. *Geological Survey of Canada Memoir*, **326**, 1–96.

Orchard, M. J. 1981. Triassic conodonts from the Cache Creek Group, Marble Canyon, southern British Columbia. *Geological Survey of Canada Paper*, **81–1A**, 357–359.

Orchard, M. J. 1983. *Epigondolella* populations and their phylogeny and zonation in the Upper Triassic. *Fossils and Strata*, **15**, 177–192, 3 plates.

Orchard, M. J. 1984a. Pennsylvanian, Permian and Triassic conodonts from the Cache Creek Group, Cache Creek, southern British Columbia. *Geological Survey of Canada Paper*, **84-1B**, 197–206, plates 22.1–2.

Orchard, M. J. 1984b. Early Permian conodonts from the

Harper Ranch Beds, Kamloops area, southern British Columbia. *Geological Survey of Canada Paper,* **84-1B**, 207–215, plate 23.1.

Orchard, M. J. 1985. Carboniferous, Permian and Triassic conodonts from the central Kootenay Arc, British Columbia: constraints on the age of the Milford, Kaslo and Slocan Groups. *Geological Survey of Canada Paper,* **85-1A**, 287–300, plates 37.1–3.

Orchard, M. J. and Struik, L. C. 1985. Conodonts and stratigraphy of Upper Paleozoic limestones in Cariboo gold belt, east–central British Columbia. *Canadian Journal of Earth Sciences,* **22**, 538–552, 2 plates.

Pessagno, E. A., Jr. and Newport, R. L. 1972. A technique for extracting Radiolaria from radiolarian cherts. *Micropalaeontology,* **18**, 231–234.

Preto, V. A. and Schiarizza, P. 1982. Geology and mineral deposits of the Eagle Bay and Fennel Formations from Adams Plateau to Clearwater, south–central British Columbia. *Geological Association of Canada, Cordilleran Section, Programme with Abstracts,* 22–24.

Ray, G. E. in press. The Hozameen Fault System and related Coquihalla Serpentine Belt of southwestern British Columbia. *Canadian Journal of Earth Sciences.*

Read, P. B. 1983. Geology, Classy Creek (104J/2E) and Stikine Canyon (104J/1W), British Columbia. *Geological Survey of Canada Open-File Map,* **940**.

Read, P. B. 1984. Klastline River (104G/16E), Ealue Lake (104H/13W), Cake Hill (104I/4W) and Stikine Canyon (104J/1E). *Geological Survey of Canada Open-File Map,* **1080**.

Read, P. B., Psutka, J. F., Brown, R. L. and Orchard, M. J. 1983. 'Tahltanian' Orogeny and Younger Deformations, Grand Canyon of the Stikine, British Columbia. *Geological Association of Canada Programme with Abstracts,* **8**, A57.

Ross, C. A. and Ross, J. R. P. 1983. Late Paleozoic accreted terranes of Western North America. In C. H. Stevens (Ed.), *Pre-Jurassic Rocks in Western North America Suspect Terranes,* Society of Economic Paleontologists and Mineralogists, Pacific Section, 7–22.

Sandberg, C. A. and Ziegler, W. 1979. Taxonomy and biofacies of important conodonts of Late Devonian *styriacus*-Zone, Unites States and Germany. *Geologica et Palaeontologica,* **13**, 173–212, 7 plates.

Schiarizza, P. 1981. Clearwater Area. In *Geological Fieldwork 1980: A Summary of Field Activities, British Columbia Department of Mines Paper,* **1981-1**, 159–163.

Schiarizza, P. 1982. Clearwater Area. In *Geological Fieldwork 1981: A Summary of Field Activities, British Columbia Department of Mines Paper,* **1982-1**, 59–67.

Smith, R. B. 1979. Geology of the Harper Ranch Group (Carboniferous–Permian) and Nicola Group (Upper Triassic) Northeast of Kamloops, British Columbia. *Unpublished M.Sc. Thesis,* University of British Columbia.

Souther, J. G. 1971. Geology and Mineral Deposits of Tulsequah map-area, British Columbia. *Geological Survey of Canada Memoir,* **362**, 1–84.

Souther, J. G. 1972. Telegraph Creek map-area, British Columbia. *Geological Survey of Canada Paper,* **71-44**, 1–38.

Struik, L. C. 1980. Geology of the Barkerville–Cariboo River area, east–central British Columbia. *Unpublished Ph.D. Thesis,* University of Calgary.

Struik, L. C. and Orchard, M. J., 1985. Upper Paleozoic conodonts from ribbon chert delineate imbricate thrusts within the Antler Formation of Slide Mountain Terrane, central British Columbia. *Geology,* **13**, 794–798.

Travers, W. B. 1978. Overturned Nicola and Ashcroft strata and their relation to the Cache Creek Group, southwestern Intermontane Belt, British Columbia. *Canadian Journal of Earth Sciences,* **15**, 99–116.

Trettin, H. P. 1980. Permian rocks of the Cache Creek Group in the Marble range, Clinton Area, British Columbia. *Geological Survey of Canada Paper,* **79-17**, 1–17.

Wardlaw, B. R. and Jones, D. L. 1981. Permian conodonts from eugeoclinal rocks of western North America and their tectonic significance. *Rivista Italiana Paleontologia,* **8**, 895–908, plate 64.

Uglow, W. L. 1922. Geology of the North Thompson Valley map-area, British Columbia. *Geological Survey of Canada Summary Report, Part A,* 72–106.

6

Conodont biostratigraphy of the Caballos Novaculite (Early Devonian–Early Mississippian), northwestern Marathon Uplift, west Texas

J. E. Barrick

Conodont faunas obtained by hydrofluoric acid processing of cherts, from bedding plane surfaces of siliceous shales, and from a single allodapic limestone unit, permit biostratigraphic subdivision and dating of the undifferentiated chert and shale member of the Caballos Novaculite. The basal chert and lower novaculite members failed to yield conodonts; by stratigraphic position they may be Late Ordovician to Early Devonian in age. The lower third of the overlying undifferentiated chert and shale member bears Early Devonian (Lochkovian?) conodonts. Middle Devonian conodonts are uncommon, but late Eifelian and late Givetian forms are present. A series of faunas ranging in age from early Frasnian (Late Devonian) to Kinderhookian (Early Mississippian) were recovered from cherts and shales at the East Bourland Mountain section.

6.1 INTRODUCTION

During the Middle Palaeozoic Era an unusual association of silica-rich strata, white novaculites, dark cherts and siliceous shales accumulated along the southern margin of North America in the depositional precursor of the Ouachita Orogenic Belt. Today, these strata are exposed only in the Marathon Basin, west Texas (Caballos Novaculite), and the Ouachita Mountain area of western Arkansas and southeastern Oklahoma (Arkansas Novaculite) (Fig. 6.1). The lithostratigraphy, petrography and sedimentology of the 'novaculitic facies' have been studied by numerous workers, but the scarcity of fossil material has prevented accurate biostratigraphic dating and correlation. As a result, the chronological relations of units within the novaculitic facies are uncertain, as are correlations with the predominantly carbonate and shale facies of the shelf margin and cratonic areas of southern North America.

Previous biostratigraphic information from the novaculitic facies consisted of a small number of conodont faunas, most of which were obtained from siliceous shales which are interbedded with cherts. These faunas were erratically distributed geographically and stratigraphically, because of the vagaries of collecting conodonts on bedding-plane surfaces of shales.

Preliminary work on the Caballos Novaculite indicates that biostratigraphically useful conodont faunas can be obtained from most cherts and some novaculites by dissolution of samples in dilute hydrofluoric acid (HF). Thus, it has been possible to sample outcrops of the novacu-

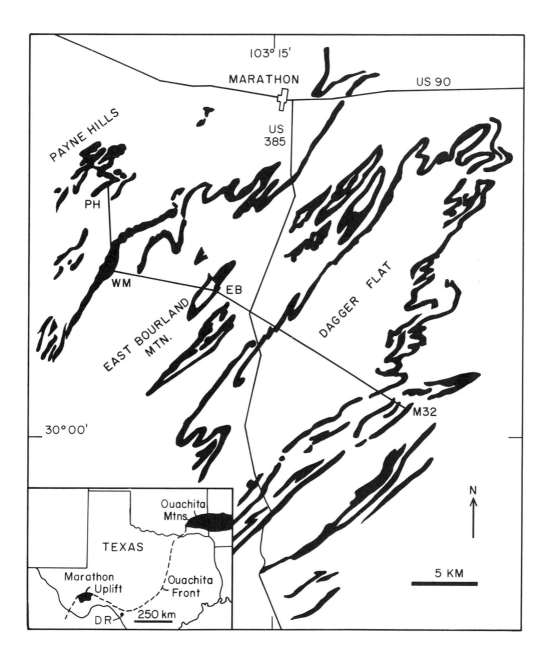

Fig. 6.1 — Sketch map of major portion of Marathon Uplift showing outcrop pattern of Caballos Novaculite (black). (After King, 1937.) The line of section is shown in Fig. 6.2. The inset map shows relation of Marathon Uplift to Ouachita Front and location of Devil's River Uplift (DR).

litic facies at closely spaced stratigraphic intervals, regardless of lithotype, and to obtain conodont elements from many of the samples. Procedures used to obtain conodonts from the cherts of the Caballos Novaculite and initial biostratigraphic results are summarized in this paper.

6.2 CABALLOS NOVACULITE

The Caballos Novaculite consists of 30–210 m of novaculite (milky-white chert) and interbedded cherts and siliceous shales which crop out in the Marathon Basin and smaller outlying areas in southwestern Texas (Fig. 6.1). The basic stratigraphic sequence and geographic variations in lithotypes have been discussed by King (1937) and McBride and Thomson (1970). Throughout the eastern part of the Marathon Basin, the Caballos Novaculite is dominated by two thick novaculite units which permit five members to be recognized (e.g. Fig. 6.2 section M 32). Typically, a thin basal unit of pale-brown chert and off-white chert with shale partings, the lower chert member, rests with sharp contact on the Upper Ordovician Maravillas Formation. The massive lower novaculite member rests either on the lower chert member or directly on the Maravillas Formation. The lower chert and shale member consists of green and grey thin chert beds with shale partings which separate the two novaculite units where the upper novaculite member is present. The massive upper novaculite is thickest in the southeastern part of the basin (120 m maximum thickness) and thins to the northeast, where it is poorly developed or absent. Where the upper novaculite is present, an upper chert and shale member forms the top of the Caballos Novaculite. In areas lacking the upper novaculite, no distinction is made between the chert and shale members (Fig. 6.2). The siliceous shales and sands of the Tesnus Formation (of Mississippian–Early Pennsylvanian age) overlie the cherts and shales of the Caballos Novaculite.

A series of papers has described in detail many aspects of the petrography of the dark cherts and novaculites but has presented differing environmental interpretations (McBride and Thomson, 1970; Folk, 1973; Folk and McBride, 1976; McBride and Folk, 1977). The novaculite members are massive to poorly bedded units composed of homogeneous milky white chert. The novaculite is almost pure microquartz (99% SiO_2) which displays a characteristic clotted or pelloidal structure in polarized light. Sparse quartz silt is the major detrital impurity. Ghosts of sponge spicules are the main recognizable biogenic component. Probable radiolarian remains occur rarely. A complex series of early formed fractures, breccias and quartz sand-filled pockets and fractures has been described.

The chert and shale members make up about one-half of the Caballos Novaculite. Rhythmically intercalated beds of green and grey chert, 1–15 cm thick, are separated by shale partings, 1–5 mm thick. Pale-brown, black, blue and various-coloured cherts are present locally. Chert beds are composed mainly of equant microcrystalline quartz in a wide variety of sizes. Several per cent of clay (chiefly illite), quartz silt and other minor impurities may be present. The grey to green siliceous shale is dominantly illite with variable amounts of microquartz. Sparse beds of chert conglomerate, chert and quartz sandstone, calcarenite and red siliceous shale occur in the chert and shale members. Radiolarians are scattered through the cherts, averaging about 1% but are locally as abundant as 15%. Palynomorphs, sponge spicules and condonts have been observed in thin sections. Sparse calcareous benthic fossils have been found in allodapic calcarenites, *Callixylon* occurs at a few localities, and *Helminthoidea*-like traces are present on some shale beds.

McBride (in McBride and Folk, 1977) has argued that the Caballos Novaculite is the result of deposition of siliceous sediments on the slopes and floor of a deep marine basin. The novaculite units represent accumulation at bathyal depths of siliceous sponge spicules plus lesser amounts of carbonate mud drifted in from a marginal carbonate shelf. Chert and shale members formed from the accumulation of siliceous oozes, dominated by radiolarians, admixed with variable amounts of terrigenous clays and rare coarser clastics introduced into the basin by turbidity currents and mass flows. In contrast, Folk (in McBride and Folk, 1977) has proposed a shallow-water origin for the

Caballos Novaculite. The novaculite was originally deposited as a mixture of carbonate pellets, sponge spicules and other opaline skeletal material in semi-restricted lagoons lacking a source of terrigenous detritus. Evaporitic conditions were attained in some peritidal flats and some subaerial exposure occurred. Chert and shale members accumulated as a clayey radio-

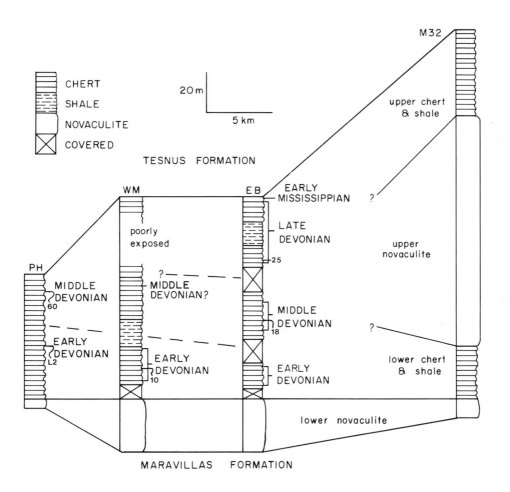

Fig. 6.2 — Stratigraphic cross-section of Caballos Novaculite showing distribution of major lithotypes and conodont faunas. The location of line of section is shown in Fig. 6.1. The lower chert member has been included in the lower novaculite. Total thickness of section WM is from King (1937, Plate 2). Section M32 is section 32 of McBride and Thomson (1970, Fig. 7), from which the description and thickness of members were taken.

larian-rich ooze in depths from the intertidal zone to perhaps 100 m.

Several palaeogeographic reconstructions of North America showing the Ouachita belt during Devonian or Early Mississippian time have been presented over the last decade (e.g. Lowe, 1975; Heckel and Witzke, 1979). Gutschick and Sandberg (1983, Figs. 5, 6, p. 82) depict the Ouachita belt in the Early Mississippian as a foreland trough situated between a converging North and South America. They postulated the existence of an elongate medial welt capped by low islands to explain the juxtaposition of shallow-water novaculite and deeper-water chert–shale lithofacies. These researches and others (e.g. Park and Croneis, 1969) agree that the predominance of non-terrigenous silica in the Ouachita belt and the shelf areas to the north represents an unusual depositional situation; silica-rich oceanic waters circulated across the southern margin of North America and siliceous organisms, sponges and radiolarians, served as the means by which the silica was deposited in areas lacking a significant influx of terrigenous clastics. The ultimate source of the silica was volcanic activity outside the area of deposition. Concentration of silica occurred as a result of dynamic upwelling of waters along the western coast of North America. Zones of high organic productivity developed, which were characterized by 'blooms' of siliceous zooplankton (Lowe, 1975).

Only radiolarians and conodonts are sufficiently abundant and widespread to permit accurate dating and correlation of the Caballos Novaculite. The radiolarians reported from the Caballos Novaculite are largely spumellarians (Aberdeen, 1940), which are presently of limited use for biostratigraphic work. Graves (1952) reported a conodont fauna 40 m above the base of the Caballos Novaculite from a carbonate unit within the lower chert and shale member in the Payne Hills (Figs. 6.1, 6.2, PH). Graes considered it to be a Late Devonian fauna, but subsequently Ziegler (1971) reinterpreted it to be Early Devonian in age. At East

Bourland Mountain (Figs. 6.1, 6.2, EB), Graves (1952) recovered species of *Palmatolepis* (Late Devonian) from bedding-plane surfaces of siliceous shales near the top of the Caballos Novaculite.

The age of the base of the Caballos Novaculite is unknown, for no biostratigraphic indicators have been reported between the top of the Maravillas Formation (Late Ordovician-–Richmondian (Berry, 1960; Bergström, 1978)) and the Early Devonian conodont fauna of Graves (1952). Workers who believe that deposition of the Ouachita facies was continuous through the Palaeozoic Era infer that Silurian, or even Late Ordovician, strata may occur in the lower part of the Caballos Novaculite (McBride and Thomson, 1970; Bergström, 1978). Lithologic evidence is equivocal; the Maravillas–Caballos contact has been interpreted as conformable (McBride and Thomson, 1970) or unconformable (Folk, in McBride and Folk, 1977, p. 1279).

The Tesnus Formation rests with apparent conformity on the Caballos Novaculite wherever seen (McBride and Thomson, 1970). Ellison (1962) reported forms of *Gnathodus* from the lower beds of the Tesnus Formation at East Bourland Mountain which he tentatively considered to be Meramecian (Late Mississippian) in age.

6.3 METHODS OF STUDY

In 1982, the present author initiated systematic collection and HF processing of chert samples from the Caballos Novaculite as part of research on Ouachita biostratigraphy funded by the Petroleum Research Fund of the American Chemical Society (Grant 13786–G2). The decision to start HF processing of cherts was based on reports that conodonts had been identified in thin sections from all rock types of the Caballos Novaculite (cherts, shales, sandstones and carbonates), except the novaculites (McBride and Thomson, 1970; McBride, pers. commun., 1981). This suggested that conodonts were sufficiently abundant in at least the chert and shale members that dissolution of small samples of chert would yield conodont elements. Because of easy land access and because previous conodont faunas had been recovered from this area, sampling of the Caballos Novaculite was restricted to the northwestern part of the Marathon Basin. Several outcrops were sampled, all of which lack the upper novaculite member (Figs. 6.1 and 6.2).

The procedure used to extract conodont elements from cherts was developed through a number of trials with the purpose of obtaining relatively well-preserved elements. When cherts are placed in cold HF solution, the micro-quartz passes slowly into solution, but calcium-bearing compounds, carbonates and phosphates, are preserved, probably by alteration to amorphous calcium fluoride or fluorite. Although carbonate microfossils such as ostracods may be exquisitely preserved (Sohn, 1956; Schallreuter, 1982), conodonts and other phosphatic material are less resistant to the HF. Elements become cracked and brittle, are often distorted and are etched or partially dissolved, even after a short period of immersion in dilute HF. The use of dilute HF (10%) and a relatively short period of immersion of the sample in the solution between washings tend to minimize the destruction of conodonts, but at the expense of extra time and labour in processing even small samples. The procedure which I use is as follows.

(1) Chert samples are crushed to pieces 1–2 cm in diameter. If any carbonate is present, it must first be dissolved out with dilute acetic or formic acid. Otherwise a large CaF_2 residue will result.

(2) Approximately 50 g of sample is immersed in 500 ml of a 10% HF solution in an HF-resistant container. Greater concentrations of HF dissolve the chert more quickly, but at the expense of conodont preservation.

(3) After 6–8 h the HF solution is decanted and saved. The sample is gently washed through sieves (stainless steel or nylon). Because of the small size of the Caballos conodonts, a 230 mesh (62 μm) sieve is used to catch the residue.

(4) The undissolved chert fragments are reimmersed in the decanted HF solution for 6–8 h and the washing is repeated. Usually the sample can be treated four to five times before the rate of dissolution diminishes appreciably. After four intervals of immersion, typically 40–50% of the sample will have dissolved.

The insoluble residues obtained are usually very small and easily examined under a binocular microscope. However, because the specimens are more brittle than is typical of conodonts, more care should be taken in handling the elements.

The number of conodont elements recovered from the Caballos Novaculite varied with rock type and stratigraphic position. Samples of the white novaculites were resistant to solution (25% dissolution) and yielded tiny residues. Only a single indeterminate simple cone was recovered from 22 samples of novaculite. The characteristic microstructure and extreme purity of the novaculite appear to be responsible for the difficulty in processing and the failure to obtain conodont faunas.

Cherts of various colours all dissolved relatively easily in dilute HF (50–60% dissolution). The size of the insoluble residue varied from very small in the purer cherts to moderate in cherts with detrital impurities, usually clays. Samples from the lower portion of the chert and shale unit yielded relatively small conodont faunas, typically from a few elements to about 50 per 100 g. In contrast, samples from the upper part of the chert and shale unit at East Bourland Mountain (Fig. 6.2) ranged in abundance to nearly 1000 elements (including fragments) per 100 g. Two-thirds of the 60 chert samples yielded at least one conodont element, but only a small portion contained biostratigraphically useful forms.

Preservation of conodont elements extracted from the Caballos cherts was variable. Simple cones, small ramiform elements and juvenile Pa elements were often exceptionally well preserved. However, larger Pa elements were sparse and, when present, were often fractured, partially dissolved or broken into unidentifiable fragments. Preservation varied from sample to sample, but no correlation with chert type is obvious. In some larger faunas, a number of extremely well-preserved Pa elements were found (Plate 6.1, sample EBC 25).

Most small conodont elements are a dark translucent brown colour. Larger elements are more white and opaque, apparently because of etching of the surface by the HF solution. It is not clear whether the brown colour of the fluoritized conodonts corresponds to the original conodont colour. The colour alteration index values of conodonts derived from carbonates of underlying strata (Upper Ordovician Maravillas Formation (Bergström, 1978)), limestone within the Caballos Novaculite, and overlying strata (Lower Pennsylvanian Dimple Limestone) range from 1 to $2\frac{1}{2}$.

Isolated conodont elements occur on bedding-plane surfaces of siliceous shales at some localities. Identification of Pa elements is difficult, however, because the upper surface of each element invariably remains embedded in the shale when it is split. An attempt was made to disaggregate the shale by conventional methods (e.g. Stoddard's solvent) but met with little success. Portions of shale bearing conodont elements were dissolved in dilute HF, but the diagnostic large Pa elements, mostly species of *Palmatolepis*, were so poorly dissolved that they were unidentifiable. Finally, the method of dissolving the conodont elements with concentrated hydrochloric acid (HCl) and making a latex cast was used with moderate success.

6.4 CONODONT FAUNAS

Conodont faunas ranging in age from Early Devonian (Lochkovian) to Early Mississippian (Kinderhookian) were recovered from the chert and shale unit of the Caballos Novaculite in the northwestern Marathon Basin. The biostratigraphic reliability of faunas from individual beds of the Caballos Novaculite depends on the sedimentary processes by which conodont elements were emplaced in the sediments. McBride (in McBride and Folk, 1977) indicates that, because the chert and shale members represent deposition of siliceous oozes and clays at relatively great depths, coarse sedimentary particles must have been introduced into the environment by mass flows and turbidity currents. The limestone beds in the Payne Hills from which Graves (1952) extracted conodonts clearly represent carbonate debris transported basinwards by turbidity currents and mass flow. As discussed below, this conodont fauna is stratigraphically admixed due to reworking of older conodont elements. McBride (pers. commun., 1981) found that conodonts in chert beds were concentrated in placers of transported heavy mineral particles which formed laminae millimetres thick. It is uncertain whether these conodont elements originated in the area of deposition or were transported into the environment by sedimentary processes. Unlike the carbonate fauna, no stratigraphic admixture of conodonts has been observed in the chert and shale units, even in closely sampled intervals.

The lower portion of the chert and shale member at all outcrops yielded condont faunas dominated by elements of simple cone species (Fig. 6.2). Elements of *Dapsilodus obliquicostatus* and *Decoriconus fragilis* are most abundant; *Panderodus unicostatus, Pseudooneotodus beckmanni* and species of *Dvorakia* and *Belodella* are also represented. Ramiform elements occur in most samples, but diagnostic Pa elements are uncommon. *Ozarkodina excavata excavata* is present in a few samples. Pb and ramiform elements such as those found in the apparatus of *O. remscheidensis* were recovered, but associated Pa elements were too fragmentary for confident identification. Juvenile I elements of *Icriodus* species occur sparingly with the coniform elements. None was large enough to be identified.

Although no diagnostic species could be identified, this faunal association suggests that the lower portion of the chert and shale member is Early Devonian in age. *Ozarkodina excavata excavata* apparently ranges no higher than the *gronbergi* Zone (Zlichovian (Klapper and Johnson, 1980)). Similarly, *Dapsilodus obliquicostatus* and *Decoriconus fragilis* have not been reported from strata younger than Early Devonian in age. The presence of *Icriodus* excludes all but a latest Silurian age assignment. The stratigraphic intervals indicated as Early Devonian on Fig. 6.2 represent strata characterized by the presence of *Dap. obliquicostatus* or *Dec. fragilis*. At the East Bourland section (EB), the Early Devonian cone fauna ranges 20 m above the base of the chert and shale unit, up to a covered interval. At section WM, the fauna occurs up to 25 m above the lower novaculite, where a barren shale and siltstone interval interrupts the chert succession.

At the Payne Hills section (Fig. 6.2, PH), no Early Devonian conodont faunas were recovered from chert beds. Approximately 27–29 m above the base of the chert and shale unit lies the 2 m thick limestone unit from which Graves (1952) recovered conodonts (PHL2). The carbonate fauna is characterized by elements of *Ozarkodina remscheidensis remscheidensis, O. excavata excavata, Decoriconus fragilis* and a *Belodella* species. Elements of other simple cone species occur: *Dapsilodus obliquicostatus, Panderodus unicostatus* and *Pseudooneotodus beckmanni. Icriodus* elements are common, but most are badly broken. A species identified as *Icriodus* n. sp. G of Klapper (in Klapper and Johnson, 1980, Plate 2, Figs. 1–3) is characteristic. A few elements of *Pedavis biexoramus* are present.

Numerous elements of older conodont species are present in the limestone beds. At least one specimen of *Icriodus woschmidti* was recovered, as well as several species of Silurian conodonts: *Pterospathodus pennatus* and *Distomodus* sp. (of Llandovery age), *Kockelella variabilis* and *Polygnathoides siluricus* (of Ludlow age; a fragment of the latter species is probably the form illustrated as *Palmatolepis* sp. by Graves (1952, Plate 80, Figs. 2, 3)), *Oulodus elegans* (late Ludlow–Pridoli age) and

Walliserodus sp. Reworked Middle to Late Ordovician coniform and ramiform elements occur in small numbers.

The youngest diagnostic species identified from the limestone fauna, *Icriodus* n. sp. G and *Pedavis biexoramus,* are characteristic of the Cordilleran *eurekaensis* Zone (Klapper, 1977), the fauna of which has been described by Murphy and Matti (1982) in central Nevada. *Ozarkodina remscheidensis remscheidensis* ranges through this zone in the Cordilleran region. However, among the numerous broken elements of the limestone fauna are two fragments of Pa elements that may represent species of *Ancyrodelloides, A. transitans* and *A. kutscheri.* If these species are present, then the fauna may be as young as the overlying Cordilleran *delta* Zone. In either case, the fauna contains no species that indicate that the limestone beds are any younger than Lochkovian in age.

Sedimentary features (McBride and Folk, 1977) and the reworked conodont fauna indicate that the limestone unit is allodapic and that the material which forms it was probably derived from an adjacent carbonate shelf region. If the carbonate detritus was transported shortly after its formation into the basin, then the lower part of the chert and shale member at the Payne Hills section is Lochkovian or older. The Early Devonian chert fauna recovered from other sections could be Lochkovian in age as well.

The cherts that overlie the Lower Devonian interval have yielded relatively few identifiable conodont elements. Typical samples bear ramiform elements, a few carminate Pa elements (*Ozarkodina?*), *Belodella* and *Dvorakia* elements, juvenile Pa elements of *Polygnathus* species and I elements of *Icriodus* species. The small size of platform elements precludes confident identification of species in most samples. No species of latest Early Devonian or earliest Middle Devonian conodonts were recovered. In one sample from East Bourland, (EBC18, 35 m above the base of the chert and shale member), a single specimen of *Tortodus kockelianus kockelianus* was found associated with a questionable specimen of *Ozarkodina biden-tata*. *Tortodus k. kockelianus* occurs only in the *kockelianus* and lowermost *ensensis* zones, of latest Eifelian age (Weddige, 1977).

At East Bourland Mountain, samples higher than the *kockelianus* sample are relatively undiagnostic, but small Pa elements similar to those of *Polygnathus linguiformis* and the *P. varcus* group range 10 m higher, up to a covered interval (Fig. 6.2). A more diagnostic late Middle Devonian fauna was found in a single chert bed 50 m above the base of the chert and shale unit at the Payne Hills section (PHC60). *Polygnathus linguiformis*, *P. xylus xylus* and an *Icriodus* species (*I. expansus*, or *I. difficilis*) occur with poorly preserved *P. varcus* group Pa elements and *Belodella* elements. *Polygnathus xylus xylus* is characteristic of the Givetian *varcus* to early Frasnian Lower *asymmetricus* zones (Klapper and Johnson, 1980).

A series of Late Devonian conodont faunas was recovered from the uppermost 35 m of the Caballos Novaculite at the East Bourland section. Three lithologic intervals comprise the upper chert and shale unit at this section (Fig. 6.2). Above a covered interval below which Middle Devonian conodonts were recovered is an 11 m sequence of interbedded green cherts and green siliceous shale (24–35 m below the top of the Caballos Novaculite). Towards the top of this interval, chert beds decrease in thickness and number. The second unit comprises 12 m (12–24 m below top) of dominantly siliceous shale with minor cherts near the base. The upper shale beds are a dark red–brown in colour. At the top of the Caballos Novaculite is a green to black chert unit 12 m thick that has only thin interbeds of shale, mostly near its base (0–12 m below the top).

An extremely abundant and unusually well-preserved Frasnian conodont fauna was recovered from a chert bed near the base of the lower chert and shale interval (EBC25; 35 m below the top). Elements of *Polygnathus decorosus*, *Icriodus symmetricus* and *Palmotolepis hassi* are common. Fragments of *Ancyrodella* species and other *Palmatolepis* species are also present among the numerous broken elements. *Palmatolepis hassi* ranges approximately from the *Ancyrognathus triangularis* through the Lower *gigas* zones (Ziegler, in Klapper and Ziegler, 1979, Fig. 5). Chert beds from the overlying several metres have produced only indeterminate juvenile forms of *Polygnathus* and *Palmatolepis* species.

Graves (1952, Plate 79) illustrated several Pa elements of *Palmatolepis* from the siliceous shales at East Bourland Mountain. At least two of the specimens (Graves, 1952, Plate 79, Figs. 1, 3) resemble *Pa. gigas,* and a third (Graves, 1952, Plate 79, Fig. 2) appears to be a specimen of *Pa. foliacea*. The latter species ranges from the upper part of the *A. triangularis* Zone into the Lower *gigas* Zone (Ziegler, in Klapper and Ziegler, 1979). Although the precise stratigraphic level of the collection was not given, it is likely that at least some of Grave's material came from this interval of interbedded cherts and shales.

The interbedded green siliceous shales near the top of the lower chert and shale interval are well exposed, and relatively abundant Pa elements of *Palmatolepis* species are present on bedding-plane surfaces. From the upper 2 m of the unit (24–26 m), at least the following two species are present: *Pa. tenuipunctata* and *Pa. quadrantinodosalobata*. This association dates the top of the interval as early Famennian, *crepida* Zone.

Conodont elements were more difficult to obtain from bedding planes in the overlying dominantly siliceous red-coloured shale unit. Two levels have yielded identifiable *Palmatolepis* species. About 3 m above the base of the shale unit (19 m), *Pa. quadrantinodosa quadrantinodosa* and *Pa. glabra lepta* occur. The association is characteristic of the Lower *marginifera* Zone (Ziegler and Sandberg, 1984). Higher in the siliceous shale unit (15–16 m), *Pa. glabra pectinata* was recovered, a species that ranges no higher than the Upper *marginifera* Zone (Ziegler and Sandberg, 1984).

Samples from the lower 10 m of the upper chert unit at East Bourland yielded no diagnos-

tic conodonts. Near the top of the unit
(1.10–2.10 m; Fig. 6.3) a conodont fauna dominated by small Pa elements of *Pseudopolygnathus marburgensis trigonicus* appears. *Palmatolepis gracilis gracilis* and possibly *Pa. g.
sigmoidalis*, *Pelekysgnathus* coniform elements
and a number of carminate Pa elements comprise the remainder of the fauna. *Pseudopolygnathus m. trigonicus* ranges from the Upper
gracilis expansa Zone into the Middle *praesulcata* Zone, late in the Famennian Stage (Ziegler
and Sandberg, 1984).

The succeeding 0.4 m of chert has yielded
small conodont faunas dominated by elements
of *Polygnathus communis communis*, various
carminate Pa elements and *P. inornatus*.

The uppermost 0.6 m of the upper chert
unit, the top of the Caballos Novaculite, has
yielded numerous small Pa elements of *Siphonodella* species (Fig. 6.3). Although most are
too small or fragmentary to assign with confi-
dence to species, the range of morphology is
like that found in *S. cooperi* and *S. duplicata*
morphotypes 1 and 2 (Sandberg *et al.*, 1978).
Only *S. cooperi* morphotype 2 can be recognized with certainty. *Polygnathus communis
communis*, various carminate Pa elements, *P.
inornatus* and a symmetrical Pa element of a

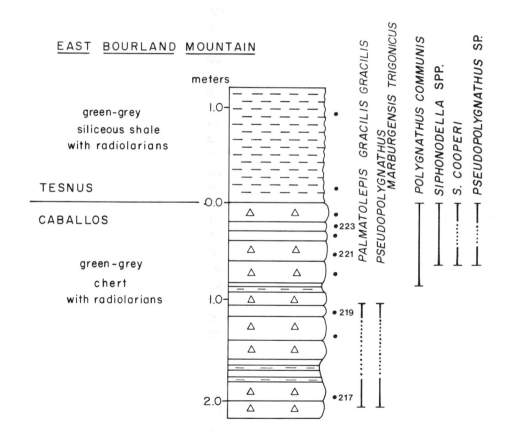

Fig. 6.3 — Distribution of important conodont species in the uppermost cherts of the Caballos Novaculite at
East Bourland Mountain (Fig. 6.1 and 6.2, EB). The dots represent conodont samples. The Devonian–Mississippian boundary lies at approximately 0.9 m below the top of the Caballos Novaculite (see text). Samples
from the lower Tesnus Formation lacked Pa elements.

Pseudopolygnathus species are also present. This fauna is no older than the Early Mississippian (Kinderhookian), Upper *duplicata* Zone, although it may range up through the Lower *crenulata* Zone.

Conodont elements are uncommon in the green radiolarian-bearing shales of the basal Tesnus Formation at East Bourland. Although fragments of ramiform elements have been seen, no Pa elements have been discovered in the lower 6 m of the Tesnus Formation. A sample at 6.7 m above the base has yielded a few small Pa elements of a *Gnathodus* species. The best-preserved specimen appears to be a Pa element of either *G. delicatus* or *G. cuneiformis,* late Kinderhookian and Osagian species.

6.5 DISCUSSION

Because the lower chert and lower novaculite members did not yield conodonts, the age of the base of the Caballos Novaculite remains uncertain. The presence of probable Early Devonian (Lochkovian) conodonts in the lower part of the overlying chert and shale member suggests that the lower chert and novaculite members may be latest Ordovician or Silurian in age. Whether an unconformity exists between the lower novaculite member and the chert and shale member, as has been proposed by McBride and Folk (1977), cannot be determined at present using conodonts.

The undifferentiated chert and shale members in the northwestern Marathon Uplift appear to represent more or less continuous deposition from Early Devonian to Early Mississippian times. The sections studied lack the upper novaculite member (Fig. 6.2) and the age of this unit remains undetermined. Lithological correlations by McBride and Thomson (1970) indicate that the upper novaculite may correlate with the middle portion of the undifferentiated chert and shale members. If this is correct, then the upper novaculite could be Middle Devonian in age.

Although there is little biostratigraphic or lithological evidence for a significant break in

deposition between the Caballos Novaculite and the Tesnus Formation at the East Bourland section, an unconformity of some magnitude may be present in the Payne Hills. Middle Devonian conodonts occur only 10 m below the top of the Caballos Novaculite and the section is atypically thin (Fig. 6.2). No conodonts have been recovered from the Tesnus Formation at this locality, but a Carboniferous radiolarian fauna characterized by latentifistulid forms is present near the base of the Tesnus Formation.

Correlation of the Caballos Novaculite with the shelf sequence in southern North America remains difficult. Ouachita strata of the Marathon Uplift comprise an allochthonous body of rock that was thrust northwards over the shelf (Ewing, 1985). The shelf region which was proximal to the area of Caballos Novaculite deposition now lies in the subsurface to the southeast, perhaps in the Devil's River Uplift, near the Texas–Mexico border (Fig. 6.1). Palaeozoic stratigraphy in the Devil's River Uplift is poorly known but, like areas farther north, Silurian and Early Devonian carbonates are unconformably overlain by Late Devonian to Early Mississippian black shale (Nicholas, 1983). Unfortunately, detailed biostratigraphic information on the subsurface Silurian to Devonian succession in the west Texas region is not readily available (see Wilson and Majewske, 1960, for a summary).

Within the Ouachita Facies, the Caballos Novaculite has been correlated with the lithologically similar Arkansas Novaculite of the Ouachita Mountains in Oklahoma and Arkansas (Fig. 6.1). Pitt *et al.* (1982) reported that L. R. Wilson obtained Silurian palynomorphs from the lower Arkansas Novaculite in Oklahoma. Hass (1951, 1956) described conodonts from the siliceous shales of the Arkansas Novaculite at several localities. Most conodonts were obtained from a middle chert and shale member and range in age from Late Devonian (Frasnian) to Early Mississippian (Osagean?). The overlying upper novaculite and upper chert and shale members were also dated as Early Mississippian in age. Like the Caballos Novaculite, too few biostratigraphic data exist to establish precise correlations within the Ouachita Facies, or with the adjacent shelf region. However, as research continues on conodont faunas of the novaculitic facies in west Texas and the Ouachita Mountain region, sufficient information should be generated to establish a framework of correlation.

6.6 ACKNOWLEDGEMENTS

Earle F. McBride, University of Texas-Austin, provided the author with unpublished data and sections, information on land access and encouragement to work on the Caballos Novaculite. Conversations with Anita Harris and Bruce Wardlaw, US Geological Survey, and Brad Robinson, ARCO Research Laboratory, were helpful in working out details of HF processing methods. David Proctor, Texas Tech University, assisted with field work and the preparation of the manuscript. David Evans, Texas Tech University, prepared the scanning electron micrographs of conodonts. Gilbert Klapper, University of Iowa, carefully reviewed the manuscript and suggested several improvements.

Acknowledgement is made to the Donors of the Petroleum Research Fund, administered by the American Chemical Society, for the support of this research (Grants PRF 13885G2 and 17283AC).

PLATE 6.1

Conodonts from the Caballos Novaculite. Specimens obtained from cherts using HF processing, except as noted. Specimens (SUI) are reposited in the paleontology collections of the Department of Geology, University of Iowa.

Pseudopolygnathus marburgensis trigonicus Ziegler 1962
Plate 6, Figs. 1, 9. Fig. 1, upper view, Pa element, specimen SUI 51688, ×70. Sample EBC217. Fig. 9, upper view, Pa element, specimen 57689, ×70. Sample EBC217.

Elictognathus sp
Plate 6, Fig. 2, Lateral view, specimen SUI 51690, ×70. Sample EBC223.

Decoriconus fragilis (Branson and Mehl 1933)
Plate 6, Fig. 3. Inner lateral view, specimen SUI 51691, ×110. Sample WMC10.

Palmatolepis gracilis gracilis Branson and Mehl 1934
Plate 6, Fig. 4. Upper view, Pa element, specimen SUI 51692, ×70. Sample EBC217.

Siphonodella sp
Plate 6, Figs. 5, 11. Fig. 5, upper view, Pa element, specimen SUI 51693, ×70. Sample EBC225. Fig. 11, upper view, Pa element, specimen 51694 ×70. Sample EBC223.

Polygnathus xylus xylus Stauffer 1940
Plate 6, Fig. 6. Upper view, Pa element, specimen SUI 51695 , ×70. Sample PMC60.

Polygnathus decorosus Stauffer 1938
Plate 6, Figs. 7, 13. Fig. 7, upper view, Pa element, specimen SUI 51696, ×70. Sample EBC25. Fig. 13, upper view of Pa element, specimen SUI 51697, ×70. Sample EBC25.

Tortodus kockelianus kockelianus (Bischoff and Ziegler 1957)
Plate 6, Fig. 8. Upper view, Pa element, specimen 51698, ×70. Sample EBC18.

Dapsilodus obliquicostatus (Branson and Mehl 1933)
Plate 6, Fig. 10. Lateral view, specimen SUI 51699, ×110. Sample WMC10.

Icriodus symmetricus Branson and Mehl 1934
Plate 6, Fig. 12. Upper view, I element, specimen SUI 51700, ×70. Sample EBC25.

Palmatolepis hassi Müller and Müller 1957
Plate 6, Figs. 14, 16. Fig. 14, upper view, Pa element, specimen SUI 51701, ×70. Sample EBC25. Fig. 16, upper view of Pa element, specimen SUI 51707, ×70. Sample EBC25.

Pedavis biexoramus Murphy and Matti 1982
Plate 6, Fig. 15. Upper view, I element, specimen SUI 51703, ×70. Limestone, sample PHL2.

Icriodus n. sp. G of Klapper (1977)
Plate 6, Fig. 17. Upper view, I element, specimen SUI 51704, ×70. Limestone, sample PHL2.

Pl. 6.1]   **Conodonts of the Caballos Novaculite**   133

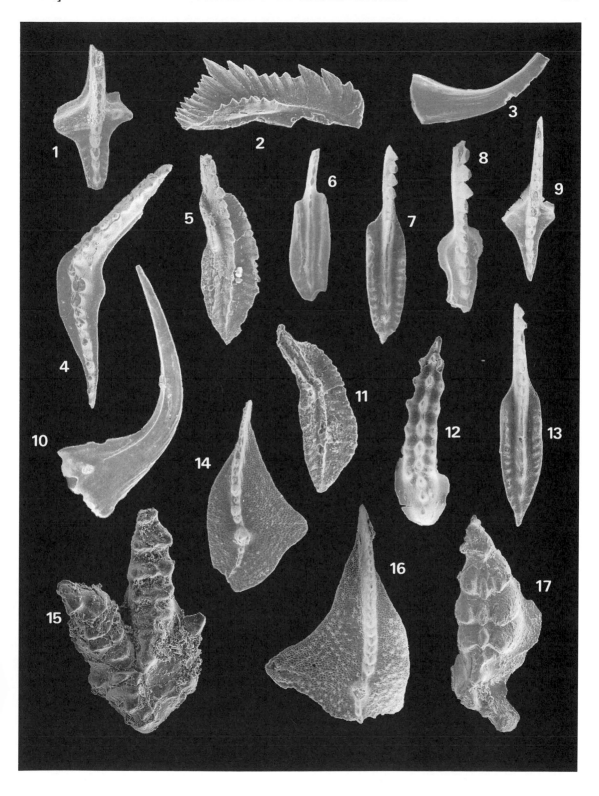

6.7 REFERENCES

Aberdeen, E. 1940. Radiolarian fauna of the Caballos Formation, Marathon Basin, Texas. *Journal of Paleontology,* **14,** 127–139, 2 plates.

Bergström, S. M. 1978. Middle and Upper Ordovician conodont and graptolite biostratigraphy of the Marathon, Texas graptolite zone reference standard. *Palaeontology,* **21,** 723–758, 2 plates.

Berry, W. B. M. 1960. Graptolite faunas of the Marathon Region, west Texas. *Publications of Bureau of Economic Geology, University of Texas,* **6005,** 1–179, 20 plates.

Bischoff, G. and Ziegler, W. 1957. Die Conodontenchronologie des Mitteldevons und des tiefsten Oberdevons. *Abhandlungen des Hesischen Landesamtes für Bodenforschung,* **22,** 1–136, 21 plates.

Branson, E. B. and Mehl, M. G. 1933. Conodonts from the Bainbridge (Silurian) of Missouri. *University of Missouri Studies,* **8,** 39–52, 1 plate.

Branson, E. B. and Mehl, M. G. 1934. Conodonts of the Grassy Creek Shale of Missouri. *University of Missouri Studies,* **8,** 171–259, 9 plates.

Ellison, S. P., Jr. 1962. Conodonts from Trans-Pecos Paleozoic of Texas. *American Association of Petroleum Geologists Bulletin,* **46,** 226.

Ewing, T. E. 1985. Westward extension of the Devil's River Uplift—Implications for the Paleozoic evolution of the southern margin of North America. *Geology,* **13,** 433–436.

Folk, R. L. 1973. Evidence for peritidal deposition of Devonian Caballos Novaculite, Marathon Basin, Texas. *American Association of Petroleum Geologists Bulletin,* **57,** 702–725.

Folk, R. L. and McBride, E. F. 1976. The Caballos Novaculite revisited: Part I, Origin of novaculite members. *Journal of Sedimentary Petrology,* **42,** 659–669.

Graves, R. W., Jr. 1952. Devonian conodonts from the Caballos Novaculite. *Journal of Paleontology,* **26,** 610–612, 3 plates.

Gutschick, R. C. and Sandberg, C. A. 1983. Mississippian continental margins of the conterminous United States. In D. J. Stanley and G. T. Moore (Eds.), *The Shelfbreak: Critical Interface on Continental Margins. Society of Economic Paleontologists and Mineralogists Special Publication,* **33,** 79–96.

Hass, W. H. 1951. Age of Arkansas Novaculite. *American Association of Petroleum Geologists Bulletin,* **35,** 2526–2541, 1 plate.

Hass, W. H. 1956. Conodonts from the Arkansas Novaculite, Stanley Shale and Jackfork Sandstone. In *Ardmore Geological Society Field Conference, Southeastern Oklahoma, Guidebook,* 25–33, 1 plate.

Heckel, P. H. and Witzke, B. J. 1979. Devonian world palaeogeography determined from distribution of carbonates and related palaeoclimatic indicators. In M. R. House, C. T. Scrutton and M. G. Bassett (Eds.), *The Devonian System, Special Papers in Palaeontology,* **23,** 99–123.

King, P. B. 1937. Geology of the Marathon Region, Texas. *U.S. Geological Survey Professional Paper,* **187,** 1–148, 27 plates.

Klapper, G. 1977. Lower and Middle Devonian conodont sequence in central Nevada; with contribution by D.

B. Johnson. *University of California-Riverside, Campus Museum Contributions,* **4,** 33–54.

Klapper, G. and Johnson, J. G. 1980. Endemism and dispersal of Devonian conodonts. *Journal of Paleontology,* **54,** 400–455, 4 plates.

Klapper, G. and Ziegler, W. 1979. Devonian conodont biostratigraphy. In M. R. House, C. T. Scrutton and M. G. Bassett (Eds.), *The Devonian System, Special Papers in Palaeontology,* **23,** 199–224.

Lowe, D. R. 1975. Regional controls on silica sedimentation in the Ouachita system. *Geological Society of America Bulletin,* **86,** 1123–1127.

McBride, E. F. and Folk, R. L. 1977. The Caballos Novaculite revisited: Part II, Chert and shale members and synthesis. *Journal of Sedimentary Petrology,* **47,** 1261–1286.

McBride, E. F. and Thomson, A. 1970. The Caballos Novaculite, Marathon region, Texas. *Geological Society of America Special Paper,* **122,** 1–129, 18 plates.

Müller, K. J. and Müller, E. 1957. Early Upper Devonian (Independence) conodonts from Iowa, Part I. *Journal of Paleontology,* **31,** 1069–1108, 7 plates.

Murphy, M. S. and Matti, J. C. 1982. Lower Devonian Conodonts (*hesperius–kindlei* Zones), Central Nevada. *University of California Publications in Geological Sciences,* **123,** 1–83, 8 plates.

Nicholas, R. L. 1983. Devil's River Uplift. In E. C. Kettenbrink, Jr. (Ed.), *Structure and Stratigraphy of the Val Verde Basin–Devil's River Uplift, Texas, West Texas Geological Society Publication,* **83–77,** 125–137.

Park, D. E., Jr. and Croneis, C. 1969. Origin of Caballos and Arkansas Novaculite Formations. *American Association of Petroleum Geologists Bulletin,* **53,** 94–111.

Pitt, W. D., Fay, R. O., Wilson, L. R. and Curiale, J. A. 1982. Geology of Pushmataha County, Oklahoma. *Llano Estacado Center for Advanced Professional Studies and Research, Studies in Natural Sciences Special Publication,* **2,** 1–101.

Sandberg, C. A., Ziegler, W., Leuteritz, K. and Brill, S. M. 1978. Phylogeny, speciation, and zonation of *Siphonodella* (Conodonta, Upper Devonian and Lower Mississippian). *Newsletters on Stratigraphy,* **7,** 102–120.

Schallreuter, R. 1982. Extraction of ostracods from siliceous rocks. In R. H. Bate, E. Robinson and L. M. Sheppard (Eds.), *Fossil and Recent Ostracods,* Ellis Horwood, Chichester, West Sussex, 169–176, 2 plates.

Sohn, I. G. 1956. The transformation of opaque calcium carbonate to translucent calcium fluoride in fossil Ostracoda. *Journal of Paleontology,* **30,** 113–114, 1 plate.

Stauffer, C. R. 1938. Conodonts of the Olentangy Shale. *Journal of Paleontology,* **12,** 411–433, 6 plates.

Stauffer, C. R. 1940. Conodonts from the Devonian and associated clays of Minnesota. *Journal of Paleontology,* **14,** 417–435, 3 plates.

Weddige, K. 1977. Die Conodonten der Eifel-Stufe im Typusgebiet und in benachbarten Faziesgebieten. *Senckenbergiana Lethaea,* **58,** 217–419, 6 plates.

Wilson, J. L. and Majewske, O. P. 1960. Conjectured Middle Paleozoic history of central and west Texas. In *Aspects of the Geology of Texas—A Symposium, Univeristy of Texas Publication,* **6017,** 65–86, 2 plates.

Ziegler, W. 1962. Taxionomie und Phylogenie Oberdevonischer Conodonten und ihre stratigraphische Bedeutung. *Abhandlungen des Hessischen Landesamtes für
Bodenforschung,* **38,** 1–166, 14 plates.

Ziegler, W. 1971. Conodont stratigraphy of the European
Devonian. In W. C. Sweet and S. M. Bergström
(Eds.), *Symposium on Conodont Biostratigraphy,*
Geological Society of American Memoir, **127,**
227–284.

Ziegler, W. and Sandberg, C. A. 1984. *Palmatolepis*-based
revision of upper part of standard Late Devonian
conodont zonation. In D. L. Clark (Ed.), *Conodont
Biofacies and Provincialism, Geological Society of
American Special Paper,* **196,** 179–194, 2 plates.

7

Lower Llanvirn–Lower Caradoc (Ordovician) conodonts and graptolites from the Argentine Central Precordillera

M. A. Hünicken and G. C. Ortega
with contributions from N. E. Vaccari and B. Waisfeld

In the Argentine Central Precordillera (Cerro Viejo, Huaco Area, San Juan Province), black shales of the Los Azules Formation (Lower Llanvirn–Lower Caradoc in age) with very important Baltoscandian region index conodonts and abundant graptolites conformably overlie limestones (and marly limestones) of the San Juan Formation with Lower Llanvirn conodonts.

Evidence is presented proving the presence of the *Eoplacognathus suecicus* Zone and the *Pygodus serra* Zone associated with important index graptolites (*Paraglossograptus tentaculatus* and *Glyptograptus teretiusculus*).

The occurrence of *Pygodus anserinus* in the upper part of this sequence (together with the latest specimens of *Glyptograptus teretiusculus* and the earliest specimens of *Nemagraptus gracilis*) is inferred.

7.1 INTRODUCTION

This work is part of a project on the biostratigraphy of the marine Ordovician rocks in the Precordillera of La Rioja, San Juan and Mendoza.

The sampled area is located at Jachál, San Juan Province, between Huaco River and a latitude 30° 10′ S (Fig. 7.1). The stratigraphic section begins with the San Juan Limestone (Kobayashi, 1937) of Arenig–Lower Llanvirn age which bears an abundant macrofauna (trilobites, brachiopods, bryozoans, sponges, etc.) and conodonts. This formation is overlain conformably by the Los Azules Shales (Harrington, 1957) of Lower Llanvirn–Lower Caradoc age with graptolites and conodonts.

The conodont elements from limestone were dissolved out of its matrix with acetic acid and those in the shales were studied using latex casts.

7.2 PREVIOUS WORK

The first find of Ordovician limestones and shales in the Precordillera was made by Stelzner (1873) and the fossils were described by Kayser (1876). The term Precordillera was proposed by Bodenbender (1902). Stappenbeck (1910) mentioned new fossiliferous localities. More recent works in the Huaco area are those by Bracaccini (1946) and Heim (1947). In the Cerro Viejo section, Borrello and Gareca (1951) described the San Juan Limestone, the Cerro Viejo Shales and the Los Azules Shales and considered them to be Llanvirn–Caradoc in age. Harrington (1957) agreed with this dating. Turner (1960) described *Glyptograptus teretiusculus Hisinger,* referring the bearing shales to the Llandeilo Series. Cuerda and Furque (1975) and Furque and Cuerda (1979) referred the limestone to the Llanvirn Series and the shales to the Llanvirn–Caradoc interval of time. Hünicken and Ortega (1980) recorded new data about the graptolite faunas.

7.3 BIOSTRATIGRAPHY

The stratigraphic distribution of important Lower Llanvirn conodonts of the San Juan Limestone and Llanvirn–Caradoc conodonts and graptolites of the Los Azules Shales is summarized on Fig. 7.2.

The ranges of the conodonts *Eoplacognathus suecicus* Bergström 1971, *Polonodus tablepointensis* Stouge 1984, *Pygodus serra* (Hadding 1913) and the graptolites *Paraglossograptus tentaculatus* (Harris), *G. teretiusculus* and *Nemagraptus gracilis* (Hall) are shown.

The Ordovician conodont sequences of the Argentine Precordillera (Hünicken *et al.*, 1982, 1985) are very similar to those of the Baltoscandian and North Atlantic regions (Serpagli, 1974; Hünicken, 1982, 1985; Hünicken and Sarmiento, 1980, 1982, 1985). For this reason, in this chapter we follow the conodont biostrati-

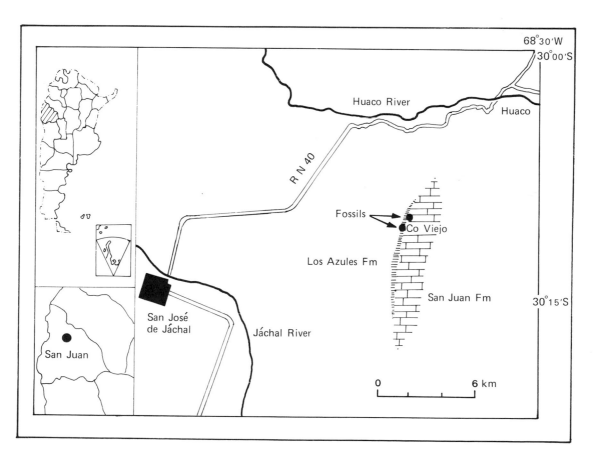

Fig. 7.1 — Location map of the Huaco region, Jáchal, San Juan Province, Argentina.

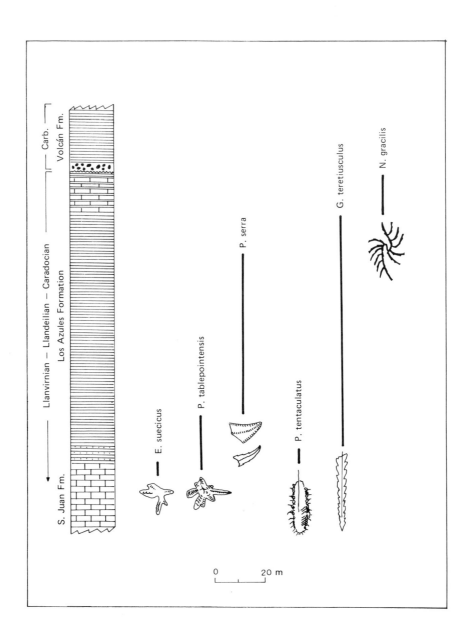

Fig. 7.2 — Stratigraphic distribution of Upper Arenig–Lower Carodoc index conodonts and graptolites within the Argentine Central Precordillera succession.

graphy of Lindström (1971), Bergström (1971), Dzik (1976, 1978) and, in particular, Löfgren (1978) and Bergström (1983).

(a) *Eoplacognathus suecicus* Zone

Löfgren (1978) considered *E. suecicus* as a zonal index for the upper Kundan and the Aserian beds (Lower Llanvirn Series of the Baltic region). In the Argentine Central Precordillera, *E. suecicus* appears in the upper beds of the San Juan Limestone (see Fig. 7.2) but it has not yet been found in the overlying shaly formation. Nevertheless, in our section, the upper boundary of the *Eoplacognathus suecicus* Zone is tentatively located at the level where *P.tablepointensis* disappears, below the first appearance of *P. serra*. The lower boundary of this zone in the San Juan Limestone remains unknown.

P. tablepointensis Stouge (1984), which occurs for the first time in the top of the San Juan Formation together with *E. suecicus*, ranges up into the lower part (20 m) of the Los Azules Formation. Also, the lower 8 m of this graptolitiferous formation have abundant specimens of *P. tentaculatus* (=*Hallograptus etheridgei*), which represent the North American *P. tentaculatus* Zone (Lower Llanvirn Series), *Tetragraptus headi* (Hall) and *Tetragraptus quadribrachiatus* Hall (Hünicken and Ortega, 1980). Immediately above, the last occurrence of *P. tablepointensis* and *Pterograptus elegans* Holm would correspond to a zone probably equivalent to part of the *Didymograptus murchisoni* Zone of northwestern Europe.

(b) *Pygodus serra* Zone

P. serra (Upper Llanvirn (Bergström, 1983) and Upper Llanvirn Series–Middle Llandeilo Series (Jaanusson, 1982, p. 8)) appears in the black shales of the Los Azules Formation, 20 m above the boundary with the San Juan Limestone. Pygodontiform and ramiform elements of *P. serra* are associated with *G. teretiusculus* (Upper Llanvirn Series–Llandeilo Series) through almost all this faunal zone. In the

Argentine Precordillera, *G. teretiusculus* appears above the upper boundary of the *Paraglossograptus tentaculatus* Zone and disappears below the first appearance of *N. gracilis*.

(c) *Pygodus anserinus* Zone

Although the occurrence of *P. anserinus* (Llandeilo Series–Lower Caradoc Series) has not been verified yet in the upper part of the formation in the studied area, this important index conodont has been found previously in the southern Argentine Precordillera, Mendoza Province (Heredia, 1982). This species is inferred to be present in the upper part of the Los Azules Formation associated with the last specimens of *G. teretiusculus* and the first occurrence of *N. gracilis*.

7.4 SYSTEMATIC PALAEONTOLOGY

All microphotographs were taken with the scanning electron microscope of the Consejo Nacional de Investigaciones Científicas y Técnicas (CONICET), Buenos Aires, Argentina.

Genus: *Eoplacognathus* Hamar 1966.

Type species: *Ambalodus lindstroemi* Hamar.

Eoplacognathus suecicus Bergström 1971, (Plate 7.1, Figs. 4–13).

Eoplacognathus suecicus Bergström, 1971, p. 141, Plate 1, Figs. 5–7, (includes synonymy through 1969).
Eoplacognathus suecicus Bergström, Löfgren, 1978, p. 59, Plate 15, Figs.9–14, 16–18 (includes synonymy through 1976, but provisionally we do not accept 1974 *Ambalodus pseudoplanus* sp.n. Viira, p. 54, Plate 6, Figs. 25, 29, 31; 1974 *Amorphognathus* sp.n. 1 Viira, Plate 7, Figs.1, 2; 1976 *Eoplacognathus pseudoplanus* (Viira), Dzik, Fig. 30g–n).
Eoplacognathus suecicus Bergström, Dzik, 1976, p. 409, Fig. 7.

Description: In our collection we have dextral (Plate 7.1, Figs. 8, 12, 13) and sinistral (Plate

7.1, Fig. 7) amorphognathiform elements. This element has four processes. The posterior one is laterally expanded, with a central row of short denticles. The posterior inner lateral process is shorter than the posterior one. The anterior process is a little shorter and thinner than the posterior one. The bifid outer lateral process has a long posterior lobe (broken in our specimens) and a short anterior lobe.

Sinistral and dextral ambalodiform elements are very similar. The majority of our specimens are dextral (Plate 7.1, Figs. 4–6, 9, 11) and only one is sinistral (Plate 7.1, Fig. 10). The Y-shaped ambalodiform elements have three subequally long processes. The gently curved anterior process makes an almost acute angle with the lateral process and the posterior one has a tendency to be wider than the others. Each process bears a central row of node-like denticles. The cusp is at the junction of the processes.

Remarks: This species is characteristic of the *Eoplacognathus suecicus* Zone, Kundan Stage and Aserian Stage (sensu Löfgren, 1978) of the Baltic region.

Occurrence and age: Top of the San Juan Formation, Cerro Viejo Section, Jáchal Department, San Juan Province. Lower Llanvirn Series.

Genus: *Polonodus* Dzik 1976.

Type species: *Ambalodus clivosus* (Viira, 1974).

Polonodus tablepointensis Stouge 1984.
(Plate 7.1, Figs. 1, 2).

Amorphognathus n.sp. Lindström, 1964, pp. 93–94, Fig. 33C.
Amorphognathus n.sp.cf. *Amorphognathus* n.sp. Lindström, 1964, Fåhraeus, 1970, p. 2065, Figs. 3A,B.
Amorphognathus variabilis Sergeeva, Fåhraeus, 1970, p.2065, Fig. 3E.
Polonodus clivosus (Viira), Dzik, 1976, p. 423, Figs. 29C,D (only).
cf. *Polonodus clivosus* (Viira), Dzik, 1976, Fig. 7, Plate 43, Fig. 1.

? Nov. gen.1 n. sp.1 Landing, 1976, p. 642, Plate 4, Fig. 20.
Polonodus ? sp. A Löfgren, 1978, pp. 76–77, Plate 16, Figs. ?9, 11, 14A,B.
Polonodus ? sp. B Löfgren, 1978, pp. 77–78, Plate 16, Figs. 7, 8, Fig. 30.
Polonodus clivosus (Viira), Löfgren, 1978, p. 76, Plate 16, Figs. 15A,B,C only.
cf. *'Amorphognathus'* n.sp. Lindström, Harris *et al.,* 1979, Plate 2, Figs. ?11,15.
Polonodus tablepointensis n.sp. Stouge, 1984, pp. 72–73, Plate 12, Fig. 13; Plate 13, Figs.1–5 (includes synonymy through 1979).

Description: The polyplacognathiform element is a conical conodont with a little cusp, a large basal cavity and four processes or lobes. The anterior process is better developed than the others. All of them are covered with concentric ridges and one or two radial rows of rounded denticles are present in each process.

The ambalodiform element has not yet been found in our section.

Remarks: *P. tablepointensis* is present in the lower and middle Table Head Formation, western Newfoundland (Stouge, 1984) and in the Lower Llanvirn Series of the Baltic region (Dzik, 1976; Löfgren, 1978; Ziegler, 1981).

Occurrence and age: Top of the San Juan Formation (Plate 7.1, Fig. 1) and the lower 20 m of the Los Azules Shales (Plate 7.1, Fig. 2), Cerro Viejo Section, Jáchal Department, San Juan Province. Lower Llanvirn Series.

Genus: *Pygodus* Lamont and Lindström 1957.

Type species: *Pygodus anserinus* Lamont and Lindström, 1957.

Pygodus serra (Hadding 1913)
(Plate 7.1, Fig. 3).

Arabellites serra Hadding, 1913, p. 33, Plate 1, Figs. 12, 13.
Periodon serra (Hadding), Lindström, 1955, p. 110, Plate 22, Figs. 17, 20, 25.
Arabellites serra Hadding, Lindström, 1960, p. 91, Fig. 7:6.

Pygodus n. sp. 2 Lindström, 1960, p. 91, Fig. 7:1.

Pygodus anserinus Lamont and Lindström, Wolska, 1961, p. 357, Figs. 4, 5.

?*Pygodus trimontis* Hamar, 1966, p. 70, Plate 7, Figs. 12, 16, 17.

Pygodus serrus (Hadding), Bergström, 1971, p. 149, Plate 2, Figs. 22, 23.

Haddingodus serra (Hadding), Viira, 1974, p. 86, Plate 11, Fig. 28.

Pygodus serrus (Hadding), Dzik, 1976, p. 443, Fig. 29a–b.

Pygodus serra (Hadding), Löfgren, 1978, p. 98, Fig. 32D–F.

Pygodus serra (Hadding), Bergström, 1983, p. 44, Fig. 3.

Description: *Pygodus serra* has pygodontiform, haddingodontiform and ramiform elements. To date, only several pygodontiform, which show three denticle rows on the upper side of the platform, and some four-branched denticulated ramiform elements (Fig. 7.2), which agree with those illustrated by Löfgren (1978, Fig. 32F) and Bergström (1983, Fig. 3), have been found. They are also very similar to those described as *Tetraprioniodus lindstroemi* by Sweet and Bergström (1962, p. 1248, Plate 170, Figs. 5, 6).

Remarks: The geographical distribution of *P. serra* includes the Baltic region, Scotland, Canada, North America, USSR, China and Australia. In the Baltoscandian region, *P. serra* ranges from the uppermost Aserian (Upper Llanvirn Series), throughout Lasnamägian to the Lower Uhakuan (Lower Llandeilo Series).

Occurrence and age: Los Azules Formation, shaly upper member (Plate 7.1, Fig. 3; Fig. 7.2), Cerro Viejo section, Jáchal Department, San Juan Province. Upper Llanvirn–Lower Llandeilo Series.

PLATE 7.1

Specimens (CORD-MP) are reposited in the Museo de Paleontología (Departmento de Geología), Universidad Nacional de Córdoba, Argentina.

Polonodus tablepointensis Stouge 1984
Plate 7.1, Figs. 1, 2. Fig. 1, upper view, specimen CORD-MP 511–1, × 75. Fig. 2, lateral view, latex cast, specimen CORD-MP 700–1, × 30.

Pygodus serra (Hadding 1913)
Plate 7.1, Fig. 3. Upper view, latex cast, specimen CORD-MP 701–1, × 75.

Eoplacognathus suecicus Bergström 1971
Plate 7.1, Figs. 4–13. Fig. 4, upper anterior view, dextral ambalodiform element, specimen CORD-MP 510–1, × 75. Fig. 5, upper postero-anterior view, dextral ambalodiform element, specimen MP 510–2, × 75. Fig. 6, antero-lateral view, dextral ambalodiform element, specimen CORD-MP 510–3, × 75. Fig. 7, upper view, posterior process of amorphognathiform element, specimen CORD-MP 510–7, × 70. Fig. 8, upper view, amorphognathiform element, specimen CORD-MP 510–8, × 75. Fig. 9, antero-lateral view, dextral ambalodiform element, specimen CORD-MP 510–4, × 75. Fig. 10, upper view, sinistral ambalodiform element, specimen CORD-MP 510–5, × 75. Fig. 11, lateral view, dextral ambalodiform element, specimen CORD-MP 510–6, × 75. Fig. 12, upper view, anterior and posterior processes of amorphognathiform element, specimen CORD-MP 510–9, × 150. Fig. 13, upper antero-lateral view, amorphognathiform element, specimen CORD-MP 510–10, × 150.

Pl. 7.1] **Lower Llanvirn — Lower Caradoc (Ordovician) conodonts** 143

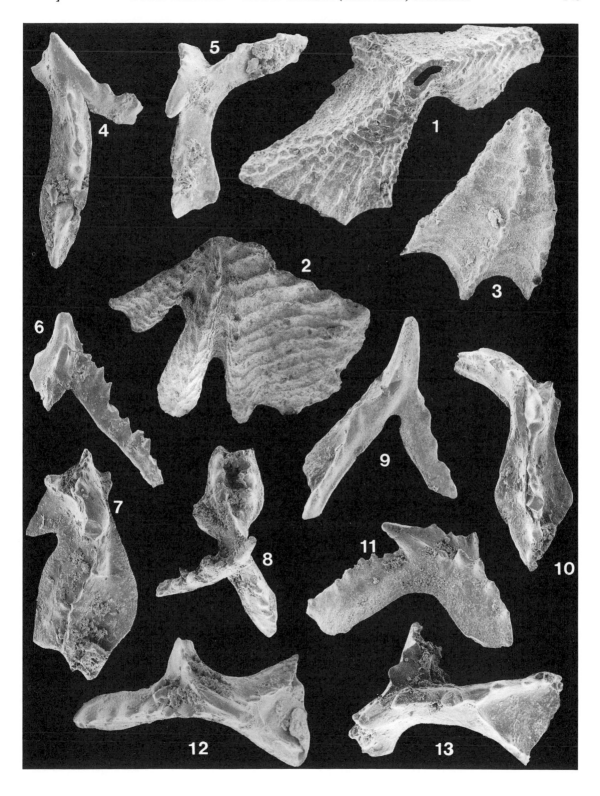

7.4 REFERENCES

Bergström, S. M. 1971. Conodont biostratigraphy of the Middle and Upper Ordovician of Europe and Eastern North America. In W. C. Sweet and S. M. Bergström (Eds.), *Symposium on Conodont Biostratigraphy, Geological Society of America Memoir*, **127**, 83–161, 2 plates.

Bergström, S. M. 1983. Biogeography, evolutionary relationships, and biostratigraphic significance of Ordovician platform conodonts. *Fossils and Strata*, **15**, 35–58.

Bodenbender, G. 1902. Contribución al conocimiento de la Precordillera de San Juan. *Boletín de la Academia Nacional de Ciencias, Córdoba, Argentina*, **17**, 203–261, 3 plates.

Borrello, A. V. and Gareca, P. G. 1951. Sobre la presencia de *Nemagraptus gracilis* (Hall) en el Ordovícico del Norte de San Juan. *Revista de la Asociación Geológica Argentina*, **6** (3), 187–193.

Bracaccini, O. 1946. Contribución al conocimiento geológico de la Precordillera sanjuanino-mendocina. *Boletín de Informaciones Petroleras (Buenos Aires)*, **23** (258), 81–105; **23** (260), 259–274; **23** (261), 361–384; **23** (262), 455–473; **23–2** (263), 22–35; **23–2** (264), 103–125; **23–2** (265), 171–192.

Cuerda, A. J. and Furque, G. 1975. Nuevos datos sobre la Paleobiogeografía de la Formación Gualcamayo, Ordovícico de la Precordillera. *Actas del Primer Congreso Argentino de Paleontología y Bioestratigrafía, Tucumán*, **1**, 49–57, 1 plate.

Dzik, J. 1976. Remarks on the evolution of Ordovician conodonts. *Acta Palaeontologica Polonica*, **21** (4), 395–455, Plates 41–44.

Dzik, J. 1978. Conodont biostratigraphy and paleogeographical relations of the Ordovician Mojcza Limestone (Holy Cross Mts., Poland). *Acta Palaeontologica Polonica*, **23** (1), 51–72, Plates 12–15.

Fåhraeus, L. E. 1970. Conodont-based correlation of Lower and Middle Ordovician strata in western Newfoundland. *Geological Society of America Bulletin*, **81**, 2061–2076.

Furque, G. and Cuerda, A. J. 1979. Precordillera de La Rioja, San Juan y Mendoza. In *Geología Regional Argentina*, Academia Nacional de Ciencias, Córdoba, **1**, 455–522.

Hadding, A. R. 1913. Undre dicellograptusskiffern i Skåne jämte några därmed ekvivalenta bildningar. *Lunds Universitets Årsskrift N. F. Avd. 2*, **9** (15), 1–90, 8 plates.

Hamar, G. 1966. The Middle Ordovician of the Oslo region, Norway. 22. Preliminary report on conodonts from the Oslo–Asker and Ringerike districts. *Norsk Geologisk Tidsskrift*, **46**, 27–83.

Harrington, H. J. 1957. Ordovician Formations of Argentina. In H. J. Harrington and A. F. Leanza (Eds.), *Ordovician Trilobites of Argentina, University of Kansas, Special Publication*, **1**, 1–59.

Harris, A. G., Bergström, S. M., Ethington, R. L. and Ross, R. J., Jr. 1979. Aspects of Middle and Upper Ordovician conodont biostratigraphy of carbonate facies in Nevada and southeast California and comparison with some Appalachian successions. In C. A. Sandberg and D. L. Clark (Eds.), *Conodont Biostrati-*

graphy of the Great Basin and Rocky Mountains, Brigham Young University Geology Studies, **26** (3), 7–43.

Heim, A. 1947. El carbón del Río Huaco. *Boletín de la Dirección de Minas y Geología, Buenos Aires*, **62**, 1–18, 2 plates.

Heredia, S. E. 1982. *Pygodus anserinus* Lamont and Lindström (conodonte) en el Llandeiliano de la Formación Ponón Trehué, Provincia de Mendoza, Argentina. *Ameghiniana, Revista de la Asociación Paleontológica Argentina*, **19** (3–4), 229–233, 1 plate.

Hünicken, M. A. 1982. La Zona do *Oepikodus evae* (Conodonte, Arenigiano Inferior) en la Formación San Juan, Quebrada de Talacasto, Departamento Ullún, San Juan, Argentina. *Actas del Quinto Congreso Latinoamericano de Geología, Buenos Aires*, **1**, 797–802.

Hünicken, M. A. 1985. Lower Ordovician conodont Biostratigraphy in Argentina. *Boletín de la Academia Nacional de Ciencias, Córdoba, Argentina*, **56** (3–4), 309–322.

Hünicken, M. A. and Ortega, G. 1980. Acerca del hallazgo de *Tetragraptus headi* (Hall) (Graptolithina) en la Formación Los Azules (Ordovícico), Departmento Jáchal, San Juan, R. Argentina. *Boletín de la Academia Nacional de Ciencias*, **53** (3–4), 343–350, 2 plates.

Hünicken, M. A. and Sarmiento, G. 1980. The Baltoscandian conodont *Prioniodus elegans* Pander (Lower Arenigian) from the San Juan Formation of the Precordillera, Guandacol River, La Rioja, R. Argentina. *Boletín de la Academia Nacional de Ciencias, Córdoba, Argentina*, **53** (3–4), 293–306, 2 plates.

Hünicken, M. A. and Sarmiento, G. 1982. La Zona Baltoscandinava de *Oepikodus evae* (Conodonte, Arenigiano Inferior) en el perfil del Río Guandacol, La Rioja, Argentina. *Actas del Quinto Congreso Latinoamericano de Geología, Buenos Aires*, **1**, 791–796.

Hünicken, M. A. and Sarmiento, G. 1985. *Oepikodus evae* (Lower Arenigian conodont) from Guandacol, La Rioja Province, Argentina. *Boletín de la Academia Nacional de Ciencias, Córdoba, Argentina*, **56** (3–4), 323–332, 2 plates.

Hünicken, M. A., Suárez Riglos, M., and Sarmiento, G. 1982. Lower Ordovician Conodonts from Argentina. *Proceedings of the Fourth International Symposium on the Ordovician System, Sundvolden, Oslo, 19–23 August 1982, Abstracts*, in *Palaeontological Contributions, University of Oslo*, **280**, 25.

Hünicken, M. A., Suárez Riglos, M., and Sarmiento, G. 1985. Conodontes tremadocianos de la Sierra de Cajas, Departamento Humahuaca, Provincia de Jujuy, R. Argentina. *Boletín de la Academia Nacional de Ciencias, Córdoba, Argentina*, **56** (3–4), 333–347, 3 plates.

Jaanusson, V. 1982. Introduction to the Ordovician of Sweden. In D. L. Bruton and S. H. Williams (Eds.), *Field Excursion Guide, Proceedings of the Fourth International Symposium on the Ordovician System, Sundvolden, Oslo, 19–23 August 1982*, in *Palaeontological Contributions, University of Oslo*, **279**, 1–10.

Kayser, E. 1876. Ueber Primordiale und Untersilurische Fossilien aus der Argentinischen Republik. *Cassel. Beiträge zur Geologie und Palaeontologie der Argen-*

tinischen Republik, 2-Palaeontologischer Theil, 1 Abteilung, Palaeontograph. Suppl., **3** (2), 1–33.

Kobayashi, T. 1937. The Cambro-Ordovician shelly faunas of South America. *Journal of Faculty of Science, University of Tokyo*, **2** (5), 369–522, 8 plates.

Lamont, A. and Lindström, M. 1957. Arenigian and Llandeilian cherts identified in the Southern Uplands of Scotland by means of conodonts, etc. *Edinburgh Geological Society Transactions*, **17**, 60–70.

Landing, E. 1976. Early Ordovician (Arenigian) conodont and graptolite biostratigraphy of the Taconic allochthon, eastern New York. *Journal of Paleontology*, **50**, 614–646.

Lindström, M. 1955. The conodonts described by A. R. Hadding, 1913. *Journal of Paleontology*, **29**, 105–111, Plate 22.

Lindström, M. 1960. A Lower–Middle Ordovician succession of conodont faunas. *21st International Geological Congress Reports*, **7**, 88–96.

Lindström, M. 1964. *Conodonts*, Elsevier, Amsterdam, 1–196.

Lindström, M. 1971. Lower Ordovician conodonts of Europe. In W. C. Sweet and S. M. Bergström (Eds.), *Proceedings of Symposium on Conodont Biostratigraphy, Geological Society of America Memoir*, **127**, 21–61, 1 plate.

Löfgren, A. 1978. Arenigian and Llanvirnian conodonts from Jämtland, Northern Sweden. *Fossils and Strata*, **13**, 1–129, 16 plates.

Serpagli, E. 1974. Lower Ordovician conodonts from Pre-cordilleran Argentina (Province of San Juan). *Bolletino della Società Paleontologica Italiana*, **13** (1–2), 17–98, Plates. 7–37.

Stappenbeck, R. 1910. La Precordillera de San Juan y Mendoza. *Anales del Ministerio de Agricultura, Sección Geología, Mineralogía y Minería*, **4** (3), 1–187, 15 plates.

Stelzner, A. 1873. Mitteilungen an Professor H. B. Geinitz. Über seine Reise durch die argentinischen Provinzen San Juan und Mendoza und die Cordillere zwischen dem 31° und 32° S. *Neues Jahrbuch für Mineralogie, Geologie und Palaeontologie*, 726–744.

Stouge, S. S. 1984. Conodonts of the Middle Ordovician Table Head Formation, western Newfoundland. *Fossils and Strata*, **16**, 1–145, 18 plates.

Sweet, W. C. and Bergström, S. M., 1962. Conodonts from the Pratt Ferry Formation (Middle Ordovician) of Alabama. *Journal of Paleontology*, **36**, 1214–1252.

Turner, J. C. M. 1960. Faunas graptolíticas de América del Sur. *Revista de la Asociación Geológica Argentina*, **14** (1–2), 5–180, 9 plates.

Viira, V. 1974. Ordovician Conodonts of the East Baltic (in Russian, with English summary). *Eesti NSV Teaduste Akadeemia Geologia Institut Valgus*, 1–142.

Wolska, Z. 1961. Konodonty z Ordowickich glazownarzutowych polski. *Acta Palaeontologica Polonica*, **6** (4), 339–365, 4 plates.

Ziegler, W. (Ed.) 1981. *Catalogue of Conodonts, IV*, Schweizerbart'sche, Stuttgart, 1–445, 40 plates.

8

Some aspects of the application of scanning electron microscopy in conodont studies

With contributions from **R. D. Burnett; and M. P. Smith**

Early scanning electron microscopy studies revealed details of surface morphology and ultrastructure of unetched fracture surfaces. The internal microstructure of conodonts may be examined by studying either fracture surfaces or polished and etched surfaces of conodonts. Etching of conodonts may alter or destroy more of the structure than is revealed. Fracture surfaces develop along preferred lines of fracture and conodonts do not fracture naturally in desired directions. Stress-induced fractures do not produce clean breaks. A new technique is described which, in addition to revealing details of the ultrastructure, also permits microprobe analysis of polished (but unetched) surfaces. Acetate peels of etched surfaces reveal internal structural detail. Cylinder etching, in combination with critical-point drying, reveals the interrelationship of constituent mineral and organic phases within conodonts.

Unlike secondary electron images, back-scattered electron imaging presents no problem with charging on points of high relief. Also, the loss of morphological detail typical of photographs of bedding-plane assemblages is not encountered with back-scattered electron imaging.

8.1 A SHORT REVIEW OF THE APPLICATION OF SCANNING ELECTRON MICROSCOPY IN CONODONT STUDIES
(by R. D. Burnett)

(a) Introduction
Possibly the earliest suggestion of the applicability of scanning electron microscopy (SEM) to general microfossil studies was that of Sandberg and Hay (1967), followed later that year by SEM illustrations of a conodont, *Palmatolepis* (Hay and Sandberg, 1967).

More comprehensively, Pietzner *et al.* (1968), in a classic early study of conodont geochemistry and ultrastructure, incorporated scan studies of overall surface morphology and of unetched fracture surfaces. However, concentrated use of transmission electron microscopy via highly complex and only partially successful preparation techniques was made to analyse ultrafine crystallite detail. The eventual move to the use of SEM techniques for microstructure examination is illustrated by Pierce and

Langenheim (1969), who describe polished and etched surfaces within *Palmatolepis* and *Polygnathus* types. They describe the textures of the compositional crystallites and propose them as distinct for each type and thus to be of taxonomic significance. Other examples of SEM-revealed detail of taxonomic interest were proposed by Pierce and Langenheim (1970), in a study of fine ornamentation (polygonal networks, striae) which show some diagnostic tendencies.

Interesting early gross examinations of conodonts by SEM include that of surficial fine features in lower Arenig forms (Lindström *et al.*, 1972) and a short examination of metamorphosed (phyllite-grade) conodonts (Schönlaub and Zezula, 1975). The success of SEM application in such work is underlined by its present ubiquitous use in illustrating biostratigraphic treatises (Müller, 1981; Sandberg and Clark, 1979).

(b) Techniques with respect to the study of the internal microstructure of conodonts

Preparation techniques for examination by SEM have been outlined by Stone (this volume). With respect to the study of internal microstructure, two schools of thought relating to techniques predominate, one in favour of examining fracture surfaces, and the other supporting the use of polished–etched sections.

Considering firstly the fracture-surface technique, this has been used extensively by Lindström and Ziegler (1971) in studying elements of *Panderodontacea*: '... 'The best surfaces for study of the internal structure were obtained by breaking conodonts. Natural fracture surfaces tend to be diagenetically recrystallized. We refrained from etching the surfaces, since this method appears to alter or destroy more of the structure than it reveals'. The technique has continued to find some applications, as typified in the examination by von Bitter and Merrill (1983) of broken denticle surfaces (cross sections) of *Ellisonia* and of broken cusps of *Ptiloconus* and *Chirognathus* (all unetched).

The argument proposed by fracture-technique advocates is typified by the ideas of Lindström and Ziegler (1971) that etching techniques could simulate structures which did not conform to the original structure, a view with a degree of truth behind it. Examples of such artefacts are rare but undoubtedly exist. The 'growth canals' reported running axially in conodonts (Barnes *et al.*, 1973b; Barnes and Slack, 1975), discovered via polishing-etching techniques, have not been re-observed by later workers and probably represent axial cavities joined by etching of their weak interconnected walls to produce one long space (Lindström and Ziegler, 1981). Barnes *et al.* (1973a) also note the existence of spheroids within white matter; these are probably 'either small fractures grown together or parting surfaces of crystallites that were artificially enhanced by (the authors') etching method' (Lindström and Ziegler, 1981).

Nonetheless, the majority of internal ultrastructure studies have applied the polishing–etching technique. Early work, following that of Pierce and Langenheim (1969), includes sectional examination of the cusps of Middle Ordovician forms (Barnes *et al.*, 1970) and polished–etched surface photographs of *Polygnathus* (Müller and Nogami, 1971). More recently Szaniawski (1982) applied the technique to illustrate inorganic and organic layering in protoconodonts. The method used in these researches is typified by that in Barnes *et al.* (1973a). These authors apply a sequence of plastic mounting, polishing, etching (35 s in 2 M hydrochloric acid, HCl) and coating, for a comparative SEM study of neurodont ultrastructure with that of hyaline and cancellate (albid) conodonts.

The arguments in support of etching–polishing are given succinctly by Barnes *et al.* (1973b) who wrote that, although fracture surfaces should permit study of unaltered structure, they present disadvantages and limitations. Conodonts do not fracture readily along all desired directions, since the arrangement of crystallites and lamellae produce preferred lines of fracture; hyaline conodonts fracture lengthwise, whereas white matter fractures transversely. Stress applied to produce alternative lines of fracture does not give a 'clean break' and could result in illusory or ill-defined structures, an argument originally levelled at polishing–etching. These researchers argue that, whilst material removed by etching is selective, structural patterns tend to be accentuated rather than falsified.

The greater volume of microstructural SEM work applying the etching–polishing technique attests to its greater applicability over fracture study. However, in applying the former technique, it is of paramount importance to be aware of the possible creation of artefacts.

8.2 A NEW TECHNIQUE FOR THE EXAMINATION OF CONODONT ULTRASTRUCTURE BY SEM
(by R. D. Burnett)

(a) Introduction
The technique outlined modifies that used by Barnes *et al.* (1973a) for studying the internal morphology of conodonts and incorporates polishing and etching of mounted specimens. Lindström and Ziegler (1971) argued that etching techniques generate structures not originally present within the fossil. However, restricting microstructure study of conodonts to fracture surfaces is impractical. It is extremely difficult to achieve precision in fracture-siting, and serial examination of individual conodonts is a near impossibility. Furthermore, comparisons of internal detail revealed by fracturing and by polishing–etching disclose no significant differences (Burnett, 1985).

(b) Technique
(i) *Initial mounting*
The conodont is manipulated into a vertical position (conventional notation) on a surface consisting of double-sided transparent adhesive tape mounted on a standard glass slide. Over the conodont is placed a brass cylinder (of dimensions 5 mm high and 4 mm in external

diameter, as used for electron-microprobe 'standards'). This is pressed down to make a closed seal with the adhesive tape. Into this cylinder, bubble-free Araldite is pipetted carefully down the side of the cylinder, to avoid both displacing the conodont and trapping air bubbles.

(ii) *Polishing*
After drying of the Araldite, the 'basal' surface of the cylinder (that originally in contact with the adhesive tape) is polished, i.e. working in a lower to upper direction. Polishing is implemented by a turntable of moderate speed in six stages.

(1) '1000 grit' paper (about 20 µm) (wet or dry Tri-M-ite 3M Scotch paper) for 20 s or until just before the desired level is attained.
(2) 14 µm diamond paste on a paper lap for 1 m.
(3) 14 µm diamond paste on a paper lap for 8 m.
(4) 6 µm diamond paste on a paper lap for 6 m.
(5) 1 µm diamond paste on a paper lap for 3 m.
(6) $\frac{1}{4}$ µm diamond paste on a velvet lap for 4 m.

The brass cylinder is mounted flush in a Perspex holder during polishing to provide a flat contact between cylinder and polishing surfaces. The diamond pastes are lanolin-based, but additional lubricant is added as a Hyprex aerosol (a thin paraffin). Insufficient lubrication causes the specimen to be torn from the holder by friction; excess lubrication causes it to become rounded.

At each polishing stage, specimens are examined in reflected light to establish that the correct (maximal) level of polish has been achieved. Between polishing grades the cylinder is washed carefully in Genklean.

(iii) *Etching*
The resulting cylinder thus contains a conodont exhibiting a flattened sectional surface continuous with that of the araldite mounting. Being flat, this surface possesses no discernible structural relief under SEM and so requires etching.

Three drops of 2 M HCl are pipetted onto the polished surface and left for an optimum 30 s. The surface is then gently flushed with distilled water and air-dried.

(iv) *Mounting*

The etched cylinder is fixed to a standard aluminium SEM stub by high-conductivity acetate-based paint (Dag). This provides good electrical continuity between stub and cylinder.

The stub surface is then coated as required. Gold–palladium, applied by sputter coater to a thickness of 250 Å, is adequate for most studies, being thick enough to prevent specimen charging but not to conceal detail. Alternatively, for qualitative X-ray analysis (energy-dispersive X-ray analysis facilities) a carbon coating is applied as gold–palladium interferes with other X-ray lines (Goldstein *et al.*, 1981).

The processes (i)–(iv) can be carried out repeatedly, to provide a serial visual record of microstructure. Coat-removal prior to repolishing is necessary is some cases, e.g. when further X-ray analyses are planned. Gold–palladium may be removed by soaking in 10% solution of $FeCl_3$ in ethanol for 6 h, according to the method of Crossmann and McCann (1979). (A word of caution, $FeCl_3$ is used for etching certain sulphides but normally as a 20% solution. Nevertheless there may be a danger of erroneous etching during removal of the gold-–palladium coating. However, although a surface may be etched after the gold is removed, the method (which essentially allows serial examination) next involves a repolishing. Thus, provided that the etching is not too deep, the next polishing phase should remove all trace of it.) Carbon coatings can best be removed by low-temperature plasma ashing for short periods.

(c) Advantages of the technique

Supplementary to revealing ultrastructure detail, the method enables microprobe analysis of polished (but not etched) conodont interiors.

Etched specimens may be used to obtain acetate peels of internal surface detail, to be carbon-shadowed for transmission electron microscope study. Finally, cylinder etching in combination with critical-point drying (Clarke, 1980) can be used to reveal the interrelationship of consti-tuent mineral and organic phases within conodonts.

8.3 THE USE OF BACK-SCATTERED ELECTRON IMAGING IN THE PHOTOGRAPHIC EXAMINATION OF BEDDING-PLANE ASSEMBLAGES
(by M. P. Smith)

Bedding-plane assemblages traditionally have been illustrated by line drawings (Schmidt, 1934), light photographs (Rhodes, 1952; von Bitter, 1976) and/or scanning electron micro-graphs using the conventional secondary-elec-tron mode (Norby, 1976; Rieber, 1980). Each of these methods has disadvantages to a greater or lesser degree. Camera lucida drawings often fail to convey the relief of the subject, there may be depth-of-field problems with light photogra-phy and there are often charging problems and inadequate depiction of detail with secondary-electron images (Fig. 8.1).

Back-scattered electron imaging has become an increasingly important petrographic technique in sedimentology (Pye and Krinsley, 1983a, 1983b, 1984; White *et al.*, 1984) and in structural and metamorphic geology (Hall and Lloyd, 1981; Lloyd and Hall, 1981). Secondary electrons are shallow low-energy emissions pro-duced by inelastic collisions in which energy is transferred from the primary-beam electrons to the atoms of the specimen. In contrast, back-scattered electrons are deeper high-energy emissions produced after elastic collisions where primary electrons interact with the speci-men but lose little of their energy prior to being scattered back out of the specimen. The back-scattered coefficient (the number of back-scat-tered electrons over the total number of prim-ary electrons) varies with the atomic number of the specimen; the greater the atomic number,

Fig. 8.1 — Secondary-electron image of a Pennsylvanian bedding plane assemblage from Bailey Falls, Illinois (×29). Specimen 5545/015 was collected by Dr. R. J. Aldridge and is deposited in the micropalaeontological collections of the Department of Geology, University of Nottingham.

Fig. 8.2 — Back-scattered electron image of the same assemblage as that in Fig. 8.1 (×28).

the higher is the back-scatter coefficient. For compounds, the coefficient is calculated by using the mean atomic number, which is based on the weight-fractions of the components (Hall and Lloyd, 1981). The higher the atomic or mean atomic number, the brighter the image appears on the screen; hence, preliminary mineral identifications may be made using back-scattered electron images.

The shale fragments on which the assemblages are usually found are trimmed and the thickness reduced as much as is safely possible. Specimens are then attached to stubs with double-sided adhesive tape prior to coating with carbon.

Apatite has a mean atomic weight of approximately 12.5 (Hall and Lloyd, 1981) and thus has a pale- to mid-gray tone on back-scattered electron images (Fig. 8.2). The background matrix is generally a slightly darker gray, probably because of the carbonaceous content of the shales. This type of image has two principal advantages over those produced by secondary electrons. Firstly, there is no problem with charging which is often seen in secondary images on points of high relief such as Pa elements. Secondly, in secondary-electron images, the elements often appear very dark gray or black. With the superimposition of charged spots onto this tone, morphological detail is often obscured. Back-scattered electron images suffer from neither of these drawbacks. Stereo images may be constructed in a manner identical with that for specimens photographed in the secondary-electron mode.

Back-scattered electron imagery overcomes all the problems previously associated with illustrating bedding-plane assemblages, namely depiction of relief, depth-of-field and charging. The technique also has applications in the photography of fused clusters and isolated elements.

8.4 ACKNOWLEDGEMENTS

The work of M. P. Smith arises from research on conodont palaeobiology funded by NERC Research Grant GR3/5105. Dr. R. J. Aldridge commented on an early version of the manuscript. Mr. A. Swift provided photograhic assistance.

8.5 REFERENCES

Barnes, C. R., Sass, D. B. and Monroe, E. A. 1970. Preliminary studies of the ultrastructure of selected Ordovician conodonts. *LIfe Science Contributions, Royal Ontario Museum*, **76**, 1–24.

Barnes, C. R., Sass, D. B. and Monroe, E. A. 1973a. Ultrastructure of some Ordovician conodonts. In F. H. T. Rhodes, (Ed.) *Conodont Paleozoology, Geological Society of America Special Paper*, **141**, 1–30.

Barnes, C. R., Sass, D. B., and Poplawski, M. L. S. 1973b. Conodont Ultrastructure: the family *Panderodontidae*. *Life Science Contributions, Royal Ontario Museum*, **90**, 1–36.

Barnes, C. R. and Slack, D. J. 1975. Conodont ultrastructure: the subfamily *Acanthodontidae*. *Life Science Contributions, Royal Ontario Museum*, **106**, 1–21.

Burnett, R. D. 1985. Conodont Diagenesis and Metamorphism: Investigations into chemo-physical aspects, with case studies from Northern England. *Unpublished Ph.D. Thesis*, Sheffield University, 1–297.

Clark, G. R. 1980, Techniques for observing the organic matrix of molluscan shells. In D. C. Rhodes and R. A. Lutz (Eds.), *Topics in Geobiology 1: Skeletal growth of aquatic organisms*, Plenum Press, New York, 607–612.

Crossman, R. S. and McCann, P. 1979. Removal of gold-palladium films from electron microscope specimens. *Micron*, **10**, 37.

Goldstein, J. I., Newberry, D. E., Echlin, P., Joy, D. C., Fiori, C. and Lifshin, E. 1981. *Scanning Electron Microscopy and X-ray Microanalysis*, Plenum Press, New York, 1–528.

Hall, M. G. and Lloyd, G. E. 1981. The SEM examination of geological materials with a semi-conductor back-scattered electron detector. *American Mineralogist*, **66**, 362–368.

Hay, W. W. and Sandberg, P. A. 1967. The scanning electron microscope, a major breakthrough for micropaleontology. *Micropaleontology*, **13**, 407–418.

Lindström, M., McTavish, R. A. and Ziegler, W. 1972. Feinstrukturelle Untersuchungen an Conodonten. II. Einige *Prioniodontidae* aus dem Ordovicium Australiens. *Geologica et Palaeontologica*, **6**, 33–43.

Lindström, M. and Ziegler, W. 1971. Feinstrukturelle Untersuchungen an Conodonten. I. Die Uberfamilie *Panderodontacea*. *Geologica et Palaeontologica*, **5**, 9–33.

Lindström, M. and Ziegler, W. 1981. Introduction to the Conodonta; morphology and composition of elements; micromorphology of elements; surface micro-ornamentation and observations in internal composition. In R. A. Robison (Ed.), *Treatise on Invertebrate Paleontology: Part W, Supplement 2, Conodonta*, Geological Society of America, Boulder, Colorado, and University of Kansas Press, Lawrence, Kansas, W41–W52.

Lloyd, G. E. and Hall, M. G. 1981. Application of scanning electron microscopy to the study of deformed rocks. *Tectonophysics*, **78**, 687–698.

Müller, K. J. 1981. Micromorphology of elements; internal structure. In R. A. Robison (Ed.), *Treatise on Invertebrate Paleontology: Part W, Supplement 2, Conodonta*, Geological Society of America, Boulder, Colorado, and University of Kansas Press, Lawrence, Kansas, W20–W41.

Müller, K. J. and Nogami, Y. 1971. Uber den Feinbau der Conodonten. *Memoirs of the faculty of Science Kyoto University Geology and Mineralogy Series*, **38**, 1–87.

Norby, R. D. 1976. Conodont apparatuses from Chesterian (Mississippian) strata of Montana and Illinois. *Unpublished Ph.D. Thesis*, University of Illinois Urbana-Champaign, 1–244, 21 plates.

Pierce, R. W. and Langenheim, R. L. 1969. Ultrastructure in *Palmatolepis* sp. and *Polygnathus* sp. *Geological Society of America Bulletin*, **80**, 1397–1400.

Pierce, R. W. and Langenheim, R. L. 1970. Surface patterns of selected Mississippian conodonts. *Geological Society of America Bulletin*, **81**, 3225–3236.

Pietzner, H., Vahl, J., Werner, H. and Ziegler, W. 1968. Zur chemischen Zusammensetzung und Mikromorphologie der Conodonten. *Palaeontolographica A*, **128**, 115–152, 10 plates.

Pye, K. and Krinsley, D. H. 1983a. Mud rocks examined by back-scattered electron microscopy, *Nature*, **301**, 412–413.

Pye, K. and Krinsley, D. H. 1983b. Interlayered clay stacks in Jurassic shales. *Nature*, **404**, 618–620.

Pye, K. amd Krinsley, D. H. 1984. Petrographic examination of sedimentary rocks in the SEM using backscattered electron detectors. *Journal of Sedimentary Petrology*, **54**, 877–888.

Rhodes, F. H. T. 1952. A classification of Pennsylvanian conodont assemblages. *Journal of Paleontology*, **26**, 886–901, 4 plates.

Rieber, H. 1980. Ein conodonten cluster aus der Grenzbitumenzone (Mittlere Trias) des Monte San Giorgio (Kt, Tessin/Schweiz). *Annalen des Naturalhistorischen Museums in Wien*, **83**, 265–274.

Sandberg, C. A. and Clark, D. L. (Eds.) 1979. Conodont Biostratigraphy of the Great Basin and Rocky Mountains, *Brigham Young University Geology Studies*, **26** (3), 1–190.

Sandberg, P. A. and Hay, W. W. 1967. Study of microfossils by means of the scanning electron microscope. *Journal of Paleontology*, **41**, 999–1001.

Schmidt, H. 1934. Conodonten-funde in ursprünglichem Zusammenhang. *Paläontologische Zeitschrift*, **16**, 76–85.

Schönlaub, H. P. and Zezula, G. 1975. Silur-Conodonten aus einer Phyllonitzone im Muralpen-Kristallin (Lungar/Salzberg). *Verhandlungen der geologischen Bundesanstal*, **4**, 253–269.

Szaniawski, H. 1982. Organic matrix structure of protoconodonts. *Abstracts*, ECOS III Publications from the Institute of Mineralogy, Palaeontology and Quaternary, Geology, University of Lund, Sweden, **238**, 22.

von Bitter, P. H. 1976. The apparatus of *Gondolella sublanceolata* Gunnell (Conodontophorida, Upper Pennsylvanian) and its relationship to *Illinella typica* Rhodes. *Life Science Contributions, Royal Ontario Museum*, **109**, 1–44.

von Bitter, P. H. and Merrill, G. K. 1983. Late Palaeozoic species of *Ellisonia* (Conodontophorida): Evolutionary and palaeocological significance. *Life Science Contributions, Royal Ontario Museum*, **136**, 1–24.

White, S. H., Shaw, H. F. and Huggett, J. J. 1984. The use of back-scattered electron imaging for the petrographic study of sandstones and shales. *Journal of Sedimentary Petrology*, **54**, 487–494.

9

Contact microradiography of conodont assemblages

R. D. Norby and M. J. Avcin

Contact microradiography has proved to be a valuable aid in locating and diagnosing buried conodont assemblages in black fissile shale of Carboniferous age. The technique also may be useful for identifying assemblages in carbonates and other rocks deposited under quiet-water conditions.

Previously used methods of mechanically splitting shale often caused fracturing among conodont elements and irregular separations which resulted in partly exposed assemblages. With the present technique, X-rays pass through an assemblage-bearing rock slab and expose film in direct contact beneath it. The differential absorption of the X-rays by the conodonts and rock matrix produces shadows on the film that results in a radiograph. Once identified on the radiograph, the assemblage can be left *in situ*, carefully excavated with a needle or chemicals, or uncovered by very careful subsplitting of the slab.

To produce a radiograph capable of the high resolution and contrast required for recognition of elements and assemblages, several interre-lated factors must be considered: kilovoltage, exposure, type of material, thickness of material and film type. The greatest resolution and contrast can be obtained by using the longest-wavelength X-rays (produced by lowest possible kilovoltage in the range 10–40 kV) that will penetrate the least-dense material (organic-rich shale is relatively transparent to X-rays), the thinnest slabs (less than 10 mm are recommended), the finest-grain film (high or maximum resolution film and plates) and the longest exposure times (several hours, depending on milliamperage). Two representative exposure-guide charts have been prepared to show the relationship of these factors.

9.1 INTRODUCTION

The basic technique of using X-ray radiography to study the internal features of fossils, or to locate and determine the spatial distribution of elements within the rock matrix, has been generally known since a year after the discovery of X-rays by Röntgen, when Brühl (1896) first indicated its usefulness in palaeontology. Subsequently, palaeontologists have used the techniques with varying success for examining vertebrates (e.g. Peyer, 1934; Stürmer, 1965, 1984; Zangerl, 1965; Zangerl and Richardson, 1963), invertebrates (e.g. Hamblin and Van Sant, 1963; Stürmer, 1970, 1984) and trace fossils (e.g. Farrow, 1966).

Contact microradiography is well known in the biological field (Bohatirchuk, 1963; Mitchel, 1963) and in industry (Kodak Ltd., 1962; Andrews and Johnson, 1963); however, its application to micropalaeontology has been limited primarily to a few studies on Foraminifera (Schmidt, 1952; Hedley, 1957; Hooper, 1959, 1965). The value of contact microradiography for conodont-assemblage studies has not been adequately described until now. Norby and Avcin (1972) first noted the usefulness of radiography to locate and diagnose conodont assemblages contained within shale slabs and both utilized the technique in their respective theses (Norby, 1976; Avcin, 1974).

Conodont assemblages can sometimes be found on bedding-plane surfaces of black fissile shale, primarily but not exclusively, of Carboniferous age (e.g. Schmidt, 1934; Scott, 1934, 1942; Du Bois, 1943; Rhodes, 1952; and numerous others). Assemblages have also been found on bedding surfaces of fine-grained bituminous or laminated argillaceous carbonates, but the occurrences are rarer (e.g. Lange, 1968; Mashkova, 1972; Scott, 1973; Briggs, *et al.*, 1983; Aldridge *et al.*, 1985). Some 20–30 reports (not listed here) have been published on bedding-plane assemblages since Schmidt (1934) and Scott (1934) independently reported on bedding-plane associations of conodonts that they firmly believed represented the remains of individual conodont animals. The term 'assemblages', as used here, means any bedding-plane association of conodont elements of natural (i.e. presumably undisturbed anatomical arrangement of a conodont apparatus), coprolitic, current-accumulated or other origins.

Most reported assemblages were found for-tuitously on weathered surfaces or, more commonly, by splitting limestone or shale slabs along bedding planes. In our studies of conodont assemblages from black fissile shales of Carboniferous age, previously used methods of obtaining the assemblages presented several problems.

(1) The procedure of mechanically separating or splitting the fine laminae of unweathered or partially weathered shale with a thin knife blade, spatula or a meat cleaver and hammer may have caused or contributed to the large amount of fracturing of the conodont elements.

(2) Assemblages often split so that individual conodont elements or parts thereof were generally divided between part and counterpart, thereby making interpretations more difficult.

(3) In some cases, only a portion of an assemblage was exposed on the split surface because of irregularities of the bedding plane, 'cleanness' of the split or slight divergence from the usual parallel orientation of the elements to the bedding surface.

(4) Gypsum commonly occurs along bedding planes of some Carboniferous black shales, thereby increasing the 'splittability' of the shale. Nevertheless, the assemblages commonly were fractured, displaced and encrusted by the growth of gypsum crystals.

These problems led us to experiment with contact microradiography. Our first attempt was based on the then unpublished work of Odum and Frost (1972), who had successfully obtained radiographs of conodonts *in situ*. Ideally, we hoped to utilize radiographs to locate assemblages in shale slabs, to identify element types, to observe their spatial arrangement and, if necessary, to excavate the assemblage by reagents or with fine needles (Norby and Avcin, 1972). In addition, we believed that the radiographs would show textual characteristics, some sedimentary structures and other information that could be used to evaluate the palaeoecology of the environment in which the conodont assemblages occur.

9.2 METHODS

Because the principles of radiography are well known, our discussion will focus on contact microradiography as applied to conodont-assemblage studies. Basic information, as well as advanced technical data on X-rays, can be found in many modern textbooks, some of which provide specific applications to sedimentology and palaeontology. Bouma (1969) is an excellent source on the basic technique as well as on most specialized aspects of radiography. Additional information was obtained from Clark (1963), Fraser and James (1969), Zangerl (1965) and *Fundamentals of Radiography* distributed by the Kodak Co. (1968). Anyone who may wish to attempt this technique safely should first read a basic reference on methods and processes involved in the use of X-ray radiation (e.g. Bouma, 1969; Clark, 1963; Brown, 1975). Particular attention should be given to proper shielding of X-ray sources and to other safety procedures necessary to safeguard the operator.

In contact microradiography a slab of potential assemblage-bearing shale is placed directly onto a sheet of high-resolution X-ray film, which is placed on a lead sheet to reduce backscattered radiation. The film and specimen are exposed together for a predetermined length of time and by a predetermined amount of soft (5–50 kV) radiation to produce a photographic record or radiograph. This radiograph is then enlarged optically from a few times to a few hundred times depending primarily on the resolution of the film.

The basic principle involved is that X-rays are differentially absorbed by materials depending on their atomic composition and their density (Kodak Co., 1968). Conodont elements, which are more dense than the enclosing shale or other matrix, absorb more radiation than the matrix does. Therefore, radiation that penetrates a shale slab will cast shadows onto the film beneath because of greater attenuation of the X-ray beam by conodont elements or any other material of greater density than shale (Fig. 9.1). Conodonts, being more opaque to the radiation, will appear as lighter areas on the radiograph, whereas the darker regions represent parts of the shale more easily penetrated by X-rays.

(a) Primary factors affecting the quality of the shadow image

The primary factors affecting the quality of the shadow image formed on the radiograph are as follows.

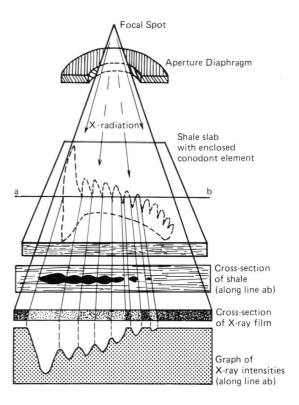

Fig. 9.1 — Diagram illustrating the penetration of a shale slab by an X-ray beam. The cross-section of the X-ray film shows the reaction of the film to the varying intensities of the X-rays which penetrate the shale slab. A graph more accurately displays the differences in X-ray intensities caused by variations in the thickness of the conodont element.

(i) Source of X-ray radiation.
(ii) Exposure.
(iii) Focus-to-film distance.
(iv) Kilovoltage.
(v) Sample thickness.
(vi) Type of film.

(i) *Source of X-ray radiation*

X-rays are produced from a modified cathode-ray tube. A current is applied to the cathode filament and, as a high potential difference is induced between the cathode and anode, electrons stream from the cathode towards the anode. The electrons strike a very small focal spot on the target metal of the anode with sufficient velocity to generate X-rays. The target is angled to project the X-rays out of the cathode-ray tube enclosure, through a port or window and onto the specimen and film.

At the present time, industrial X-ray machines available in most geology departments are capable of producing high-resolution radiographs. A Picker Hotshot 110 kV portable X-ray unit in the Department of Geology, University of Illinois at Urbana-Champaign, was made available to us by G. de V. Klein. The focal spot on this unit is very small (0.5 mm), which is necessary to obtain high-resolution radiographs. The voltage range is continuously adjustable from 10 kV to 110 kV while the amperage control provides a continuous range from 0 mA to 3.5 mA. Contact microradiography uses voltages of 50 kV or less that produce 'soft' long-wavelength radiation. This unit is equipped with a beryllium window that facilitates transmission of such long-wavelength low-kilovolt radiation.

(ii) *Exposure*

Exposure, the total amount of radiation to which a sample is subjected at a specific kilovoltage, is directly proportional to the period of radiation times the milliamperage. This product is usually expressed in milliampere seconds (mA s) or in milliampere minutes (mA min).

An increase in one factor must be accompanied by a proportional decrease in the other to provide the same exposure. Zangerl (1965) noted that the quality of radiographs is slightly different if high milliamperages (of the order of 60 mA) and short exposures were used as opposed to low milliamperages and longer exposures. Zangerl favoured the low milliamperages (± 5 mA) and longer exposure times. Stürmer (1984) routinely used a milliamperage of 25 mA and very long exposures of up to several hours. For the Picker Hotshot, the maximum recommended milliamperage of 3.5 mA was utilized for all exposures, which reduced the exposure time to the minimum allowable by the exposure relationship. Our exposure times usually varied between 2 min and 30 min, although we also tried times ranging up to 6 h. As in light photography, an improperly exposed film will result in an image with less contrast and detail than a properly exposed film. Interestingly enough, the slower-speed industrial films show an increase in contrast as the density of the radiograph increases (Fraser and James, 1969). With a more intense light source, these 'darker' or slightly overexposed radiographs are still readable and provide the desired greater contrast and resolution.

(iii) *Focus-to-film distance*

A change in the focus-to-film distance affects the intensity of the radiation reaching the specimens and film. This relationship between distance and intensity is known as the inverse square law, because the intensity varies inversely as the square of the focus-to-film distance. For example, in our studies, early trial radiographs were made at a focus to film distance of 70 cm. Later this distance was reduced to 44 cm. The intensity at 44 cm was 2.53 (inverse of $44^2/70^2$) times the intensity at 70 cm. If the intensity at 70 cm produced properly exposed radiographs, a reduction to 0.395 of the intensity (mA) or exposure time at 44 cm would be necessary to prevent overexposed radiographs. In contact microradiography, other scientists have used focus-to-film distances from 8 cm to 30 cm (Bouma, 1969). Stürmer (1984), in his ultra-high-resolution microradiographic work, used a focus-to-film distance of 50 cm. The primary reason for using the shorter focus-to-film distances is to gain shorter exposure times. Theoretically, a decrease in this distance should result in a very slight loss in sharpness and a very slight increase in magnification, because more X-rays are at angles of less than 90° to the specimens and film. Any increase in magnification is negligible and, for exploratory radiographs, no noticeable loss of sharpness was observed from similarly exposed radiographs taken at the two different focus-to-film distances. For retakes, the sample can be positioned at a point to receive the X-rays that are more nearly normal to the plane of the specimen and film. No noticeable change in contrast occurred with the reduction in the focus-to-film distance.

(iv) *Kilovoltage*

One of the most important factors affecting image quality in X-ray radiography is kilovoltage. A change in kilovoltage changes wavelength and thereby the penetrating power of the beam. An increase in kilovoltage produces radiation of a shorter, more penetrating wavelength. In addition, X-rays with longer, less penetrating wavelengths are still present at the higher kilovoltage, but they are present in greater intensity. Thus, the total intensity of the beam is greatly increased.

The effect of different kilovoltages on conodont elements is illustrated in Fig. 9.2. Graph A shows the X-ray intensity pattern at 20 kV which theoretically will result in a poorly exposed negative of the particular conodont-bearing shale slab. Graph B shows the intensity pattern for the same slab at 40 kV and at the same exposure (mA min) that was used at 20 kV. The overall intensity is increased, possibly resulting in an overexposed negative and, more importantly, the contrast has been greatly decreased, as the intensity peaks and valleys are nearly the same. This loss of detail is due to the increased penetrating power of the shorter-wavelength X-rays and lesser attenuation of the X-rays by the conodont element. In graph C, the kilovoltage remains at 40; the intensity of the X-rays has been reduced by decreasing the exposure, but the percentage differences in the intensities from peaks to valleys remain the same as they were in graph B. The contrast in graph C is about the same as that in graph B and neither of these has the high contrast present in graph A. Some resolution is also lost at the

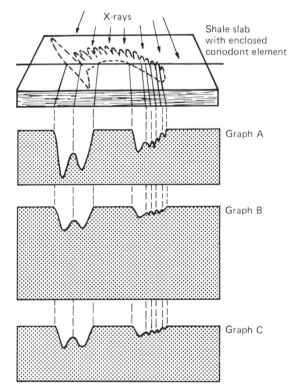

Fig. 9.2 — Diagrams showing the effect of kilovoltage and exposure on subject contrast. Graph A shows a hypothetical intensity pattern at 20 kV. Graph B shows a higher intensity pattern at 40 kV with the same exposure (mA min) used in graph A. The contrast in graph B has been reduced relative to graph A. In graph C, the intensity pattern at 40 kV is reduced by decreasing the exposure, but the percentage differences in intensity (and contrast) remain the same as they were in graph B. The intensity increases towards the top of each graph.

higher kilovoltages. Therefore, the lowest possible kilovoltage should be used to obtain the highest subject contrast. For every kilovoltage value, however, there is a maximum thickness of shale of a given density that the X-rays can penetrate. X-rays generated at a given kilovoltage cannot penetrate shale slabs with thicknesses greater than the maximum, regardless of the intensity (mA) or exposure time. Shale slabs with thicknesses just less than the maximum will require the longest exposure times, but these slabs will show the highest possible subject contrast at the given kilovoltage.

In our study, the kilovoltages usually ranged between 20 kV and 30 kV but occasionally values over 50 kV and a low of 10 kV were used. This range is comparable with the 25–45 kV range used by Stürmer (1984).

(v) *Sample thickness*
The densities of conodonts in a particular shale are nearly constant, but the ratio of thickness of a conodont element relative to the thickness of shale varies greatly. This ratio is very important in determining the amount of subject contrast between the two. The thickness of conodont elements lying on a bedding plane varies but usually does not exceed 0.1 mm with a maximum limit of about 0.5 mm. In a very thin slab (1 mm), a conodont element with a maximum thickness of 0.1 mm will occupy 10% of the thickness, whereas the same element in a 5 mm slab will occupy only 2% of the thickness. Therefore, the absorption effect of the conodont element relative to the matrix is lessened and the contrast between the conodont element and the matrix is reduced in the thicker slab. This is graphically portrayed in Figs. 9.3(a)–(c). In these figures, a constant of 20 kV was used and exposure was increased to compensate for the increased thickness of the shale slab. If a higher kilovoltage had been used for the 3 mm slab, even less contrast between the shale matrix and the conodont element would result than that portrayed in Fig. 9.3(c). Also, lower kilovoltage with compensating increased exposure for the 1 mm slab would have resulted in even more contrast than that shown in Fig 9.3(a).

The thickness of shale slabs X-rayed for this study ranged between about 1 mm and 10 mm. In a shale slab 10 mm thick exposed to 50 kV radiation, large Pa elements of platform apparatuses and robust elements of *Idioprioniodus* were still recognizable on radiographs; however, only a few larger Pb and ramiform elements were recognized as parts of a mature assemblage (the elements were only faintly visible). Elements and assemblages can be recognized more easily if the matrix is very homogeneous. With longer exposure times, shale thicknesses of 15 mm or 20 mm could be radiographed but, in general, the conodonts will produce too little contrast and most assemblages may go unnoticed. For practical work, assemblages may be recognized in shale slabs up to 10 mm with a kilovoltage of 35 kV and an exposure of 50 mA min on Kodak M film.

(vi) *Type of film*
Radiographic film is produced by several manufacturers (e.g. Gevaert, Ilford and Kodak) for industrial and medical use. These films vary in speed, grain size and contrast. Film is available in several convenient sizes and packaged in boxes of separate film sheets, in boxes of film sheets with protective interleaves of paper or in a 'ready pack' in which the film is enclosed with interleaves of black paper and sealed in a light-protective envelope. Kodak M film in a 'ready pack' was used in most of our studies because it is a fine-grained high-contrast film which permitted good resolution of the conodonts. The slow to medium speed resulted in long (by medical standards) but reasonable exposure times.

For best results, the shale sample must be in very close contact with the film. When using small light-weight shale pieces and 'ready packs', extra care must be taken to make sure that wrinkles or thin air pockets do not cause a gap between the film and specimen.

The double emulsion of M film created some minor problems for assemblages not centred directly under the X-ray beam. The radiation passed through the specimen and film at an angle, producing a double image: one image in the upper emulsion layer and one in the lower emulsion. These two images are slightly offset (the amount of offset depends on the angle of incident X-rays), which results in a slightly blurred image if examined or printed at high magnifications. Close examination of conodont elements in Plate 9.1, Fig. 1b, reveals these slightly blurred images of an *Idioprioniodus healdi* assemblage that was positioned near the edge of a sheet of M film. Single emulsion layer film such as Kodak R film can be used to solve the problem. However, the film speed is only

about one-quarter as fast as M film, which increases the exposure by a factor of approximately 4. Stürmer (1984) used either Kodak High Resolution Film or Agfa Scientia Film 23 D 56 at exposure times of 5 min to 5 h at 25 mA and 25–45 kV. However, Stürmer achieved his best resolution with Kodak Maximum Resolution Plates or Agfa Millimask Plates at exposure times of 2–20 h at 25 mA and 25–45 kV.

In our studies, M film was adequate for exploratory shots whereas R film was used only for radiographs of partially exposed assemblages and for previously radiographed assemblages which showed unusual characteristics. The slight increase in resolution alone does not justify the use of R film as a general practice. For publication prints, however, R film or one

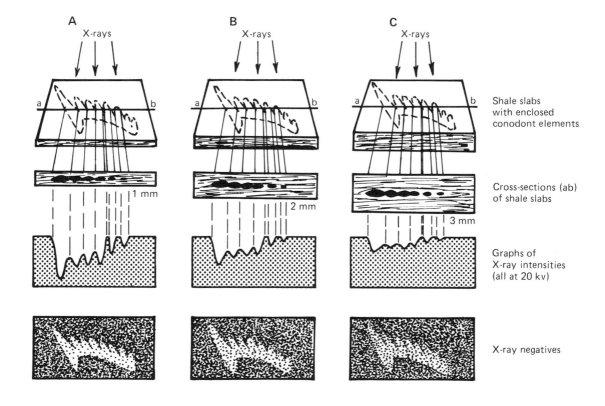

Fig. 9.3 — Diagrams illustrating the reduced contrast between the shale matrix and the conodont element with increased thicknesses of shale. Exposure is successively increased from A to B to C to provide sufficient intensity to penetrate shale slab. Intensity increases towards top of graph.

of the higher-resolution films or plates is recommended. At the time of our initial studies, we were not aware of the availability of the high-resolution films and plates as used by Stürmer. We believe that the higher resolution offered by these films and plates would produce prints superior to those illustrated in Plate 9.1 and also would allow for greater magnification if needed.

We used normal photographic enlarging techniques to print the radiographs, but it was difficult to obtain adequate prints within a grey tone while preserving the contrast of the radiograph. Also, print exposure times for some of the denser radiographs ranged up to 10 min. Zangerl (1965) noted that some of the problems of printing could be solved with a method called Logetronography (Elmer and Dwin, 1957). Bouma (1969) described a machine that automatically corrects for density variations of film. The conventional exposure lamp is replaced by a high-intensity cathode-ray tube that produces a small spot of light. This spot of light is projected through the radiograph and scanned in a series of approximately 300 overlapping linear lines to cover the entire radiograph. According to Zangerl (1965, p. 313), 'Good logEtronic prints not only reproduce the very finest shadow detail visible on the original radiogram but actually show details so minute that they are invisible to the human eye on the original radiogram.' Stürmer (1984, Figs. 16A, B) illustrated the superior quality of a radiograph print made by the Logetronic process compared with a print made in the conventional manner. The equipment necessary to undertake this electronic contrast controlled printing is not generally available but radiographs could be processed commercially by this technique.

(b) Excavation technique

After identifying a buried assemblage on a radiograph, we marked its position on the slab. The slab was then trimmed to a 3–4 cm square blank with the assemblage approximately centred on the blank. The blank was radiographed a second time to obtain a better understanding of the position and character of the assemblage. If an assemblage appeared to be in good conditon, it was embedded in paraffin so that only one face of the blank was exposed. The blank was immersed in a 5% solution of sodium hypochlorite. At this concentration, the surface of the blank was sufficiently disaggregated after 12 h to permit removal of a thin layer of sediment. This was accomplished with fine synthetic fibre brushes and dissecting needles. The process was repeated until the assemblage was partially exposed, at which point the concentration of the sodium hypochlorite was decreased to permit slower excavation. We discovered that many of the elements had actually been fractured *in situ* probably prior to or during lithification of the shale. Although fracturing may not occur *in situ* in all buried assemblages, the use of the sodium hypochlorite solution or any other chemical solution as an aid in excavation has a tendency to expand or separate the fine shale laminae or clay particles and to disrupt and fracture any contained conodont elements and assemblages as they became exposed.

In partially buried assemblages and some completely buried ones as well, conodont elements can be carefully excavated by using a fine dissecting needle (and steady hands) with the radiograph as a guide. Several assemblages have been successfully excavated in this manner but patience is necessary. In a few cases, the radiograph may provide enough information on the assemblage to make excavation unnecessary. If the elements need to be visually examined, the slab may be very carefully subsplit along planes of fissility; however, this may result in some damage to the assemblage. If the assemblage is not found, the splits may be reradiographed, the assemblage relocated and the conodont elements excavated with a needle.

9.3 RADIOGRAPH EXPOSURE GUIDE

The two exposure guides illustrated in Figs. 9.4 and 9.5 were produced from data taken from general exploratory radiographs and some systematically controlled radiographs. Ideally, a step wedge of the rock material to be studied should be radiographed at varying exposures and kilovoltages to produce a guide chart. Because each X-ray unit has slightly different characteristics, a guide chart should be made even when similar materials, kilovoltage values and films are being used.

For each radiograph, the exposure and thickness of the shale slab were plotted on a semilogarithmic graph. The kilovoltages used were listed beside each point. Next the relative

density of each radiograph was determined visually with a microscope and fluorescent light source. More precise densities can be determined with a densitometer. Those radiographs that were slightly dark (denser) but readable determined the points used for the ideal isopotential line. Occasionally, lighter and darker radiographs were used to interpolate an ideal point. For each isopotential line, a zone of readable radiographs exists which extends about halfway to the next adjacent isopotential line on each side.

The isopotential lines on the graph (Fig. 9.4)

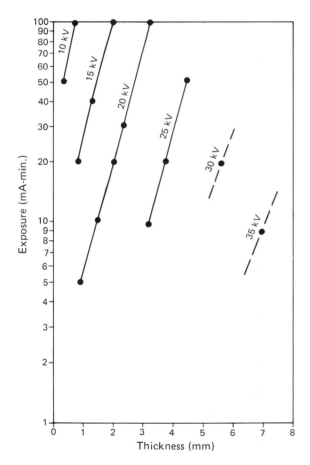

Fig. 9.4 — Exposure guide for brownish-black 'paper' shales from the Stonehouse Canyon Member of the Tyler Formation (Mississippian age), central Montana (Norby, 1976) using Kodak M film and a focus-to-film distance of 44 cm.

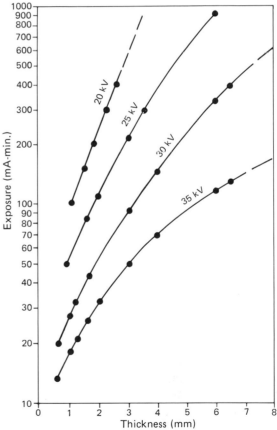

Fig. 9.5 — Exposure guide for black fissile shale from the Modesto Formation (Pennsylvanian age), northern Illinois (Avcin, 1974) using Kodak R film and a focus-to-film distance of 44 cm.

for Kodak M film are approximately straight in the utilized portion of the graph. At low kilovoltages, the thicker specimens will require much longer exposures (more than 100 mA min, which is above our utilized area). Therefore, above 100 mA min the isopotential line may tend to curve more towards the left and to become asymptotic to a thickness line, as suggested by Fraser and James (1969). This occurs because a maximum threshold thickness exists above which the X-rays at a given kilovoltage cannot penetrate. For example, in one of our longest exposures of 2100 mA min at 20 kV (a point that falls outside the graph in Fig. 9.5), the resultant radiograph (R film) for a Pennsylvanian black shale slab 6.0 mm thick was readable but a little light, whereas the image of a 6.5 mm slab on the same radiograph was virtually transparent. A proper exposure for the 6.0 mm slab would probably be 4000–8000 mA min. At this exposure the 6.5 mm slab may or may not show some detail. Thus, 6–7 mm is about the maximum thickness of Pennsylvanian black shale that 20 kV can penetrate. For Mississippian black shale, radiation of 20 kV can probably penetrate a thickness greater than 6–7 mm. Therefore, one must be careful in extending the 'straight' isopotential lines to determine exposure guides, as the isopotentail lines may curve with very long exposures. The graph (Fig. 9.5) for Kodak R film also suggests some opposite curvature in the 30 kV and 35 kV isopotential lines. Additional test shots are needed to extend these lines to determine proper exposures for both graphs.

Brownish-black papery shale from the Tyler Formation (of Upper Mississippian age) of central Montana (Norby, 1976) was utilized for producing the guide (Fig. 9.4) for Kodak M film. The guide (Fig. 9.5) for the Kodak R film was based on black fissile shale from the Modesto Formation (of Pennsylvanian age) of northern Illinois (Avcin, 1974). In our general comparison of the two black shales using Kodak M film, the exposure times were about twice as long for the Pennsylvanian shale with all other factors remaining the same. This is logical, as

the density of the Pennsylvanian shale is nearly twice that of the Mississippian shale. This emphasises the necessity for making individual exposure guides for the specific material being studied, over the range of kilovoltages to be used. Depending on kilovoltage, exposure values for a carbonate may be from two to five times as long as for a Pennsylvanian shale (Fraser and James, 1969).

9.4 OTHER SPECIALIZED TECHNIQUES

A few additional specialized techniques are described here briefly for conodont researchers who may want to experiment with them. Hooper (1965) described point-projection radiography and was able to produce enlarged images of internal structures of foraminifera without destroying the specimen. Point-projection X-ray microscopy does not seem to have wide applications. Some of this same information can be obtained from transmission electron microscopy, scanning electron microscopy and scanning transmission electron microscopy techniques which are more widely known and which require equipment more readily available.

Infrared (Rolfe, 1965a) and ultraviolet radiation techniques (Rolfe, 1965b) might prove to be important in examining surficial natural bedding-plane assemblages for evidence of soft-body structures. We have had no experience in these specialized areas.

9.5 SUMMARY AND DISCUSSION

In contact microradiography, an X-ray unit capable of generating soft radiation in the range 5–50 kV is necessary. We commonly used values between 20 kV and 30 kV, which resulted in moderate exposure values for shale thickness up to 8 mm with Kodak M film. Our kilovoltages are comparable with those of Stürmer (1984) who used values between 25 kV and 45 kV. To obtain the best contrast and resolution in the radiographs, the lowest possible kilovoltage that can penetrate the slab should be used.

In our studies, a shale thickness of 10 mm was a practical upper limit. Large conodonts probably could be recognized in shales up to 20 mm thick if very long exposures were used with kilovoltage (approximately 40 kV) just above the threshold of penetration. However, the chances of identifying an assemblage are greatly reduced. Ideally, a 20 mm slab could be split to 10 mm or less and the pieces radiographed separately.

Zangerl (1965) believed that the milliamperage should be low (±5 mA), which is typical of many industrial units. Stürmer (1984), however, was able to obtain superior radiographs using 25 mA. As the millamperage apparently is not a factor in quality, X-ray units with water-cooled anodes that will tolerate the heat produced at higher milliamperages can be used to shorten exposure times.

The focus-to-film distance should be kept short (7.5–30 cm is recommended by Bouma (1969)); however, we used 44 cm and Stümer (1984) used 50 cm.

A high-resolution fine-grained film, preferably with a single emulsion layer such as Kodak R, or any ultrahigh-resolution film, should be used. Finer-grained films invariably are slower in speed and require longer exposure times.

It is very important to have very good contact between the film and the sample. In some cases, a film holder may be appropriate for single sheets of film or a special light-tight box that holds both the film and the sample in direct contact.

Better radiographs can be obtained with very long exposure times. Stürmer (1984) used exposures up to 30 000 mA min (20 h at 25 mA).

A considerable amount of experience is needed in working with these long exposures. To make the necessary exposure guide, each successive radiograph must be checked as exposure times are gradually increased; otherwise, much time may be wasted.

All the factors summarized previously are important, but to differing degrees. Those which affect resolution and contrast most are film type, kilovoltage and slab thickness. As an example, the figures (radiograph positives) in Plate 9.1 illustrate typical resolution and contrast of assemblages at various thicknesses of slabs and varying kilovoltages; all utilized Kodak M film. Plate 9.1, Figs. 2, 4 and 7, show the high resolution which occurs in slabs 1 mm thick taken at 10–15 kV. Some loss in resolution occurs with increased thickness (2–3 mm) and higher kilovoltages (20 kV) as shown in Plate 9.1, Figs. 1 and 3. Plate 9.1, Figs. 5 and 6, show a greater loss of detail and contrast in 4–5 mm slabs taken at 20–25 kV.

This radiographic treatment has proved to be useful in determining the elemental arrangement of buried assemblages and in locating positions of hidden elements in partially exposed assemblages on shale bedding planes. The process is moderately time-consuming but no more so than mechanically splitting and visually examining bedding-plane surfaces. It is a technique not for use routinely on any sample, but only on samples known or expected to contain conodont assemblages. We have not attempted to use this technique on carbonates, bentonites or other possible assemblage-bearing rocks but, if conodonts are abundant and physical evidence of a very quiet-water environment exists, then the technique should be attempted.

PLATE 9.1

All figures are radiograph positives produced with the Picker Hotshot X-ray machine at a focal-to-film distance of 44 cm. All radiographs were made on Kodak M film and printed with normal enlarging techniques. The enlarger diaphragm was fully open ($f = 4.5$) to cut down the enlarging exposure time; times were still several minutes long with considerable amount of 'dodging' to obtain a reproducible print on low- to medium-contrast printing paper.

Symbols used to identify element types follow or are modified from Sweet (1981). Assemblage specimens (prefix 57P- or 62P-) are reposited in the Illinois State Geological Survey micropaleontological collections.

All slabs with figured assemblages are from field samples H-B-1-A or H-B-1-B of the Tyler Formation, Upper Mississippian age, central Montana (Norby, 1976), except slab shown in Fig. 6 which is from a black shale bed at the top of the Modesto Formation, Pennsylvanian age, northern Illinois (Avcin, 1974).

Idioprioniodus healdi (Roundy 1926)
Plate, 9.1, Figs. 1–3, buried bedding-plane assemblage. Fig. 1a, scattered natural(?) assemblage, slab is broken near left edge and additional elements may be lost, 2.0–2.5 mm slab radiographed at 20 kV and 25 mA min, specimen 62P-603, ×5.3. Fig. 1b, left part of assemblage, specimen 62P-603, ×10.6. Fig. 2, partial assemblage of six elements, 1.2 mm slab radiographed at 15 kV and 35 mA min, specimen 62P-605, ×10.6. Fig. 3, natural(?) assemblage of eight elements, Sa element is present in slab 5 mm to right, others not obseved, 2.0–2.8 mm slab radiographed at 20 kV and 18 mA min, specimen 62P-604, ×10.6.

Kladognathus? sp.
Plate 9.1, Figs. 4a, 4b, partially buried bedding-plane assemblage. Fig. 4a, elements in lighter area in upper left are exposed, 0.8 mm slab radiographed at 10 kV and 70 mA min, specimen 62P-751A, ×10.6. Fig. 4b, elements in lighter area at top are exposed, 0.9–1.1 mm slab radiographed at 15 kV and 35 mA min, specimen 62P-751B, ×10.6.

Lochriea commutata? (Branson and Mehl 1941)
Plate 9.1, Fig. 5, buried bedding-plane assemblage. 4.2 mm slab radiographed at 25 kV and 25 mA min, specimen 62P-313, ×10.6.

Streptognathodus? sp.
Plate 9.1, Fig. 6 , buried bedding-plane assemblage. 4.6 mm slab radiographed at 20 kV and 180 mA min, specimen 57P-500, ×10.6.

Gnathodus bilineatus (Roundy 1926)
Plate 9.1, Fig. 7, partially buried assemblage. Specimen has been completely excavated, only Pa element on left, Sc elements on bottom and *Idioprioniodus healdi* element at top were visible before excavation, 0.8 mm slab radiographed ar 10 kV and 70 mA min, specimen 62P-16A, x10.6.

Pl. 9.1]   **Contact microroadiography of conodont assemblages**   165

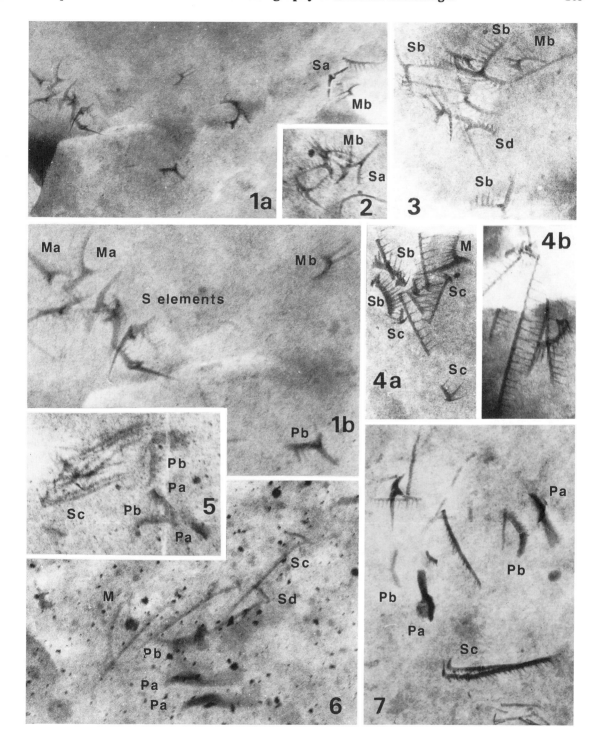

9.6 REFERENCES

Aldridge, R. J., Briggs, D. E. G. and Smith, M. P., 1985. The structure and function of panderodontid conodont apparatus. In R. J. Aldridge, R. L. Austin, and M. P. Smith (Eds.), *Fourth European Conodont Symposium (ECOS IV), Abstracts*, Private publication, University of Southampton, 1.

Andrews, K. W. and Johnson, W., 1963. Applications of microradiography of iron and steel. In G. L. Clark, (Ed.) *Encyclopedia of X-rays and Gamma Rays*, Reinhold, New York, 618–621.

Avcin, M. J. 1974. Des Moinesian conodont assemblages from the Illinois Basin. *Unpublished Ph.D. Dissertation*, University of Illinois, Urbana-Champaign, 1–151, 5 plates.

Bohatirchuk, F. 1963. Magnified X-ray images in medicine and biology (micro-macroradiography and historadiography). In G. L. Clark (Ed.), *Encyclopedia of X-rays and Gamma Rays*, Reinhold, New York, 567–571.

Bouma, A. H. 1969. *Methods in the Study of Sedimentary Structures*, Interscience, New York, 1–458.

Branson, E. B. and Mehl, M. G. 1941. New and little known Carboniferous conodont genera. *Journal of Paleontology*, **15**, 97–106, 1 plate.

Briggs, D. E. G., Clarkson, E. N. K. and Aldridge, R. J. 1983. The conodont animal. *Lethaia*, **16**, 1–14.

Brown, J. G. 1975. *X-rays and their Applications*, Plenum, New York, 1–258.

Brühl, 1896. Ueber Verwendung von Röntgen'schen X-Strahlen zu palaeontologisch-diagnostischen Zwecken. *Verhandlangen der Berliner Physiologischen Gesellschaft, Archiv für Anatomie und Physiologie, Physiologische Abteilung Jahrgang 1896*, 547–550.

Clark, G. L. (Ed.) 1963. *Encyclopedia of X-rays and Gamma Rays*, Reinhold, New York, 1–1149.

Du Bois, E. P. 1943. Evidence of the nature of conodonts. *Journal of Paleontology*, **17**, 155–159, 1 plate.

Elmer, G. J. and Dwin, R. C. 1957. Logetronography. *American Journal of Roentgenology*, **78**, 124–133.

Farrow, G. E. 1966. Bathymetric zonation of Jurassic trace fossils from the coast of Yorkshire, England. *Palaeogeography, Palaeoclimatology, and Palaeoecology*, **2**, 103–151.

Fraser, G. S. and James, A. T. 1969. Radiographic exposure guides for mud, sandstone, limestone, and shale. *Illinois State Geological Survey Circular*, **443**, 1–20.

Hamblin, W. K. and Van Sant, J. 1963. Radiography in paleontology, In G. L. Clark (Ed.), *Encyclopedia of X-rays and Gamma Rays*, Reinhold, New York, 684–686.

Hedley, R. H., 1957. Microradiography applied to the study of Foraminifera. *Micropaleontology*, **3**, 19.

Hooper, K. 1959. X-ray absorption techniques applied to statistical studies of Foraminifera populations. *Journal of Paleontology*, **33**, 631.

Hooper, K. 1965. X-ray microscopy in morphological studies of microfossils. In B. Kummel and D. Raup (Eds.), *Handbook of Paleontological Techniques*, Freeman, San Francisco, Californa, 320–326.

Kodak, Co. 1968. *The Fundamentals of Radiography*, Radiography Markets Division, Eastman Kodak Co., Rochester, New York, 1–76.

Kodak Ltd. 1962. Contact microradiography. *Kodak Data Sheet*, **IN-12**, 1–2.

Lange, F. G. 1968. Conodonten-Gruppenfunde aus Kalken des tieferen Oberdevon. *Geologica et Palaeontologica*, **2**, 37–57, 6 plates.

Mashkova, T. V. 1972. *Ozarkodina steinhornensis* (Ziegler) apparatus, its conodonts and biozone. *Geologica et Palaeontologica*, **SB1**, 81–90, 2 plates.

Mitchel, G. A. G., 1963. Microradiography in biology. In G. L. Clark (Ed.), *Encyclopedia of X-rays and Gamma Rays*, Reinhold, New York, 607–618.

Norby, R. D. 1976. Conodont apparatuses from Chesterian (Mississippian) strata of Montana and Illinois. *Unpublished Ph.D. Dissertation*, University of Illinois, Urbana-Chanpaign, 1–294, 21 plates.

Norby, R. D. and Avcin, M. J. 1972. Techniques for the study of conodont assemblages (Abstract). *Geological Society of America Abstracts with Programs*, **4**, 340–341.

Odum, I. E. and Frost, S. H. 1972. Use of radiography in the study of microfossils in black shales (Abstract). *Geological Society of America Abstracts with Programs*, **4**, 342.

Peyer, B. 1934. Über die Röntgenunterschungen von Fossilien, hauptsächlich von Vertebraten. *Acta Radiologica*, **15**, 364–379.

Rhodes, F. H. T. 1952, A classification of Pennsylvanian conodont assemblages. *Journal of Paleontology*, **26**, 886–901, 4 plates.

Rolfe, W. D. E. 1965a. Uses of infrared rays. In B. Kummel and D. Raup (Eds.), *Handbook of Paleontological Techniques*, Freeman, San Francisco, California, 345–350.

Rolfe, W. D. E. 1965b. Uses of ultraviolet rays. In B. Kummel and D. Raup (Eds.), *Handbook of Paleontological Techniques*, Freeman, San Francisco, California, 350–360.

Roundy, P. V. 1926. The microfauna in Mississippian formations of San Saba County, Texas. U.S. *Geological Survey Professional Paper*, **146**, 1–63, 4 plates.

Schmidt, H. 1934. Conodonten-Funde in ursprunglichem Zusammenhang. *Paläontologische Zeitschrift*, **16**, 76–85.

Schmidt, R. A. M. 1952. Microradiography of microfossils, *Science*, **115**, 94.

Scott, H. W. 1934. The zoological relationship of the conodonts. *Journal of Paleontology*, **8**, 448–455, 2 plates.

Scott, H. W. 1942. Conodont assemblages from the Heath Formation, Montana. *Journal of Paleontology*, **16**, 293–300, 4 plates.

Scott, H. W. 1973. Conodontochordata from the Bear Gulch Limestone (Namurian, Montana). *Publications of the Museum, Michigan State University, Paleontological Series*, **1**, 81–100, 3 plates.

Stürmer, W. 1965. Röntgenaufrahmen von einigen Fossilien aus dem Geologischen Institut der Universität Erlanger–Nürnberg. *Geologische Blätter für Nordost-Bayern und angrenzende Gebiete*, **15**, 217–223.

Stürmer, W. 1970. Soft parts of cephalopods and trilobites. Some surprising results of X-ray examinations of Devonian slates. *Science*, **170**, 1300–1302.

Stürmer, W. 1984. Interdisciplinary palaeontology. *Interdisciplinary Science Reviews*, **9**, 1–14.

Sweet, W. C. 1981. Macromorphology of elements and apparatuses. In R. A. Robison (Ed.), *Treatise on*

Invertebrate Paleontology, Part W, Supplement 2, Conodonta, Geological Society of America, Boulder, Colorado, and University of Kansas Press, Lawrence, Kansas, W5–W20.

Zangerl, R. 1965. Radiographic techniques. In B. Kummel and D. Raup (Eds.), *Handbook of Paleontological Techniques*, Freeman, San Francisco, California, 305–320.

Zangerl, R. and Richardson, E. S., Jr. 1963. The paleoecological history of two Pennsylvanian black shales. *Fieldiana, Geology Memoirs*, **4**, 1–352, 55 plates.

10

Morphometrics and the analysis of shape in conodonts

N. MacLeod and T. R. Carr

Quantitative methods of morphological analysis have not been applied extensively to the study of conodonts. Nevertheless, many aspects of conodont palaeobiology are best approached from a quantitative point of view. Application of many currently available techniques of quantitative data-acquisition and analysis can contribute substantially to the investigation of conodont systematics, biostratigraphy, palaeoecology, palaeobiogeography and phylogenetics. In addition, quantitative techniques may be used to study species-specific developmental processes in conodonts. A generalized analysis of quantitative morphology for the conodont species *Neogondolella polygnathiformis* indicates that size-dependent patterns of morphologic variation differ for carina and platform regions of the phenotype. Such variations may reflect differing modes of formation for these two aspects of generalized conodont morphology.

10.1 INTRODUCTION

The fundamental unit in palaeontology is the taxonomic species, the existence of which forms the basis of all other aspects of palaeontologic inquiry (e.g. biostratigraphy, palaeobiology and phylogeny). In modern terms, taxonomic species are regarded as individuals rather than collections and are recognized by the possession of a series of unique attributes (Ghiselin, 1964; Hull, 1976). In the study of palaeontological materials, the taxonomist is usually restricted to attributes which are morphological in character. Therefore, the study of palaeontology is to a large extent the study of morphology.

Analysis of the morphology of palaeontological specimens can be approached from either a qualitative or a quantitative point of view. The qualitative analysis of species has historical precedence, with many centuries of biological and palaeontological tradition behind it. Qualitative characterization of morphology also has several practical advantages, including economy of effort and the ability to define a very large number of morphological features regardless of their complexity. Nevertheless, the ambiguities and deficiencies inherent in the qualitative approach to taxonomy can be demonstrated by means of the 'complete' diagnosis for *Declinognathodus noduliferus*. The diagnoses reads as follows (Sweet, in Ziegler, 1975): 'Platform long, lanceolate, posteriorly pointed; inner and outer parapets ornamented with regularly spaced alternating nodes; outer row of nodes continues anteriorly as a thin blade with laterally compressed, short, sharp-pointed, discrete denticles; outer side of platform set with one or more (usually three) nodes which extend in a row posteriorly from the junction of the blade and platform; oral surface of the platform transversed by a deep, median, longitudinal trough, continuous the length of the platform; aboral attachment scar slender to moderately flaring.'

Such a qualitative characterization of an important taxon such as *D. noduliferus,* even when amended with additional qualitative observations, may place severe constraints on our ability to address questions of taxonomic, biogeographical and ecological variation in morphology within such a widely dispersed taxon. In addition, qualitative characterization of morphology increases the difficulty of assessing the nature of evolutionary relationships between *D. noduliferus* and other taxa that have been postulated as being closely related (e.g. *Gnathodus girtyi simplex*). In general, the qualitative approach limits us to qualitative discussion, which leads to inconsistency in usage and forms the basis for exhausting and untestable arguments between workers. The quantitative characterizaton of morphology possesses its own distinct advantages. These include increased precision and objectivity in character definition, and an explicit recognition of the degree and limits of morphological variability of single characters or entire organisms.

A quantitative morphometric analysis typically requires large numbers of specimens and a multitude of morphological measurements. These morphological variables will, to a greater or lesser degree, display some type of variation both within a single taxon and between taxa. Subsequent to data collection, numerical methods of data analysis may be used to reduce this variation to more easily interpretable patterns. These patterns may then be compared with quantitative patterns obtained from other samples or used to test various hypotheses. As can be readily seen from this general description, a primary disadvantage of the quantitative approach is that it often places such stringent demands on our materials (e.g. availability and quality of specimens) that many taxa cannot be adequately treated by these techniques.

The collection of quantitative data has been traditionally ignored in palaeontology because of the time consumed in gathering and analysing such data. The current availability of image-analysis equipment and small powerful computers, however, provides the means to gather and analyse large amounts of morphological data within relatively short time frames. Despite these recent technological advances, many palaeontologists remain unimpressed by the quantitative approach to morphological analysis. Indeed, for large differences in morphology between taxa (e.g. generic or familial comparisons), the 'eye-ball test', as employed by the experienced taxonomist, remains unrivalled. However, as differences in morphology become smaller and more subtle, the inaccuracies and inconsistencies of our qualitative taxonomy overwhelm our individual or, more importantly, our collective taxonomic abilities. We then enter the realm of endless taxonomic debate.

The rationale of this paper is not to argue for the superiority of either the qualitative or the quantitative approach. Rather, we propose that the nature of the palaeontological problem to be investigated will, in many cases, constrain the choice of approach to the characterization of morphology. Since the methods of quantitative morphometric analysis are unfamiliar to a large number of conodont workers, a survey of different approaches to quantitative morphological analysis which may prove useful is included. In addition, we discuss quantitative measurement strategies, and the hardware and software available for their rapid data collection. We also provide a brief survey of the different methods for analysing quantitative morphological data and some of the applications of such data.

10.2 ACQUISITION OF MEASUREMENTS

The success of quantitative analysis of morphology is dependent on several aspects of the data-collection process. First, the mechanics of the process (hardware and software) should ease data acquisition and analysis. Second, the data should have a high degree of accuracy with a minimum of error introduced by image distortion (electronic or optical) and inconsistent specimen orientation. Finally, the measurements must be suitable for the question to be addressed. These three aspects of the acquisition of morphological data should determine the selection of data-acquisition strategies and the mechanics of data collection.

Traditional data-acquisition strategies have concentrated on the measurement of such aspects of morphology as maximum lengths, widths and diameters (Fig. 10.1). Examples of application of traditional strategies to the measurement of conodont morphology include the work of Barnett (1971), Croll and Aldridge (1982), Croll et al. (1982) and Murphy and Cebecioglu (1984). One problem with this type of measurement strategy, for example, is that it collects data in a highly biassed unsystematic manner and can leave large portions of the morphology relatively unsampled. This may lead to difficulties when a change in morphology, pertinent to the problem under investi-

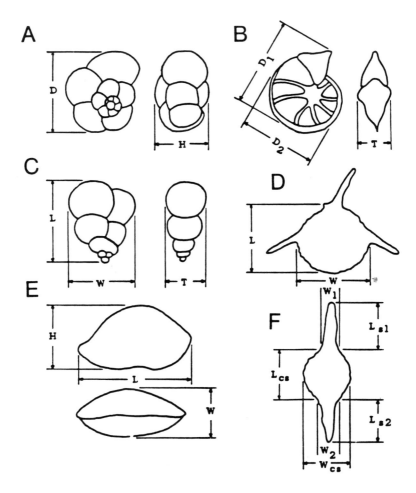

Fig. 10.1 — Traditional microfossil measurement strategies for (A)—(C) foraminifera, (E) ostracods and (D), (F) radiolaira.

gation, is underrepresented or represented by an inherently biassed set of measurements. Conclusions derived from such data will probably be incomplete at best.

One straightforward method of data collection which overcomes many of these difficulties is to sample the entire form as an outline. The outline strategy uses mathematical formulations of open or closed curves to represent the boundaries of regions within an object or the boundary between an object and its surroundings. The simplest outline data-acquisition method consists of the digitized outline of the region or object. The advantage of outline analysis is that the data can be obtained from any recognizable region or object and will represent that outline in a systematic and unbiassed manner. A major disadvantage is that outline analysis largely ignores relationships between biologically homologous points or landmarks, resulting in the absence of one-to-one correspondence with homologous points on different specimens (i.e. identical recognizable landmarks such as a basal pit). In addition, taxonomically important landmarks located in the interior of a specimen will not be recorded by the outline strategy (e.g. the cusp of *Palmatolepis* or *Neogondolella*).

As an alternative to morphological data acquisition via form outlines, a series of easily recognizable landmarks may be designated and the distances between landmarks measured. Similar to the collection of outline data, the collection of landmark distances is flexible, allowing the investigator to choose between analysis of the complete form or any part thereof. As opposed to the outline strategy, a landmark strategy can obtain data from points located within the outline of the form. In addition, landmark analysis can obtain distance data between biologically homologous points. Therefore, landmark analysis may facilitate an explicit study of evolutionary transformations when morphometric analysis is accompanied by hypotheses of phylogenetic ancestry and descent. Disadvantages of the landmark approach to data collection include, in many taxa, a relative lack of easily recognizable topological landmarks that can be located with certainty (e.g. simple cones). In addition, the number of possible measurements between landmark points increases dramatically as the number of designated landmarks increases. Also, many of the measurements between landmarks may represent redundant information, leading to overdetermination within the landmark–distance data sets, and subsequent complication of the data analysis.

One landmark–distance measurement selection strategy that overcomes many of these problems is the so-called 'truss analysis' (Strauss and Bookstein, 1982). Truss analysis specifies that measured distances between landmarks be taken in such a way as to organize the morphology into a series of adjacent quadrilateral cells, each of which is further constrained or 'stiffened' by the measurement of the diagonal distances between opposite corners (Fig. 10.2). There are several advantages in using the truss–analysis method. First, the truss–analysis strategy allows for precise and redundant location of the relative positions of the landmark points, resulting in accurate form reconstruction from a set of simple distance measurements. Second, the truss measurements, because of their organization as a series of independent quadrilateral cells, contain a large amount of systematic information concerning localized morphological variation over the entire form.

If the quantitative analysis of morphology is to have more than a minor place in palaeontology, the physical acquisition of data (be it traditional, outline or landmark) must be rapid and accurate. We are in various stages of developing hardware and software systems to make acquisition and analysis of quantitative data a realistic adjunct to, or alternative for, qualitative analysis of morphology. What follows is a brief outline of the systems currently under development.

We have approached the problem of collecting landmark data from microfossils such as conodonts and radiolaria by producing a video image from either an optical microscope or an

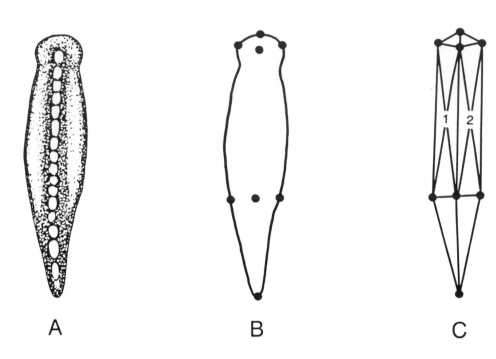

Fig. 10.2 — Conodont 'truss analysis' measurement network: (A) plan view of generalized conodont; (B) outline of conodont with selected landmarks; (C) network of distance measurements between landmarks used to quantify conodont morphology, (truss quadrilateral cells indicated by 1 and 2; see text for discussion).

electron microscope (Fig. 10.3). The image may be analysed in real time or stored on a laser video disc for later reference. Stored images are catalogued and accessed by a specialized database management system installed on a small microcomputer. Measurement of the video image is accomplished by superimposing a cursor image from a digitizing tablet on the video monitor and recording the (x, y) coordinates of the predetermined landmarks. The landmark coordinates may then be stored on floppy disc for subsequent analysis. The storing of landmark locations minimizes space requirements and permits the generation of a large number of possible distance measurements using a simple computer program.

Outline data are collected in a semiautomatic manner using image enhancement of a digitized video image and a grey-level edge detection algorithm. The coordinates of the outline are obtained automatically and stored for later analysis.

Two problems which plague the collection of quantitative morphological data are maintaining consistent orientation between specimens and removing optical and electronic distortion. Consistent orientation can be maintained with a universal stage and proper orientation protocol (e.g. MacLeod and Carter, 1984). Distortion (both vertical and horizontal) should be checked periodically. We have found that an 'English Finder' is an excellent scale to use to check for distortion over the entire field of view.

10.3 TECHNIQUES OF MORPHOLOGICAL DATA ANALYSIS

Techniques of numerical data analysis which can be applied to quantitative morphology range from simple graphical procedures, such as the construction of frequency histograms or scatter plots, to mathematically sophisticated multivariate methods designed to assess inter-actions between many variables simulta-neously. Although it may be unnecessary to have an intimate familiarity with the mathemat-ics used in the various quantitative procedures, it is recommended that one obtain at least a qualitative understanding of the procedures which are to be employed.

Currently, a wide range of quantitative tech-niques are available for the purpose of morpho-logical data analysis. These techniques may be divided into two basic groups: (1) methods designed to analyse landmark data sets and (2) methods designed to analyse form outlines. A possible third group consisting of 'hybrid' tech-niques which attempt to treat both landmarks and outlines simultaneously does exist but, as yet, these 'hybrid' techniques have had only limited application.

(a) Landmark techniques

The most widely used group of quantitative techniques, multivariate morphometrics, is a heterogeneous assemblage of numerical pro-cedures, loosely held together by the fact that each is designed to analyse multivariate data sets. Typically, these procedures have enjoyed wide application outside the field of morpho-metrics. Indeed, many were first developed by statisticians as generalized extensions of more traditional univariate methods to the multi-variate case. Included under the heading of

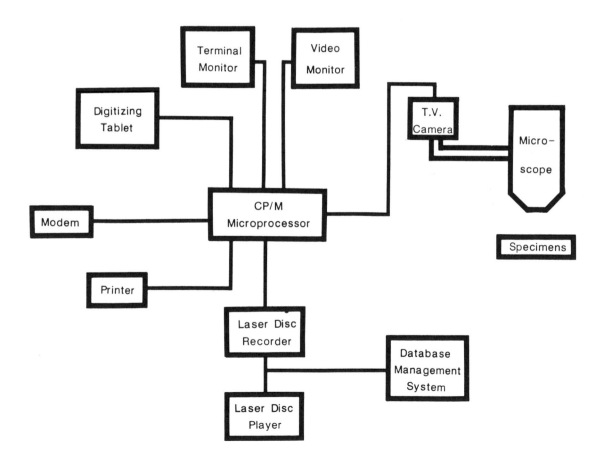

Fig. 10.3 — Schematic diagram of data acquisition system used in this study. See text for discussion.

multivariate morphometric techniques are principal-component analysis (Morrison, 1976), factor analysis (Harman, 1976), cluster analysis (Sneath and Sokal, 1973), linear discriminant analysis (Lachenbruch and Goldstein, 1979), canonical variate analysis (Campbell and Atchley, 1981) and mulitvariate analysis of variance (Cooley and Lohnes, 1971). The common goal of each of these procedures is that of dimensionality reduction, which is to say the summarization of complex multidimensional relationships between variables (*R* mode) or individuals (*Q* mode) through the creation of a relatively small number of latent vectors which correspond to the major directions of variation within the data set. A number of excellent descriptions of these multivariate techniques are available for both general and mathematically oriented audiences, including those given by Reyment *et al.* (1984), Neff and Marcus (1980), Gnanadesikan (1977) and Davis (1973). The reader is directed to these references for specific information regarding multivariate analytical techniques.

Multidimensional scaling (Torgerson, 1952; Gower, 1966) belongs to a group of multivariate analytical procedures known as 'ordination techniques' (Everitt, 1978). The purpose of multidimensional scaling is graphically to depict multivariate relationships between individuals within a low-order spatial coordinate system. Computationally similar to principal-component analysis, this technique is not usually included in discussions of multivariate morphometric procedures but can be of considerable value in identifying redundant measurements within a multivariate landmark–distance data set and in reconstructing 'average' or representative distributions of landmark points for use in intragroup and intergroup comparisons.

In order to illustrate the application of multidimensional scaling to reconstructing morphology, a series of eight landmarks were located on each of 19 specimens of *Neogondolella polygnathiformis* (Triassic age, Glenn Shale, east–central Alaska). A pair of simple bivariate plots of the reconstructed two-dimensional distributions of the eight landmark points (Fig. 10.4) shows the power of the technique. Fig. 10.4(a) represents the 'best-fit' landmark distribution using all 28 possible interpoint distances, whereas Fig. 10.4(b) represents the 'best-fit' distribution of landmarks when seven (possibly redundant) interpoint distances have been removed from the data set. The similarity between the two reconstructions indicates that the original data set of 28 interpoint distances was overdetermined, a situation which could decrease the effectiveness of many of the more traditional multivariate morphometric techniques mentioned above.

Removal of the seven designated distances effectively reduces the overdetermination of the data set while preserving the spatial constraints on landmark location which define the morphology of this species.

(b) Outline techniques

The tangent-angle method of outline analysis was originally proposed by Bookstein (1978) as a way of explicitly comparing form outlines. Conceptually, the technique defines the shape of a closed curve by relating the angle formed by a line drawn from the centre of the curve to an outline point and an arbitrarily defined baseline (the radius vector), to the angle formed by a line tangent to the curve at the point of the radius vector and another arbitrarily defined baseline (the tangent angle; see Fig. 10.5(a)). Once a series of outline points has been located in this manner, the shape of the outline may be graphically portrayed as a curvilinear function on a plot of radius vector (0–360°) versus tangent angle (0–360°) (Fig. 10.5(b)). Tangent-angle shape functions derived in this manner may be compared directly with one another or with shape functions derived from a series of 'end-member' shapes (i.e. triangles, squares, pentagons, etc.).

Another largely underexploited data-analysis technique which could be applied to outlines of organic forms is fractal analysis. Computer graphics specialists have used fractals to imitate a wide variety of organic morphologies, but the

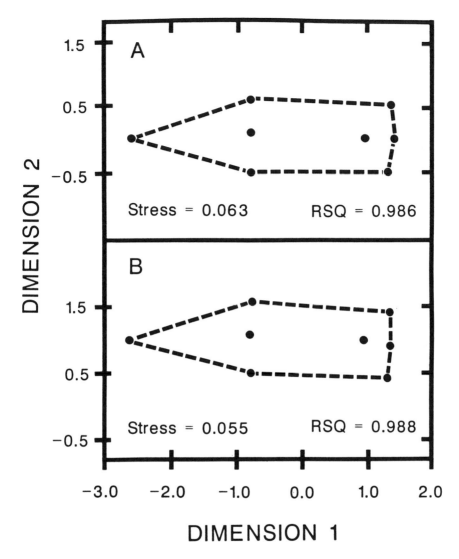

Fig. 10.4 — Comparison of results of multidimensional scaling of landmark–distance measurement networks for *N. polygnathiformis:* (A) 28 possible distances between eight landmarks used in analysis; (B) 21 distances used in analysis. Stress and *RSQ* values indicate relative deviation of scaled distances from input distances, (see text for discussion).

application of fractal analytic techniques to descriptive morphometrics is only now beginning. Richardson (1961) first recognized that for many types of closed curves there exists a linear relationship between the logarithm of the perimeter and the logarithm of the unit of measurement (step size) used to determine that perimeter. This relationship is independent of the method of measurement chosen and so may be regarded as an intrinsic characteristic of the curves in question. Mandelbrot (1967) later argued that this linear relationship was an expression of the inherent dimensionality of the curve and suggested a method whereby the fractal dimension of any closed curve could be calculated. The fractal methodology (Mandelbrot, 1967, 1983) offers new ways of analysing very complex biologic shapes and will provide new insights into the entire problem of form characterization.

    Fourier analysis has a long and distinguished record of service in the geophysical sciences where it is most often employed in the spectral analysis of waveform data. Application of Four-

ier techniques to closed curves representing the outlines of fossil forms was first attempted by Kaesler and Waters (1972). In what has become the standard radial Fourier procedure in palaeontological applicatons, (x, y) coordinate data from points located on the form outline are first transformed into polar coordinates using the centre of the form as the origin of the

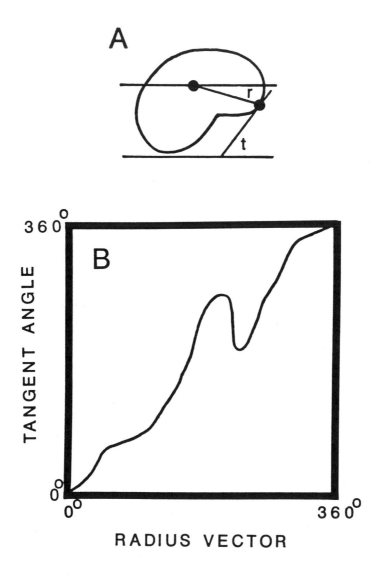

Fig. 10.5 — Geometric relationship between components of tangent-angle characterization of form outlines: (A) location of outline point with respect to its radius vector and tangent angle; (B) resultant tangent-angle form outline.

coordinate system. The outline may then be expressed as a complex cosine function which may then be broken down into a series of harmonic components using the Fourier expansion. This harmonic series corresponds to a group of abstract shape-components and these harmonic shapes are related back to the original closed curve by an amplitude and a phase angle.

Both harmonic amplitude and phase angle are necessary shape descriptors in radial Fourier analysis (Christopher and Waters, 1974); however, recent practitioners of this analytical technique have concentrated on form characterization by harmonic amplitude only (Younker and Ehrlich, 1977; Healy-Williams and Williams, 1981; Healy-Williams, 1983). Form analysis via radial Fourier techniques may be accomplished for single specimens or samples drawn from fossil populations. In this latter application, harmonic amplitude data for each specimen in the sample are grouped together and assessed for pattern formation via the analysis of frequency histograms. Radial Fourier analysis is not applicable to forms in which any radius drawn from the centre crosses the outline at more than one point. This is a serious limitation in the analysis of conodonts with long processes (e.g. *Pterospathodus* and *Eoplacognathus*).

Another method of Fourier analysis which has recently received attention in the morphometric literature is the so-called elliptic Fourier analysis (Ferson *et al.*, 1985; Kuhl and Giardina, 1982). In this variation of radial Fourier analysis, the form outline is defined by two separate cosine functions which correspond to form variation about the *x* and *y* coordinate axes. Rather than reducing an entire closed curve into a standard Fourier harmonic series, elliptical Fourier analysis performs a Fourier expansion on the *x* and *y* component shape functions independently, resulting in the definition of a series of ellipses which are spatially arranged within the *x–y* plane. The shape of these ellipses is analogous to the radial Fourier harmonic amplitudes, and their positions within the *x–y* plane are roughly analogous to phase

angles. These ellipse distributions, when added together in a cumulative manner represent a least-square approximation of the outline of the original form. One distinct advantage of the elliptical Fourier method over standard radial Fourier analysis is that the results are invariant under rotation of the original outline.

Finally, eigenshape analysis, a relatively new technique developed by Lohmann (1983), represents an attempt to apply eigenanalysis numerical procedures (a basic concept in the landmark-oriented multivariate morphometric family of techniques) to outline analysis. Basically, eigenshape analysis begins with the collection of (*x, y*) coordinate locations for a series of evenly spaced points located on the outline of a closed curve. From these data a shape function is calculated which expresses the net angular change in direction at each of 128 steps along the outline (see Zahn and Roskies, 1972). Once this shape function has been obtained for each specimen in the sample, a pairwise comparison of all shape functions is made and the results quantified in terms of a similarity matrix. Standard eigenalysis is then applied to this similarity matrix, resulting in the generation of a series of latent shape functions representing the major orthogonal shape components present in the sample. These latent shape functions are termed eigenshapes. Each original shape function may then be compared with the latent eigenshapes in order to obtain a shape distribution of forms within the sample. Alternatively, the eigenshapes may be summed together to obtain a 'best-fit' representation of the average shape of all specimens comprising the sample. This averaged shape representation may then be used in intragroup or intergroup comparisons. Outline data from 19 specimens of *N. polygnathiformis* were used in an eigenshape analysis to generate an 'average' shape characterization for this species (Fig. 10.6).

(c) Hybrid techniques
Resistant-fit theta-rho analysis (RFTRA) may be considered a hybrid morphometric technique under the present classification, in that it

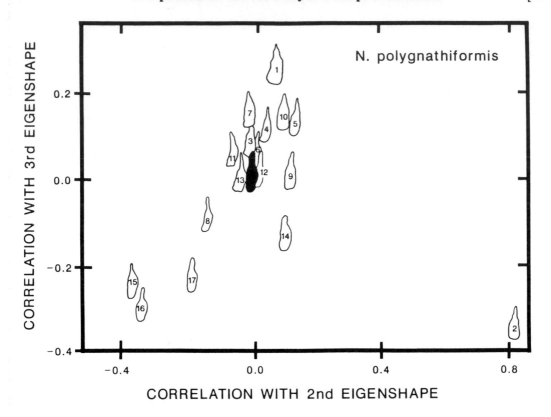

Fig. 10.6 — Bivariate distribution of *N. polygnathiformis* shape outlines on the second and third eigenshape axes. The solid form in the centre of plot represents an 'averaged' eigenshape for entire sample.

incorporates aspects of both landmark and outline analysis into its procedure. RFTRA was developed by Siegel and Benson (1982) as a way of studying morphologic transitions between two forms. RFTRA traces its conceptual ancestry back to a suggestion originally made by Sneath (1967) concerning the application of least-squares trend-surface analysis to the study of morphometric grid patterns of the kind used by Thompson (1942).

RFTRA begins with the collection of outline and landmark data for two forms. Landmarks may be located interior to or exterior to the form outline, with the only requirement being that they be topologically homologous for the forms under consideration. These data are then subjected to a series of point-for-point pairwise comparisons which, through a series of least-

squares and repeated median algorithms, are used to arrive at a set of transfer functions. These transfer functions specify outline rotations, size changes and point translations such that one form may be efficiently mapped into the other according to a set of implicit optimality criteria. Such smooth mapping transformations are of great heuristic value in understanding the significance of morphological change in a number of contexts.

10.4 APPLICATIONS OF QUANTITATIVE MORPHOLOGY

Quantitative approaches to morphological characterization and analysis are primarily used to investigate patterns of variation in morphology. These investigations may be carried out in either an exploratory mode (e.g. examination of a single sample for overt clustering of individual morphotypes within shape space, identification of the gradient-oriented pattern of continuous morphological variation) or a discriminant mode (e.g. examination of hypotheses of group distinctiveness). Both modes of data analysis have direct application in the fields of systematics, biostratigraphy, palaeobiogeography and phylogeny. As an example, in biostratigraphy (a field whose analytical methodology has traditionally been based on the qualitative assessment of morphological variation and which has been slow to take advantage of quantitative data-analysis approaches), Reyment (1980) describes a method whereby subtle variations in the gross morphology of a Cretaceous foraminifer may be used as a continuous correlation tool for basin-wide comparisons within discrete biostratigraphic zones.

In addition to the analysis of overt patterns of morphologic variation in space and time, quantitative techniques may also be used to investigate the nature of species-specific developmental processes. These applications include the analysis of variations in ontogenetic timing of developmental events (heterochrony), description of size-dependent shape transformations (allometry) and assessment of both size-dependent and size-independent interactions between morphological characters (morphological integration).

The analysis of allometry attempts to relate variation in morphology of an organism to variation in overall body size. Since the time of Galileo, body size has been recognized as a primary determinant of the functional aspects of organ design. The explicit analysis of the effect of size changes on gross morphology has only recently commanded the widespread attention of biologists. Huxley (1924) and Teissier (1931) were responsible for expressing allometry in quantitative terms in the form of the second-order power function:

$$y=bx^a \qquad (10.1)$$

where y is the size of a particular organ, b the size of organ when it appears in ontogeny, x the total body size and a the percentage rate of growth. Note that, when $a=1$, any change in x will engender a linearly proportional change in y. This is a special case of allometry termed 'isometry' in which the shape of a component morphological structure remains constant regardless of variation in body size. In cases in which $a>1$ or $a<1$ (termed positive and negative allometry, respectively), the size of the structural component changes at a progressively greater or lesser rate with respect to body size, resulting in an alteration of component shape. In addition, the allometric growth equation indicates that the growth rate of any component of the morphology is simultaneously proportional to (1) a specific constant which is characteristic of the organ in question, (2) the size of the organ at any instant during ontogeny and (3) a general growth factor which is dependent on age and environment and may be considered constant for all parts of the body (Huxley, 1972). Computationally, the analysis of allometry is simplified if the allometric growth equation (10.1) is transformed into the logarithmic expression

$$\log y = a \log x + \log b \qquad (10.2)$$

In this form the expression has the familiar $y=ax+b$ notation of a linear function whose slope a and y intercept b may be estimated via regression analysis.

Examination of patterns of allometry exhibited by an entire phenotype is often a useful way of summarizing the overall patterns of morphological change exhibited by a sample of specimens. The data necessary to perform these calculations are simply the measured distances between a series of common landmarks and an estimate of the size of each individual in the sample. (Note that individual size should be estimated using a multivariate size index such as the mean of all distances measured on the individual or the individual's score on the first principal component.) Once these data have been obtained, a regression analysis of the logarithm of distance versus the logarithm of body size for the sample will determine the allometric growth rate (slope of the regression line) and initial distance between landmarks at the onset of ontogeny (y intercept of regression line). Conformance of the data as a whole to the allometric growth model may be assessed by testing the significance of the regression.

An allometric analysis was performed on the landmark–distance matrix of 21 measurements obtained from the upper surface of the conodont species *N. polygnathiformis* (Fig. 10.7). The regressions for each of the 21 analyses were significant with well over 90% of the observed variation being explained by the calculated regression lines (Fig. 10.8). Thus, size-dependent variations in morphology for this conodont species conform well to the allometric model.

Apart from the simple characterization of size-dependent shape variation, the regression equations may also be used to predict morphology at a series of sizes and to construct a putative growth series. This is accomplished by

calculating the predicted separation between landmarks at a series of sizes and then using multidimensional scaling to estimate the two-dimensional configuration of the landmarks for each hypothetical size morph. This type of generalized strategy may be used to reconstruct the landmark configurations of different species in order to hold size constant during interspecific comparisons of morphology.

In addition to analysing size-dependent morphological variation with bivariate regression methods, morphological variation may also be studied by examining multivariate data sets for interactions existing between internal components of the form. These techniques also have the added advantage, in many cases, of being able to partition morphological variation into a series of mutually uncorrelated components. The investigation of such patterns of correlation between morphological variables has been termed the analysis of morphological integration (Olsen and Miller, 1959). Such correla-

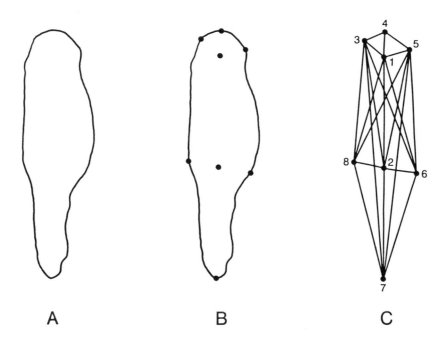

A   B   C

Fig. 10.7 — Landmark–distance measurement network used in the multivariate analysis of morphology in *N. polygnathiformis:* (A) outline of a typical specimen; (B) outline with designated landmarks; (C) interlandmark measurement network, (individual measurements are identified according to spanned landmarks in Table 10.1).

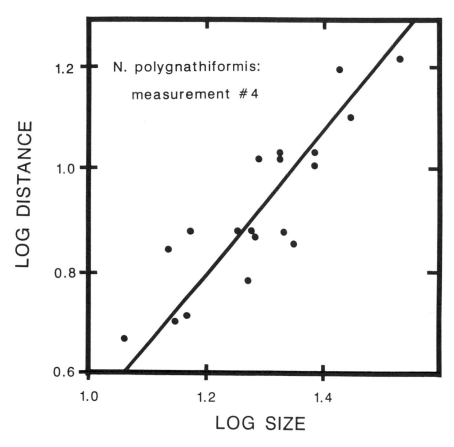

Fig. 10.8 — Example of logistic regression analysis used to estimate size-dependent (allometric) components of shape variation in *N. polygnathiformis*. 94.64% of the variation present in the sample is accounted for by the major axis regression. (log $y = 1.517x - 1.305$).

tions between different morphological variables probably arise from a combination of functional constraints on various aspects of the phenotype and/or patterns of gene distribution imposed on the organism by phylogenetic ancestry (Mayr, 1969).

Patterns of morphological integration among landmark–distance variables were analysed by Wright (1934), using the method of path coefficients and by Olsen and Miller (1959) using a cluster analysis technique. More recent studies of morphological integration, however, have usually employed principal-component analysis (MacLeod, 1984, 1985) or factor analysis (Gould, 1967; Gould and Garwood, 1969)

for the purpose of pattern recognition among morphological variables.

The application of morphological integration analysis to conodont studies may be demonstrated using the 21 landmark–distance measurements from the 19 specimens of *N. polygnathiformis* considered in the previous discussion of allometry in this species. These 21 distance measurements were transformed to logarithms and then submitted to a principal-component routine. (Note that, since principal-component analysis will only reveal linear relationships between variables, and we already know that, for this species, size-dependent shape variation is non-linear (allometric), such a transformation is appropriate and will optimize the power of the analysis.)

The eigenvectors extracted from the matrix of correlations between the landmark–distance variables partition the variation present within the data set into a series of independent components. For most organisms which exhibit ontogenetic size increase, the aspect of development which contributes most to variation between individuals within the sample is size. Size is primarily extracted by the first eigenvector, which usually summarizes the contribution of each individual landmark–distance variable to size characterization (Jolicoeur and Mosimann, 1960). Such is the case with the first eigenvector

extracted from the *N. polygnathiformis* data set which accounts for 80.3% of the observed variation between individuals. The relationship between the first eigenvector and size may also be assessed by plotting the scores for each individual on the first eignevector against an independent measure of individual size (see Fig. 10.9).

If the first eigenvector is determined to be a 'size' index, then the remaining eigenvectors must express different (although not necessary size-independent) components of variation between individuals. Patterns of covariation between variables detected through the analysis of the distribution of variable loadings on these

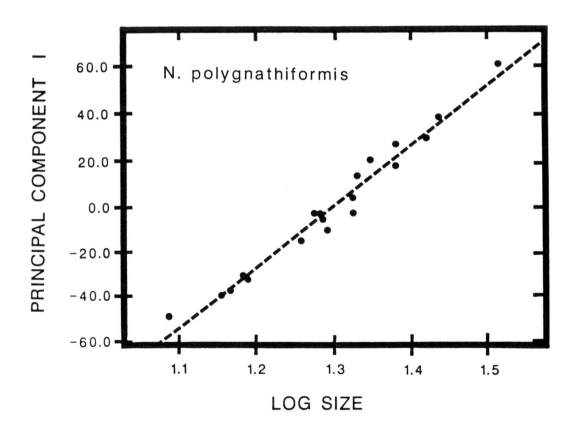

Fig. 10.9 — Relationship between individual score on the first-principal component (eigenvector) axis and an independent measure of individual size for a sample of 19 specimens of *N. polygnathiformis* ($r=0.9846$). Individual size was quantified by determination of the mean of all 21 distance measurements taken on each specimen.

remaining eigenvectors are usually interpreted as indications of the type and degree of morphological integration exhibited by the phenotype.

A plot of the pattern of landmark–distance variable loadings for the *N. polygnathiformis* data set upon the second and third eigenvectors (which explain 6.65% and 3.67% of the observed variation within the sample, respectively) reveals the presence of two distinct groups of integrated variables (Fig. 10.10). These groups are distinguished entirely by their scores on the second eigenvector. Reference to Table 10.1 and Fig. 10.7 reveals that the variables

Table 10.1 — Measurements used to quantify morphology in *N. polygnathiformis*. Refer to Fig. 10.7 for landmark location.

Measurement	Landmarks
1	1-2
2	1-3
3	1-4
4	1-5
5	1-6
6	1-8
7	2-3
8	2-5
9	2-6
10	2-7
11	2-8
12	3-4
13*	3-6
14	3-7
15*	3-8
16	4-5
17*	5-6
18	5-7
19*	5-8
20	6-7
21	7-8

which compose each group bear a relatively consistent morphological relationship to one another, in that one group consists of distance measurements between landmarks which are sensitive to variation in the direction parallel to the carinal axis, while the other group of variables is confined to the measurement of shape variation in the direction transverse to the carinal axis.

Additional information concerning the interpretation of these two putative character complexes is gained by comparing the allometric growth coefficients for all 21 measurements, after separating these estimates of size-dependent morphological variation into 'axial' and 'transverse' groups (see Table 10.2). Once again, a clear distinction between variation in the transverse group (which is typified by significant positive allometry) and the axial group (which is predominantly isometric) is expressed in the range and average values of the growth coefficients of both putative groups. The second eigenvector then may be interpreted as a 'shape' index, which expresses the contribution of the size-dependent rate of change in each measured variable per unit size increase, to overall morphological variation.

The extent of these and other patterns of morphological variation in conodonts needs to be investigated. Unfortunately, the inadequate breadth and depth of the data analysed in the *N. polygnathiformis* example do not warrant anything more than speculative interpretation at present. However, the difference in the allometric growth rates and patterns of size-dependent integration between morphological landmark–distance variables observed between the axial and transverse regions in this species do raise a number of interesting points. The growth patterns observed in the carina of *N. polygnathiformis* may be similar to those of many blade-like elements and even coniforms. If this is true it may indicate that these morphological components share a common mode of formation, such as secretion in epithelial cavities (Bengston, 1983). However, the contrasting patterns of morphological variation found in the transverse elements of the platform may reflect a different mode of formation (e.g. secretion after partial extrusion). These hypotheses will

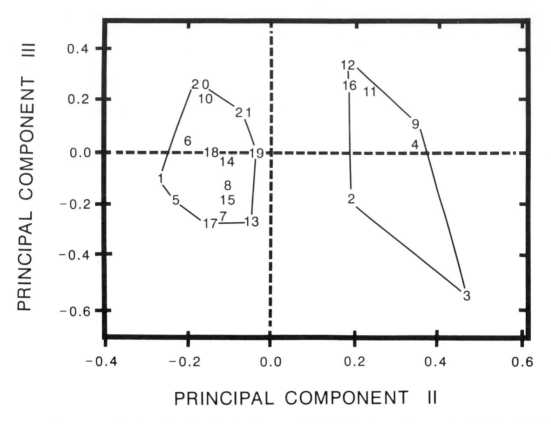

Fig. 10.10 — Bivariate plot of landmark–distance variable distribution on the second and third principal component (eigenvector) axes. See text for discussion.

take far more detailed analyses to resolve, but they may only be effectively addressed via the application of quantitative methods of analysis.

10.5 SUMMARY

Because of the inherent nature of fossil materials, the primary data base of palaeontology is morphology. Up to the present, the overwhelming majority of palaeontological studies have employed a qualitative approach to the analysis of morphological variation. Quantitative data-analysis procedures have played a limited role in the development of palaeontology as a scientific discipline, because of the relative lack of appropriate data-acquisition technology and the sheer volume of calculations necessary for the numerical analysis of all but the smallest

quantitative data sets. Fortunately, these traditional limitations have now largely been overcome, and quantitative morphological analysis will undoubtedly play a large role in future palaeontological research.

For the most part, discussions of quantitative versus qualitative approaches to morphological analysis which have appeared in the palaeontological literature have been written from the standpoint of methodological advocacy. Such a limited point of view serves only to increase polarization within the palaeontological community and completely misses the essential point that the nature of the questions to be addressed and the quality of the available materials should be the primary determinants of methodology. Simple qualitative procedures (the 'eye-ball' test) are perfectly adequate for analysing gross morphological similarities or differences between individuals and groups of specimens. Indeed, the astonishing number of successes wrought in over 250 years of qualitative palaeontology should humble even the most vitriolic numerical taxonomist. Nevertheless, qualitative analyses can only yield qualitative results. In many areas of contemporary palaeontological research, more exact and objective methods are desired.

Quantitative analysis of morphology may be divided into two basic approaches: the analysis of form outlines, and the analysis of landmark–distances. The analysis of form outlines satisfies the need for quantification in morphological analysis and can serve to summarize rapidly the nature of morphological variation within a sample, to compare similarities and differences between individuals or groups and to determine representative sample shapes.

However, most of the approaches to outline analysis fail to incorporate rigorous hypotheses of formal biological homology between forms into their methodological structure. This limits the interpretability of the results of these analytical methods in many palaeontological contexts. Finally, the collection and analysis of landmark–distance data are adequate for the complete range of morphological analysis between individuals or groups and implicitly incorporates the concept of biological homology into each data set but is itself often limited by the availability of sufficient numbers of landmarks spread over the entire form for reasonably complete morphological characterization.

The fact that conodonts represent an extinct group restricts our knowledge of them to their morphology and their distribution in the stratigraphic record. In spite of our limited biological knowledge of this group, conodonts have become one of the most important biostrati-

Table 10.2 — Separation of estimates of size-dependent morphological variation into axial and transverse groups.

Axial group		Transverse Group	
Measurement	Allometric growth coefficient	Measurement	Allometric growth coefficient
1	1.051	2	1.316
5	0.960	3	1.597
6	1.091	4	1.517
7	0.943	9	1.146
8	1.094	11	1.061
10	1.021	12	1.604
13	0.943	16	1.372
14	0.897		———
15	0.941		
17	1.063		
18	1.042		
19	1.096		
20	1.186		
21	0.879		
Mean 1.015		Mean 1.373	

graphic indicator fossils for the Palaeozoic Era and Triassic System. Quantitative analysis of overt patterns of morphological variation of conodont elements may provide improved systematic, biostratigraphical, palaoenvironmental and palaeobiogeographical intepretations and further enhance their value to the geologist. In addition, quantitative analysis of shape as applied to species-specific developmental processes of conodont morphology (e.g. heterochrony, allometry and morphological integration) may provide insight into the biological functions and relationships of this enigmatic group.

10.6 ACKNOWLEDGEMENTS

The authors would like to thank Craig Suchland for advice in adapting many of the numerical data analytical programs discussed herein to the IBM 4341 mainframe computer at ARCO's research laboratory in Plano, Texas, Harold Hietala for many helpful discussions on the application of statistical techniques to morphometric analysis and Roberta Baker for typing manuscript. During the course of these studies, N. MacLeod was supported by National Science Foundation Grant EAR8305894. We also thank ARCO for supporting this work and permitting its publication.

10.7 REFERENCES

Barnett, S. G. 1971. Biometric determination of the evolution of *Spathognathodus remsheidensis:* A method for precise intrabasinal time correlations in the northern Appalachians, *Journal of Paleontology,* **45,** 274–300.

Bengston, S. 1983. A functional model for the conodont apparatus. *Lethaia,* **16,** 38.

Bookstein, F. L. 1978. The measurement of biological shape and shape change. *Lecture Notes in Biomathematics,* **24,** Springer–Berlin, 1–191.

Campbell, N. A. and Atchley, W. R. 1981. The geometry of canonical variates analysis. *Systematic Zoology,* **30,** 268–280.

Christopher, R. A. and Waters, J. A. 1974. Fourier series as a quantitative descriptor of microspore shape. *Journal of Palaeontology,* **48,** 697–709.

Cooley, W. W. and Lohnes, P. R. 1971. *Multivariate Data Analysis,* Wiley, New York, 1–364.

Croll, V. M. and Aldridge, R. J. 1982. Computer applications in conodont taxonomy: 2. Classification methods. Computer Application in Geology I & II,

Miscellaneous Paper **14,** Geological Society of London. 247–261.

Croll, V. M., Aldridge, R. J. and Harvey, P. K. 1982. Computer applications in conodont taxonomy: 1. Characterization of blade elements. Computer Application in Geology I & II, Miscellaneous paper **14,** Geological Society of London, 237–246.

Davis, J. C. 1973. *Statistics and Data Analysis in Geology,* Wiley, New York, 1–550.

Everitt, B. 1978. *Graphical Techniques for Multivariate Data,* Heineman, London, 1–117.

Ferson, S., Rohlf, J. F. and Koehn, R. K. 1985. Measuring shape variation in two-dimensional outlines. *Systematic Zoology,* **34,** 59–68.

Gnanadesikan, R. 1977. *Methods for Statistical Data Analysis of Multivariate Data,* Wiley, New York, 1–311.

Ghiselin, M. T. 1964. *The Triumph of the Darwinian Method,* University of Chicago Press, Chicago, Illinois, 1–287.

Gould, S. J. 1967. Evolutionary patterns in pelycosaurian reptiles; a factor analytic study. *Evolution,* **21,** 385–401.

Gould, S. J. and Garwood, R. A. 1969. Levels of integration in mammalian dentitions: an analysis of correlations in *Nesophontes micrus* (Insectivoria) and *Oryzomys couesi* (Rodentia). *Evolution,* **23,** 276–300.

Gower, J. C. 1966. Some distance properties of latent root and vector methods used in multivariate analysis. *Biometrika,* **53,** 325–328.

Harman, H. H. 1976. *Modern Factor Analysis* (3rd edition), University of Chicago Press, Chicago, Illinois, 1–474.

Healy-Williams, N. 1983. Fourier shape analysis of *Globorotalia truncatulinoides* from Late Quaternary sediments in the southern Indian Ocean. *Marine Micropaleontology,* **8,** 1–15.

Healy-Williams, N. and Williams, D. F. 1981. Fourier analysis of test shape of planktonic foraminifera. *Nature,* **289,** 485–487.

Hull, D. L. 1976. Are species really individuals? *Systematic Zoology,* **85,** 174–191.

Huxley, J. S. 1924. Constant differential growth-rates and their significance. *Nature,* **114,** 895–896.

Huxley, J. S. 1972. *Problems of Relative Growth* (reprint of 1932 edition), Dover Publications, New York, 1–311.

Jolicoeur, P. and Mosimann, J. E. 1960. Size and shape variation in the painted turtle. *Growth,* **24,** 339–354.

Kaesler, R. L. and Waters, J. A. 1972. Fourier analysis of the ostracod margin. *Geological Society of America Bulletin,* **83,** 1169–1178.

Kuhl, F. P. and Giardina, C. R. 1982. Elliptic Fourier features of a closed contour. *Computer Graphics and Image Processing,* **18,** 236–258.

Lachenbruch, P. A. and Goldstein, M. 1979. Discriminant analysis. *Biometrics,* **35,** 69–86.

Lohmann, G. P. 1983. Eigenshape analysis of microfossils: a general morphometric procedure for describing changes in shape. *Mathematical Geology,* **15,** 659–672.

MacLeod, N. 1984. Morphological integration in Mesozoic

Radiolaria. *Geological Shape Analysis Conference, Abstracts,* Woods Hole Oceanographic Institution.

MacLeod, N. 1985. Analysis of mophologic integration and the recognition of character complexes in *Perispyridium, Pachyoncus,* and *Parvincingula* (Radiolaria): a comparative study. *Abstracts with Programs, (Geological Society of America)* **17** (6).

MacLeod, N. and Carter, J. L. 1984. A method for obtaining consistent specimen orientations for use in microfossil biometric studies. *Micropaleontology,* **30,** 306–310.

Mandelbrot, B. B. 1967. How long is the coast of Britain? Statistical self similarity and fractional dimension. *Science,* **155,** 636–638.

Mandelbrot, B. B. 1983. *The Fractal Geometry of Nature,* Freeman, San Francisco, California, 1–468.

Mayr, E. 1969. *Principles of Systematic Zoology,* McGraw-Hill, New York, 1–428.

Morrison, D. F. 1976. *Multivariate Statistical Methods* (2nd edition), McGraw-Hill, New York, 1–338.

Murphy, M. A. and Cebecioglu, M. K. 1984. The *Icriodus steinachensis* and *I. claudie* lineages (Devonian conodonts), *Journal of Paleontology,* **58,** 1399–1411.

Neff, N. A. and Marcus, L. F. 1980. *A Survey of Multivariate Methods for Systematics,* privately published, New York, 1–241.

Olsen, E. C. and Miller, R. L. 1959. *Morphological Integration,* University of Chicago Press, Chicago, Illinois, 1–317.

Reyment, R. A. 1980. *Morphometric Methods in Biostratigraphy,* Academic Press, New York, 1–175.

Reyment, R. A., Blacksmith, L. and Campbell, L.,

1984. *Multivariate Morphometrics* (2nd edition), Academic Press, New York, 1–233.

Richardson, L. F. 1961. The problem of contiguity: an appendix of statistics of deadly quarrels. *General Systems Yearbook,* **6,** 139–187.

Siegel, A. F. and Benson, R. H. 1982. A robust comparison of biological shapes. *Biometrics,* **38,** 341–350.

Sneath, P. H. A. 1967. Trend-surface analysis of transformation grids. *Journal of the Zoological Society of London,* **151,** 65–122.

Sneath, P. H. A. and Sokal, R. R. 1973. Numerical taxonomy. *Systematic Zoology,* **31,** 113–135.

Strauss, R. E. and Bookstein, F. L. 1982. The truss: body form reconstructions in morphometrics. *Systematic Zoology,* **31** (2), 113–135.

Teissier, G. 1931. Researches morphologiques et physiologiques sur la croissance des insects. *Travaux de la Station Biologie Roscoff,* **9,** 27–238.

Thompson, D'A. W. 1942. *On Growth and Form* (2nd edition), Cambridge University Press, Cambridge, Cambridgeshire, 1–1116.

Torgerson, W. S. 1952. Multidimensional scaling. I. theory and method. *Psychometrika,* **17,** 401–419.

Wright, S. 1934. The method of path coefficients. *Annals of Mathematical Statistics,* **5,** 161–215.

Younker, S. L. and Ehrlich, R. 1977. Fourier biometrics: harmonic amplitudes as multivariate shape descriptions. *Systematic Zoology,* **26,** 336–342.

Zahn, C. T. and Roskies, R. Z. 1972. Fourier descriptors for plane closed curves. *IEEE Transactions — Computers,* **C-21,** 269–281.

Ziegler, W. (Ed.) 1975. *Catalogue of Conodonts, II.* Schweizerbart'sche, Stuttgart, 1–404.

11

Application of conodont colour alteration indices to regional and economic geology

G. S. Nowlan and C. R. Barnes

The fact that conodonts vary in colour has been known for over 40 years, but only within the last decade has a formal conodont colour alteration index (CAI) been developed and applied to solving problems in regional and economic geology. The scale of CAI 1–8 allows assessment of the burial temperature of conodont-bearing strata up to 600 °C. CAI data are therefore complementary to various other techniques used in establishing thermal maturity.

The many applications of CAI are reviewed with examples from recent investigations. Regional and basinal studies using CAI data are valuable for hydrocarbon exploration. Zones of immature, mature and supramature strata can be identified, as shown recently for parts of the Canadian Arctic Islands, the Appalachian basins in eastern Canada and the Canning Basin of western Australia, and in the recognition of overthrust hydrocarbon plays in the eastern and western USA.

CAI data have been applied to various facets of mineral exploration. The thermal effects of mineralized intrusions can be mapped. Data are being applied to several Mississippi Valley-type deposits where the temperature of the ore-bearing fluids was significantly higher than the host strata.

Regional tectonic interpretations can be assisted through detailed CAI studies which may help to define aspects of geothermal and deformational history. Multiple thermal events have been identified in studies in Tasmania, in the Canning Basin of western Australia and in the Canadian Appalachian Orogen. In the southern Appalachians, CAI data have supported models of thin-skinned tectonics by establishing burial temperatures in both allochthonous and subjacent autochthonous strata. The thermal effects of major and minor intrusive events have been documented in several studies.

The geological significance of hotspots such as Iceland and Yellowstone is becoming appreciated and their ancient traces defined. CAI data have defined thermal anomalies in Ordovician strata in southern Quebec and northwestern Newfoundland which are interpreted as the tracks of Mesozoic hotspots now located at the Great Meteor and Madeira hotspots, respectively. The recognition of hotspot tracks is shown to be highly significant for hydrocarbon exploration in interior basins or along continental margins. Most kimberlites appear to be related to the passsage of hotspots and the delineation of the tracks by CAI data can be useful in diamond exploration.

11.1 INTRODUCTION

The early realization that coal is a sensitive indicator of thermal events and that oil and gas are generated by heating organic matter (Rogers, 1860) led to the development of techniques for interpretation of past thermal history. Later work (e.g. White, 1915) related the distribution of grades of coal to the distribution of petroleum. Subsequently, thermal maturation indicators were used routinely in the search for gas and oil. Techniques commonly in use today include palynomorph colour and translucency, vitrinite reflectance and fluorescence microscopy (see below). Epstein et al. (1977) added a new technique, that of assessment of colour in conodonts.

The fact that conodonts vary with colour has been known for some time (Ellison, 1944; Lindström, 1964), but these colour differences have only recently been placed in a systematic context. Epstein et al. (1977) compiled field data from the Appalachian Basin of the eastern USA which, combined with experimental data, showed that colour alteration in conodonts is time and temperature dependent. The change in colour is related to the progressive and irreversible alteration of trace amounts of organic matter within the conodont elements. The sequential colour changes were assigned to a numerical colour alteration index (CAI) 1–8 scheme (Table 11.1)corresponding to colours from pale yellow (CAI 1) through light to dark brown to black (CAI 5) and from black to grey through white to clear (CAI 8). CAI values 1–5 were calibrated with temperature ranges and with other indices of organic metamorphism. Recently, thermal boundaries for CAI 6–8 have been established by Rejebian (1985) (A. G. Harris, pers. commun., 1985) and these values are included in Table 11.1.

The main practical application of CAI values is the compilation of regional maps showing thermal trends and anomalies. Several regional compilations of CAI data have been completed recently. Data on the Appalachian Basin of the eastern USA have been supplied by Epstein et al. (1977), Harris et al. (1978) and Harris (1979). The Valley and Ridge Province of the southern Appalachians has been assessed by Harris and Milici (1977) and a reassessment of the structure of the Allegheny Frontal Zone has been supplied by Perry et al. (1979), based on material from the Bane Dome. Perry et al. (1983) used CAI data in a study of the structure, burial history and petroleum potential of the frontal thrust belt and adjacent foreland in southwest Montana.

In Canada, Mayr et al. (1978) briefly discussed conodont CAI data from deep wells on Bjorne Peninsula, southwest Ellesmere Island, Arctic Archipelago. Legall et al. (1982) made a comprehensive study of conodont and acritarch CAIs in southern Ontario and adjacent Quebec. Nowlan (1983) used structural hypotheses to explain a pronounced local CAI anomaly in Upper Ordovician strata of the Appalachian Orogen. Nowlan and Barnes (in press) have recently compiled all available CAI data for the Canadian Appalachians and adjacent marginal basins.

Other regional compilations have been completed for Ordovician rocks of the Caledonides in Scandinavia and the British Isles (Bergström, 1980; Oliver et al., 1984; Aldridge, 1986), Silurian rocks of the Oslo region in Norway (Aldridge 1984) and Palaeozoic strata of the Canning Basin of Australia (Nicoll and Gorter, 1984a, 1984b).

11.2 METHODS

Assessment of samples for CAI is a simple and inexpensive procedure. The best way to obtain consistent values is to assemble a set of standards with conodonts of different sizes and morphologies representing each CAI division. For each assessment, several specimens should be compared directly with the standard set because size and thickness of specimens affect their translucency. In our experience it is best to determine CAI values from the lightest parts of specimens but avoiding specimens that are strongly compressed or extremely thin walled.

Most conodont elements have areas of white matter, usually within the denticles; these are known as albid conodonts to distinguish them from hyaline or fibrous conodonts. The latter tend to give higher CAI values than albid conodonts from the same sample. Some Ordovician fibrous conodonts show a wide range of colours in the same sample and are thus unreliable for CAI determinations. Accordingly, CAI values based solely on hyaline or fibrous conodonts should be treated with caution.

Mayr *et al.* (1978) presented evidence to suggest that CAI values may be affected by host-rock composition. They showed that CAI generally increased with depth in a deep drill hole but that a reversal of this trend occurred through a transition from black shale to carbonate sediments. They reported anomalously high values in samples of black shale and, although stratigraphic mixing of chips cannot be ruled out, they suggested that host-rock composition may affect CAI determinations by as much as one unit. Such complications can be

Table 11.1 — Correlation of selected indices of organic metamorphism

	Conodonts			Palynomorphs		Vitrinite	
CAI	Temperature range (°C)	Approximate overburden (m)	AAI	Translucency index AMOCO	Amount of carbon in kerogen (wt. %)	Reflectance	Amount of fixed carbon (wt. %)
1	< 50–80	< 1200	1	1–5	< 82	< 0.8	< 60
1½	50–90	1200–2400	2–3	5–upper 5	81–84	0.70–0.85	60–65
2	60–140	2400–3600	4–5	5–6	81–87	0.85–1.30	65–73
3	110–200	3600–5500		Upper 5–6	83–89	1.40–1.95	73–84
4	190–300	5500–8000	Black to disintegrated	6	84–90	1.95–3.60	84–95
5	300–400	8000+		Upper 6–7	90+	3.60+	95+
6	350–435						
6½	425–500						
7	480–610						
8	600+						

Approximate overburden thicknesses for CAI values are examples only; they are not universally applicable, but are based on work in the Appalachian Basin, USA (Harris *et al.*, 1978) and Michigan Basin, Canada (Legall *et al.*, 1982). The acritarch alteration index (AAI) correlation is after Legall *et al.* (1982); other palynomorph and vitrinite correlations are after Epstein *et al.* (1977). The temperature ranges for CAI 6–8 are based on Rejebian (1985) (Dr. Anita G. Harris, U.S. Geological Survey, written commun., 1985). This table is modified from Nowlan and Barnes (in press). Conodonts with CAI 1 have a colour equivalent to Munsell soil colour pale yellow (2.5Y7/4 to 8/4); those with CAI 1½ (very pale brown 10YR7/3 to 10YR8/4), 2 (brown to dark brown 10YR4/2 to 7.5YR3/2, 3 Very dark grayish brown 10YR4/2 to dark reddish brown 5YR 2.5/2) 4 black, 5YR/2.5/1 to 10YR2.5/1), 5 (black, 7.5YR2.5/0 to 2.5YR2.5/0). Information derived from Epstein *et al.* (1977, Table1).

avoided by using samples of a single lithological type (e.g. carbonates) only.

Data obtained should be recorded in detail. Sample number(s), age(s), stratigraphic unit, locational information (e.g. latitude and longitude) and a locality description should be supplied for each CAI determination. Such attention to detail permits future evaluation of the same samples, which may be important if other features of conodonts, such as surface crystallinity, prove useful in thermal metamorphism studies. Finally, the data may be plotted on maps of a suitable scale and isopleths drawn to reveal the regional pattern of thermal maturation.

11.3 OBJECTIVES

Conodont CAI values provide a useful tool for the evaluation of thermal maturity levels in sediments. The distribution patterns of CAI values may be used to interpret local and regional geothermal histories of basins. Conodont CAI has some advantages over other commonly employed indicators of thermal maturation. Firstly, it provides a consistent response because all specimens are composed of the same type of material and were formed by closely related organisms. This is not true for vitrinite reflectance where the materials used may have various origins. Thus a single method can provide reliable readily comparable values for strata of Cambrian–Triassic age (the stratigraphic range of conodonts). Secondly, conodonts provide data over a considerably greater temperature range than do other methods, surviving up to about 600 °C. This feature, combined with the resistance of conodont elements to weathering, means that conodonts provide information under conditions that would have destroyed other organic matter.

Analysis of CAI patterns permits assessment of the thermal history of a sedimentary basin, providing indications of regions and stratigraphic intervals suitable for the preservation of oil and gas. This type of study usually involves a large region. Conodont CAI also has

application in the solution of more local problems, such as the identification of buried intrusions which may be useful in the localization of certain types of mineral deposit. The local thermal effect of a small volcanic plug on CAI values has been well described by Nicoll (1981). CAI may also be used to identify local tectonic effects which cause heat flow, such as initiation of rifting or major overthrusting with resulting deep tectonic burial. Regional thermal anomalies have also been related to the passage of hotspots (Legall *et al.*, 1982; Nowlan and Barnes, in press). Attempts have also been made to relate conodont CAI data to fluid-inclusion temperatures for ore-bearing fluids in Mississippi Valley-type base metal deposits (Nowlan and Barnes, in press).

Thus, conodont CAI data can be used in hydrocarbon assessments, in mineral exploration and in unravelling tectonic history. They provide a valuable tool for other specialists who require a knowledge of the thermal history of their sampled strata, such as palaeomagnetists, isotope geochemists and fluid-inclusion specialists. The following sections will endeavour to elucidate some examples of these applications.

11.4 CAI AND HYDROCARBON EXPLORATION

In both the prediction and the discovery of petroleum fields, explorationists assess a large number of geological criteria. These include factors such as suitable source rocks, potential reservoir horizons, structural and stratigraphic traps, thermal history and maturation levels, size and complexity of deposits, and quality and type of hydrocarbons. Conodonts are widely used in biostratigraphic and biofacies studies for assessing local and regional subsurface stratigraphy. Within the last decade, their utility has been expanded significantly to provide data on thermal maturation levels and geothermal history in sedimentary basins. As noted above, this potential was known for several decades but was not systematically studied until the benchmark study of Epstein *et al.* (1977).

It must be emphasized that, although thermal maturation is one of the most critical factors in hydrocarbon exploration, there is a range of techniques in addition to CAI which are used to determine such levels (Héroux *et al.*, 1979). These include total organic carbon, vitrinite reflectance, clay mineral diagenesis, thermal alteration index (based on palynomorphs), carbon preference index, kerogen hydrogen-to-carbon ratio, gravity (API), bitumen-to-organic-carbon ratios, time–temperature index (TTI), as well as actual down-hole temperature data from exploration wells. Typically, no single technique provides a complete and unambiguous assessment, and thus a range of studies is preferred. Vitrinite reflectance, for example, is difficult to obtain from strata older than Late Silurian because land plants first developed in the Silurian Period. Alteration indices based on colour changes in palynomorphs are more applicable in low-temperature regimes (less than 100 °C). Conodont CAI values are less sensitive than palynomorphs at low temperatures but permit assessments up to 600 °C. Thermal maturation levels can therefore best be determined by using a range of methods which can be correlated (Table 11.1). Although only recently applied to problems of hydrocarbon exploration, there have already been some valuable results from CAI studies.

Anomalously low CAI values were reported by Harris (1979) from tectonic windows in the southern Appalachians. The values indicated that, although most of the surface rocks were highly deformed and metamorphosed, this alteration was obtained largely prior to their emplacement as allochthons. The underlying rocks, exposed in the tectonic windows, yield low CAI values, indicating limited burial thicknesses and temperatures. These data provided further support for models of thin-skinned tectonics for the southern Appalachians (Harris and Milici, 1977). It also encouraged overthrust plays in hydrocarbon exploration in that region. Major gas discoveries, in particular, have resulted from this reinterpretation of the geological structure and the recognition that some of the underlying autochthonous rocks had not experienced burial temperatures in excess of those of the liquid window. A similar important line of CAI study involving the recognition of ancient hotspot tracks is covered in a separate section (11.7) below and also has considerable significance for hydrocarbon exploration.

A more typical exploration situation is where maturation levels increase with depth. A critical aspect of drilling, particularly in expensive off-shore or Arctic drilling programs, is to determine when the drill has passed through the liquid window (Fig. 11.1) with little chance of encountering oil deposits below. Such a progressive increase in CAI from 1 to 2 and from 2 to 4 was present in two exploration wells on Bjorne Peninsula, Ellesmere Island, Canadian Arctic Archipelago (Mayr *et al.*, 1978). Whereas the wells provided important subsurface stratigraphic data, the final 1000 m of the Panarctic Tenneco *et al.* CSP Eids M-66 well was through supramature strata and was therefore unlikely to yield liquid hydrocarbons.

The geothermal history of a sedimentary basin is commonly complex (e.g. Cercone, 1984). As pointed out below (in Section 11.6), basic heat flow from the continental crust will vary and may be supplemented by regional increases due to rifting, tectonism, metamorphism, plutonism and sedimentary burial. Many other factors, such as differential thermal conductivity of the strata and overpressuring, produce local variations in maturation levels. This emphasizes the need to secure a regional coverage of CAI values, including extension into the subsurface, to obtain a realistic assessment of the geothermal history. Once obtained, and particularly when combined with data from other techniques, it is possible to define areas of potential hydrocarbon generation and preservation (Table 1; Fig. 11.1). Immature strata may yield hydrocarbons received through migration. Mature strata may have sourced hydrocarbons. Supramature strata will contain only dry gas, unless oil migrated in during a later cooler period. The maturation level indicated from analysis of the oil must therefore be compared with the maturation level of the host strata.

The timing of thermal events is most significant with respect to the particular regional or stratigraphic plays being considered. With the possibility of thermal influences from hotspot tracks, caution must be exercised in preparing simple basin history models based largely on constant heat flow and progressive sedimentary burial. This is particularly true in using the increasingly popular TTI developed by Lopatin in 1971 (e.g. Waples, 1980; Issler, 1984; Cercone, 1984). This index is based on the concept that chemical reaction rates affecting organic material will double for every 10 °C increase in temperature or with a doubling of exposure

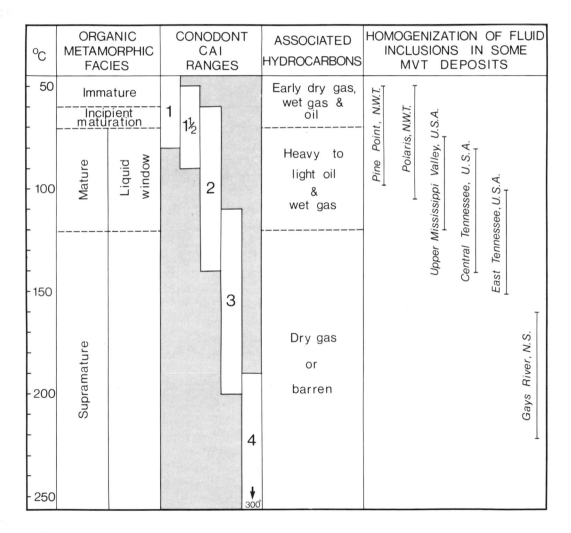

Fig. 11.1 — Correlation of CAI values with stages of hydrocarbon generation and homogenization temperatures of fluid inclusions in selected Mississippi Valley-type (MVT) lead–zinc deposits: N.W.T., Northwest Territories; N.S., Nova Scotia. CAI temperature ranges are from Epstein *et al.* (1977). Organic metamorphic facies and associated hydrocarbons columns are based on information presented by Legall *et al.* (1982). Homogenization temperatures of fluid inclusions are from Roedder (1976) except those for the Gays River deposit which are from Akande and Zentilli (1983).

time (Hood *et al.*, 1975). This means that a similar level of maturation can be achieved by both a high temperature acting for a short time and a low temperature effective over a long period. For applications of this TTI, information on the temperature levels attained and the time and duration of the thermal events are all important. Conodont CAI data can obviously assist in this regard. Sequential sampling for conodonts throughout a region is commonly required to distinguish between multiple thermal events (Burrett, 1984). Prospective areas for hydrocarbon exploration were so identified in the CAI studies for eastern Canada by Legall *et al.* (1982) and Nowlan and Barnes (in press). Similar specific observations were made for the Canning Basin, Western Australia, by Nicoll and Gorter (1984a, 1984b) and for overthrust plays in the Laramide Thrust Belt of Montana by Perry *et al.* (1983).

11.5 CAI AND MINERAL EXPLORATION

Conodont CAI values have several potential applications in the search for mineral deposits, particularly with respect to the identification of buried intrusions and the assessment of both timing and temperature of emplacement of ore-bearing fluids.

Regional patterns of CAI data may serve to outline the limits of buried intrusions, particularly where intrusions are suspected, but not known at the surface. The intrusions may reflect the location of such deposit types as porphyry copper and skarns. CAI data compiled for the Chaleur Bay region of Quebec and New Brunswick by Nowlan and Barnes (in press) were compared with patterns of thermal maturity obtained by Duba and Williams-Jones (1983) based on illite crystallinity, organic-matter reflectance and isotopic techniques. The patterns compare quite closely and may be interpreted to show the outlines of buried intrusions. The occurrence of porphyry copper deposits in the area is established (Béland, 1958) and thermal maturity data may serve to localize areas suitable for exploration; however, this will require

more detailed and systematic sampling in the future.

Nowlan and Barnes (in press) suggested the possibility that conodont CAI data be used to detect the temperature of ore fluids at the time of emplacement. As conodonts are readily obtained from carbonate rocks, they suggested that CAI data could be readily used in Mississippi Valley-type deposits. In regions of low thermal maturity (CAI 1–2), samples taken in close proximity to ore may yield elevated CAI values related to the passage of ore fluids. The reported temperatures of emplacement for several Mississippi Valley-type deposits are known from work on fluid inclusions (Roedder, 1976). Temperatures range considerably from 50–90 °C to 100–150 °C and some have an upper range limit of over 200 °C (Fig. 11.1). These are within the range of conodont CAI detectability but, if the fluids cooled rapidly, they may not have affected the conodont CAI.

Preliminary studies to test this hypothesis have been undertaken by one of us (G.S.N.) and Dr. D. F. Sangster of the Geological Survey of Canada. Conodonts from strata enclosing the Daniels Harbour deposit in western Newfoundland and the Polaris deposit on Little Cornwallis Island, Northwest Territories, have been assessed for CAI values. In the case of Daniels Harbour the background temperature for the region is too high, obscuring any possible relationship between local CAI values and ore-fluid emplacement. At the Polaris deposit, no effect on CAI has been detected that could be ascribed to proximity of hot ore-bearing fluids. Samples from within 50 cm of ore are of CAI 1 or slightly higher (1+ of Nowlan and Barnes (in press)). Fluid-inclusion data for the Polaris deposit indicate a temperature range 52–105 °C (Roedder, 1976) which theoretically could produce CAI values of $1\frac{1}{2}$–2. It is possible that the duration of elevated thermal conductivity was too short and produced no effect. In contrast, the small volcanic plug reported by Nicoll (1982) did have an appreciable effect, although admittedly it had a much higher internal temperature. Work continues on this problem and

one of the areas which may provide some defini-
tive answers is Tennessee (Fig. 11.1) where
Mississippi Valley-type deposits are known
across a belt that ranges from CAI 1 to CAI 5.

It is expected that other applications will be
found for CAI data in the field of mineral
exploration (e.g. sedimentary copper deposits),
particularly as more regional maps are completed. The potential application of CAI studies
in diamond exploration is discussed under
Section 11.7. In the short term it is important
that CAI values be well intercalibrated with
other thermal maturation indices and that they
be repeatedly tested in the case histories of
several mineral deposits and districts.

11.6   CAI AND REGIONAL TECTONIC INTERPRETATION

Conodont CAI data may be used in the interpretation of tectonic events in orogenic belts,
particularly in placing constraints on the depth
of burial. In evaluating the geothermal history
of a sedimentary basin, the general level of heat
flow with time is important but usually difficult
to establish. Certain typical heat-flow values
may have to be assumed (Epstein *et al.*, 1977;
Legall *et al.*, 1982), based on a knowledge of
known or assumed crustal thickness, structure
and deformational history. Regional heat flow
will be lower in an intracratonic basin than for a
newly rifted continental margin. Increases in
heat flow may be generated by later orogenic or
intrusive events which may be of regional or
local effect. Examples from recent studies
include the eastern Michigan Basin of southwestern Ontario (Legall *et al.*, 1982) and the
Canning Basin of Western Australia (Nicoll and
Gorter, 1984a, 1984b). In the former, the
intracratonic Michigan Basin was developed
on a continental crust 40 km thick and
CAI isopleths show only broad and gradual
change across the edge of the basin. Variation in
thermal maturity is most likely in response to

variation in depth of burial. In the Canning
Basin, the initial heat-flow regime was
influenced by rifting and graben development in
Late Devonian–Early Carboniferous times and
locally affected by Late Triassic–Early Jurassic
basic intrusives. This is locally overprinted by a
Miocene thermal event which produced the
Fitzroy Lamproites in the Lennard Shelf and
adjacent Fitzroy Graben. Nicoll and Gorter
(1984a, 19184b) have shown that the thickness
of CAI intervals in the Canning Basin is variable; the CAI 1 interval varies from 800 m to
2300 m, whereas the thickness of higher CAI
intervals is considerably less (e.g. 400–700 m for
CAI 2). They speculated that the variation may
be the result of differences in the local insulating
effects of thick shale, quartz sandstone or evaporite units to give a thick blanket of strata of
CAI 1 over strata with variable CAI which
reflect the complexities of basement rifting and
variable heat flow.

In tectonically complex areas, it may be
possible to determine the effects of tectonic
rather than sedimentary burial. For example, it
may be possible to determine the previous
extent and thickness of allochthonous strata.
Nowlan and Barnes (in press) have shown that
the allochthonous strata in the Taconic Orogen
of western Newfoundland were probably never
very thick or much more extensive than they are
now. Such analyses provide information only on
the structural thickness and do not reflect the
total stratigraphic thickness of the allochthonous strata.

Analysis of conodont CAI data from structurally deformed belts presents greater
problems than those studies conducted in a
single basin because the diversity of factors
influencing the geothermal gradient is greater.
Each locality must be treated on its own merits
and assessed in terms of the tectonic framework. General regional isopleth maps may yield
relatively little useful information because of
the geological complexity of strata intervening
between any two points. The construction of
isopleths must be undertaken with careful consideration of the geology and of the possibility

of multiple thermal events. Conodont CAI data nevertheless have many applications at the detailed level. To date, most studies of CAI patterns cover large areas, mainly because each study is usually the pioneering one for the area in question. Detailed studies of conodont CAI values over small areas may produce valuable results in orogenic regions. For example, the discovery of anomalously low values in tectonically deformed areas suggests the existence of a tectonic window (Harris, 1979) or requires some further explanation. Nowlan (1983) encountered such an anomaly in Upper Ordovician strata of the Matapedia area, Quebec, and provided structural models to explain the anomaly. Such local observations place constraints on the thermal history of the rock unit under study and may even provide significant insights into the tectonic history of the area. The CAI data of Burrett (1984) showed that Early Devonian orogenic events in eastern and western Tasmania, both previously attributed to the Tabberabberan Orogeny, must belong to two separate and distinct orogenic phases.

Although it is difficult to interpret precisely the conodont CAI data from orogenic belts, it is possible to draw tentative conclusions in some cases. For example, it may be possible to determine the limits of buried intrusions which, as noted above, could prove valuable in mineral exploration. Aldridge (1984) concluded that patterns of conodont CAI values in Silurian strata of the Oslo region, Norway, are closely related to the proximity of Permian intrusions. Nowlan and Barnes (in press) concluded that buried intrusions were the likely cause of variable CAI values in Lower Palaeozoic strata in the Chaleur Bay region of Quebec and New Brunswick.

It may be possible to define major fault lineaments using conodont CAI patterns, particularly if the fault was a line of increased thermal activity or the junction between two thermally distinct areas. Overthrusting should also have an influence on local thermal maturation levels and conodont CAI patterns may help in identification of thrust faults where the

tectonically emplaced overburden is of significant thickness.

11.7 CAI AND HOTSPOTS

One geological phenomenon which has received relatively little consideration as a factor in generating thermal anomalies in the geological record is that of hotspots. These are the surface expression of mantle plumes. They are commonly developed as volcanic centres (e.g. Hawaii, Iceland and Yellowstone) and as physiographic swells typically 1200 km in diameter. There are some 40 known active hotspots at present, the surface features of which occupy 10% of the Earth's surface (Burke and Wilson, 1976; Crough, 1979, 1983; Morgan, 1982, 1983; Vink *et al.*, 1985). Their locations have become better defined from recent satellite data on the Earth's surface topography and gravity anomalies.

The origin and evolution of mantle plumes are still under debate but there is considerable evidence (reviewed in the papers cited above) to indicate that they are initiated close to the core–mantle boundary and remain stable in a relatively fixed position for many tens of millions of years. Although fixed geographically, they may be pulsative in their upwelling velocities, with consequent pulsative surface volcanism and thermal swelling. Sleep (1984), however, has argued that a heterogeneous mantle structure may provide an alternative model to that of mantle plumes.

The relatively fixed positions of plumes contrast with the active motion of the lithospheric plates. As such plates move over hotspots, the surface effects generated by the hotspots also migrate and produce a trace that may be preserved in the migrating lithosphere (Fig. 11.2). Such features may include a chain of seamounts (dead volcanic mountains) such as the Emperor–Hawaiian Seamounts associated with the Hawaiian hotspot in the Pacific (Burke and Wilson, 1976) and the New England Seamounts associated with the Great Meteor Hotspot in the Atlantic (Crough, 1979, 1983; Morgan,

1982, 1983; Duncan, 1984). As the lithospheric plate moves over the plume, the hotspot swell develops and then fades as the lithosphere migrates away and cools. The vertical changes in elevation may be over 1 km (Crough, 1983) and may have a significant effect on the preserved stratigraphic record for a submerged oceanic seamount or for uplifted continental crust, where significant local erosion would occur together with probable deposition in flanking basins.

With information on the spreading rates and directions of lithospheric plates and knowledge of the present sites of hotspots, the traces of the assumed fixed hotspots have been determined or predicted (Crough, 1981, 1983; Morgan, 1982; Thompson and Morgan, 1984). These traces plot the paths of some of the present hotspots back to their Mesozoic positions. Some (e.g. the Great Meteor Hotspot) originated under continental crust and passed beneath the continental margin and out into the ocean basin. Thus the area of the Earth's surface which has been affected by the present 40 hotspots is considerable. The effects of hotspots should therefore be expected, and possibly predicted, in the ancient stratigraphic record. CAI data can be of value in determining thermal anomalies produced by hotspots and, together with conodont biostratigraphic data, may reveal unconformities related to erosion on swells. Regional CAI data should indicate those areas which have been uplifted, as opposed to those areas flanking a hotspot trace which may have undergone deep burial upon receipt of the erosional products of the swell. In terms of the

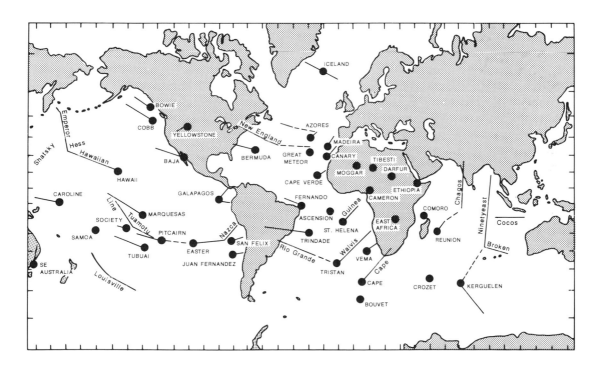

Fig. 11.2 — Global distribution of hotspots (dots) and hotspot traces (lines). Sources include Morgan (1982), Burke and Wilson (1976) and Jarrard and Clague (1977). Some traces are well documented and are labelled with the name of the volcanic feature; others, such as that of Bermuda, are hypothetical showing the predicted path if the hotspot had remained active for a long period. The dashed lines connect trace segments in gaps where no trace should be observed, usually because of ridge movement over hotspots. (From Crough, 1983.)

geological significance of determining ancient hotspot tracks, such information can aid in hydrocarbon and diamond exploration and in interpreting plate tectonic events, configurations and motions.

Legall *et al.* (1982) interpreted a regional thermal anomaly in eastern Ontario and southern Quebec, centred near Montreal, to be the result of a Cretaceous hotspot, now located in the eastern North Atlantic as the Great Meteor Hotspot. Conodont CAI values of 2 in the upper Ottawa Valley region and in most of southwestern Ontario increase progressively to over 4 in the Montreal region. The Cretaceous Monteregian intrusions of southern Quebec, dated at 118–136 Ma (Eby, 1984), the White Mountain Igneous Province in New England, with three clusters of ages at 230 Ma, 200–156 Ma and 124–100 Ma (Foland and Faul, 1977), and the New England Seamounts at 103–82 Ma (Duncan, 1984) were viewed by Crough (1981), Morgan (1982) and Duncan (1984) as generated by and defining the trace of the Great Meteor Hotspot. Alternative interpretations have been offered by McHone (1981) and Eby (1984). Other alkaline intrusions may be associated with tranform fault during episodes of plate-motion change (Jansa and Pe-Piper, 1985). The eastern extent of the CAI-based thermal anomaly recognized by Legall *et al.* (1982) has been documented by Nowlan and Barnes (in press), who advocate a hotspot interpretation. The limited northeastward extent of the anomaly, like the discontinuous line of intrusives, may be a reflection of the pulsative nature of the upwelling plume.

A hotspot interpretation for a thermal anomaly in northwest Newfoundland was also offered by Nowlan and Barnes (in press). The general area of Labrador and the Labrador Sea to the north was apparently the trace of two or three hotspots, including those now located at the Azores and Maderia (Crough, 1981; Duncan, 1984; Vink *et al.*, 1985). Jurassic intrusions along the Labrador coast and the presence of the Newfoundland Seamounts provide supporting evidence for such Mesozoic hotspot traces.

A highly significant aspect of such hotspot interpretations is that a continental margin affected by such thermal events (up to CAI 3–4) could have generated hydrocarbons from source beds. The off-shore Atlantic margin is currently one of the most actively explored of Canada's frontier petroleum provinces, following the 1979 discovery of the Hibernia oil field on the Newfoundland Grand Banks (Arthur *et al.*, 1982; Powell, 1982; Procter *et al.*, 1984). Space limitations do not allow development of the full details here and a separate paper on hotspots and hydrocarbons is in preparation. It suffices to note that, in a hydrocarbon exploration play along a margin affected by hotspots, the exact timing of the thermal event in relation to the age of the source rocks and the development of appropriate reservoir and sealing horizons are critical. If the margin is affected after the appropriate mix of sediments has accumulated, then the hotspot can clearly generate hydrocarbons. However, the temperatures along the trace may exceed those of the oil window, causing migration to flanking regions, and only gas, if any hydrocarbons at all, will occur in the zone of the trace. Furthermore, the thermal swell associated with the passage of the hotspot will complicate oil migration and lead to local unconformities or crustal flexures. The hotspot should also have a significant effect on evaporite deposits, encouraging diapiric rise, such as those characteristic of the Late Triassic–Early Jurassic of the Atlantic and Gulf coasts of North America. Similarly, Nowlan and Barnes (in press) noted the areas of the St. Lawrence Lowlands which were apparently affected by the Great Meteor Hotspot and this range of immature to supramature regions should be taken into account in future exploration of Lower Palaeozoic strata in the Quebec and Anticosti basins.

In addition to aiding hydrocarbon exploration, the recognition of ancient hotspot traces can be important in diamond exploration. The traces of hotspots are characterized by alkaline intrusions, probably generated from a mantle source (Crough *et al* 1980). Some include kimberlites which may have potential for diamonds. Kimberlites are known from several places on the Canadian Shield and sparse diamonds are known from glacial till, but no commercial diamond-bearing kimberlites have yet been found. Hotspot traces may provide a useful predictive model to define narrow belts for detailed diamond exploration. Crough *et al.* (1980) showed that most known post-Triassic kimberlites formed within 5° of a mantle hotspot. Their statistical analysis demonstrated that this kimberlite–hotspot correlation is significant above the 90% level. Jaques *et al.* (1984) and Nicoll and Gorter (1984a, 1984b) report diamond-bearing pipes and associated leucite lamproites which were generated by a Miocene thermal event in Western Australia and which are targets for diamond exploration. The thermal event may have been related to the passage of Western Australia over a mantle plume.

11.8 SUMMARY

The pioneering work of Epstein *et al.* (1977) provided a calibrated scheme of conodont CAI, from CAI 1 to CAI 8. Conodont CAI data provide information on the thermal maturation level of conodont-bearing strata which can be used in assessment of many problems in regional and economic geology. The data obtained are complementary to the many other techniques used for thermal maturation studies. Particular advantages of conodont CAI data are the considerable temperature range over which they are useful (up to 600 °C) and the fact that all conodonts are of similar compositions (unlike vitrinite material) and thus react consistently to thermal events.

CAI data can be employed as a tool in hydrocarbon exploration, determining levels of thermal maturity in surface and subsurface strata. Zones of immature, mature and supramature strata can be defined, permitting assessment of sedimentary basins for hydrocarbon potential. Conodont CAI data from the over thrust belt of the southern Appalachians

Harris, 1979) showed that, although surface rocks were highly deformed, underlying strata preserved in tectonic windows were only slightly thermally altered. This discovery encouraged hydrocarbon exploration and some major gas discoveries have resulted. CAI data are used to determine the increase in thermal alteration with depth, thus providing useful limits on the oil window during drilling. Regional compilation of data leads to the definition of zones with the best potential for hydrocarbon generation and preservation. Studies of CAI data which outline prospective areas for hydrocarbon exploration have recently been carried out for parts of the Canadian Arctic Islands (Mayr *et al.*, 1978), part of the Michigan Basin in Canada (Legall *et al.*, 1982), the Laramide Thrust Belt of Montana (Perry *et al.*, 1983), the Canning Basin of Western Australia (Nicoll and Gorter, 1984a, 1984b) and the foreland basins of the Appalachian Orogen (Nowlan and Barnes, in press).

Conodont CAI data are also valuable in mineral exploration. Studies are being made of their applicability in measuring the thermal effects of ore-bearing fluids, particularly in Mississippi Valley-type deposits. These should reveal information on the temperature and timing of emplacement of ore bodies. It is proposed that similar studies be carried out on sedimentary copper deposits. The limits of buried intrusions which may host porphyry copper or skarn deposits can also be outlined using CAI data. It is also proposed that CAI data be used in the search for diamond-bearing kimberlite pipes associated with hotspot tracks.

Regional tectonic interpretations can benefit from evaluation of conodont CAI data which provide insights into the geothermal and deformational history of a region. In the southern Appalachians, for example, CAI data have supported models of thin-skinned tectonics (Harris and Milici, 1977; Harris, 1979) by defining burial temperatures in both autochthonous and allochthonous strata. Multiple thermal events have been demonstrated using CAI data in Tasmania (Burrett, 1984), the Canning Basin of

Western Australia (Nicoll and Gorter, 1984a, 1984b) and the Appalachian Orogen of eastern Canada (Nowlan and Barnes, in press). The recognition of thermal events can have implications for the timing of orogenic events; for example, Burrett (1984) demonstrated that Early Devonian orogenic events, previously considered as part of a single orogeny in Tasmania, must in fact represent two distinct orogenic phases. Regional heat flow will be greater in a rifted continental margin than in an intracratonic basin and thus CAI data may contribute to the understanding of basin development. In tectonically complex areas, it may be possible to determine the effects of tectonic burial caused, for example, by the structural thickness of allochthonous slices (Nowlan and Barnes, in press). Conodont CAI data may also be used to define the thermal effects of both major and minor intrusive events.

Conodont CAI data have defined thermal anomalies in Ordovician strata of southern Quebec and northwestern Newfoundland which are interpreted as having been caused by the passage of the North American continent over Mesozoic hotspots now located at the Great Meteor and Maderia hotspots respectively (Nowlan and Barnes, in press). The geological significance of hotspots, which are the surface expression of mantle plumes, has only recently received attention, but their effect on heat-flow regimes must be considerable. As mobile lithospheric plates move over a hotspot, local swells and intrusions occur which, with time, form a trace on the surface of the continent. The recognition of hotspot tracks is highly significant for hydrocarbon exploration in interior or marginal basins. CAI data, combined with conodont biostratigraphic data, can define thermal anomalies as well as unconformities related to erosion on the swells. Regional CAI data should also indicate areas flanking a hotspot track now deeply buried by erosional products from the swell. The recognition of hotspot tracks aids in hydrocarbon and diamond exploration and in interpreting the configuration and motion of lithospheric plates

and related tectonic events.

It is predicted that CAI data will be applied to an increasingly broad range of geological problems, following the pioneer projects of the past few years. In addition to the regional and economic applications, CAI data will be used in other disciplines which require supporting burial temperature information, such as palaeomagnetism, isotope geochemistry and fluid-inclusion studies. The next decade will see conodont specialists collaborating with a wide spectrum of other geoscientists in the resolution of diverse geological problems.

11.9   ACKNOWLEDGEMENTS

The authors thank Dr. Anita G. Harris (US Geological Survey) for providing a comprehensive set of conodont CAI standards and for a very helpful review of this manuscript. Dr. H. Williams (Memorial University of Newfoundland) is also thanked for his thoughtful review of an earlier version of the manuscript. Dr. D. F. Sangster (Geological Survey of Canada) is acknowledged for his advice on Mississippian Valley-type lead–zinc deposits. C. R. Barnes is grateful for continuing financial support for conodont research from the Natural Sciences and Engineering Research Council of Canada.

11.10   REFERENCES

Akande, S. O. and Zentilli, M. 1983. Genesis of the lead–zinc mineralization at Gays River, Nova Scotia, Canada. In G. Kisvarsanyi, S. K. Grant, W. P. Pratt and J. W. Koenig (Eds.), *Proceedings of International Conference on Mississippi Valley Tytpe Lead–Zinc Deposits*, University of Missouri-Rolla Press, 546–557.

Aldridge, R. J. 1984. Thermal metamorphism of the Silurian strata of the Olso region, assessed by conodont colour. *Geological Magazine*, **121**, 347–349.

Aldridge, R. J. 1986. Conodont palaeobiology and thermal maturation in the Caledonides. *Journal of the Geological Society, London*, **143**, 177–184.

Arthur, K. R., Cole, D. R., Henderson, G. G. L. and Kushnir, D. W. 1982. Geology of the Hibernia discovery. In M. T. Halbou (Ed.), *The Deliberate Search for the Subtle Trap, American Association of Petroleum Geologists Memoir*, **32**, 181–196.

Béland, J. 1958. Oak Bay area, electoral districts of Matapedia and Bonaventure. *Quebec Department of*

Mines, *Geological Surveys Branch, Preliminary Report*, **375**, 1–12.

Bergström, S. M. 1980. Conodonts as paleotemperature tools in Ordovician rocks of the Caledonides and adjacent areas in Scandinavia and the British Isles. *Geologiska Föreningens i Stockholm Förhandlingar*, **102**, 377–392.

Burke, K. C. and Wilson, J. T. 1976. Hot spots on the Earth's surface. *Scientific American*, **235**, 46–57.

Burrett, C. 1984. Early Devonian deformation and metamorphism of conodonts in western Tasmania — economic, theoretical and nomenclatural implications. In P. W. Baillie and P. L. F. Collins (Eds.), *Mineral Exploration and Tectonic Processes in Tasmania, November, 1984, Abstract Volume and Excursion Guide*, 14–17.

Cercone, K. R. 1984. Thermal history of the Michigan Basin. *American Association of Petroleum Geologists Bulletin*, **68**, 130–136.

Crough, S. T. 1979. Hotspot epeirogeny. *Tectonophysics*, **61**, 321–333.

Crough, S. T. 1981. Mesozoic hotspot epeirogeny in eastern North America. *Geology*, **9**, 2–6.

Crough, S. T. 1983. Hotspot swells. *Annual Review of Earth and Planetary Sciences*, **11**, 165–193.

Crough, S. T., Morgan, W. J. and Hargraves, R. B. 1980. Kimberlites, their relation to mantle hotspots. *Earth and Planetary Science Letters*, **50**, 260–274.

Duba, D. and Williams-Jones, A. E. 1983. The application of illite crystallinity, organic matter reflectance, and isotopic techniques to mineral exploration: a case study in southwestern Gaspe, Quebec. *Economic Geology*, **78**, 1350–1363.

Duncan, R. A. 1984. Age progressive volcanism in the New England seamounts and the opening of the central Atlantic Ocean. *Journal of Geophysical Research*, **89** (B12), 9980–9990.

Eby, G. N. 1984. Geochronology of the Monteregian Hills alkaline igneous province, Quebec. *Geology*, **12**, 468–470.

Ellison, S. P. 1944. The composition of conodonts. *Journal of Paleontology*, **18**, 133–140.

Epstein, A. G., Epstein, J. B. and Harris, L. D. 1977. Conodont color alteration — an index to organic metamorphism. *U.S. Geological Survey Professional Paper*, **995**, 1–27.

Foland, K. A. and Faul, H., 1977. Ages of the White Mountain intrusive — New Hampshire, Vermont and Maine, USA. *American Journal of Science*, **277**, 888–904.

Harris, A. G. 1979. Conodont color alteration, an organo-mineral metamorphic index, and its application to Appalachian Basin geology. In P. A. Scholle and P. R. Schluger (Eds.), *Aspects of Diagenesis, Society of Economic Paleontologists and Mineralogists Special Publication*, **26**, 3–16.

Harris, A. G., Harris, L. D. and Epstein, J. B. 1978. Oil and gas data from Paleozoic rocks in the Appalachian Basin: maps for assessing hydrocarbon potential and thermal maturity (conodont color alteration isograds and overburden isopachs). *U.S. Geological Survey Map*, **I-917-E**.

Harris, L. D. and Milici, R. C. 1977. Characteristics of thin-skinned style of deformation in the southern Appalachians, and potential hydrocarbon traps. *US Geological Survey Professional Paper*, **1018**, 1–40.

Héroux, Y., Chagnon, A. and Bertrand, R. 1979. Compilation and correlation of major thermal maturation indicators. *American Association of Petroleum Geologists Bulletin*, **63**, 2128–2144.

Hood, A., Gutjahr, C. C. M. and Heacock, R. L. 1975. Organic metamorphism and the generation of petroleum. *American Association of Petroleum Geologists Bulletin*, **59**, 986–996.

Issler, D. R. 1984. Calculation of organic maturation levels for offshore eastern Canada — implications for general application of Lopatin's method. *Canadian Journal of Earth Sciences*, **21**, 477–488.

Jansa, L. F. and Pe-Piper, G. 1985. Early Cretaceous volcanism on the northeastern American margin and implications for plate tectonics. *Geological Society of America Bulletin*, **96**, 83–91.

Jaques, A. L., Webb, A. W., Fanning, C. M., Black, L. P., Pidgeon, R. T., Ferguson, J., Smith, C. B. and Gregory, G. P. 1984. The age of the diamond-bearing pipes and associated leucite lamproites of the West Kimberley region, Western Australia. *Bureau of Mineral Resources Journal of Australian Geology and Geophysics*, **9**, 1–7.

Jarrard, R. D. and Clague, D. A. 1977. Implications of Pacific islands and seamount ages for the origin of volcanic chains. *Review of Geophysics and Space Physics*, **15**, 57–76.

Legall, F. D., Barnes, C. R. and Macqueen, R. W. 1982. Thermal maturation, burial history and hotspot development, Paleozoic strata of southern Ontario–Quebec, from conodont and acritarch colour alteration studies. *Bulletin of Canadian Petroleum Geology*, **29**, 492–539, 4 plates (imprint 1981).

Lindström. M. 1964. *Conodonts*, Elsevier, Amsterdam, 1–196.

Mayr, U., Uyeno, T. T. and Barnes, C. R. 1978. Subsurface stratigraphy, conodont zonation, and organic metamorphism of the Lower Paleozoic succession, Bjorne Peninsula, Ellesmere Island, District of Franklin. *Geological Survey of Canada Paper*, **78–1A**, 393–398.

McHone, J. C. 1981. Comment on 'Mesozoic hotspot epeirogeny in eastern North America'. *Geology*, **9**, 341–342.

Morgan, W. J. 1982. Hotspot tracks and the opening of the Atlantic and Indian oceans. In C. Emiliani (Ed.), *The Sea, 7, The Ocean Lithosphere*, Wiley, New York, 443–487.

Morgan W. J. 1983. Hotspot tracks and the early rifting of the Atlantic. *Tectonophysics*, **96**, 123–139..

Nicoll, R. S. 1981. Conodont colour alteration adjacent to a volcanic plug, Canning Basin, Western Australia. *Bureau of Mineral Resources Journal of Australian Geology and Geophysics*, **6**, 265–267.

Nicoll, R. S. and Gorter, J. D. 1984a. Conodont colour alteration, thermal maturation and the geothermal history of the Canning Basin, Western Australia. *Australian Petroleum Exploration Association Journal*, **24**, 243–258.

Nicoll, R. S. and Gorter, J. D. 1984b. Interpretation of additional conodont colour alteration data and the thermal maturation and geothermal history of the Canning Basin, Western Australia. In P. G. Purcell (Ed.), *The Canning Basin, W. A., Proceedings of a Geological Society of Australia–Petroleum Exploration Society of Australia Symposium, Perth, 1984*, 411–425.

Nowlan, G. S. 1983. Biostratigraphic, paleogeographic, and tectonic implications of Late Ordovician conodonts from the Grog Brook Group, northwestern New Brunswick. *Canadian Journal of Earth Sciences*, **20**, 651–671, 3 plates.

Nowlan, G. S. and Barnes, C. R. In press. Thermal maturation of Paleozoic strata in eastern Canada from conodont colour alteration index (CAI) data with implications for burial history, tectonic evolution, hotspot tracks and mineral and hydrocarbon potential. *Geological Survey of Canada Bulletin*, **367**.

Oliver, G. J. H., Smellie, J. L., Thomas, L. J., Casey, D. M., Kemp. A. E. S., Evans, L. J., Baldwin, J. R. and Hepworth, B. C. 1984. Early Palaeozoic metamorphic history of the Midland Valley, Southern Uplands––Longford-Down massif and the Lake District, British Isles. *Transactions of the Royal Society of Edinburgh: Earth Sciences*, **75**, 245–258.

Perry, W. J., Jr., Harris, A. G. and Harris, L. D. 1979. Conodont-based reinterpretation of the Bane Dome – structural reevaluation of Allegheny Frontal Zone. *American Association of Petroleum Geologists Bulletin*, **63**, 647–654.

Perry, W. J., Jr., Wardlaw, B. R., Bostick, N. H. and Maughan, E. K. 1983. Structure, burial history and petroleum potential of Frontal Thrust Belt and adjacent foreland, southwest Montana. *American Association of Petroleum Geologists Bulletin*, **67**, 725–743.

Powell, T. G. 1982. Petroleum geochemistry of the Verrill Canyon Formation: a source for Scotian Shelf hydrocarbons. *Bulletin of Canadian Petroleum Geology*, **30**, 167–179.

Procter, R. M., Taylor, G. C. and Wade, J. A. 1984. Oil and natural gas resources of Canada 1983. *Geological Survey of Canada Paper*, **83–31**, 1–59.

Rejebian, V. A. 1985. *B.A. Thesis*, Princeton University.

Roedder, E. 1976. Fluid inclusion evidence on the genesis of ores in sedimentary and volcanic rocks. In K. H. Wolf (Ed.), *Handbook of Strata-bound and Stratiform Ore Deposits, I: Principles and General Studies; 2: Geochemical Studies*, Elsevier, Amsterdam, 67–110.

Rogers, H. D. 1860. On the distribution and probable origin of the petroleum or rock oil of Pennsylvania, New York and Ohio. *Proceedings of the Philosophical Society of Glasgow*, **4**, 355–359.

Sleep, N. H. 1984. Tapping of magmas from ubiquitous mantle heterogeneities: an alternative to mantle plumes? *Journal of Geophysical Research*, **89** (B12), 10029–10041.

Thompson, G. A. and Morgan, W. J. 1984. Introduction and tribute to S. Thomas Crough 1947–1982. *Journal of Geophysical Research*, **89** (B12), 9869–9872.

Vink, G. E., Morgan, W. J. and Vogt, P. R. 1985. The Earth's hotspots. *Scientific American*, **252**, 50–58.

Waples, D. W. 1980. Time and temperature in petroleum formation: application of Lopatin's method to petroleum exploration, *American Association of Petroleum Geologists Bulletin*, **64**, 916–926.

White, D. 1915. Some relations in origin between coal and petroleum. *Washington Academy of Science Journal*, **5**, 189–212.

12

Contact metamorphism of conodonts as a test of colour alteration index temperatures

H. A. Armstrong and M. R. Strens

Temperature ranges for conodont colour alteration index (CAI) are based upon open-air heating runs. To test the applicability of these for limestones buried under thin cover, heat-flow theory has been used to predict the country-rock temperatures away from a basaltic dyke. Theoretical and CAI temperatures are similar, suggesting with the present degree of accuracy that the experimental data can be applied to this geological situation.

12.1 INTRODUCTION

Of the many methods of assessing the maturation of sedimentary rocks, the conodont colour alteration index (CAI) has proved particularly successful and has been widely applied in Palaeozoic carbonate platform areas. The pioneer work of Epstein *et al.* (1977) allowed the temperature ranges 50–300 °C to be derived for CAI values of 1–5. Their experimental work was based upon isolated conodont elements and thus there is the question as to whether these can be applied to geological situations.

The change in colour of conodonts in areas adjacent to igneous intrusions has long been known (Ellison, 1944; Sweet and Bergström, 1966; Nicoll, 1981). Nicoll analysed conodonts in traverses away from a small leucite lamproite plug and concluded that the magma had exceeded 600 °C (CAI 7–8+) and that the thermal effects of the intrusion extended only 1 m into the country rock. His temperature estimates confirmed thin-section studies which suggested a liquid melt temperature of 750 °C.

In this chapter, we document conodont colour changes away from a basaltic dyke and correlate these with theoretically derived country-rock temperatures. Preliminary data suggest that the heating effects of the dyke extended more than 3 m into the country rock, that the contact temperature of the dyke did not exceed 700 °C (CAI 6) and that a correlation exists between numerically and experimentally derived temperatures. We conclude that with the present degree of accuracy the temperature values for CAI 3–6 can be applied to rocks buried under thin cover.

12.2 NUMERICAL MODEL

For an intrusive sheet under thin cover, solidification proceeds inwards from both contacts until the temperature at the centre of the intrusion begins to fall (Jaeger, 1968). For times greater than this, no simple heat-flow theory exists and the progress of cooling must be followed numerically. Jaeger (1957, 1968) calculated temperature distributions across an intrusion and into the country rock for a number of examples. Figs. 12.1 and 12.2 illustrate the temperature across a basaltic dyke 9 m thick. In this example, allowance is made for a latent heat of 100 cal g^{-1}, it is assumed that the dyke has a fixed melting point of 1000 °C, an intrusion temperature of 1000 °C and that the thermal properties of the dyke and country rock are the same (Jaeger, 1968). Time (Fig. 12.1) is represented on each curve by a Fourier number

τ which is proportional to kt/a^2, where k is the thermal diffusivity of the rock and has been approximated to 6.6×10^{-7} cm^2 s^{-1}, t is the time in seconds and a is the linear dimension of the dyke. Solidification will be complete when the Fourier number is 0.69, a real time of 0.63 years.

The time–temperature points for the sampled distances away from the dyke have been plotted on the Arrhenius diagram first published by Epstein *et al.* (1977) which also illustrates their experimental data for the CAI boundaries (Fig. 12.3). Heat-flow theory predicts (Fig. 12.3) that conodonts 3 m away from the dyke should have a CAI value of 3, conodonts 2 m away from the dyke should have a CAI value of 4 and for conodonts at a distance less than 1 m away, the CAI values should be greater than 5.

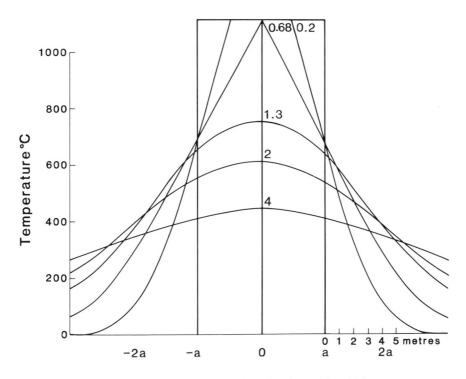

Fig. 12.1 — Temperature variation across a cooling intrusive sheet of 9 m thickness. This numerical model (after Jaeger, 1968) allows for a latent heat of 100 cal g^{-1}. The numbers on the curves are Fourier numbers which are equivalent to $\tau/k = t$ s, where τ is the Fourier number and k is the thermal diffusivity of basalt (approximately 6.6×10^{-7} cm^2 s^{-1}). The calculations assume a fixed melting point of 1000 °C with intrusion at 1000 °C, similar thermal properties for the intrusion and bedrock and a thermally insulated country rock.

12.3 TEST

To test these predictions, conodont sampling was undertaken at intervals away from the Beadnell dyke in northeastern Northumberland (Fig. 12.4). This Permo-Carboniferous dyke is 9 m thick adjacent to the transect and intrudes upper Viséan (Brigantian) limestones of the upper Middle Limestone Group. All the samples underwent standard processing techniques for conodonts using 10% acetic acid and bromoform separation. Residues were dried at 30 °C. CAI values are the minimum for the sample and were taken at the edges of the posterior processes of the Sc elements of *Gnathodus* sp.. Yields were generally low, an average of 20 specimens per kilogram but provided sufficient specimens for this preliminary analysis. The limestone host rock is a homogeneous dark-grey micrite.

Samples from the baked country rock 0.02 m from the edge of the dyke yielded conodont elements with a CAI of 6 falling to CAI 3 at 2 m (Fig. 12.5) and possibly CAI 2 at 3 m.

12.4 DISCUSSION

Conodonts from the Sandbanks Limestone (Fig. 12.4) 65 m away from the dyke record CAI 1 and this is taken as the local background. This value indicates that the host did not reach tem-

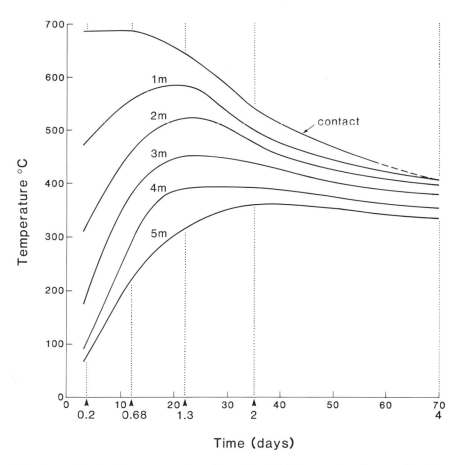

Fig. 12. 2 — Generalized cooling curves, replotted from Fig. 12.1. The curves bear distances from the contact. The vertical lines are Fourier numbers.

peratures above 80 °C, which at an average geothermal gradient of 33 °C km^{-1} is equivalent to less than 3 km depth of burial.

The data presented suggests that the Beadnell dyke affected the country rocks for some 3 m adjacent to it. At the outcrop, these effects can only be seen at the contact where the limestone is baked in a zone 25 mm wide. CAI and numerical modelling suggest that this zone never exceeded a temperature of 700 °C.

Our model only allows for a heating time which, although short, was sufficient for the conodont colours to equilibrate. From the model, sample BDR3 at 1 m from the contact exceeded temperatures of 550 °C for approximately 384 h, sufficient time to produce a CAI of 5. Sample BDR4 at 2 m from the contact exceeded temperatures of 500 °C for 312 h, again sufficient to produce a CAI value of 4. At 3 m (sample BDR5) a temperature of 480 °C

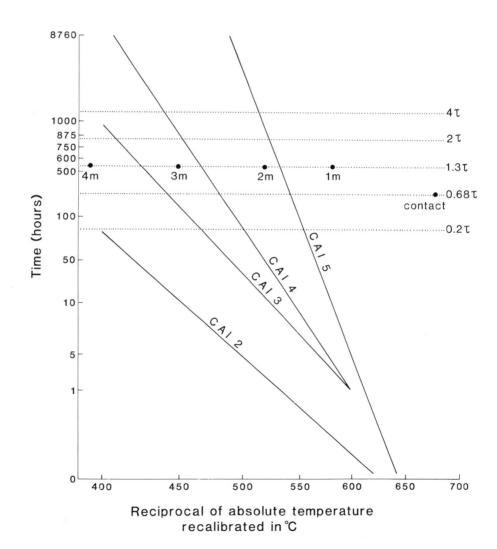

Fig. 12.3 — Predicted CAI values at given distances away from the dyke. The maximum time which each locality exceeded the plotted temperature is taken from Fig. 12.2 and plotted on an Arrhenius diagram. The boundaries of the CAI fields are from Epstein *et al.* (1977).

was maintained for 456 h, producing a theoretical CAI of 3. In comparison with the actual CAI values (Fig. 12.5), those at 2 m and 3 m do not conform to the predicted values; those in sample BDR4, 2 m away, are lower than predicted. Conodonts from sample BDR5, 3 m away from the dyke, proved difficult to assess. Specimens show a range of CAI values from 2 to 4, embracing the expected value of 3. Until further specimens are available, the CAI data from this sample are considered to be dubious.

12.5 CONCLUSIONS

From the values discussed, it may be concluded that the conodonts metamorphosed and equilibrated in a relatively short period of time. In addition, actual CAI values of 5 and less than 3 fall within the temperature fields predicted by the model and experimental work of Epstein *et al.* (1977). It is difficult to assess the causes of anomalous CAI values at 2 m and 3 m away from the dyke. It must be emphasized that the model represents an oversimplification of the

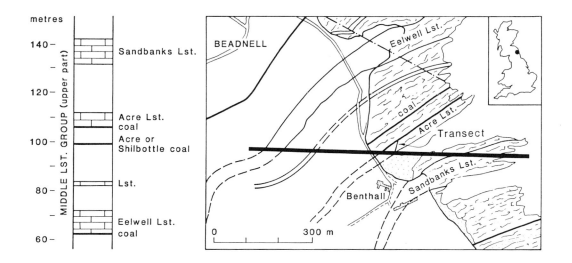

Fig. 12.4 — Sample locality and general geology map.

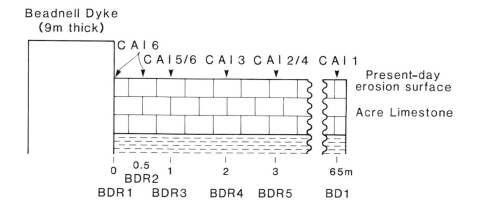

Fig. 12.5 — Actual CAI values adjacent to the Beadnell dyke. BDR and BD numbers are samples, stored in the micropalaeontological collections, Department of Geology, University of Newcastle upon Tyne.

conditions pertaining during the intrusion of the dyke. The model demands a simple and predictable cooling of the dyke and cannot take into consideration the effects of water circulation or volatiles in the system. The behaviour of these in particular may be critical in retarding (Epstein *et al.*, 1977) or enhancing (Wardlaw and Harris, 1984) the carbonization process. In our system, it appears that carbonization was retarded in a zone at 2–3 m. Harris (pers. commun. 1985) has also found this zone of anomalous CAI values in her many studies of dykes. This suggests that the phenomenon may occur widely and because of the apparent lowering of CAI values may be associated with water circulation close to the dyke. Using heat-flow modelling and closer sampling, it is hoped to refine CAI temperature boundaries.

12.6 ACKNOWLEDGEMENTS

The paper was critically read by Dr. R. J. Aldridge and Dr. A. G. Harris. We gratefully acknowledge Christine Jeans for drafting the diagrams and Elizabeth Walton who typed the manuscript. Dr. Armstrong acknowledges the receipt of a University of Newcastle upon Tyne research grant and Dr. Strens the continuing support of the NERC.

12.7 REFERENCES

Ellison, S. 1944. The composition of conodonts. *Journal of Palaeontology*, **18**, 133–140.

Epstein, A. G., Epstein, J. B. and Harris, L. D. 1977. Conodont color alteration — an index to organic metamorphism. *U.S. Geological Survey Professional Paper*, **995**, 1–27.

Jaeger, J. C. 1957. The temperature in the neighbourhood of a cooling intrusive sheet. *American Journal of Science*, **255**, 306–318.

Jaeger, J. C. 1968. Cooling and solidification of igneous rocks. In H. H. Hess and A. Poldervaart (Eds.), *Basalts: The Poldervaart Treatise on Rocks of Basaltic Composition*, **2**, Interscience, New York, 503–536.

Nicoll, R. S. 1981. Conodont colour alteration adjacent to a volcanic plug, Canning Basin, Western Australia. *Bureau of Mineral Resources Journal of Australian Geology and Geophysics*, **6**, 265–267.

Sweet, W. C. and Bergström, S. M. 1966. Ordovician conodonts from the Penobscott County, Maine. *Journal of Paleontology*, **40**, 151–154.

Wardlaw, B. R. and Harris, A. G. 1984. Conodont-based thermal maturation of Paleozoic rocks in Arizona. *American Association of Petroleum Geologists Bulletin*, **68**, 1101–1106.

13

Conodont alteration in metamorphosed limestones from northern Hungary, and its relationship to carbonate texture, illite crystallinity and vitrinite reflectance

S. Kovács and P. Árkai

Results are summarized of micropalaeontological, carbonate texture and metamorphic petrological investigations carried out on northern Hungarian Palaeozoic–Mesozoic rocks forming the innermost part of the western Carpathians.

Grades of diagenesis and incipient regional metamorphism were distinguished using illite crystallinity (IC) and vitrinite reflectance (R) as well as the mineral parageneses present in intercalated metavolcanics. When these parameters are compared with the colour alteration indices (CAIs) and carbonate-textural characteristics, the following conclusions are drawn.

(1) The low-temperature boundary of incipient metamorphism (very-low-grade, anchizone: IC=$0.37°$ (2θ); $R_{max} \approx 3.5\%$) is characterized by CAI values of about 5 or more and by the initiation of recrystallization (deformation) of carbonates.

(2) In epizonal metamorphism (low-grade, greenschist facies) (IC<$0.25°$ (2θ); R_{max}>6%), intense recrystallization obscures the original biogenic sedimentary features of the carbonates and CAI values of 6–7 are common.

(3) Local deviations from the general correlation indicate heterogeneous (independent) variations of the chemical and physical conditions.

(4) Conodont and carbonate textural alterations are of great value in establishing the boundary between diagenesis and the beginning of metamorphism but do not show a sharp difference between the anchizone and epizone.

13.1 INTRODUCTION

Palaeozoic–Mesozoic terrains in northern Hungary have been the subject of detailed investigations carried out within the framework of the Nationwide Type Section Programme and the remapping of the Aggtelek–Rudabánya Mountains, both sponsored by the Hungarian Geological Survey. As a byproduct of conodont biostratigraphical research and connected microfacies studies, metamorphic alteration of conodonts and limestones has been studied and compared with the results of metamorphic petrological investigations.

(a) Conodont alteration

It has been well known since the 1970s that conodonts are excellent indicators of organic metamorphism. The scale of conodont colour alteration index (CAI) was originally published by Epstein *et al.* (1977) and is based on the chemical transformation of the organic matter which was originally present in small quantities in the conodonts. The CAI scale has eight grades and was calibrated by heating experiments under atmospheric pressure. For CAI values of up to 5, the scale was correlated with vitrinite reflectance values and colour changes of palynomorphs. It was found that the colour corresponding to the value of CAI 5 forms above 300 °C (without any pressure changes). The temperature intervals for the values >5 have only recently been established (see Nolan and Barnes, this volume, Table 11.1).

Conodont colour alterations up to CAI 5, controlled by laboratory experiments, are already well known. They are valid for burial metamorphism and have become widely used to determine organic-matter maturity in the past few years. However, in Alpine-type folded orogenic belts, where fluid and oriented (stress) pressures play a very important role, these alterations are not as yet well understood. To our knowledge, conodonts of higher CAI value and their other alteration effects (recrystallization and formation of new minerals) have been reported only by Schönlaub and Zezula (1975), Schönlaub *et al.* (1976), Schönlaub (1979) and Schönlaub *et al.* (1980) from the Austrian Alps. Deformed and recrystallized conodonts have also been figured by Kuwano (1979) from the Mikabu greenstones, Japan, and by Kozur and Mock (1977) from the Bükk Mountains, northern Hungary, both of Triassic age. Mock (1980) also emphasized that conodonts are good indicators of incipient metamorphism.

(b) Metamorphic petrology

In metasedimentary sequences, the illite crystallinity (IC), the lattice parameter (b_0), the coal rank (vitrinite reflectance R) and the textural characteristics have been the main criteria for classifying the rocks into diagenetic zone, anchizone and epizone in the sense of Kübler (1968, 1975).

In the last 15 years, it has become obvious that there are no overall valid general quantitative relationships between illite crystallinity, vitrinite reflectance, mineral parageneses and textural characteristics (Wolf, 1975; Frey *et al.*, 1980; Árkai, 1983). This is why the grade determination of incipient metamorphic rocks should be based on the comparative evaluation of all the parameters mentioned above, comparing them with conodont and carbonate texture alterations.

13.2 GEOLOGICAL SETTING

The simplified geological map (Fig. 13.1) shows the surface outcrops of Palaeozoic and Mesozoic rocks in northern Hungary. They record, at least from what has been proved biostratigraphically, continuous marine sedimentation from the middle Devonian Period until the end of the Jurassic Period (cf. Kovács and Péró, 1983; Bérczi-Makk and Pelikán, 1984; Kozur, 1984). Silurian and Lower Devonian limestones are known only from olistoliths (Kovács and Péró, 1983). This area, forming geographically the innermost part of the west Carpathians, is characterized by complex Alpine-type nappe tectonics. The different nappes are of different metamorphic grades, ranging from diagenetic zone to epizone (greenschist facies, biotite zone) (Árkai, 1977, 1983)). There is no evidence of Variscan metamorphism; intra-Carboniferous tectogenesis seems to be excluded by the presence of continuous Upper Palaeozoic and Triassic marine sedimentation (Árkai, 1983; Kovács *et al.*, 1983). The area considered consists of a number of major tectonic units (Grill *et al.*, 1984, in preparation; Kovács and Péró, 1983; Pelikán, in press; the characterization of metamorphism after Árkai 1983 and Árkai and Kovács, in press). In the Aggtelek–Rudabánya Mountains (from north to south, Fig. 13.1) there are the following tectonic units:

(1) Silicicum. Non-metamorphosed Upper

Permian–Jurassic rocks (Silica nappe s. s.: Triassic rocks of mostly carbonate platform facies: Bódva nappe, Triassic rocks of

mostly red deep-water facies; some parts of the latter reached the anchizone).

(2) Meliaticum. Anchizonal to epizonal meta-

Fig. 13.1 — A simplified geological map of Palaeozoic–Mesozoic terrains in northern Hungary and adjacent Czechoslovakia: 1, Palaeozoic; 2, Triassic; 3, Jurassic; 4, Neogene plus Quaternary. Towns are in black.

morphosed oceanic Triassic and Jurassic rocks (glaucophanitic greenschist facies).

(3) Tornaicum. Anchizonal to epizonal metamorphosed Triassic rocks of gray basinal facies.

(4) Bükkium s. l. in the following areas:

(a) Szendrö Mountains. Epizonal metamorphosed Middle Devonian–Middle Carboniferous rocks (greenschist facies, chlorite (biotite) isogrades).

(b) Uppony Mountains. Transitional anchizonal–epizonal metamorphosed Silurian–Middle (and Upper?) Carboniferous rocks.

(c) Bükk Mountains where the following subdivision is recognized:

(i) North Bükk anticline. Non-metamorphosed to anchizonal metamorphosed Middle–Upper Carboniferous to Middle Triassic rocks.

(ii) Bükk Plateau unit. Anchizonal-–lower epizonal metamorphosed Triassic rocks.

(iii) Little Plateau unit. Non-metamorphosed Triassic rocks.

(iv) South Bükk unit. Non-metamorphosed to anchizonal metamorphosed Jurassic rocks.

(Note that, from the Bükk Mountains only the Bükk plateau unit, having conodont-bearing basinal facies, is considered in this chapter.)

13.3 METHODS

(a) Conodont alteration

During the conodont investigations of the northern Hungarian Palaeozoic and Triassic rocks the following empirical scale (see Kovács, in press; Kovács, 1983), was established (Table 13.1) (the correspondence of the colours of this scale in the range CAI<5 to the scale of Epstein *et al.* (1977) is only partial, i.e. the correlation is approximate only).

In all the investigated samples with CAI>5, not only did the colour of the conodonts change but also they were recrystallized and mostly deformed.

To check the degree of recrystallization, a series of SEM photographs was taken from conodonts of all CAI values and of all tectonic units at standard magnifications (1000× and 3000x). The pictures (Plates 13.3 and 13.4) were always taken from similar smooth parts of the specimens, namely from the anterior part of the blade of the Pa elements. To distinguish recrystallized surfaces from corroded surfaces, it is necessary to use a scanning electron microscope.

(b) Carbonate texture alteration

On the basis of studies of northern Hungarian Palaeozoic and Triassic limestones, three stages of textural alteration caused by oriented pressure (stress) were distinguished. The observed

Table 13.1 — Calcius for CAI 1–8.

CAI	Colour
1	Yellowish white, china white, glossy
1.5	Pale brown, translucent
2	Brownish grey, greyish brown, waxy
3	Pale grey, matt
4	Darker grey, matt, the tips of the teeth are white
5	Black, matt
6	Gray
7	Opaque white
8	Crystal clear, glassy

limestone textures, according to increasing deformation, are classified into Types A, B and C (Plates 13.1 and 13.3).

Type A. No changes are observed in the original texture, disregarding the occasional neomorphic recrystallization of the matrix.

Type B. Weak preferred orientation and incipient foliation in the matrix are developed. Allochemical components (bioclasts and intraclasts) are still recognizable. However, they are flattened into the plane of the foliation.

Type C. Allochemical components, with the exception of large echinoderm fragments, are indistinguishable from the recrystallized matrix. A homogeneous sparite texture with a preferred orientation is formed (metasparite).

The formation of textures with a preferred orientation (foliation) perpendicular to the stress direction was caused by the excellent translational properties, pressure-solution and reprecipitation of calcite, which results in the extension of its crystals in one direction. The deformation of conodonts is connected with these textural alterations.

Adding 'oriented' or 'unoriented' textures, Friedman's (1965) terminology of recrystallized carbonates can be extended to express the effect of dynamometamorphism.

(c) Illite crystallinity, the lattice parameter b_0 and vitrinite reflectance

In the northern Hungarian Palaeozoic–Mesozoic rocks, illite crystallinity was measured on X-ray diffractograms, using the widths 2θ in degrees of the first (10 Å) peak of illite–muscovite, at half-height. Measurements were carried out on random powder preparations of whole rock and acid-insoluble-residue samples as well as on size fractions with a grain size of less than 2 μm, according to the method of Kübler (1968, 1975). With increasing grade (temperature) of metamorphism the illite crystallinity increases because of aggradation of illite with decrease in H_2O and uptake of K^+ in the interlayer space. In order to compensate for instrumental effects, illite crystallinity standards 32, 34 and 35, kindly provided by Professor Kübler, were used for the linear calibration by the least-squares method.

In this way, Kübler's anchizone correlates roughly with the pumpellyite–prehnite–quartz facies (the temperature of the diagenetic zone–anchizone boundary is generally somewhat higher than the boundary between the laumontite–prehnite–quartz and pumpellyite––prehnite–quartz facies (see Kisch, 1974; Stadler, 1979).

However, the illite crystallinity boundaries of Kübler's anchizone, calibrated in this way, cannot be applied mechanistically to sequences differing strongly in lithology. As the effects of detrital white mica were detected even in the < 2 μm fractions, the actual illite crystallinity boundaries will vary depending on the ratio of inherited to newly formed muscovite and illite and also on the grain size and degradation state of detrital muscovite, etc. For example, in the 'less mature' northern Hungarian Palaeozoic––Mesozoic rocks, somewhat lower illite crystallinity boundaries correspond to the anchizone, correlated with the pumpellyite–prehnite––quartz facies in metabasites and with the vitrinite reflectance ranges $R_{max}=3.5$–6.0% and $R_{random}=3.0$–5.0% (Árkai, 1983). These R data are average values of the appropriate boundaries given by Weber (1972), Wolf (1972), Kisch (1974), Teichmüller et al. (1979) and Frey et al. (1980) and are somewhat higher than those proposed by Kübler (1975) and Kübler et al. (1979), i.e. $R_{random}=2.7$–4.1%.

The parameters $b_0 \approx 6 \times d(060, \overline{3}31)$ of illite––muscovite, which correlate with the $Al \leftarrow (Fe^{2+}, Mg^{2+})$ isomorphic substitution in the octahedral positions of the white mica lattice, were applied to the characterization of pressure conditions during metamorphism according to Sassi (1972), Guidotti and Sassi

(1976) as well as of Padan *et al.* (1982). Padan *et al.* (1976) extended Sassi's b_0 method to the high-temperature part of the anchizone. The b_0 boundaries of the low-to-intermediate and intermediate-to-high pressure-ranges are about 9.000 Å and 9.035 Å, respectively. The semi-quantitative pressure calibration (Table 13.2) was based on the statistical evaluation of b_0 parameters, as well as on the pressure-sensitive mineral parageneses of the intercalated metabasites (for details and exact descriptions of the methods see Árkai, 1983; Árkai and Kovács, in press).

The measurements of vitrinite reflectance R_{max}, R_{min} and R_{random} were carried out by Dr. Z. A. Horváth (Laboratory for Geochemical Research, Hungarian Academy of Sciences) by means of a Reichert Zetopan-Pol microscope, using the reflectance glass prism series of Bituminous Coal Research Inc. as reference standards, in oil immersion, at a wavelength of 548 nm.

13.4 COMPARISON OF CONODONT AND CARBONATE TEXTURE ALTERATIONS IN DIAGENETIC ZONE, ANCHIZONE AND EPIZONE WITH PETROGRAPHIC CHARACTERISTICS

Table 13.2 summarizes the data obtained from the northern Hungarian Palaeozoic and Triassic rocks, affected by various degree of Alpine-type regional dynamothermal metamorphism. Boundaries of diagenetic zone, anchizone and epizone are given by illite crystallinity (IC) and vitrinite reflectance (R) values, and conodont and carbonate texture alterations are incorporated in this scheme. They show the following zonal distribution.

(a) Diagenetic zone
(IC>0.37; R_{max}<3.5%)
All CAI values below 5 fall into this zone and partly also the value 5. No conodont recrystallization (not to be confused with corrosion!) or deformation occurs. The carbonate texture is always A type.

(b) Anchizone
(IC=0.37–0.25; R_{max}=3.5–6.0%)
In this zone a CAI value of 5 is most common but, depending on various local pressure and temperature conditions, CAI values of 6 and 7 may also occur. The conodonts are remarkably different from those of the diagenetic zone and are always recrystallized; also, depending on the local pressure conditions, they may be either non-deformed or deformed. The texture of the carbonate host-rocks is always oriented and may be either B or C type.

(c) Epizone (greenschist facies)
(IC 0.25; R_{max}>6.0%)
The CAI value is usually 6 or 7, but a CAI value of 5 may also occur. The conodonts are strongly recrystallized and more or less deformed. The carbonate texture is always C type and well oriented; the original allochemical components have completely vanished, i.e. these rocks should be classified as marbles.

(d) Mesozone (amphibolite facies)
Rocks of this metamorphic grade are not present in northern Hungary, and so they were not involved in our investigations. Conodonts of the highest CAI value (CAI 8) have been reported from metamorphosed rocks (amphibolite facies) from the middle Austrian alpine nappe system of the eastern Alps (Schönlaub, 1979; Epstein *et al.*, 1977).

Table 13.3 shows the conodont and carbonate texture alterations in the tectonic units of northern Hungary. Only conodont-bearing units (except the North Bükk anticline) are considered, (for their metamorphic grade and brief characterization, see above). The fluid pressure has been determined from the b_0 lattice parameters and pressure-sensitive mineral assemblages of metavolcanics, and the temperature from illite crystallinity and vitrinite reflectance values (cf. Árkai 1983; Árkai and Kovács, in press). In this table we demonstrate that conodont and limestone texture alterations were strongly influenced by the different pres-

Table 13.2 — Correlation of conodont and carbonate texture alterations with petrographic characteristics in the diagenetic zone, anchizone and epizone.

Zones in metasedimentary rocks	Mineral facies in metabasites	Illite crystallinity (2θ (deg))		Vitrinite reflectance (%)		CAI	Conodont Recrystallization	Carbonate texture Deformation
		Whole rock*	Fraction $<2\ \mu m$†	R_{max}	R_{random}			
Diagenetic zone	Diagenesis and zeolite (laumontite–prehnite–quartz) facies	>0.37	>0.34	<3.5	<3.0	1–5	—	A
Anchizone	Pumpellyite–prehnite-quartz facies	0.25–0.37	0.25–0.34	3.5–6.0	3.0–5.0	5–(6–7)	±	B–C
	Pumpellyite–actinolite facies						+	
Epizone	Greenschist facies	<0.25	<0.25	>6.0	>5.0	(5–)6–7	+	C

*Random powder preparations.
†Sedimented (highly oriented) preparations.

Table 13.3 — Conodont and carbonate texture alterations in the tectonic units of northern Hungary, together with the average pressure and temperature indices of regional dynamothermal metamorphism.

Tectonic unit	Age	Pressure P (kbar)	Temperature T (°C)	CAI	Conodont Recrystallization	Conodont Deformation	Carbonate texture
Szendrö Mountains	Devonian–Carboniferous	≈ 3	≈ 400	(5)6–7	+	+	C
Uppony Mountains	Silurian–Carboniferous	≈ 2.5	≈ 350	5	+	–	B–C
Bükk Mountains (only Bükk plateau nappe)	Triassic	≈ 3	≈ 350	5–7	+	±	B–C
Aggtelek–Rudabánya Mountains, Tornaicum	Triassic	≈ 7	≈ 350	6–7	+	+	B–C
Aggtelek–Rudabánya Mountains, Silicicum	Triassic	Indeterminable (diagenetic)	<200	1–5	–	–	A

sure and temperature conditions in the different tectonic units.

13.5 DISCUSSION OF CONODONT ALTERATIONS

In the diagenetic zone, where alterations are determined largely by temperature changes and fluid pressure does not play an important role (apart from compaction), only the colour of the conodonts changes. However, it is presumably also influenced by the colour of the host-rock (Schönlaub et al., 1980).

During anchizonal and epizonal regional dynamothermal metamorphism, alterations of conodonts are connected to the textural and mineral paragenesis alteration of the host-rocks. Here, not only does their colour change (CAI>5) but also they become recrystallized and often deformed, in parallel with the formation of an oriented texture of the enclosing limestones.

Deformation of conodonts is especially common in medium- and high-pressure metamorphosed rocks (see the tectonic unit 'Tornaicum', in Table 13.3). It is linked to the formation of rock cleavage in the host-rock and is promoted by the good translational ability of the enclosing calcite (see Plate 13.5, Fig. 2). Conodonts are not always as ductile as it seems at first glance under the normal optical microscope; their deformation occurs frequently as a sum of minor displacements along cracks, as can be seen from the scanning electron micrographs (Plate 13.5, Figs. 4, and 9).

Recrystallization of conodonts shows a sharp difference with respect to the diagenetic zone. However, we could not observe a large difference between the anchizone and epizone. It is presumably also explained by the different pressure and temperature conditions. Together with recrystallization and cracking of conodonts, new minerals may also form on their surfaces (Schönlaub and Zezula, 1975; Flajs and Schönlaub, 1976; Schönlaub et al., 1980; Lelkes-Felvári et al., 1984).

Epstein et al. (1977) noted that the transition from CAI 5 to CAI 7 may have been very short-lived, because they were reported even from the same bed. According to our observations, however, this transition is rather long-lived because these values, depending on the local pressure and temperature conditions, occur throughout the anchizone and epizone. Furthermore, black, gray and white colours may occur not only just in the same bed but also often within the same specimen; the recrystallized conodont material loosened along minor cracks looks paler than it really is.

A similar phenomenon indicating the local (microscopic-scale) variations in pressure field (mainly due to stress) was found by Árkai et al. (1981) describing the quasi-'coexistence' of anthracite and graphite. In slates and metacherts of the Uppony Palaeozoic, as anthracite with $R_{random}=4.43–6.29\%$ is frequent in the intercleavage (massive) parts of the rocks, it was transformed into graphite along the cleavage planes because of the local increase in shearing stresses.

13.6 CONCLUSIONS

The conclusion from our observations is that conodont and carbonate texture alterations are useful tools for recognizing the boundary between diagenesis and incipient metamorphism. However, the boundary between anchizonal and epizonal metamorphism cannot be unambiguously recognized simply by these methods, because these alterations are strongly influenced by local pressure and temperature conditions.

Apart from their significance in metamorphic studies, conodonts are of great stratigraphic usefulness in marbles having a C-type texture; they are generally the only microfossils preserved in them, because all calcareous microfossils determinable in thin section have disappeared as a result of complete recrystallization.

PLATE 13.1

These illustrations show A- and B-type limestone textures.

Plate 13.1, Fig. 1. Non-metamorphosed Triassic carbonate platform facies, with an A-type texture; algal (dasycladacean) bioorthosparite, grainstone, ×10. Wetterstein Limestone, Silica nappe, locality Alsóhegy, sample 6-1972.

Plate 13.1, Fig. 2. Non-metamorphosed Triassic pelagic basinal facies, with an A-type texture; radiolarian filamentous biomicrite, wackestone, ×33. Bódvalenke Limestone, Bódva nappe, locality Bódvalenke, sample 2.

Plate 13.1, Fig. 3. Weakly metamorphosed Triassic carbonate platform facies with a B-type texture; biointraorthosparite, grainstone; intraclasts (coated grains) show a preferred orientation and are flattened into the plane of foliation, ×10. Berva Limestone, Bükk Mountains, locality Felnémet (Berva Quarry), sample F5 (From F. Veledits.)

Plate 13.1, Fig. 4. Weakly metamorphosed Triassic pelagic basinal facies with B-type texture; radiolarian biomicrite, wackestone; the rock shows a preferred orientation and calcified radiolarian tests are flattened into the plane of foliation, ×33. Pötschen Limestone, Torna nappe, locality Nagykö at Hidvegardo, sample H-26.

Pl. 13.1]   **Conodont alteration in limestones from northern Hungary**   219

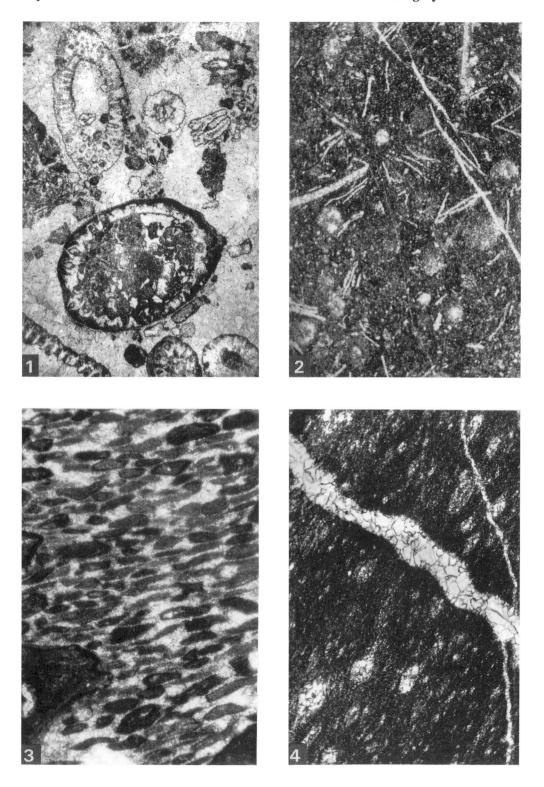

PLATE 13.2

These illustrations show C-type limestone textures (marbles).
The original allochemical components (bioclasts, intraclasts) have completely disappeared as a result of metamorphic recrystallization. The textural type (which is classified according to Friedman (1965) with the exception of specifying sparite as 'metasparite' to indicate its metamorphic origin) in all figures is non-equigranular xenotopic porphyrotopic metasparite with a preferred orientation.

Plate 13.2, Fig. 1. Metamorphosed Triassic carbonate platform facies, orientation from lower left to upper right, ×33. Metamorphosed Steinalm Limestone, Torna nappe, locality Esztramos Hill, sample Esz1.

Plate 13.2, Fig. 2. Metamorphosed Triassic pelagic basinal facies, orientation from lower left to upper right, ×33. Unnamed Ladinian pelagic limestone from the same locality as Fig. 1, sample Esz23/a.

Plate 13.2, Fig. 3. Metamorphosed, strongly recrystallized Palaeozoic carbonate platform facies, orientation from lower right to upper left, ×33. Carboniferous Upper Rakaca Marble Formation, Szendrö Mountains, locality Rakacaszend, sample Szrö8.

Plate 13.2, Fig. 4. Metamorphosed Palaeozoic basinal facies, orientation from lower right to upper left, ×33. Verebeshegy Limestone Member of the Carboniferous Upper Rakaca Marble Formation, Szendrö Mountains, locality Rakaca, sample Szrö2.

Pl. 13.2]   **Conodont alteration in limestones from northern Hungary**   221

PLATE 13.3

These illustrations show the surfaces of Triassic conodonts with different CAI values.
The specimens will be deposited at the Museum of the Hungarian Geological Survey (at present they are stored at the North Hungarian Department of the Survey).

Gondolella sp. aff. *G. bulgarica* (Budurov and Stefanov)
Plate 13.3, Figs. 1, 2. CAI 1.5, the surface of the Pa element is completely smooth, ×750, (Fig. 1),×2250 (Fig. 2). Silica nappe, diagenetic zone, locality Baradla Cave, sample of red brachiopod limestone, specimen 1.

Gondolella foliata inclinata Kovacs
Plate 13.3, Figs. 3, 4. CAI 4.5, the striations on the Pa element are partly original and partly due to some corrosion,×750 (Fig. 3),×2250 (Fig. 4). Szölösardö unit (frontal slice of Silica nappe), diagenetic zone, borehole Szölösardö 1, 365.03–365. 20 m, sample of Nadaska Limestone.

Gondolella excelsa (Mosher)
Plate 13.3, Figs. 5, 6. CAI 6–7, the surface of the Pa element is completely and strongly recrystallized, ×750 (Fig. 5),×2250 (Fig. 6). Torna nappe, lower part of epizone, locality Esztramos Hill, sample Esz18, specimen 2.

Pl. 13.3]   **Conodont alteration in limestones from northern Hungary**   223

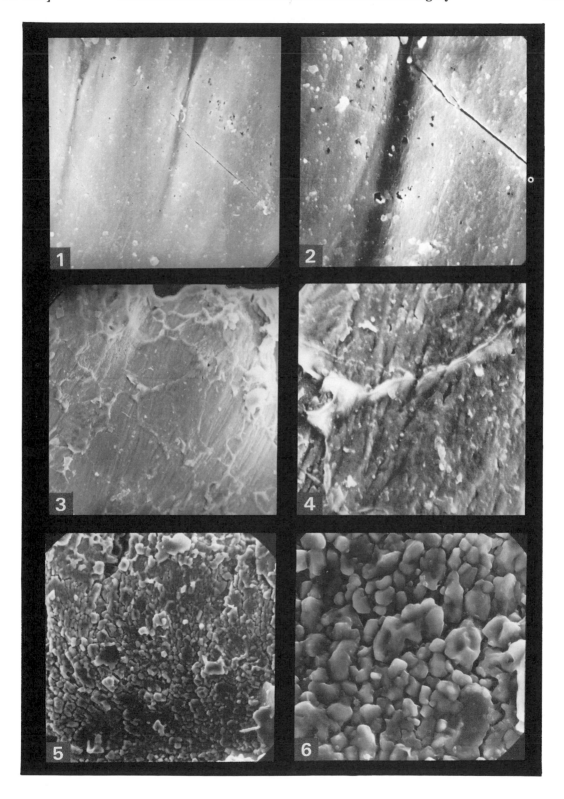

PLATE 13.4

These illustrations show the surfaces of metamorphosed Palaeozoic conodonts (Figs. 1–4) and the formation of new minerals during metamorphism (Figs. 5 and 6).
The specimens will be deposited at the Museum of the Hungarian Geological Survey (at present they are stored at the North Hungarian Department of the Survey).

Palmatolepis gracilis gracilis Branson and Mehl.
Plate 13.4, Figs. 1, 2. CAI 5, Pa element, ×750 (Fig. 1), ×2250 (Fig. 2). Upper part of anchizone, locality Uppony Mountains, sample U16/a.

Gnathodus bilineatus bilineatus Roundy
Plate 13.4, Figs. 3, 4. CAI 6–7, Pa element, ×750 (Fig. 3), ×2250 (Fig. 4). Lower part of epizone, locality Szendrö Mountains, sample Szrö94.

Icriodus symmetricus Branson and Mehl
Plate 13.4, Figs. 5, 6. CAI 6–7, authigenic quartz crystal formed on the posterior extremity of an I element, ×60 (Fig. 5), ×150 (Fig. 6). Locality Szendrö mountains, sample Szrö49.

Pl. 13.4]   **Conodont alteration in limestones from northern Hungary**   225

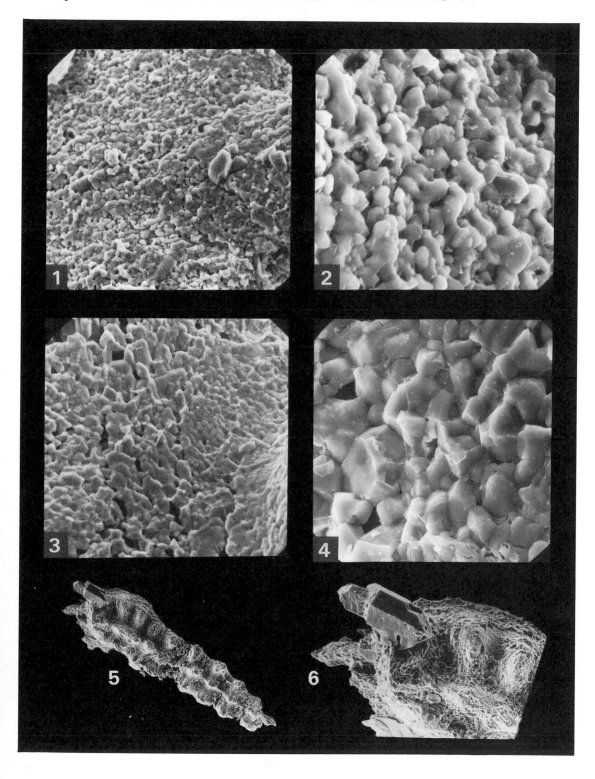

PLATE 13.5

These illustrations are of deformed conodonts.
The specimens will be deposited at the Museum of the Hungarian Geological Survey (at present they are stored at the North Hungarian Department of the Survey).

Gondolella navicula Huckriede
Plate 13.5, Figs. 1a, 1b. Fig. 1a, upper view, Pa element 'sandwiched' along parallel shear planes, ×440. (Fig. 1b), lower view, Pa element 'sandwiched' along parallel shear planes, ×44. Torna nappe, locality Hidvégardó Ruda-oldal, sample H7.

Gondolella sp
Plate 13.5, Fig. 2. Fragment of the posterior portion of Pa element showing step-like shift along parallel shear planes following the cleavage of the host rock, ×66. Torna nappe, locality Esztramos Hill, sample Esz19.

Gondolella navicula Huckriede
Plate 13.5, Figs. 3a, 3b. Pa element bent in a crook-like form with minor subparallel fractures, ×62 (Fig. 3a), ×66 (Fig. 3b). Torna nappe, locality Hidvégardó Ruda-oldal, sample H10.

Gondolella constricta Mosher and Clark
Plate 13.5, Fig. 4, Pa element, ×52. Torna nappe, locality Esztramos Hill, sample Esz18 .

Gondolella navicula Huckriede
Plate 13.5, Figs. 5a, 5b. Lower lateral views, Pa element, ×66. Torna nappe, locality Hidvégardó Ruda-oldal, sample H10.

Paragnathodus nodosus (Bischoff)
Plate 13.5, Fig. 6. Upper view, Pa element, ×52. Locality Szendrö Mountains, Rakacaszend, sample Szrö38.

Lochriea commutata (Branson and Mehl)
Plate 13.5, Fig. 7. Upper view, Pa elements, ×66. Locality Szendrö Mountains, Rakaca, sample Szrö1.

Idiognathoides noduliferus noduliferus Ellison and Graves
Plate 13.5, Fig. 8. Upper view, Pa element with the free blade bent in a semicircular arc, ×66. Locality Szendrö Mountains, Rakacaszand, sample Szrö24.

Idiognathiodes noduliferus noduliferus Ellison and Graves
Plate 13.5, Figs. 9a, 9b. Fig. 9a, upper view, Pa element, ×44. Fig. 9b, lateral view, Pa element, ×44. Locality Szendrö Mountains, Rakacasz and, sample Szrö24.

Ancyrodella sp
Plate 13.5, Fig. 10, Upper view, Pa element, ×36. Locality Szendrö Mountains, Szendrö, Nagy-Somos-tetö, sample Szrö88.

Pl. 13.5]　　Conodont alteration in limestones from northern Hungary　　227

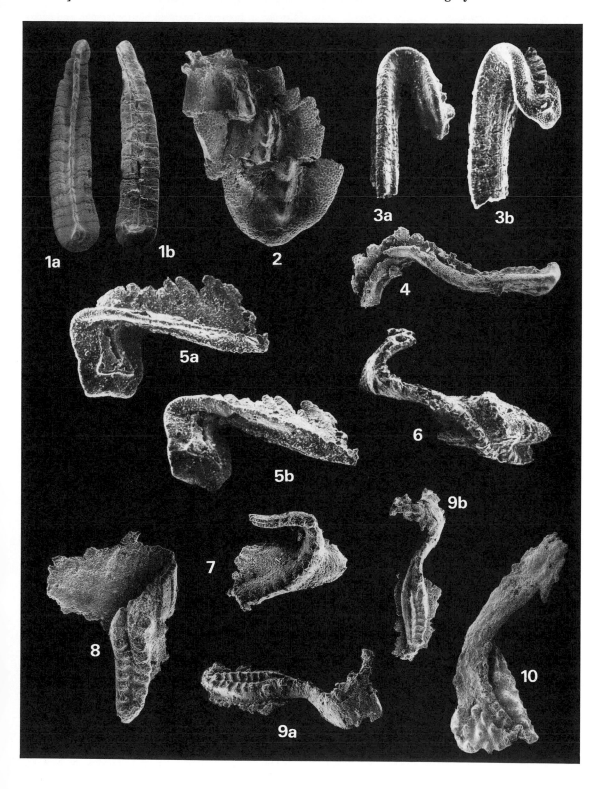

13.7 REFERENCES

Árkai, P. 1977. Low-grade metamorphism of Palaeozoic sedimentary formations of the Szendrö Mountains (NE Hungary). *Acta Geologica Academiae Scientiarum Hungaricae*, **21** (1–3), 53–80.

Árkai, P. 1983. Very low- and low-grade Alpine regional metamorphism of the Paleozoic and Mesozoic formations of the Bükkium, NE Hungary. *Acta Geologica Hungarica*, **26** (1–2), 83–101.

Árkai, P., Horváth, Z. A. and Tóth, M. 1981. Transitional very low- and low-grade regional metamorphism of the Palaeozoic formations, Uppony Mountains, NE Hungary: mineral assemblages, illite crystallinity, b_0 and vitrinite reflectance data. *Acta Geologica Academiae Scientarum Hungaricae*, **24** (2–4), 265–294.

Árkai, P. and Kovács, S. In press. Diagenesis and regional metamorphism of the Mesozoic Aggtelek–Rudabánya Mts. (NE Hungary). *Acta Geologica Hungarica*.

Bérczi-Makk, A. and Pelikán, P. 1984. Jurassic formations from the Bükk Mountains (in Hungarian with English summary). *A Magyar Állami Földtani Intézet Évi Jelentése az 1982 (Annual Report of the Hungarian Geological Institute from 1982)*, 137–166.

Epstein, A. G., Epstein, J. B. and Harris, L. D. 1977. Conodont color alteration—an index to organic metamorphism. *U.S. Geological Survey Professional Paper*, **995**, 1–27.

Flajs, G. and Schönlaub, H. P. 1976. Die biostratigraphische Gliederung des Altpaläozoikums am Polster bei Eisenerz (Nördliche Grauwackenzone, Österreich). *Verhandlungen der Geologischen Bundesanstalt*, **1976** (2) 257–303, 4 plates.

Frey, M., Teichmüller, M., Teichmüller, R., Mullis, J., Künzi, B., Breitschmid, A., Gruner, U. and Schwizer, B. 1980. Very low-grade metamorphism in external parts of the Central Alps: illite crystallinity, coal rank and fluid inclusion data. *Eclogae Geologicae Helvetiae*, **73**, 173–203.

Friedman, G. M. 1965. Terminology of crystallization textures and fabrics in sedimentary rocks. *Journal of Sedimentary Petrology*, **35** (3) 643–655.

Grill, J., Kovács, S., Less, G. Y., Piros, O., Réti, Z. S., Róth, L., Szentpétery, I., Borka, Z. S., Gyuricza, G. Y. and Sádi, L. In preparation. Geological monograph of the Aggtelek–Rudabánya Mts.

Grill, J. Kovács, S., Less, G. Y., Réti, Z. S., Róth, L. and Szentpétery, I. 1984. Geology and evolution of the Aggtelek–Rudabánya Mts. (in Hungarian). *Földtani Kutatás*, **24** (4) 49–56.

Guidotti, C. V. and Sassi, F. P. 1976. Muscovite as a petrogenetic indicator mineral in pelitic schists. *Neues Jahrbuch für Mineralogie, Abhandlungen*, **127**, 97–142.

Kisch, H. J. 1974. Anthracite and meta-anthracite coal ranks associated with 'anchimetamorphism' and 'very-low-stage' metamorphism, I–III. *Koninklijke Nederlandse Akademie van Wetenschappen, Proceedings, Series B*, **77** (2) 81–118.

Kovács, S. 1983. Results of conodont investigations in Hungary until 1981 (in Hungarian, with English summary). *Discussiones Palaeontologicae*, **30**, 73–111.

Kovács, S. In press. Conodont-biostratigraphic and microfacies investigations in the northeastern part of the Rudabánya Mts. *A Magyar Állami Földtani Intézet Évi Jelentése az 1984 (Annual Report of the Hungarian Geological Institute from 1984)*.

Kovács, S., Kozur, H. and Mock, R. 1983. Relations between the Szendrö-Uppony and Bükk Palaeozoic in the light of the latest micropalaeontological investigations. *A Magyar Állami Földtani Intézet Évi Jelentése az 1981 (Annual Report of the Hungarian Geological Institute from 1981)*, 155–175.

Kovács, S. and Péró, Cs. 1983. Report on Stratigraphical Investigation in the Bükkium (northern Hungary). In F. P. Sassi and T. Szederkényi (Eds.), *IGCP Project 5 Newsletter*, **5**, 58–65.

Kozur, H. 1984. New Radiolarian taxa from the Triassic and Jurassic. *Geologische und Paläontologische Mitteilungen der Universität Innsbruck*, **13** (2), 49–88.

Kozur, H. and Mock, R. 1977. Conodonts and holothurian sclerites from the Upper Permian and Triassic of the Bükk Mts. (north Hungary). *Acta Universitatis Szegediensis, Acta Mineralogica et Petrographica*, **23**, 109–126, 3 plates.

Kübler, B. 1968. Evaluation quantitative du métamorphisme par la cristallinité de l'illite. *Bulletin Centre Recherche, Pau, SNPA*, **2** (2), 385–397.

Kübler, B. 1975 Diagenese-anchimétamorphisme et métamorphisme. *Publication de l' Institut National de la Recherche Scientifique, Pétrole, Quebec*.

Kübler, B., Pittion, J.-L., Héroux, Y., Charollais, J. and Weidmann, M. 1979. Sur le pouvoir reflecteur de la vitrinite dans quelques roches du Jura, de la molasse et des Nappes préalpines, helvetiques et penniques. *Eclogae Geologicae Helvetiae*, **72** (2), 343–373.

Kuwano, Y. 1979. Triassic Conodonts from the Mikabu Greenrocks in Central Shikoku. *Bulletin of the National Science Museum, Series C (Geology and Paleontology)*, **5** (1), 9–24, 4 plates.

Lelkes-Felvári, Gy., Kovács, S. and Majoros, Gy. 1984. Lower Devonian pelagic limestone in borehole Kékkút 4, Bakony Mts. *A Magyar Állami Földtani Intézet Évi Jelentése az 1982 (Annual Report of the Hungarian Geological Institute from 1982)*, 289–315.

Mock, R. 1980 Novel knowledge and some problems as regards the geology of the inner west Carpathians (in Hungarian with English abstract). *Földtani Kutatás*, **23** (3), 11–15.

Padan, A., Kisch, H. J., Shagam, R. 1982. Use of lattice parameter b_0 of the dioctahedral illite/muscovite for the characterization of P/T gradients of incipient metamorphism. *Contributions to Mineralogy and Petrology*, **79**, 85–95.

Pelikán, P. In press. Structural outline of the Bükk Mts. (in Hungarian). *Általános Földtani Szemle, Budapest*.

Sassi, F. P. 1972 The petrologic and geologic significance of the b_0 value of potassic white micas in low-grade metamorphic rocks. An application to the eastern Alps. *Tschermaks Mineralogische und Petrographische Mitteilungen*, **18**, 105–113.

Schönlaub, H. P. 1975. Das Paläozoikum im Österreich. *Abhandlungen der Geologischen Bundesanstalt*, **33**, 124, 6 plates.

Schönlaub, H. P., Exner, Ch. and Nowotny, A. 1976. Das

Altpaläozoikum des Katschberges und seiner Umgebung (Österreich). *Verhandlungen der Geologischen Bundesanstalt,* (2), 115–145, 3 plates.

Schönlaub, H. P., Flajs, G. and Thallmann, F. 1980 Conodontenstratigraphie am Steirischen Erzberg (Nördliche Grauwackenzone). *Jahrbuch der Geologischen Bundesanstalt,* **123** (1), 169–229, 7 plates.

Schönlaub, H. P. and Zezula, G. 1975. Silur-Conodonten aus einer Phyllonitzone im Muralpen-Kristallin (Lungau/Salzburg). *Verhandlugen der Geologischen Bundesanstalt,* 253–269, 3 plates.

Stalder, P. J. 1979. Organic and inorganic metamorphism in the Taveyannaz sandstone of the Swiss Alps and equivalent sandstones in France and Italy. *Journal of Sedimentary Petrology,* **49**, 463–482.

Teichmüller, M., Teichmüller, R. and Weber, K. 1979. Inkohlung und Kristallinität vergleichende Untersuchungen im Mesozoikum und Palaozoikum von Westfalen. *Fortschritte in der Geologie von Rheinland und Westfalen,* **27**, 201–276.

Weber, K. 1972. Kristallinität des Illits in Tonschiefern und andere Kriterien schwacher Metamorphose im nordöstlichen Rheinischen Schiefergebirge. *Neues Jahrbuch für Geologie und Paläontologie, Abhandlungen,* **141**, 333–363.

Wolf, M. 1972. Beziehungen zwischen Inkohlung und Geotektonik im nördlichen Rheinischen Schiefergebirge. *Neues Jahrbuch für Geologie und Paläontologie, Abhandlungen,* 141, 222–257.

Wolf, M. 1975. Über die Beziehungen zwischen Illit-Kristallinität und Inkohlung. *Neues Jahrbuch für Geologie und Paläontologie, Monatshefte,* **1**, 1–64.

14

The electron spin resonance technique in conodont studies

Z. Belka, K. Miaskiewicz and T. Zydorowicz

Conodonts from Ordovician to Triassic age deposits have been investigated by electron spin resonance. This technique permits detection of the type and relative number of paramagnetic centres having different absorption signals. Conodonts exhibit signals induced by natural background radiation. The intensity of the signals is dependent upon the radiation dose received by the sample as well as upon the temperature.

Electron spin resonance is not recommended for dating the age of conodonts but can be used for determining the age of completion of the latest heating event (i.e. most frequently the time of the last upheaval of the host-rocks). Because there is a correlation between the conodont colour alteration index (CAI) and the electron spin resonance spectral structure, it seems that this method may be applied to testing CAI values.

14.1 INTRODUCTION

Electron spin resonance (ESR) (or electron paramagnetic resonance) is a branch of spectroscopy in which microwave-frequency radiation is absorbed by molecules, ions or atoms possessing unpaired electrons. This technique enables detection of the presence and relative abundance of various paramagnetic centres. These centres include radicals, vacant lattice points and transition metals.

ESR spectroscopy utilizes the different energy states which arise from the interaction of the magnetic moment of an unpaired electron with a static magnetic field, and the transition between them upon the application of microwave radiation.

The basic equation is

$$h\nu = g\mu_B B$$

where ν is the fixed frequency of the microwave radiation, B is the strength of the static magnetic field at resonance, μ_B is the electron Bohr magneton, and g is the Landé factor, which is a unique property of the electron as a whole. If the electron spin is the only source of the magnetism, then $g = 2.0023$. In atoms or ions an orbital angular momentum may be also present, in which case g is different from the free spin value. The g factor is measured from the spectra by the use of standards (commonly diphenylpicrylhydrazyl, with $g = 2.0036$) and the relationship

$$g = g_{standard} \frac{B_{standard}}{B}$$

When nuclei with spin value *I* not equal to 0 are present in paramagnetic molecules, the electron magnetic moment can interact with nuclear magnetic moments. This interaction splits the ESR signal on some lines (so-called hyperfine splitting). The magnitude of the splitting is described by the hyperfine splitting constant *A* which is measured directly from the spectra by the separation of the component peaks in terms of magnetic field units (millitesla (mT) or gauss (G)). The ESR spectra are commonly presented as derivatives, e.g. the first derivative of the absorption curve is plotted against the magnetic field strength *B* (Fig. 14.1).

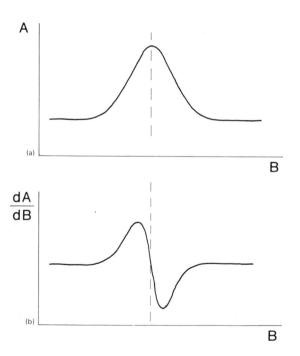

Fig. 14.1 — The shape of a typical ESR spectrum presented as (a) absorption curve and (b) derivative curve.

The *g* factor and hyperfine coupling constant *A* are generally anisotropic, with three principal values along three orthogonal axes (i.e. they are second-rank tensors). In gaseous and liquid samples the anisotropy is averaged, but in solid samples the anisotropy significantly alters the shape of the ESR spectrum.

Comprehensive reviews of ESR spectroscopy have been published by Atherton (1973), Drago (1977) and Symons (1978). ESR spectroscopy has occasionally been applied to geological materials. Ikeya and Miki (1980) reported that only a few geological materials are suitable for dating by ESR, provided that the materials possess defects and traps of high thermal stability, e.g. apatites. ESR spectroscopy has been used to date stalactites (Ikeya, 1975), animal and human bones (Ikeya and Miki, 1980), and flints (Robins *et al.*, 1978; Garrison *et al.*, 1981). As a dating technique, it has been applied to conodonts by Morency *et al.* (1970). In addition, ESR spectroscopy has been used to derive information on the formation of manganese nodules (Wakeham and Carpenter, 1974) and the distribution of Mn^{2+} in carbonates (Schindler and Ghose, 1970; Wildeman, 1970; Low and Zeira, 1972). Cubbitt and Wilkinson (1976) used ESR spectroscopy to classify Kansas shales, and Gilinskaya *et al.* (1982) have classified Siberian apatites. The potential of ESR spectroscopy in studies of clay minerals has been discussed by Friedlander *et al.* (1963).

Conodonts consist of carbonate apatite, approximately of the composition of the mineral francolite (Pietzner *et al.*, 1968). In addition to the mineral phase, conodonts contain an organic soft tissue characterized by cells, fibrous material and lumina (cf. Fåhraeus and Fåhraeus-van Ree, 1985). It is assumed that paramagnetic centres in conodonts can be related to both the inorganic and the organic materials. When the paramagnetic centres consist of organic free radicals and/or electron traps, the relative number of the centres is dependent upon external factors, such as the total radiation or temperature.

The purpose of this study is to determine the dependence of the presence and intensity of conodont ESR signals upon sample age in addition to temperature and the micromorphological composition. Since heat alters conodont colour (Epstein *et al.*, 1977) and other investigations (cf. Hutton and Troup, 1966) have shown that the colour of many materials is related to paramagnetic centres, we have also compared the conodont ESR spectra with their colour alteration index (CAI) values.

The study was performed on 17 conodont samples (see Table 14.1) which range in age from Ordovician to Triassic and were obtained from various localities in Poland, North America, Germany and Svalbard. The results presented are preliminary and indicate only qualitative analyses of the spectra.

14.2 EXPERIMENTAL CONDITIONS

The conodont preparation was identical with that for stratigraphic work. The minimum recommended weight of the conodont sample is 5 mg, i.e. about 300 conodont elements of various growth stages.

The ESR spectra were recorded on an SE/X 2542 spectrometer (Radiopan, Poland), using the X band with 100 kHz modulation. All spectra were recorded using the same modulation amplitude (0.1 mT) and microwave energy. This procedure does not damage the conodont material.

14.3 SPECTRAL STRUCTURE

Four distinct signals can usually be observed in the ESR spectra of conodonts (Fig. 14.2). The signals have been labelled A, B, C and D in order of descending *g* factor values (Table 14.2). In addition, some spectra indicate hyperfine splitting, with a peak split into four component peaks (indicated by asterisks (*) in Fig.

Table 14.1 — Conodont samples used in this study.

Sample	Locality	Age	Conodont CAI
1	Erratic boulder from Poland	Arenig	1
2	Mojcza, Holy Cross Mountains, Poland	Llanvirn	1
3	Borehole Solarnia IG-1, depth of 845 m, Silesia, Poland	Givetian	4
4	Amsdell Creek, New York, USA	Frasnian	3
5	Borehole ZMZ-23, depth of 281 m, Cracow Upland, Poland	Frasnian	3.5
6	Harz Mountains, Germany	Famennian	4
7	Plucki, Holy Cross Mountains, Poland	Frasnian	2
8	Dule, Holy Cross Mountains, Poland	Famennian	2
9	Borehole BK-318, depth of 662 m, Cracow Upland, Poland	Famennian	2.5
10	Borehole B-1, depth of 940 m, Cracow Upland, Poland	Tournaisian	2
11	Galezice, Holy Cross Mountains, Poland	Viséan	2
12	Lower Silesia, Poland	Ladinian	1
13	Strzelce Opolskie, Lower Silesia, Poland	Anisian	1
14	Strzelce Opolskie, Lower Silesia, Poland	Anisian	1
15	Strzelce Opolskie, Lower Silesia, Poland	Anisian	1
16	Gasiorowice, Lower Silesia, Poland	Anisian	1
17	Botneheia, Isfjorden, Svalbard	Ladinian	2

14.2). The splitting is characterized by a hyperfine splitting constant A of 2.31 mT, and an intensity ratio of 1:3:3:1.

The structure of signals B, C and D varied between samples, whereas signal A, if present, was always identical in shape and showed a very marked g factor anisotropy.

14.4 SOURCES OF SIGNALS

Our preliminary investigations have not allowed identification of the specific paramagnetic centres within the conodonts. The data can only indicate possible signal sources, and therefore the signal source remains an open question.

Peak asymmetry, as shown by signal A, is related to the mineral content of the conodont.

Signals with g factors very similar to those of peaks B and C were noted in the spectra obtained from the apatites from various metamorphic rocks of Siberia (Gilinskaya et al., 1982). The signals were reported to have been produced by the radicals SO_3^- and PO_3^{2-}. Therefore, signal B is probably derived from the SO_3^- radical, the signal of which has a congruent g factor. PO_3^{2-} radicals originate from damage of the apatite structure when uranium ions replace calcium ions (Gilinskaya and Shcherbakova, 1975; Gilinskaya et al., 1982), the concentration of the radicals being

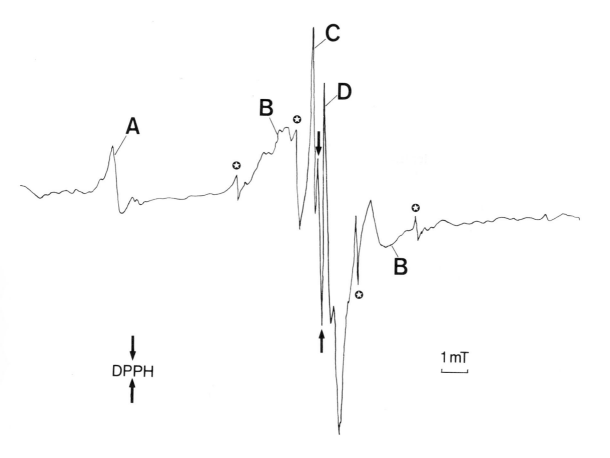

Fig. 14.2 — ESR spectrum obtained for conodonts of the Amsdell Creek locality (sample 4), to show the recognized signals A, B, C and D present in the ESR spectra of conodonts: *, an additional signal split on four component peaks; DPPH, diphenylpicrylhydrazyl.

Table 14.2 — g Factor values of signals present in conodont ESR spectra.

Signal	g factor
A	2.051 ± 0.001
B	2.004 ± 0.001
C	2.000 ± 0.001
D	1.999 ± 0.001

proportional to the uranium content. It is unlikely that signal C could originate from PO_3^{2-} radicals, because uranium was not detected in the conodont crowns, and the presence of uranium in the conodont bases is doubtful (Pietzner *et al.*, 1968).

We attribute signal D, the only centre which is stable to 1000°C (cf. table 14.3), to the presence of organic matter.

A signal split into four component peaks (see Fig. 14.2) indicated the presence of three equivalent nuclei with a spin value I of $\frac{1}{2}$, the nucleus ^{19}F being a probable source.

14.5 SIGNAL INTENSITY AS A FUNCTION OF TIME AND TEMPERATURE

Geological dating by ESR spectroscopy is based on the general principle that the signals emitted by certain geological materials (bones, flints, apatites, etc.) are proportional to the total radiation received. By assuming a constant radiation dose, the signals are therefore proportional to the duration of radiation.

Conodonts contain very low concentrations of radioactive elements. Their ESR spectra were therefore induced by natural background radiation, i.e. cosmic rays and radiation from radioactive elements present in the host-rocks. The ESR spectra obtained from the conodonts investigated do not indicate such a simple dependence. It appears that the spectral structure is influenced by both time and temperature

and that the observed signal intensity is dependent upon both the radiation dose received by the sample and the temperature.

(a) Influence of time
Although a dependence of the signal intensities on the sample age was not observed, the spectral structure enables a subdivision into three groups to be made. Spectra obtained from Triassic conodonts are distinct in that signal A is not exhibited. Only signal B is present, with signals C and D being either very weak or absent (Fig. 14.3). The second category of spectra consists of Devonian and Carboniferous conodonts. This category is characterized by the presence of peak B, and also signals A, C and D which are usually strong although, in some samples, signals A or C may be absent. The absence of the signals is more a function of heating than time.

The third group consists of the spectra of Ordovician conodonts in which signals C and D are very weak relative to the strong signals of peaks A and B (see Figs. 14.3 and 14.4).

(b) Temperature influence
Two samples, samples 1 and 4, were heated in open air from 100–1000°C for a period of 0.5–96 h. The resultant thermal annealing caused changes in the intensity of particular signals, with eventual disappearance of signals except peak D, the intensity of which had almost doubled by a temperature of 1000°C (cf. Table 14.3 and Fig. 14.4). Because thermal annealing generates the decay of signals A, B and C, the intensity of these signals in altered conodonts with an advanced thermal history will be proportional to the time of the last geological upheaval and not to the total age of the sample. For such conodonts, the intensity of signal D will always be greater than that produced by the background radiation during the total geological age of the sample.

Signal A, absent in the spectra obtained from Triassic conodonts (even those which are unaltered), occurs in the spectra of Palaeozoic samples (cf. Figs. 14.3 and 14.5). It appears that radiation persisting for longer than 225 ma was necessary to produce the paramagnetic centres responsible for signal A. Therefore, all Palaeo-zoic conodonts which do not emit peak A must have been heated at a time after the Carboniferous Period.

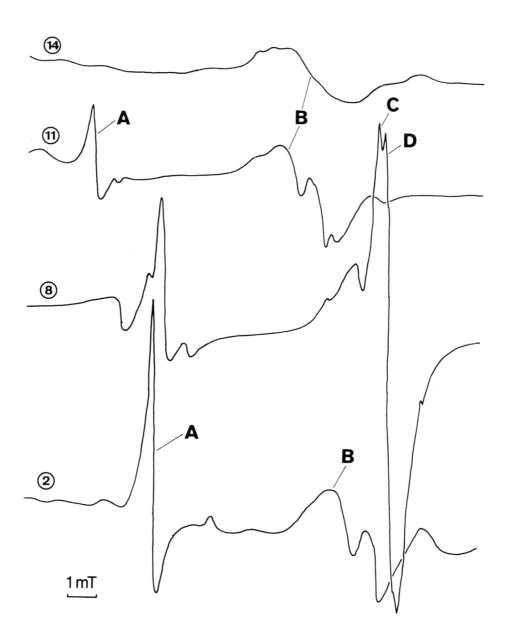

Fig. 14.3 — ESR spectra for conodonts from Triassic, Carboniferous, Devonian and Ordovician deposits, showing a dependence of the spectrum structure upon the age of sample. The circled numbers denote the investigated conodont samples (see Table 14.1).

Table 14.3 — Changes in the intensity of ESR signals present in the spectra of conodonts, resulting from the influence of temperature (data based on the experimental heating of two conodont samples (samples 1 and 4) to 100 °C,150 °C and 200 °C for 8 h, to 500 °C for 3 h, and to 1000 °C for 6 h.

	Change in intensity at the following temperatures				
Signal	100 °C	150 °C	200 °C	500 °C	1000 °C
A	No change	↓	↓	0	0
B	No change	No change	No change	0	0
C	↑	↑	↑	0	0
D	No change	No change	No change	↑	↑

↓ , decrease; ↑ , increase; 0, absent.

14.6 COMPARISON OF ESR SPECTRA WITH CONODONT CAI VALUES

Although a precise correlation between the intensity of ESR signals and conodont CAI values has not been recorded, nevertheless the ESR spectra of conodonts with an identical CAI value do show some similarities. These similarities are in accord with data which we obtained from the experimental heating of conodonts. The spectra of conodonts which have a CAI value of 1 show signals C and D to be always weaker than signal B, or to be absent, e.g. the Triassic samples. With higher CAI values, an increase in intensity of peaks C and D is initially observed. Signal C decays in conodonts with a CAI of 3.5 or higher, whereas both signals D and B are strong (Fig. 14.5).

The structure of signal B shows the strongest correlation with CAI value. The observation is surprising as experimental heating of conodonts does not indicate that signal B is responsible for conodont colour. This is why signal B decays at a temperature of 500 °C. Upon closer examination, signal B is shown to be a band consisting of two weakly separated signals, indicated B′ and B″ (cf. Fig. 14.6). With increasing CAI values, an increase in the intensity of signal B″ is noted in relation to signal B′. Only one of the investigated samples, sample 9, departed from this trend, by exhibiting a signal B structure similar to that of conodonts with a CAI of 3.5 in spite of the sample colour corresponding to CAI 2.5.

It appears that the absence of a precise quantitative correlation between the intensity of ESR signals and the conodont CAI value is a result of the thermal instability of the paramagnetic centres responsible for signals A, B and C (cf. Table 14.3), coupled with the fact that the intensity of the ESR signal is a combined function of both the radiation time and the thermal history of the conodonts.

14.7 EFFECT OF MICROMORPHOLOGICAL COMPOSITION

Pietzner *et al.* (1968) showed that there are differences in the chemical compositions of the conodont crown and base. The basal filling is more fine grained and contains a higher concentration of trace elements; for example, yttrium can be enriched in the base by a factor of 20 greater than that in the crown.

Only in one sample investigated (sample 7, from the Frasnian deposits of the Holy Cross Mountains) were a large number of conodonts observed in which the crown was attached to the base. The sample was divided into two portions, one with conodont crowns only, and the other in which crowns were attached to bases. Both portions were investigated by ESR spectroscopy. The portion containing crowns attached to the bases yielded spectra with signals which were all stronger than the portion comprising crowns alone; the intensities of signals A, B and C were twice as strong, and signal D was five times as strong. The greater organic content in the basal filling is one reason for the marked increase in the intensity of signal D.

Therefore it is important to investigate conodont crowns alone when using ESR spectroscopy. This is of additional importance, as the concentration of trace elements in the bases is most probably due to later (post-mortem?) inputs (Pietzner *et al.*, 1968).

14.8 APPLICATION

ESR spectroscopy was proposed as a dating technique for conodonts by Morency *et al.* (1970). The technique consists of subjecting the

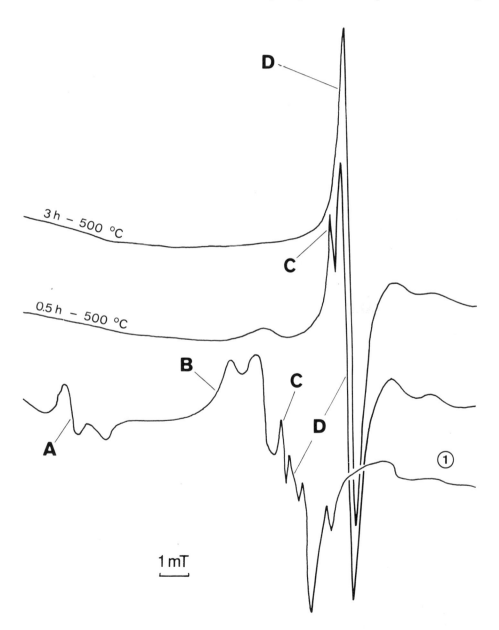

Fig. 14.4 — ESR spectrum for Ordovician conodonts (sample 1) and its changes after heat treatment.

conodonts to small doses of laboratory γ-radiation in order to establish the dependence of signal growth upon the radiation dosage. If the signal growth is linear, then it is possible to extrapolate to obtain the geological age of the sample if the total radiation dose is known (cf. Zeller *et al.*, 1967).

This investigation, however, has shown

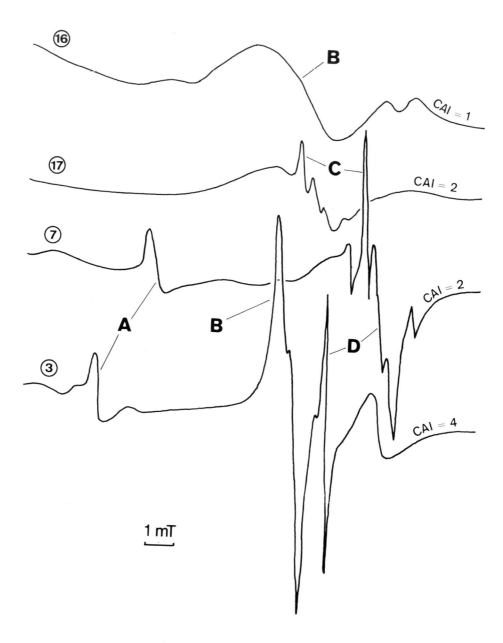

Fig. 14.5 — ESR spectra of conodonts with different CAI values. Note the distinct increase in signals B and D for conodonts with a maximum thermal history. The circled numbers denote the investigated conodont samples (see Table 14.1).

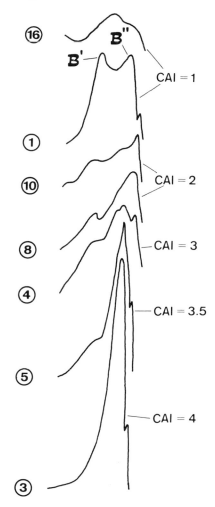

Fig. 14.6 — Correlation of the structure of signal B with conodont CAI. The circled numbers denote the investigated conodont samples (see Table 14.1).

radiation (cf. Morency *et al.*, 1970). The effect was observed when sample 1 had been subjected to a γ-radiation dose of 0.17×10^6 R. After the radiation, signals A, B and C disappeared whilst signal D increased by a factor of almost 4. According to Vaz and Zeller (1966), the decrease in signal intensity in response to additional radiation occurs with saturation of paramagnetic centres. In this case, the number of unpaired electrons responsible for the ESR signals tends to decrease.

As a consequence, we do not recommend ESR spectroscopy as a technique for age-determination of conodonts. It is recommended instead that ESR spectroscopy be used for dating the last heating event, i.e. most frequently the age of the completion of the last upheaval of the conodonts.

ESR spectra can yield useful information about the heat treatment of conodonts, especially in cases of heating to less than 200 °C for very short periods of time, and in which no colour changes have been produced in the conodonts.

ESR spectroscopy has a potential application in the testing and correlation of CAI values of conodonts from different host-rocks, or in situations where the conodont colour has been modified by the lithology, diagenesis and/ or weathering.

14.9 ACKNOWLEDGEMENTS

We would like to thank Dr. J. Dzik and Dr. K. Zawidzka for providing some conodont samples for this study. Thanks are also due to Professor Dr. A. Radwański and Dr. A. Gize who critically read the manuscript and offered helpful suggestions.

that, for the 17 conodont samples studied, the signal intensity is a function of both time and, most definitely, temperature. As a heating event, even of short duration, produces a decay of signals A, B and C, with a simultaneous growth of signal D, then any dates obtained from the ESR spectra of conodonts can be very misleading. In addition, during studies of Early Palaeozoic conodonts, another difficulty has appeared. This difficulty is a decrease in signal intensity as a result of the additional laboratory

14.10 REFERENCES

Atherton, N. M. 1973. *Electron Spin Resonance*, Halsted Press, London, 1–430.
Cubbitt, J. M. and Wilkinson, J. G. 1976. A classification study of shales in Kansas using electron-spin resonance spectroscopy. *Journal of Mathematical Geology*, **8** (3), 337–344.
Drago, R. S. 1977. *Physical Methods in Chemistry*, W. B. Saunders, New York, 1–660.

Epstein, A. G., Epstein, J. B. and Harris, L. D. 1977. Conodont color alteration — an index to organic metamorphism. *U.S. Geological Survey Professional Paper*, **995**, 1–27.

Fåhraeus, L. E. and Fåhraeus-van Ree, G. E. 1985. Histomorphology of soft tissue conodont matrix. In R. J. Aldridge, R. L. Austin, and M. P. Smith (Eds.), *Fourth European Conodont Symposium (ECOS IV)*, *Abstracts*, Private publication, University of Southampton, 10–11.

Friedlander, H. Z., Saldick, J. and Frink, C. R. 1963. Electron spin resonanace spectra in various clay minerals. *Nature*, **199** (4888), 61–62.

Garrison, E. G., Rowlett, R. M., Cowan, D. L. and Holroyd, L. V. 1981. ESR dating of ancient flints. *Nature*, **290**, (5801), 44–45.

Gilinskaya, L. G., Gerasimov, E. K., Sukhoverkhova, M. B. and Pospelova, L. N. 1982. ESR spectra of apatites from metamorphozed rocks (in Russian). *Proceedings of the USSR Acadamy of Sciences, Geological Series*, **3**, 100–112.

Gilinskaya, L. G. and Shcherbakova M. J. 1975. Investigations of structural damage in apatite by electron spin resonance spectroscopy (in Russian). In *Physics of apatite*, Nauka, Novosibirsk, 7–63.

Hutton, D. R. and Troup, G. J. 1966. Paramagnetic resonance centres in amethyst and citrine quartz. *Nature*, **211**, (5049), 621.

Ikeya, M. 1975. Dating a stalactite by electron paramagnetic resonance. *Nature*, **225**, (5503), 48–50.

Ikeya, M. and Miki, T. 1980. Electron spin resonance dating of animal and human bones. *Science*, **207** (4434), 977–979.

Low, W. and Zeira, S. 1972. ESR spectra of Mn^{2+} in heat-treated aragonite. *American Mineralogist*, **57** (7–8), 1115–1124.

Morency, M., Emond, P. L. and von Bitter, P. H. 1970. Dating conodonts using electron spin resonance: a possible technique. *Kansas Geological Survey Bulletin*, **199** (1), 17–19.

Pietzner, H., Vahl, J., Werner, H. and Ziegler, W. 1968. Zur chemischen Zusammensetzung und Mikromorphologie der Conodonten. *Palaeontographica A*, **128**, 115–152, 10 plates.

Robins, G. V., Seeley, N. F., McNeil, D. A. C. and Symons M. R. C. 1978. Identification of ancient heat treatment in flint artefacts by ESR spectroscopy. *Nature*, **276** (5689), 703–704.

Schindler, P. and Ghose, S. 1970. Electron paramagnetic resonance of Mn^{2+} in dolomite, and magnesite, and Mn^{2+} distribution in dolomites. *American Mineralogist*, **55** (11–12), 1889–1896.

Symons, M. R. C. 1978. *Chemical and Biochemical Aspects of Electron Spin Resonance Spectroscopy*, Van Nostrand Reinhold, New York, 1–185.

Vaz, J. E. and Zeller, E. J. 1966. Thermoluminescence of calcite from high gamma radiation dose. *American Mineralogist*, **51**, 1156–1166.

Wakeman, S. and Carpenter, R. 1974, ESR spectra of marine and freshwater manganese nodules. *Chemical Geology*, **13**, (1), 39–47.

Wildeman, T. R. 1970. The distribution of Mn^{2+} in some carbonates by electron paramagnetic resonance. *Chemical Geology*, **5** (3), 167–177.

Zeller, E. J., Levy, P. W. and Mattern, P. L. 1967. Geologic dating by electron spin resonance. *Radioactive Dating and Methods of Low-Level Counting, Proceedings of a Symposium*, International Atomic Energy Agency, Vienna, 531–540.

15

Temperature as a factor affecting conodont diversity and distribution

J. E. Geitgey and T. R. Carr

Recent studies involving oxygen-isotope measurements of conodonts indicate that such analyses may provide a primary marine palaeotemperature signal. Examination of all $\delta^{18}O$ results at present reported, for samples of Ordovician–Pennsylvanian age, allows for the compilation of a Palaeozoic marine palaeotemperature curve which is consistent with previously described qualitative interpretations of Palaeozoic palaeoclimatology.

There is a marked degree of correspondence between the isotopically and qualitatively inferred palaeotemperture record and intervals of fluctuation of conodont diversity and distribution. In particular, during intervals of inferred higher or increasing palaeotemperature (i.e. Ibexian, Mohawkian, Late Devonian and Triassic intervals), diversity and/or distribution appear to have been relatively high. Conversely, during intervals of lower or decreasing palaeotemperature (i.e. Whiterockian, Cincinnatian, Carboniferous and Permian intervals), diversity and/or distribution were reduced.

The observed correspondence suggests that temperature was an important environmental variable which affected conodonts. It is proposed that conodont diversity was to some degree controlled by fluctuations in the volume of thermally suitable habitat available for exploitation by the group. Thus, with increased temperatures and a decreased latitudinal temperature gradient, more habitat was available, allowing for greater diversity to develop as predicted by the species–area effect.

15.1 INTRODUCTION

Since conodonts were first described by Pander (1856), they have proved to be excellent biostratigraphic tools, and have become increasingly useful in palaeoenvironmental interpretations and as indices of thermal metamorphism. Recent work in the area of conodont geochemistry suggests that equally valuable information concerning various aspects of Palaeozoic and Triassic palaeobiology, palaeoecology, palaeoclimatology and palaeo-oceanography may also be obtainable from analyses of conodont elements.

The reason why geochemical analyses of conodonts appear to be such a powerful tool is the fact that conodonts possessed skeletal elements composed of the carbonate–apatite francolite (Pietzner *et al.*, 1968), which contains a host of trace elements. Theoretical and empirical studies suggest that apatitic minerals remain essentially inert within a low-temperature inorganic diagenetic environment after initial precipitation (Winter, *et al.*, 1940; Brodskii and Sulima, 1953; Urey *et al.*, 1951; Tudge, 1960). In addition, the consistent experimental results from geochemical analysis of conodonts suggest that conodont apatite does possess primary marine isotopic and trace-element significance (Holser *et al.*, 1982; Kolodny *et al.*, 1983; Wright *et al.*, 1984; Geitgey, 1984). Most of the previous work which has sought to obtain a record of variations in marine chemistry through the geological record has relied upon geochemical analyses of carbonates. However, carbonates are less stable chemically than phosphates, and reliable results are generally limited to rocks of Cretaceous age and younger. Data obtained from older rocks are more suspect. Clearly, the potential is to utilize geochemical analyses of Palaeozoic and Triassic conodonts as a means of obtaining reliable marine geochemical data for samples much older than the carbonate samples typically analysed.

The recent interest in the area of conodont geochemistry, and specifically trace-element and isotopic studies, was largely spurred by Jack Kovach, who initiated research into a variety of geochemical analysis of conodonts. Listed below is a summary of the literature concerned with geochemical studies of conodonts and some of the potential applications which might be made of such data.

(1) Neutron activation analysis (Bradshaw *et al.*, 1973) revealed differences in isotopic content between lamellar and fibrous conodonts.

(2) Strontium concentration and isotopic analyses have been reported by Kovach (1980, 1981a, 1981b), who suggested that such studies might permit the construction of a more refined curve depicting secular variations in the $^{87}Sr/^{86}Sr$ composition of the marine environment through time. Additionally, such studies might prove useful for purposes of correlation, particularly for global correlations between dissimilar faunas.

(3) The potential for radiometrically dating conodonts has been discussed by Sachs *et al.* (1980), who used fission-track dating, and also by Kovach and Zartman (1981), who attempted U–Th–Pb dating. Although refinement of the procedures would probably be required, both papers concluded that reasonable dates should be obtainable.

(4) Studies dealing with neodymium isotopic analyses of conodonts have been published by Kovach and Futa (1983), Wright *et al.* (1984) and Shaw and Wasserburg (1985). Such work establishes the possibility of measuring secular variations of neodymium in the oceans through time. Kovach and Futa (1983) point out that, since the $^{143}Nd/^{144}Nd$ ratio has been shown to vary between oceanic basins in the modern marine realm, analyses of conodonts may allow for enhanced palaeocirculation pattern and palaeogeographical reconstructions.

(5) Wright *et al.* (1984) reported rare-earth element analyses for a large number of Palaeozoic and Triassic conodont samples. Of particular interest are their measurements of cerium, which they suggest may be interpretable in terms of anoxic events.

One final area of interest, which will serve as the subject of the rest of this chapter, is oxygen-isotope analysis of conodonts. Recently, researchers at the Hebrew University in Jerusalem, Israel, have rekindled interest in the area of phosphate oxygen-isotope palaeothermometry for the purpose of obtaining primary marine palaeotemperature data. Their work has included analyses and interpretations of fish-bone apatite (Kolodny *et al.*, 1983), phosphorite rocks (Shemesh *et al.*, 1983) and conodonts (Luz *et al.*, 1984). In their papers, these researchers argue that it has been empirically demonstrated that rapid equilibration can occur between isotopic oxygen in biogenic phosphate and the ambient marine waters (Koshland and Clarke, 1953; Cohn, 1953; Stein and Koshland, 1952; Dahms and Boyer, 1973). That observation, coupled with the previously discussed knowledge that apatites may remain essentially inert after initial precipitation, led Kolodny *et al.* (1983, p. 399) to suggest that biogenic apatites should be ideally suited to isotopic and trace-element studies. Since $\delta^{18}O$ analyses for the purpose of obtaining marine palaeotemperature data have for the most part been performed on less stable carbonate samples, reliable data have been limited to relatively younger samples. However, in light of the apparent stability of conodont apatite, one should be able to measure primary $\delta^{18}O$ values from conodonts and thus make marine palaeotemperature inferences for samples much older than the carbonates typically analysed.

The following sections will outline briefly the experimental procedure required to perform $\delta^{18}O$ analyses of conodonts and then examine the palaeotemperature significance of the $\delta^{18}O$ measurements at present available. That data set will form the basis of a discussion regarding the apparent correspondence between the isotopically and qualitatively inferred palaeotemperature record and fluctuations in the diversity and distribution of conodonts throughout their record.

15.2 METHODS

Because oxygen occurs in both PO_4 and CO_3 sites within the apatite lattice, the analytical technique must isolate the oxygen bound to the phosphorus and ignore the exchangeable CO_3 oxygen. The analytical technique, which involves a series of dissolutions and precipitations, was developed by Tudge (1960), was outlined by Kolodny *et al.* (1983) and Shemesh *et al.* (1983) and was detailed in Geitgey (1984). A brief outline of the procedure follows.

In order to obtain accurate results of the $\delta^{18}O$ analyses, 10 mg of phosphatic material is required. This translates to approximately 1000 to 6000 conodont elements per sample, depending upon the size of the individual elements involved. After hand picking, the samples are processed according to an involved chemical procedure. The final step in the chemical procedure is the precipitation of the oxygen-bearing $BiPO_4$ residue. Subsequently, the $BiPO_4$ is fluorinated, and the oxygen converted to CO_2 and analysed mass-spectrometrically.

The precision of this technique, as applied to conodont analyses, was discussed at some length by Luz *et al.* (1984, p. 6) and was estimated to be 0.5‰. All the results discussed in this chapter are reported in δ notation, with respect to standard mean ocean water.

Because of the large quantity of conodonts at present required by $\delta^{18}O$ analysis, it is hoped that a future improvement will be to decrease the required sample size below 10 mg. An additional improvement to the technique would be to increase analytical precision beyond 0.5%.

15.3 ISOTOPICALLY INFERRED MARINE PALAEOTEMPERATURE RECORD

Table 15.1 lists all currently available $\delta^{18}O$ analyses of thermally unaltered Ordovician conodonts and represents most of the data base for this report. Analysed samples range in age from Ordovician to Pennsylvanian. At present, no data are available for the Silurian, Permian or Triassic Periods. All $\delta^{18}O$ analyses were performed upon samples collected from the North American midcontinent area and thus represent a low-latitude palaeotemperature record.

Because of possible secular variations in the $\delta^{18}O$ composition of ocean waters through time (which must be addressed in order to calculate an absolute palaeotemperature), the discussion is restricted to relative palaeotemperature 'trends' suggested by the isotopic data. With that caveat, one needs only to remember that, the greater the $\delta^{18}O$ value, the 'cooler' is the inferred palaeotemperature. Such an interpretation is almost certainly overly simplistic, as it ignores the potential effects of variations in salinity and $\delta^{18}O$ composition of the marine environment and may or may not be supported by further work.

Luz et al. (1984) and Geitgey (1984) have argued, on the basis of several lines of evidence, that the $\delta^{18}O$ values measured from conodonts are indeed interpretable as a palaeotemperature record. Those arguments are not repeated here. However, as evidence of the palaeotemperature significance of the isotopic data, the $\delta^{18}O$ record is compared in Fig. 15.1 with two qualitative interpretations of palaeotemperature fluctuation for the Palaeozoic Era. The curve to the left is a qualitative record of the distribution and abundance of evaporite versus

Table 15.1 — Currently available $\delta^{18}O$ analyses of thermally unaltered conodonts.

Sample	Formation	Locality	Age	$\delta^{18}O$
From Luz et al. (1984)				
UG-10	Manitou	Colorado	Ibexian	15.1
72SB-196	Joins	Oklahoma	Lower Whiterockian	16.3
72SE-222	McLish	Oklahoma	Whiterockian	16.9
72SE-456	McLish	Oklahoma	Whiterockian	17.5
72SE-545	McLish	Oklahoma	Whiterockian	17.4
72SJ-13	Bromide	Oklahoma	Whiterockian	17.7
66KK-26	Ashlock	Kentucky	Maysvillian	21.8
66KK-30	Ashlock	Kentucky	Maysvillian	17.9
From Geitgey (1984)				
60G7-22	Point Pleasant Ls.	Kentucky	Edenian	16.0
60G10-1	Point Pleasant Ls.	Kentucky	Edenian	17.3
60G10-1	Point Pleasant Ls.	Kentucky	Edenian	17.7
60G10-1	Point Pleasant Ls.	Kentucky	Edenian	17.0
60G11-1	Point Pleasant Ls.	Kentucky	Edenian	17.8
60G11-1	Point Pleasant Ls.	Kentucky	Edenian	17.2
60G12-5	Point Pleasant Ls.	Kentucky	Edenian	17.7
60G12-5	Point Pleasant Ls.	Kentucky	Edenian	18.2
83GA-1	Kope	Kentucky	Maysvillian	17.5
83GA-1	Kope	Kentucky	Maysvillian	17.0
Previously unreported				
60T6-5	Point Pleasant Ls.	Kentucky	Edenian	16.0
60T6-5	Point Pleasant Ls.	Kentucky	Edenian	15.3

glacial deposits (Frakes, 1979, Fig. 4–2). While not a palaeotemperature record *per se*, such a record should have palaeotemperature significance. The right-hand curve, also taken from Frakes (1979, Fig. 9–1), is his qualitative interpretation of actual palaeotemperature fluctuations through the geological record. Illustrated between those two curves is the $\delta^{18}O$ record as determined by analyses of conodonts. Examination of these curves suggests that there appears to be a reasonable degree of correspondence between the palaeotemperature fluctuations qualitatively described by Frakes and the $\delta^{18}O$ curve. In particular, there is a trend of decreasing temperatures through the Ordovician Period, with a short-lived Mohawkian-age warming. During Middle–Late Devonian times, palaeotemperatures increased to a maximum, then declined through the Mississippian Period and remained low through the Pennsylvanian Period. Although such comparisons are at this

time fraught with uncertainties (i.e. the actual palaeotemperature significance of the curves and the paucity of data), the preliminary correspondence between the curves suggests that the $\delta^{18}O$ record does possess palaeotemperature significance.

Another observation is the degree of correspondence between the isotopically inferred palaeotemperature record and fluctuations in sea level. Fig. 15.2 depicts fluctuations in Ordovician $\delta^{18}O$ versus the (upper) Vail *et al.* (1977) sea-level curve, and the sea-level curve (lower) published by Fortey (1984). These comparisons appear to indicate sympathetic fluctuations between $\delta^{18}O$ and sea level for much of the Ordovician Period, with lower temperatures corresponding to lower sea level and vice versa. This suggests some type of linkage mechanism between the two, such as albedo affects associated with relative sea-level changes. On the assumption that the correspondence is real,

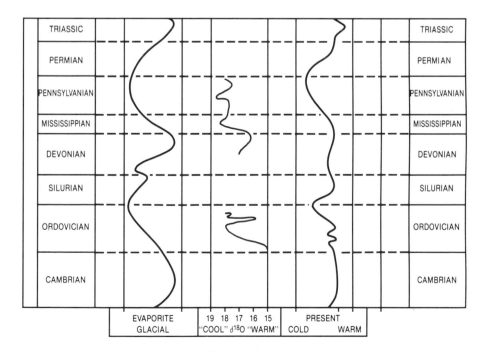

Fig. 15.1 — Comparison of isotopically inferred palaeotemperature record (centre curve) versus qualitatively inferred palaeotemperature curves (evaporite versus glacial, left-hand curve, after Frakes (1979, Fig. 4-2); inferred palaeotemperature record, right-hand curve, after Frakes (1979, Fig. 9-1)).

$\delta^{18}O$ analyses may also aid in the process of refining sea level fluctuation curves for the Palaeozoic Era.

Because it has been exceedingly difficult to infer palaeotemperatures and/or palaeotemperature fluctuations reliably for much of the geological record, relatively few studies have addressed the potential palaeoecological effects of this important environmental variable. A notable exception has been the $\delta^{18}O$ studies of Cretaceous and younger foraminifera, particularly since the advent of the Deep Sea Drilling Project. That work has provided an increasingly accurate record of marine palaeotemperatures for the past 100 Ma and has allowed for many questions regarding palaeoclimatology, palaeoceanography, and palaeoecology to be addressed (e.g. Savin, 1977, 1982). The hope is that the $\delta^{18}O$ record provided by conodonts will produce palaeotemperature data that can be employed to

elucidate various aspects of conodont palaeobiology.

15.4 ORDOVICIAN CONODONT DIVERSITY AND DISTRIBUTION VERSUS PALAEOTEMPERATURE

The most extensive discussions of temperature as a factor affecting conodont distribution have been made by those researchers who have examined Ordovician provincialism (*e.g.* Sweet *et al.*, 1959; Barnes *et al.*, 1972; Sweet and Bergström, 1974, 1984). These workers suggested that latitudinal temperature differences played a role in the development of a low-latitude, warm-water or North American midcontinent fauna, distinct from a coeval high-latitude, cold-water or North Atlantic fauna (see Fig. 15.3).

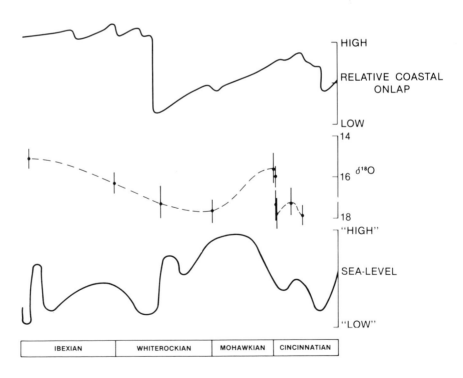

Fig. 15.2 — Comparison of the Palaeozoic $\delta^{18}O$ record with relative fluctuations in sea level as proposed by upper curve (Vail *et al.*, 1977, Fig. 1) and lower curve (Fortey, 1984, Fig. 3).

Sweet and Bergström (1974) observed that members of the low-latitude warm-water fauna invaded the higher-latitude North Atlantic Province during intervals when warmer waters spread polewards, as inferred by the presence of Bahama-type carbonates.

As an alternative to the temperature hypothesis, Barnes *et al.* (1972) and Barnes and Fåhraeus (1975) suggested that Ordovician conodont provincialism may have been related to tectonic setting. In particular, they suggested that members of the Midcontinent Province occupied an epieric sea environment, whereas the North Atlantic fauna inhabited the deeper waters of continental margins. Sweet and Bergström (1974) and Bergström and Carnes (1976) subsequently pointed out that an attempt to relate conodont distribution to particular palaeotectonic regimes was probably overly simplistic. Specifically, these researchers pointed out that to characterize all members of the cold-water fauna as deeper-water or 'eugeosynclinal' was misleading, as a number of the cold-water genera had their greatest abundance in epicontinental sediments of the Baltic Shield and Appalachian shelf deposits. Bergström and Carnes (1976) noted that the Middle Ordovician conodont association which they obtained from relatively shallow-water high-latitude deposits at the Lunne section in Sweden is most closely related to coeval conodonts from relatively deeper-water rocks of the Appalachians and not to coeval shallow-water deposits. Baltoscandia was at approximately 60°S during the Early and Middle Ordovician Period (Bergström and Noltimier, 1982). Bergström and Carnes (1976) suggested that members of the high-latitude fauna submerged as they migrated towards the equatorially located Appalachian region, in order to maintain an optimal thermal

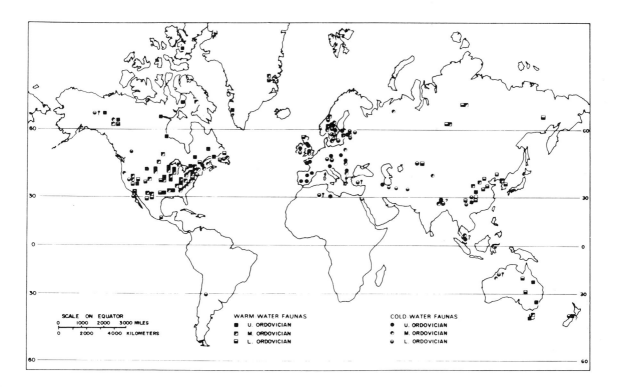

Fig. 15.3 — Localities where warm-water versus cold-water Ordovician conodonts have been collected. (From Sweet and Bergström (1984, Fig. 1).

environment (i.e. 'cooler' waters). The observation of Bergström and Carnes suggests that water temperature was probably more important in controlling the habitat available for occupation by the cold-water fauna than simply tectonic setting. Thus, because suitably 'cooler' temperatures were present both at high latitudes in relatively shallow waters and in deeper waters at lower latitudes, members of the North Atlantic or cold-water province are found in a variety of tectonic environments. However, because members of the North American Midcontinent Province were more thermophilic than the cold-water fauna, their distribution appears to be related to low-latitude shelf settings, where warmer waters were developed. However, at the macroevolutionary level, conodonts were thermophilic and could not tolerate the much 'cooler' marine palaeotemperatures which would have been present at the highest latitudes. This is supported by the general absence of conodonts from Palaeozoic rocks deposited at high palaeolatitudes (Nicoll, 1976).

Regarding conodont diversity in the Ordovician Period, Clark (1983) and Sweet (1985) demonstrated that conodont diversity reached a peak during the Early Ordovician Period which was never to be exceeded (Fig. 15.4). Following the Early Ordovician (Ibexian) peak, diversity decreased to a low point near the Whiterockian–Mohawkian boundary, recovered to a degree in the Mohawkian Epoch, dropped off in the late Mohawkian Epoch and remained low through the Cincinnatian Epoch. Sweet and Bergström (1974) observed that conodont stocks which survived into the Silurian Period were derived from the warm-water Midcontinent fauna. Such an observation is paradoxical in that marine palaeotemperatures were most probably reduced during the Late Ordovician Period, as evidenced by extensive continental glaciation on Gondwana at that time (Beuf et al., 1971; Harland, 1972; Frakes, 1979). Thus, it appears more reasonable to expect that the cold-water fauna would have survived the Late Ordovician refrigeration at the expense of the warm-water fauna. The

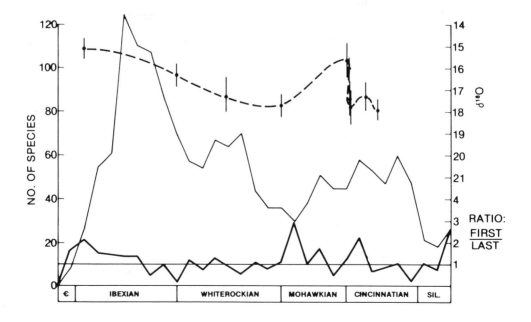

Fig. 15.4 — Ordovician $\delta^{18}O$ record versus specific conodont diversity as measured by Sweet (1985, Fig. 7).

isotopic record demonstrates an apparent correspondence between isotopically inferred temperature and diversity (Fig. 15.4) (Sweet, 1985). The peak in diversity during the Ibexian Epoch corresponds to apparently higher temperatures, whereas the Whiterockian–Mohawkian diversity 'low' corresponds to inferred lower temperatures. As temperatures apparently warmed through the Mohawkian Epoch, conodont diversity began to increase again, with another major diversity 'low' occurring across the Ordovician–Silurian boundary. Palaeotemperatures apparently dropped in the late Mohawkian–early Cincinnatian interval and, although no $\delta^{18}O$ data are available for latest Ordovician times, the record of Gondwana glaciation suggests that palaeotemperatures were reduced through that interval. Overall, even though the $\delta^{18}O$ record is rather sparse at this time, one can suggest that conodont diversity appears to have been higher during intervals of higher or increasing temperature (Ibexian Epoch and Mohawkian Epoch) and lower during intervals of inferred lower or decreasing temperature (Whiterockian Epoch and Cincinnatian Epoch). The observation that conodont diversity appears to have fluctuated sympathetically with the $\delta^{18}O$ record was first made and discussed by Geitgey (1984, 1985).

When the previously discussed observations of other workers are combined with the isotopic data presented, the following picture emerges. During the Early Ordovician Period, conodont diversity reached a peak, and palaeogeographic distribution was apparently widespread, as evidenced by the presence of a 'cold-water' fauna at palaeolatitudes of 60° S. During that time, the isotopically inferred palaeotemperature record suggests that "warm" marine temperatures were developed. Through the rest of the Ordovician Period, conodont diversity and palaeotemperature appear to have fluctuated sympathetically. Diversity was highest during intervals of increasing or higher palaeotemperature, and lowest during decreasing or lower temperatures. During the Cincinnatian Epoch, conodont diversity decreased, and distribution

apparently became more restricted (Baltoscandia moved to a low-latitude position), and temperatures decreased, culminating in a major glaciation on Gondwana. Paradoxically, although temperatures decreased in the later Ordovician Period, the conodont stocks which survived the terminal Ordovician refrigeration were not members of the cold-water North Atlantic fauna, but principally members of the warm-water Midcontinent fauna.

15.5 POST-ORDOVICIAN RECORD OF CONODONT DIVERSITY AND DISTRIBUTION VERSUS PALAEOTEMPERATURE

On the basis of the suggestion that conodont diversity was higher during intervals of increasing or higher marine temperature, an examination of the post-Ordovician record was made. Because $\delta^{18}O$ data are currently unavailable, the Silurian Period is excluded from the discussion.

Fig. 15.5 is a plot of the number of conodont genera present through time (Clark, 1983) versus the $\delta^{18}O$ record. When the Middle Devonian–Pennsylvanian interval is examined, conodont diversity appears to correlate with palaeotemperature. Specifically, the Late Devonian peak in conodont diversity correlates with an inferred peak in palaeotemperature. Diversity then dropped through the Mississippian Period and remained low through the Pennsylvanian Period. Correspondingly, palaeotemperatures dropped from a Late Devonian peak to a Mississippian low and fluctuated at relatively 'cooler' temperatures through Pennsylvanian times.

Although no $\delta^{18}O$ data are currently available from conodonts of Permian or Triassic age, temperature effects upon conodont diversity and distribution have been suggested by Nicoll (1976). That paper discussed the absence or scarcity of conodonts of Permian age in Austra-

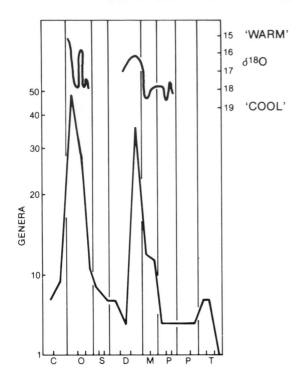

Fig. 15.5 — Palaeozoic $\delta^{18}O$ record versus conodont diversity at the generic level as measured by Clark (1983, Fig. 2).

tion occurred on Gondwana during Permo-Carboniferous times, suggesting the presence of lower marine temperatures, particularly in the southern hemisphere. At some point during the Permian Period, temperatures began to rise and this trend apparently continued into the Triassic Period.

The preceding reconstruction of palaeotemperature fluctuation, and its apparent correspondence to conodont diversity and distribution, is consistent with the observations made for the Ordovician Period and the Devonian–Pennsylvanian interval.

lia and suggested that this was due to the presence of 'cooler' marine temperatures in the region at that time.

Fig. 15.6, from Nicoll's paper, demonstrates the differences in global distribution of Permian and Triassic conodonts. As is seen in Fig. 15.6, Early Permian conodonts are found near, or north of, the Permian equator. By Late Permian times, occurrences range from 30° N to 40° S. In the Triassic Period, conodont distribution ranged from 60° N to 70° S. Nicoll further noted that conodont abundance similarly expanded from the Permian Period to the Triassic Period (See Fig. 15.5).

In examining the factors which might account for the observed distributions of Permian and Triassic conodonts, Nicoll concluded that fluctuations in climatic conditions were probably responsible. He observed that glacia-

15.6 MARINE TEMPERATURES AND CONODONT DIVERSITY: AN ORDOVICIAN MODEL

A model of a causal relationship between inferred marine palaeotemperature and conodont diversity for the Ordovician Period must also incorporate the biogeographical selectivity of the Late Ordovician extinction. In particular, while there is a general decrease in conodont diversity at the close of the Ordovician Period, there is preferential extinction among the conodonts of the colder-water North Atlantic Province compared with conodonts of the warmer-water Midcontinent Province. The apparent paradox that a colder environment leads to the demise of the colder-water conodont fauna, whereas the warmer-water fauna, although decreased in diversity, survived into the Silurian Period to provide the stock for the reradiation of post-extinction conodont faunas must be incorporated into any model for Ordovician conodont-diversity patterns.

In order to incorporate the observed biogeographical characteristics of Ordovician conodont-diversity patterns, it is postulated that, as marine temperatures decreased in the Late Ordovician Period, the latitudinal temperature gradient increased and the thermally suitable area available for habitation was compressed

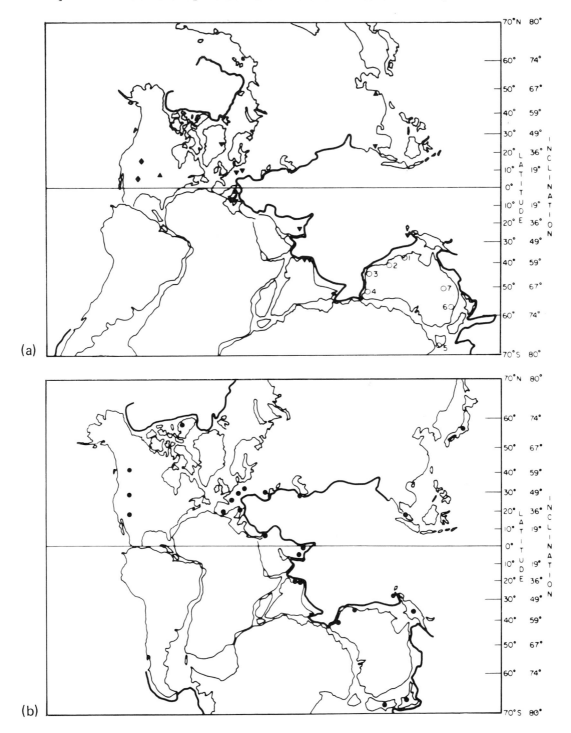

(a)

(b)

Fig. 15.6 — (a) Global distribution of Permian conodont localities: ▲, Early Permian; ▼, Late Permian; ◆, both; o, sampled Australian basins. (From Nicoll (1976, Fig. 2).) (b) Global distribution of Triassic conodont localities. (From Nicoll (1976, Fig. 3).)

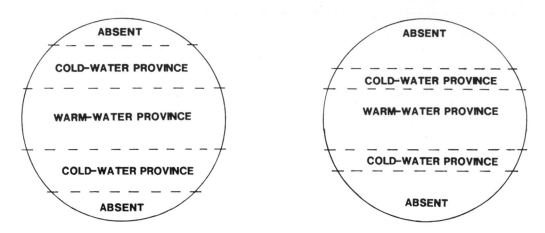

Fig. 15.7 — Schematic model of a preferential decrease in habitable area for conodonts of the colder-water North Atlantic Province compared with the warmer-water Midcontinent Province due to a decrease in temperature and increase in thermal gradient during the Late Ordovician Period.

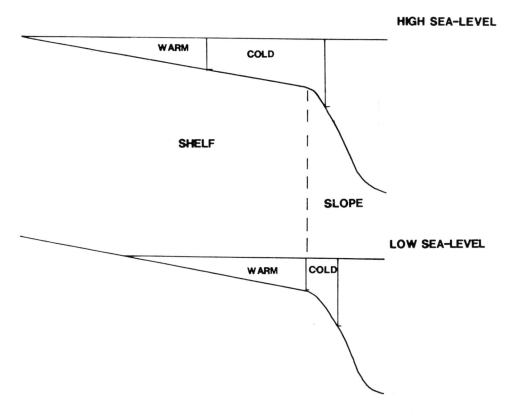

Fig. 15.8 — Schematic model of a preferential decrease in habitable area on tropical shelves for conodonts of the colder-water North Atlantic Province compared with the warmer-water Midcontinent Province due to an eustatic decrease in sea level during the Late Ordovician Period.

towards the equator (Fig. 15.7). The effect of a general contraction of habitable area on diversity can be approached by extrapolation of the theory of island biogeography (MacArthur and Wilson, 1967) into an evolutionary time frame (e.g. Simberloff, 1974; Sepkoski, 1976). Although not the only explanation, a species–area relationship is consistent with observed changes in inferred marine palaeotemperatures and conodont diversity throughout the Palaeozoic Era and Triassic Period. The strength of the relationship is not unduly surprising, since temperature appears to be a major determinant of geographical range in marine species (Stehli *et al.*, 1972; Jablonski *et al.*, 1985).

Sweet and Bergström (1984) suggested that, during the Late Ordovician Period, much of the North Atlantic Province was at approximately 30° S latitude, a somewhat higher palaeo-latitude than the equatorial Midcontinent Province. It is proposed that, as marine temperature fell during the Late Ordovician Period, the latitudinal temperature gradient increased, and the area thermally suitable for conodonts was compressed towards the equator. Although the conodont fauna of the North Atlantic Province preferred somewhat 'cooler' temperatures relative to the North American Midcontinent Province, the lower thermal limit tolerable to the fauna would have contracted closer to the equator. If the thermal gradient increased coincident with a decrease in temperature, a greater areal restriction would have been placed on the North Atlantic fauna relative to the continental glaciation on Gondwana. The decrease in sea level provided additional restriction of habitable shelf area available to conodonts. Because of the probable positions of the Midcontinent and North Atlantic faunas on the tropical shelves (Bergström and Carnes, 1976), a relative sea-level drop would have tended to place greater areal restriction on the North Atlantic fauna, which appears to have occupied the deeper shelf to slope environments (Fig. 15.8). The combi-nation of relatively greater latitudinal and depth restriction might then explain the preferential extinction of conodonts of the North Atlantic Province.

15.7 SUMMARY

(1) Because of the chemical stability of phosphate, it appears that biogenic apatites, such as conodont elements, preserve original marine isotopic and trace-element signatures. Thus, geochemical analyses of biogenic apatites should provide valuable information for samples much older than have typically been analysed. This geochemical data should be useful in addressing a variety of geological and palaeontological questions.

(2) Specifically, the reasonable and consistent experimental results obtained from $\delta^{18}O$ analyses of conodonts indicates that primary marine palaeotemperature data may be obtained.

(3) A preliminary comparison of the diversity and distribution of conodonts, with the isotopically and qualitatively inferred palaeo-temperature record, suggests some degree of correspondence. In particular, it appears that, during intervals of higher or increasing palaeotemperature (i.e. Ibexian, Mohawkian, Late Devonian and Triassic times), conodont diversity and/or distribution were relatively higher. Alternatively, during intervals of lower or decreasing palaeotemperature (i.e. Whiterockian, Cincinnatian, Carboniferous and Permian times), conodont distribution and/or diversity were relatively low.

(4) From observations on diversity and distribution of conodonts together with the inferred palaeotemperature record, temperature appears to have been an important environmental variable which affected the conodonts. It is suggested that temperature fluctuations influenced the diversity and distribution of conodonts by changing the

volume of habitable marine ecospace available for exploitation. Changes in habitable ecospace due to temperature limitations were thus an important determinant for conodont diversity and biogeography. Specifically, a schematic model is proposed to explain overall Ordovician conodont diversity patterns and, in particular, a preferential decrease in diversity and subsequent extinction of the colder-water North Atlantic conodont fauna in the Late Ordovician Period.

15.8 ACKNOWLEDGEMENTS

The research presented here was initiated by the senior author as a M.Sc. Thesis carried out under the supervison of Dr. Walter C. Sweet, whose constructive discussion on many of the topics reported here is gratefully acknowledged. Special thanks go to Dr. Yehoshua Kolodny, the Hebrew University, Jerusalem, Israel, for performing the $\delta^{18}O$ analyses and aiding in the interpretation of the results. Drs. S. Bergström, K. Foland, J. Kovach and S. Savin are acknowledged for their comments on many of the ideas presented here.

Financial support was provided by the Friends of Orton Hall Fund, the Chevron Thesis Support Fund, the Appalachian Basin Industrial Associates, the Geological Society of America (Grant 3147–83) and the Petroleum Research Fund as administered by the American Chemical Society.

We also thank ARCO for supporting this work, for allowing its publication and for the drafting of figures.

15.9 REFERENCES

Barnes C. R. and Fåhraeus, L. E. 1975. Provinces, communities, and the proposed nektobenthic habit of Ordovician conodontophorids. *Lethaia*, **8**, 133–149.

Barnes, C. R., Rexroad, C. B., and Miller, J. F. 1972. Lower Paleozoic Provincialism. In F. H. T. Rhodes (Ed.), *Conodont Paleozoology, Geological Society of America Special Paper*, **141**, 157–190.

Bergström, S. M. and Carnes, J. B. 1976. Conodont biostratigraphy and paleoecology of the Holston Formation (Middle Ordovician) and associated strata in eastern Tennessee. In C. R. Barnes (Ed.), *Conodont Paleoecology, Geological Society of Canada Special Paper*, **15**, 27–58.

Bergström, S. M. and Noltimier, H. C. 1982. Latitudinal positions of Ordovician continental plates based on palcomagnctic cvidcncc. In D. L. Bruton and S. H. Williams (Eds.), *Proceedings of the International Symposium on the Ordovician System Abstracts, Palaeontological Contributions from the University of Oslo*, **280**.

Beuf, S., Biju-Duval, B., de Charpal, O., Rognon, P., Gariel, O. and Bennacef, A. 1971. *Les Gres du Paleozoique Inferieur au Sahara (Sedimentation et Discontinuites; Evolution Structurale d'un Craton)*, Editions Technip, Paris, 1–464.

Bradshaw, L. E., Noel, J. A. and Larson, R. J. 1973. In F. H. T. Rhodes (Ed.), *Conodont Paleozoology, Geological Society of America Special Paper*, **141**, 67–84.

Brodskii, A. I. and Sulima, L. V. 1953. Isotope exchange of oxygen in solutions of phosphate acid. *Doklady Akademii Nauk*, **92**, 589.

Clark, D. L. 1983. Extinction of conodonts. *Journal of Paleontology*, **57** (4), 652–661.

Cohn, M. 1953. A study of oxidative phosphorylation with $\delta^{18}O$-labeled inorganic phosphate. *Journal of Biological Chemistry*, **201**, 735–750.

Dahms, A. S. and Boyer, P. D. 1973. Occurrence and characteristics of $\delta^{18}O$ exchange reactions catalyzed by sodium- and potassium-dependent adenine triphosphatases. *Journal of Biological Chemistry*, **248**, 3155.

Fortey, R. A. 1984. Global earlier Ordovician transgressions and regressions and their biological implications. In D. L. Bruton (Ed.), *Aspects of the Ordovician System, Palaeontological Contributions from the University of Oslo*, **295**, 37–50.

Frakes, L. A. 1979. *Climates Throughout Geologic Time*, Elsevier, Amsterdam, 1–310.

Geitgey, J. E. 1984. Significance and interpretation of oxygen-isotope analyses of Ordovician conodonts and other phosphatic materials. *Unpublished M.Sc. Thesis* Ohio State University, Columbus, Ohio, 1–116.

Geitgey, J. E. 1985. Correspondence between Ordovician conodont diversity and isotopically inferred marine paleotemperatures. *Geological Society of America Abstracts with Programs*, **17** (2), 92.

Harland, W. B. 1972. The Ordovician ice age. *Geological Magazine*, **184**, 1860.

Holser, W. T., Kovach, J. and Wright-Clark, J. 1982. Geochemistry of conodont apatite. In L. Jeppsson and A. Lofgren, (Eds.), *Third European Conodont Symposium (ECOS III)*, *Abstracts*, Private publication. University of Southampton, 14.

Jablonski, D., Flessa, K. W. and Valentine, J. W. 1985

Biogeography and paleobiology. *Paleobiology,* **11,** 75–90.

Kolodny, Y., Luz, B. and Navon, O. 1983. Oxygen Isotope variations in phosphates of biogenic apatites, I. Fish bone apatite — rechecking the rules of the game. *Earth and Planetary Science Letters,* **64,** 405.

Koshland, D. E. and Clarke, E. 1953. Mechanisms of hydrolysis of adenosinetriphosphate catalyzed by lobster muscle. *Journal of Biological Chemistry,* **205,** 917–924.

Kovach, J. 1980. Variations in the strontium isotope composition of seawater during Paleozoic time determined by analysis of conodonts. *Geological Society of America Abstracts with Programs,* **12** (7), 465.

Kovach, J. 1981a. Variation in strontium isotopic composition of Paleozoic oceans. *US Geological Survey Cross Section,* **12,** 19–20.

Kovach, J. 1981b. The strontium content of conodonts and possible use of the strontium concentrations and strontium isotopic composition of conodonts for correlation purposes. *Geological Society of America Abstracts with Programs,* **13** (6), 285.

Kovach, J. and Futa, K. 1983. Conodonts as possible indicators of variations in the neodymium isotopic composition of Paleozoic seawater. *Geological Society of America Abstracts with Programs,* **15** (4), 247.

Kovach, J. and Zartman, R. E. 1981. U–Th–Pb dating of conodonts. *Geological Society of America Abstracts with Programs,* **13** (6), 285.

Luz, B., Kolodny, Y. and Kovach, J. 1984. Oxygen isotope variations in phosphate of biogenic apatites, III. Conodonts. *Earth and Planetary Science Letters,* **69,** 255.

MacArthur, R. H. and Wilson, E. O. 1967. *The Theory of Island Biogeography,* Princeton University Press, Princeton, New Jersey.

Nicoll, R. S. 1976. The effect of Late Carboniferous–Early Permian glaciation on the distribution of conodonts in Australia. In C. R. Barnes (Ed.), *Conodont Paleoecology, Geological Society of Canada Special Paper,* **25,** 263–278.

Pander, H. 1856. *Monographie der Fossilen Fische des silurischen Systems der russisch-baltischen Gouvernements,* Akademie de Wissenschaften, St. Petersburg, 1–91, 9 plates.

Pietzner, H., Vahl, J., Werner, H. and Ziegler, W. 1968 *Zur chemischen Zusammensetzung und Mikromorphologie der Conodonten.* Palaeontographica, **128,** 115152, 10 plates.

Sachs, H. M., Denkinger, M., Bennett, C. L. and Harris, A. G. 1980. Radiometric dating of sediments using fission tracks in conodonts. *Nature,* **228,** 358–361.

Savin, S. M. 1977. The history of the Earth's surface temperatures during the past 100 million years. *Annual Review of Earth and Planetary Sciences,* **5,** 314–355.

Savin, S. M. 1982. Stable isotopes in climatic reconstruction. In *Climate in Earth History,* National Academy, Washington, DC, 164–171.

Sepkoski, J. J., Jr. 1976. Species diversity in the Phanerozoic: species–area effects. *Paleobiology,* **2,** 298–303.

Shaw, H. F. and Wasserburg, G. J. 1985. Sm–Nd in marine carbonates and phosphates: implications for Nd isotopes in seawater and crustal ages. *Geochimica et Cosmochimica Acta,* **49,** 503–518.

Shemesh, A., Kolodny, Y. and Luz, B. 1983. Oxygen isotope variations in phosphate of biogenic apatites, III. Phosphorite rocks. *Earth and Planetary Science Letters,* **64,** 405–416.

Simberloff, D. 1974. Permo-Triassic extinctions: effects of area on biotic equilibrium. *Journal of Geology,* **82** 267–274.

Stehli, F. G., Douglas, R. G. and Kafescioglu, I. A. 1972 Models for the evolution of planktonic foraminifera. In T. J. M. Schopf (Ed.), *Models in Paleobiology,* Freeman–Cooper, San Francisco, California, 116–128.

Stein, S. S. and Koshland, D. E. 1952. Mechanism of action of alkaline phosphatose. *Archives of Biochemistry and Biophysics,* **39,** 229–230.

Sweet, W. C. 1985. Conodonts: Those fascinating little whatzits. *Journal of Paleontology,* **59** (3), 485–494.

Sweet, W. C. and Bergström, S. M. 1974. Provincialism exhibited by Ordovician conodont faunas. *Society of Economic Paleontologists and Mineralogists Special Publications,* **21,** 189–202.

Sweet, W. C. and Bergström, S. M. 1984. Conodont provinces and biofacies of the Late Ordovician. In D. L. Clark (Ed.), *Conodont Provinces and Provincialism, Geological Society of America Special Paper,* **196,** 69–87.

Sweet, W. C., Turco, C. A., Warner, E., Jr. and Wilkie, L. C. 1959. The American Upper Ordovician Standard, I. Eden conodonts from the Cincinnati region of Ohio and Kentucky. *Journal of Paleontology,* **33,** 1029–1068.

Tudge, A. P. 1960. A method of analysis of oxygen isotopes in ortho-phosphates — its use in the measurement of paleotemperatures. *Geochimica et Cosmochimica Acta,* **18,** 81–93.

Urey, H. C., Lowenstam, H. A., Epstein, S. and McKinney, C. R. 1951. Measurement of paleotemperatures of the Upper Cretaceous of England, Denmark and the southern United States. *Geological Society of America Bulletin,* **62,** 399–416.

Vail, P. R., Mitchum, R. M. and Thompson, S. 1977. Seismic stratigraphy and global changes of sea-level, Part 4: Global cycles of relative changes of sea-level. In C. E. Payton (Ed.), *Seismic Stratigraphy-Applications to Hydrocarbon Exploration, American Association of Petroleum Geologists Memoir,* **26,** 83–98.

Winter, E. R. S., Carlton, M. and Briscoe, H. V. A. 1940. The interchange of heavy oxygen between water and inorganic oxygen. *Journal of the Chemical Society,* **31,** 131.

Wright, J., Seymour, R. S. and Shaw, H. F. 1984. REE and Nd isotopes in conodont apatite: variations with geological age and depositional environment. *Geological Society of America Special Paper,* **196,** 325–340.

16

Conodont chemostratigraphy across the Cambrian–Ordovician boundary: western USA and southeast China

J. Wright, J. F. Miller and W. T. Holser

The trace-element composition of biogenic apatite contains a detailed record of perturbations of sea-water chemistry that is useful for regional and intercontinental correlation; such correlations are referred to as chemostratigraphy. Detailed studies of conodont and some inarticulate brachiopod apatite from sections spanning the Cambrian–Ordovician boundary show that strata can be correlated by secular variations of trace elements. These strata consist of marine carbonates deposited in different tectonic settings: the Texas strata were deposited in a marginal cratonic sea, the Oklahoma strata in an aulacogen and the Utah strata on a carbonate platform. The strata in southeast China were deposited in slightly deeper water on a continental shelf. Precise biostratigraphic correlation is controlled by important conodont and trilobite lineages and by well-defined eustatic events. The biostratigraphic horizons are used as the basis for correlation of the chemical signatures. Chemical data from conodont and inarticulate brachiopod samples in the western USA show excellent correlation of secular variations of the cerium anomaly in rare-earth element patterns. A metal-rich zone occurs at the base of the Late Cambrian *Cordylodus proavus* conodont zone and coincides with the first occurrence of *Eurekia apopsis* trilobites in western USA sections. A similar geochemical signal is recognized in the section from China and was used to predict the correlative horizon in China.

16.1 INTRODUCTION

Previous work has shown that carbonate apatite of conodonts, inarticulate brachiopods and fish all retain a greatly enriched but unfractionated geochemical signature of the sea-water in which the skeletal material was deposited (Wright and Holser, 1981; Wright-Clark, 1982; Wright et al., 1984; Shaw and Wasserburg, 1985; Keto and Jacobson, 1985; Elderfield and Pagett, in press; Staudigel et al., in press). We report the first detailed chemostratigraphic study of trace-element variations through a time interval encompassing the Cambrian–Ordovician boundary. The strata investigated consist of well-exposed marine carbonates deposited in different tectonic settings in the western USA (Miller *et al.*, 1982) and southeast China (Lu and Lin, 1983). Fig. 16.1 shows the locations of sections on a reconstruction of plate positions during Late Cambrian times (Scotese *et al.*, 1979). The Lange Ranch section is in the Llano Region of Texas and was deposited near the oceanic margin of a shallow cratonic sea. The Chandler Creek section in the Wichita Mountains of Oklahoma was deposited in an aulacogen. The Lava Dam sections in the House Range, Utah, were deposited on a carbonate platform. The Dadoushan section in western Zhejiang Province, China, represents relatively deep-water deposition on a continental shelf. Table 16.1 lists the intervals at which each section was sampled and includes the stratigraphic position of each sample as measured from the base of Miller's sections.

The present chapter deals with a portion of a broader study of the trace-element geochemistry of Palaeozoic to Recent biogenic apatite. The Cambrian–Ordovician boundary

interval was chosen for detailed study for two reasons. First, Miller (1984) proposed that two worldwide eustatic events (possibly related to glaciation in Gondwanaland) strongly affected the evolution of contemporary conodonts and other marine invertebrates during this boundary interval. It seemed likely that these eustatic events might also have produced geochemical changes in seawater that were recorded in conodonts and other skeletal apatite. Second, the IUGS International Working Group on the Cambrian–Ordovician Boundary is in the process of standardizing the base of the Ordovician System by considering the definition of a boundary point in a stratotype section. Geochemical research that might shed new light on problems of correlation and on environmental interpretations is therefore particularly timely.

16.2 SAMPLE PREPARATION

In the present study we used a composite of all taxa of either conodont elements or inarticulate brachiopod shells recovered from each sample.

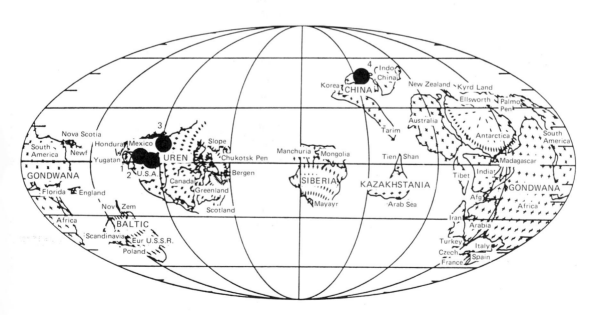

Fig. 16.1 — Palaeogeographical reconstruction of continental positions for middle Late Cambrian (Franconian Age). Mollweide equal-area projection. Sample locations are as follows: 1, Texas; 2, Oklahoma; 3, Utah; 4, China. (After Scotese *et al.* (1979).)

Table 16.1 — Locations and chronostratigraphic units from which samples were collected.

Sample	System	Series or stage	Location	Formation
JW3-9Br	Ordovician	'Ibexian' Series*	Lange Ranch, Texas	Tanyard Formation 1517 ft
JW3-8Br	Ordovician	'Ibexian' Series*	Lange Ranch, Texas	Wilberns Formation 1512 ft
JW3-5MEI-WU;5NS-WU; 5Br-NASA; 5MEI-NASA; 5NS-NASA	Ordovician	'Ibexian' Series*	Lange Ranch, Texas	1497 ft
JW3-4Br-WU; 4Br-NASA	Ordovician	'Ibexian' Series*	Lange Ranch, Texas	1492 ft
JW3-3Br	Ordovician	'Ibexian' Series*	Lange Ranch, Texas	1487 ft
JW3-24NS-WU; 24Br-NASA; 24NS-NASA; 24MEI-NASA; 24MEI/cl-NASA	Ordovician	'Ibexian' Series*	Lange Ranch, Texas	1472 ft
JW3-23NS-WU; 23NS-WU†; 23Br-NASA; 23MEI-NASA	Ordovician	'Ibexian' Series*	Lange Ranch, Texas	1457 ft
JW3-17Br-WU; 17Br-NASA	Ordovician	'Ibexian' Series*	Lange Ranch, Texas	1412 ft
JW3-17A-Br-WU	Ordovician	'Ibexian' Series*	Lange Ranch, Texas	1411.5 ft
JW2-9.5A-Br; 9.5B-Br‡	Cambrian	Trempealeauan Stage	Lange Ranch, Texas	Wilberns Formation, 1400.5 ft
JW2-9Br; 9Br†	Cambrian	Trempealeauan Stage	Lange Ranch, Texas	>1400 ft
JW2-8Br-WU; 8Br-WU†; 8Br-NASA	Cambrian	Trempealeauan Stage	Lange Ranch, Texas	<1400 ft
JW2-25Br-NASA	Cambrian	Trempealeauan Stage	Lange Ranch, Texas	1132 ft
JW3-46Br	Ordovician	'Ibexian' Series*	Chandler Creek, Oklahoma	Signal Mountain Limestone, 1910 ft
JW3-34Br-WU; 34Br-WU†; 34Br-NASA; 34NS-NASA; 34MEI-NASA	Ordovician	'Ibexian' Series*	Chandler Creek, Oklahoma	1687 ft
JW3-31Br	Ordovician	'Ibexian' Series*	Chandler Creek, Oklahoma	1529 ft
JW3-29Br	Ordovician	'Ibexian' Series*	Chandler Creek, Oklahoma	1512 ft
JW3-28Br	Ordovician	'Ibexian' Series*	Chandler Creek, Oklahoma	1508 ft
JW3-27Br	Ordovician	'Ibexian' Series*	Chandler Creek, Oklahoma	1500 ft
JW3-26Br	Ordovician	'Ibexian' Series*	Chandler Creek, Oklahoma	1497 ft
JW3-25Br; 25Br†	Cambrian	Trempealeauan Stage	Chandler Creek, Oklahoma	1490 ft
JW2-37Br	Cambrian	Trempealeauan Stag	Chandler Creek, Oklahoma	1427 ft
JW3-75Br‡	Ordovician	'Ibexian' Series*	House Range, Utah	Fillmore Formation, 2291 ft
JW3-73Br	Ordovician	'Ibexian' Series*	House Range, Utah	2271 ft
JW3-71Br	Ordovician	'Ibexian' Series*	House Range, Utah	House Limestone, 2246 ft
JW3-67Br	Ordovician	'Ibexian' Series*	House Range, Utah	2150 ft
JW3-65Br	Ordovician	'Ibexian' Series*	House Range, Utah	2125 ft
JW3-64Br	Ordovician	'Ibexian' Series*	House Range, Utah	2109 ft
JW3-58Br	Ordovician	'Ibexian' Series*	House Range, Utah	2101 ft
JW3-57Br; 57Br†	Ordovician	'Ibexian' Series*	House Range, Utah	2080 ft
JW3-59Br	Ordovician	'Ibexian' Series*	House Range, Utah	2048 ft
JW3-54Br	Ordovician	'Ibexian' Series*	House Range, Utah	1710 ft
JW3-52Br	Ordovician	'Ibexian' Series*	House Range, Utah	Notch Peak Formation, 1601 ft
JW3-50Br‡	Ordovician	'Ibexian' Series*	House Range, Utah	1592 ft
JW3(2)-55Br	Cambrian	Trempealeauan Stage	House Range, Utah	1583 ft 4 in
JW2-54A-Br; 54B-Br; 54B-Br†	Cambrian	Trempealeauan Stage	House Range, Utah	1582 ft 8 in
JW2-45Br†	Cambrian	Trempealeauan Stage	House Range, Utah	Orr Formation, 287 ft
JW2-40Br†	Cambrian	Trempealeauan Stage	House Range, Utah	95 ft
JW3-101Br; 101Br†	Ordovician	Lower Ordovician Series	Dadoushan, southeast China	Yinchupu Formation, 269 ft
JW3-100Br	Ordovician	Lower Ordovician Series	Dadoushan, southeast China	197 ft
JW2-105Br; 105Br†	Cambrian	Upper Cambrian Series	Dadoushan, southeast China	Siyangshan Formation, 184 ft
JW2-104A-Br; 104A-Br†;104B-Br‡	Cambrian	Upper Cambrian Series	Dadoushan, southeast China	154 ft
JW2-103Br; 103Br†	Cambrian	Upper Cambrian Series	Dadoushan, southeast China	151 ft
JW2-102Br;2-102Br†	Cambrian	Upper Cambrian Series	Dadoushan, southeast China	144 ft
JW2-101Br‡	Cambrian	Upper Cambrian Series	Dadoushan, southeast China	133 ft
JW2-100Br; 100Br†	Cambrian	Upper Cambrian Series	Dadoushan, southeast China	125 ft

† Weight corrected to average CaO content of 52 wt. %.

‡ Inarticulate brachiopod sample.

* Ibexian was introduced as a series name for the Lower Ordovician in North America by Hintze (1982). We place the name, Ibexian, in quotation marks because it may not be possible to use it as a series name. Ibexian may be preoccupied by previous use as the Ibex Substage of LeMone (1975) or as the Ibex Member of the Ely Springs Dolostone of Budge and Sheehan (1980).

The use of composite samples minimizes metabolic effects and gives an average of the seawater environment in which the organisms were deposited. The carbonate matrix of samples was removed by dissolution in 10% acetic acid. Fossil apatite was further concentrated in the heavy liquid tetrabromoethane. Tests for chemical contamination by acetic acid or tetrabromoethane (Wright, 1985; Shaw, 1984) show that heavy liquids may remove part of the trace elements present but do not cause fractionation. Fig. 16.2 compares the rare-earth element patterns of samples of conodonts prepared three ways: hand picked without heavy-liquid concentration, concentrated with tetrabromoethane, or concentrated with methylene iodide. Use of either heavy liquid reduces the absolute concentration of rare-earth elements, but the rare-earth element patterns look the same, i.e. no fractionation occurs. Isotopic tracer studies of conodonts separated mechanically compared with those separated chemically with acetic acid also show removal or exchange of rare earths, but no fractionation (Shaw,. 1984). Comparison of bromine contents of samples analysed with and without the use of tetrabromoethane (Table 16.2) indicate that bromine is added during heavy-liquid separation and does not occur naturally in the fossil apatite. It is advisable to run similar tests for chemical contamination when using other reagents, e.g. sodium hypochlorite, which also might affect elemental concentrations.

After this two-step separation process, conodont elements or inarticulate brachiopod shells were hand picked from the heavy fraction of the insoluble residue using pure alcohol as a wetting agent. Each sample of fossil material was cleaned ultrasonically in pure alcohol to remove any adhering material, particularly clays and iron minerals. Each sample was then filtered with a millipore system and whole fossils or fragments were hand picked again to ensure that extraneous material was not included in the final portion to be analysed. Samples were placed in ultrapure quartz glass tubes and sealed by fusing the glass in a flame. Sample weights

were determined by weight difference of the empty and filled quartz tubes prior to sealing. The sensitivity of the balance for each weight is ± 0.01 mg. For very small samples, e.g. less than 0.20 mg, this may be a significant source of error. The analytical data for very small conodont samples have been normalized to a nominal CaO concentration for apatite of 52 ± 4 wt. % (Pietzner *et al.*, 1968; Deer *et al.*, 1966). These normalized values are listed in Table 16.2 after the set of values calculated using the measured weight. Normalization to a nominal CaO concentration does not change the shape of the rare-earth element pattern or the value of the cerium anomaly.

16.3 ANALYTICAL PROCEDURES

Instrumental neutron activation analysis was used to determine trace-element contents of samples. Instrumental neutron activation analysis is a quantitative application of γ-ray spectroscopy. It is a non-destructive and highly sensitive technique by which concentrations of many elements may be determined simultaneously. A general description of this technique has been given by Laul (1979). Modifications to standard procedures which facilitate analysis of small microfossil samples include the use of high-purity quartz tubes, high neutron flux ($(1-5) \times 10^{18}$ neutrons cm^{-1}) and long counting times (8–12 h). Samples were analysed at either the Johnson Space Center, National Aeronautics and Space Administration, Houston, Texas, or at Washington University, St. Louis, Missouri, using the procedures of Jacobs *et al.* (1977). Recent modifications at Washington University include the use of the TEABAGS computer program (Lindstrom and Korotev, 1982) and new fly-ash standards (Korotev, in press). Interlaboratory bias was negligible between the two laboratories. Analytical uncertainties were lower at Washington University because of a higher neutron flux and longer counting times. Fig. 16.2 illustrates the consistency of analyses for splits of the same sample analysed at each laboratory. Small variations in neodymium between the laboratories result from the differences in counting times and neutron fluxes.

Trace-element data for Cambrian–Ordovician boundary samples are given in Table 16.2.

All trace elements are in parts per million (ppm) except gold which is in parts per billion (ppb). At the range and level of elemental concentrations reported here, the 1σ (one standard deviation) error based on counting statistics is

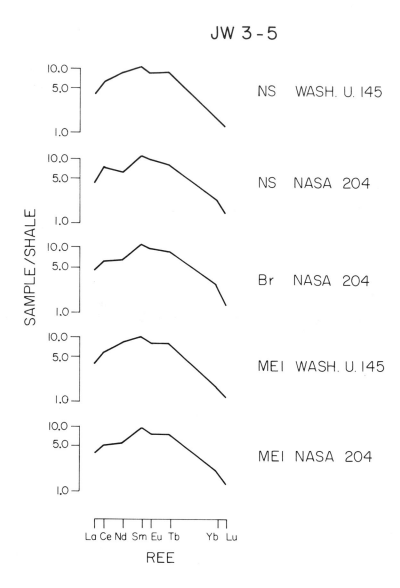

Fig. 16.2 — Comparison of rare-earth element patterns of samples: Br, separated with tetrabromoethane; MEI, separated with methylene iodide; NS, not separated with heavy liquid. They also show the variation in rare-earth element analyses between different instrumental neutron activation analysis laboratories: WASH U, Washington University, St. Louis, Missouri; NASA, National Aeronautics and Space Administration, Johnson Space Center, Houston, Texas. Concentrations normalized to shales are indicated on a logarithmic scale on the vertical axis. See text for further explanation.

Table 16.2 — Chemical data on conodonts from Cambrian–Ordovician boundary sections.

Texas	JW3-9Br	JW3-8Br	JW3-5MEI-WU	JW3-5NS-WU	JW3-5Br-NASA	JW3-5MEI-NASA
Trace elements						
Sc	6	0.4	0.9	1.2	0.8	0.7
Cr	66	220	69	70	13	12
Co	0.7	1	0.5	0.6	1.1	0.8
Ni	230	660	260	210	ND	100±60
Zn	100	56	65	68	38	32
As	ND	ND	ND	ND	ND	ND
Br	27	12	3	3	0.5	ND
Sr	9800	7000	6000	6400	5600	4600
Sb	0.04±0.08	0.01±0.28	0.2±0.1	0.1±0.1	ND	ND
La	200	100	130	130	150	130
Ce	620	270	420	460	450	380
Nd	590	190	280	300	220	190
Sm	130	43	57	62	65	54
Eu	18	7	10	11	12	10
Tb	13	3.7	6.9	7.7	7.6	6.5
Yb	7	3	5.1	5.1	8.5	6.4
Lu	0.9	0.2±0.1	0.5	0.6	0.6	0.6
Au	8±6	8±11	−3±5	4±4	6.9	5.6
Th	77	47	47	51	61	37
U	9.4	11	9.5	9.8	12	11
Mass (mg)	1.01	0.14	0.46	0.29	1.10	0.98

Table 16.2 — *continued.*

Texas	JW3-5NS-NASA	JW3-4Br-WU	JW3-4Br-NASA	JW3-3Br	JW3-24NS-WU	JW3-24Br-NASA
Trace elements						
Sc	1	2.5	3.2	0.31	4.7	2.9
Cr	23	17	24	110	67	ND
Co	2	0.3	1.4	0.7	0.5	1.4
Ni	ND	50±40	ND	350	210	NA
Zn	41	63	26	51	56	70
As	ND	ND	ND	ND	ND	NA
Br	ND	34	7.3	8.8	3.8	0.6
Sr	4200	5300	4700	5900	19100	6500
Sb	ND	0.01±0.1	ND	0.1±0.1	0.0±0.1	ND
La	130	150	160	60	180	160
Ce	550	530	600	180	540	480
Nd	210	350	350	120	370	180
Sm	63	74	84	24	65	120
Eu	12	12	15	4.2	10	9.5
Tb	7.1	8.5	9.7	2.8	7.5	6.3
Yb	7.3	6.2	8.5	1.7	5.8	5.9
Lu	0.6	0.7	0.8	0.2	0.7	0.5
Au	6.6	5±5	5.5	2±3	7±6	ND
Th	53	59	63	22	65	35
U	11.	8.5	9	7.7	12	13
Mass (mg)	0.55	0.82	0.95	0.44	0.42	0.94

Table 16.2 — *continued.*

Texas	JW3-24NS-NASA	JW3-24MEI-NASA	JW3-24MEI/cl-NASA	JW3-23NS-WU	JW3-23NS-WU*	JW3-23Br-NASA
Trace elements						
Sc	4.3	2.5	1.8	7.2	5.9	9.2
Cr	ND	ND	ND	78	64	55
Co	2	ND	2.4	1.4	1.2	3.1
Ni	NA	NA	NA	290	240	ND
Zn	61	120	140	90	74	ND
As	NA	NA	NA	ND	ND	ND
Br	ND	ND	0.2±0.1	ND	ND	6.7
Sr	5100	5700	4900	9000	7400	10100
Sb	ND	ND	ND	0.3±0.3	0.2±0.3	ND
La	200	170	110	260	210	250
Ce	650	470	390	880	730	1020
Nd	260	200	180	760	630	650
Sm	140	110	74	140	120	140
Eu	12	10	7.6	18	15	18
Tb	9.3	6.9	5.3	13	11	16
Yb	9.3	8.1	7.9	9.3	7.7	15
Lu	0.7	0.6	0.6	0.9	0.8	1.4
Au	ND	ND	ND	10±20	8±15	8.3
Th	58	23	30	93	76	106
U	14	16	13	11	8.9	20
Mass (mg)	0.60	0.55	0.23	0.15	0.12	0.20

Table 16.2 — *continued.*

Texas	JW3-23MEI-NASA	JW3-17Br-WU	JW3-17Br-NASA	JW3-17A-BR	JW2-9.5A-Br	JW2-9.5B-Br
Trace elements						
Sc	6.9	0.6	0.9	1.1	5.3	1.6
Cr	28	2.3	24	2±1	36	32
Co	1.6	0.5	2±1	0.7	2.3	0.8
Ni	120±70	20±10	ND	10±30	170	190
Zn	60	73	240	77	180	86
As	ND	ND	ND	3±2	8.6	ND
Br	1.2	18	1.2	2.7	4.2	ND
Sr	9100	5800	4200	5400	4700	2000
Sb	ND	0.06±0.03	ND	0.06±0.04	0.5±0.3	0.3±0.2
La	220	78	87	140	280	530
Ce	700	230	290	440	1290	1920
Nd	440	150	190	300	900	1430
Sm	120	32	37	65	180	250
Eu	16	5.1	7.1	10	31	40
Tb	9.9	3.7	5.2	7.5	21	30
Yb	11	2.5	3.5	4.4	11	15
Lu	1	0.3	0.7	0.5	1.4	1.7
Au	4.5	3±3	1.9	1±4	14±8	20±10
Th	59	14	15	18	50	10
U	9.8	5	7.6	6.2	18	41
Mass (mg)	0.78	1.08	0.53	0.81	0.21	0.51

Table 16.2 — *continued.*

Texas	JW2-9Br	JW2-9Br*	JW2-8Br-WU	JW2-8Br-WU*	JW2-8Br-NASA	JW2-25Br
Trace elements						
Sc	4.8	3.6	5.6	4.6	4.3	18
Cr	340	260	63	52	41	77
Co	4.3	3.2	7.4	6.1	12	31
Ni	1130	840	210	170	ND	ND
Zn	220	160	190	160	110	130
As	ND	ND	ND	ND	50±30	ND
Br	240	180	530	430	1.4	32
Sr	5100	3800	5200	4300	4000	2100
Sb	0.9	0.7	0.3±0.4	0.3±0.4	1.1±0.6	ND
La	580	430	260	230	230	410
Ce	2350	1740	960	790	1030	1930
Nd	1520	1130	640	530	270	860
Sm	330	240	130	110	130	220
Eu	51	38	20	17	24	44
Tb	39	29	14	12	14	28
Yb	25	18	7.8	6.4	12	24
Lu	3	2.2	0.8	0.7	0.9	2.3
Au	20±30	10±30	9±18	7±18	18	10
Th	140	110	26	21	33	81
U	22	16	13	11	11	25
Mass (mg)	0.05	0.04	0.11	0.09	0.33	0.20

Table 16.2 — *continued.*

Oklahoma	JW3-46Br	JW3-34Br-WU	JW3-34Br-WU*	JW3-34Br-NASA	JW3-34NS-NASA	JW3-34MEI-NASA
Trace elements						
Sc	0.4	2.7	3.3	2.3	5	1.6
Cr	17	24	29	NA	NA	NA
Co	0.4	0.5	0.5	1.8	26	18
Ni	40±50	50±90	60±90	NA	NA	NA
Zn	32	80	97	60	190	90
As	ND	ND	ND	NA	NA	NA
Br	12	24	29	0.7	ND	ND
Sr	5900	12000	14500	3500	3600	3300
Sb	0.0±0.2	0.4±0.8	0.5±0.8	ND	ND	ND
La	37	98	120	100	210	92
Ce	110	290	350	270	490	230
Nd	74	160	190	180	ND	ND
Sm	15	32	39	47	38	19
Eu	2.4	4.9	5.9	6.7	13	4.4
Tb	1.5	2.2	2.7	2.8	5.1	2.6
Yb	0.6	1.3	1.6	3.2	ND	ND
Lu	0.06±0.03	0.3	0.3	ND	ND	ND
Au	6±5	ND	ND	ND	ND	ND
Th	17	11	13	10	29	24
U	4.2	3.0	3.6	ND	ND	ND
Mass (mg)	0.73	0.17	0.21	0.46	0.23	0.45

Table 16.2 — *continued.*

Oklahoma	JW3-31Br	JW3-29Br	JW3-28Br	JW3-27Br	JW3-26Br	JW3-25Br
Trace elements						
Sc	0.4	1.2	0.7	0.5	2.1	2.7
Cr	5.4	11	7	0.5±3.4	6	13
Co	0.4	0.9	1.6	0.7	1	1.2
Ni	30±40	50±30	30±90	10±70	−30±70	0±120
Zn	79	64	69	90	65	80
As	ND	ND	10	ND	ND	ND
Br	10	14	27	12	17	53
Sr	5500	4900	5400	5300	4300	3700
Sb	0.1±0.1	0.1±0.1	0.5	0.02±0.35	0.2±0.3	0.3±0.7
La	26	91	66	38	84	160
Ce	98	380	220	93	250	860
Nd	65	190	110	59	150	610
Sm	12	33	20	10	26	110
Eu	1.8	5.6	3.5	1.8	4.3	19
Tb	0.9	3.1	1.6	0.7	2.5	10
Yb	0.5	1.7	0.5	0.3±0.2	1.1	2.8
Lu	0.04±0.04	0.2	0.06±0.04	0.1±0.1	0.2±0.1	0.3
Au	ND	−1±3	ND	ND	1±4	10±10
Th	11	34	13	15	32	38
U	2.1	3.5	2	1.4±0.9	3.8	6
Mass (mg)	1.07	0.77	0.35	0.35	0.51	0.17

Table 16.2 — *continued.*

Oklahoma	JW3-25Br*	JW2-37Br	Utah JW3-75Br	JW3-73Br	JW3-71Br
Trace elements					
Sc	3.1	2.1	1.5	0.1	0.4
Cr	15	13	13	15	19
Co	1.4	0.8	1.7	0.5	1.4
Ni	0±100	30±100	10±30	56	30±70
Zn	92	100	61	39	61
As	ND	ND	ND	ND	6±9
Br	61	20	15	12	36
Sr	4300	6000	1600	18400	10700
Sb	0.4±0.7	0.4±0.9	0.1±0.1	0.07±0.10	0.1±0.3
La	190	55	140	42	47
Ce	1000	220	290	100	86
Nd	700	190	170	59	52
Sm	130	37	28	9.8	9.6
Eu	22	6.2	5.6	2	1.7
Tb	12	2.4	3.5	1	1.2
Yb	3.2	1.0	2.7	0.3±0.2	0.7
Lu	0.4	0.2±0.1	0.3	0.05±0.03	0.1±0.1
Au	20±10	ND	-1±5	ND	ND
Th	44	27	24	8.9	12
U	7	3.2	4.8	0.2±0.7	5.8
Mass (mg)	0.20	0.12	1.80	0.61	0.45

Table 16.2 — *continued*.

Utah	JW3-67Br	JW3-65Br	JW3-64Br	JW3-58Br	JW3-57Br	JW3-57Br*
Trace elements						
Sc	7	0.2	0.09	0.5	0.5	0.6
Cr	21	46	28	18	7±5	9±5
Co	2.9	0.6	0.2±0.1	0.5	0.8	1.1
Ni	50±130	140	71	30±20	10±70	20±70
Zn	79	ND	ND	65	90	120
As	ND	ND	ND	ND	ND	ND
Br	23	4	4.3	12	15	19
Sr	7200	10100	11500	9200	9300	12100
Sb	0.2±0.4	0.05±0.08	0.01±0.10	0.03±0.05	0.9±0.8	1.2±0.8
La	97	36	18	55	45	59
Ce	220	85	42	150	110	150
Nd	130	46	22	99	50	65
Sm	20	7.9	4	18	12	15
Eu	4	1.6	0.8	3.1	2.5	3.3
Tb	3.3	0.9	0.5	2.3	1.9	2.5
Yb	2.2	0.8	0.3	1.3	0.7	0.9
Lu	0.6	0.08	0.02±0.03	0.1	0.07±0.06	0.09±0.06
Au	ND	ND	ND	3±4	ND	ND
Th	9.5	3.8	1.8	13	8.7	11
U	1.5	1.3	1.1	1.8	5.9	7.7
Mass (mg)	0.27	0.97	0.89	0.97	0.18	0.23

Table 16.2 — *continued.*

Utah	JW3-59Br	JW3-54Br	JW3-52Br	JW3-50Br	JW2-55Br	JW2-54A-Br
Trace elements						
Sc	0.4	0.4	0.09	0.3	0.6	1.1
Cr	2±2	12	15	6.1	21	160
Co	2.3	1.3	0.7	3.7	1.4	2.5
Ni	0±60	0±60	40±20	−9±40	60	460
Zn	84	47	ND	ND	29	50
As	ND	11	5.3	37	12	53
Br	49	8.8	5.6	39	67	67
Sr	7100	8700	4900	3200	2900	2300
Sb	0.1±0.2	0.09±0.27	0.02±0.09	0.3	0.06±0.06	0.1±0.1
La	71	45	23	130	150	300
Ce	160	77	55	240	360	560
Nd	100	43	34	160	220	370
Sm	16	6.8	6.2	28	39	92
Eu	3.3	1.3	1.1	7.9	6.8	13
Tb	1.9	0.7	0.6	3.0	4.1	8.4
Yb	1.0	0.5	0.2±0.1	1.0	2.4	6
Lu	0.2	0.08	ND	0.2	0.2	0.7
Au	2±5	ND	ND	5±6	3±5	7±4
Th	2.4	8.8	8.3	7.1	54	21
U	2.1	1.1	0.7±0.4	5.7	5.3	19
Mass (mg)	0.74	0.49	0.99	0.44	0.69	0.63

Table 16.2 — *continued.*

Utah	JW2-54B-Br	JW2-54B-Br*	JW2-45Br	JW2-40Br	China	JW3-101Br
Trace elements						
Sc	14	10	0.5	0.4		0.7
Cr	103	71	10	4		38
Co	6.2	4.3	1.4	2±1		2.3
Ni	200±100	160±90	20±30	30±30		50±100
Zn	150	100	46	76		80
As	66	46	ND	ND		ND
Br	99	69	12	4±3		7.9
Sr	5000	3500	2400	1200		2700
Sb	0.7	0.5	0.01±0.05	0.00±0.06		0.6
La	330	230	210	170		73
Ce	670	480	460	360		180
Nd	520	360	300	260		190
Sm	96	67	50	39		40
Eu	18	13	8.6	10		7.2
Tb	10	7.2	5	4.9		4.5
Yb	5.2	3.6	2.6	1.8		1.3
Lu	0.8	0.6	0.3	0.3		0.2±0.1
Au	87	60	8±7	3.5		16±8
Th	52	36	17	8.2		23
U	14	9.5	7.3	9.3		1.5
Mass (mg)	0.05	0.04	1.97	0.92		0.08

Table 16.2 — *continued.*

China	JW3-101Br*	JW3-100Br	JW2-105Br	JW2-105Br*	JW2-104A-Br	JW2-104A-Br*
Trace elements						
Sc	0.9	0.5	1.2	1.6	1	0.9
Cr	51	62	570	760	540	490
Co	3.1	1.9	37	50	3.6	3.2
Ni	70±140	210	1710	2200	1820	1630
Zn	110	58	2170	2900	140	120
As	ND	ND	55	73	ND	ND
Br	11	10	25	33	35	31
Sr	3600	3100	2700	3600	3800	3400
Sb	0.8	0.4	1.6	2.1	0.9	0.8
La	98	270	200	260	310	280
Ce	240	620	410	540	830	740
Nd	250	330	290	390	660	590
Sm	54	64	56	74	0.8	0.7
Eu	9.7	12	9.8	13	27	24
Tb	6.1	8.5	6.7	8.9	18	16
Yb	1.8	7.6	2.7	3.6	7.2	6.5
Lu	0.3±0.2	0.9	0.4	0.5	0.7	0.7
Au	20±10	10±9	22	29	40±20	30±20
Th	31	31	32	42	76	68
U	2.0	22	2.4	3.2	13	11
Mass (mg)	0.11	0.34	0.07	0.09	0.05	0.05

Table 16.2 — *continued*.

China	JW2-104B-Br	JW2-103Br	JW2-103Br*	JW2-102Br	JW2-102Br*	JW2-101Br
Trace elements						
Sc	0.5	0.7	0.6	1	0.9	0.6
Cr	18	210	180	450	390	43
Co	1.8	4.6	4	7.4	6.4	1.9
Ni	110±90	660	570	1420	1230	160
Zn	150	56	49	ND	ND	ND
As	ND	ND	ND	10±5	9±5	ND
Br	42	250	210	33	29	62
Sr	2100	1700	1500	2000	1700	1400
Sb	0.5	0.8	0.7	1	0.8	0.6
La	450	290	250	520	450	320
Ce	1280	610	520	1300	1130	800
Nd	1050	340	290	840	730	550
Sm	230	70	61	160	140	95
Eu	47	12	11	26	23	17
Tb	28	8.6	7.5	22	19	11
Yb	11	8.1	7	17	15	6.4
Lu	1	0.9	0.8	2	1.8	0.6
Au	21	2±15	2±15	20±10	21	10±10
Th	61	52	45	87	75	75
U	19	48	42	21	18	27
Mass (mg)	0.32	0.18	0.16	0.23	0.20	0.31

Table 16.2 — *continued.*

China	JW2-100Br	JW2-100Br*
Trace elements		
Sc	1.1	2.1
Cr	490	910
Co	7.4	14
Ni	1520	2810
Zn	ND	ND
As	ND	ND
Br	83	150
Sr	1100	2040
Sb	1.4	2.6
La	790	1470
Ce	420	790
Nd	250	460
Sm	47	86
Eu	8.5	16
Tb	5.4	10
Yb	3.6	6.7
Lu	0.5	1
Au	52	96
Th	36	67
U	25	46
Mass (mg)	0.08	0.15

ND, not detected; NA, not analysed.

All trace-element concentrations are in parts per million, except gold which is in parts per billion.

* Weight corrected to average CaO content of 52 wt. %.

† Upper limit value, where $1\sigma > 50\%$.

as follows: lanthanum, samarium, europium, thorium < 2%; scandium, cerium, terbium, ytterbium, lutetium < 5%; strontium, neodymium, uranium < 10%; chromium, cobalt, nickel, arsenic, zinc, bromine < 20%; antimony, gold < 30%. Where the precision is low, i.e. $1\sigma > 50\%$, the upper limit values are listed in Table 16.2 (In Figs. 16.5–16.8, these values are given a different symbol (see figure captions).) A complete listing of data including exact 1σ error has been given by Wright (1985).

16.4 BIOSTRATIGRAPHY AND CHRONOSTRATIGRAPHY

We have chosen a stratigraphic interval that has been carefully studied and constrained on the basis of conodonts and trilobites in the western USA (Miller *et al.*, 1982; Miller, 1984). Fig. 16.3 illustrates correlation of strata in Texas, Oklahoma and Utah with biostratigraphic and chronostratigraphic units. Correlation horizons are based on a combination of trilobite and conodont biostratigraphic zones and subzones currently recognized in the western USA. Not all these horizons have been identified in southeast China. The base of the *Missisquoia depressa* trilobite subzone is at present regarded as the Cambrian–Ordovician boundary in the USA. There is as yet no international agreement on the biostratigraphic position of the Cambrian–Ordovician boundary and therefore no globally accepted chronostratigraphic classification of this interval. This problem is being studied by the IUGS Working Group on the Cambrian–Ordovician Boundary. Chemical events observed as signatures in this study may be useful in redefining the position of this chronostratigraphic horizon.

In the Jiangshan area of Zhejiang Province, southeast China, the Cambrian–Ordovician boundary has been placed at the base of the *Hysterolenus* trilobite zone (Lu and Lin, 1983). We have arbitrarily used the Cambrian–Ordovician boundary as defined in each of these regions as the primary correlation horizon for all four sections. Other biostratigraphic hori-

zons used for correlation purposes in the USA. are the bases of trilobite and/or conodont zones or subzones as indicated in Fig. 16.3. The bases of the *Missisiquoia depressa* trilobite subzone and the *Missisquoia typicalis* trilobite subzone = the *Clavohamulus elongatus* conodont subzone correlate within a hiatus in the Texas section. A broken line across the hiatus in Figs. 16.4–16.8 connects the samples on either side of the erosional surface. The hiatus interval is represented by similar broken lines connecting correlative samples in the Oklahoma and Utah sections in these figures.

16.5 RESULTS AND DISCUSSION: CERIUM ANOMALY

Rare-earth elements are normalized to a well-characterized standard to remove the odd–even effect of elemental abundances. We use the North American Shale Composite Standard (Haskin *et al.*, 1968) to normalize the rare-earth elements in our samples. Then the rare-earth elements are plotted with the concentrations on a logarithmic scale versus the atomic number on a linear scale. This type of plot permits visual representation of significant variations. One of these variations is the variation in cerium relative to neighbouring elements lanthanum and neodymium. In some samples, cerium concentrations are higher (enriched) and, in other samples, cerium concentrations are lower (depleted) than lanthanum and neodymium concentrations. The cerium variation can be isolated by giving a mathematical value to its position on the line between lanthanum and neodymium, by the formula

$$Ce_{anom} = \log\left(\frac{3Ce_n}{2La_n + Nd_n}\right),$$

where n is the shale-normalized value of each rare earth (from Elderfield and Greaves, 1982). Comparison with analyses of fish debris in modern sediments from different environments shows that Ce_{anom} values < -0.10 indicate a

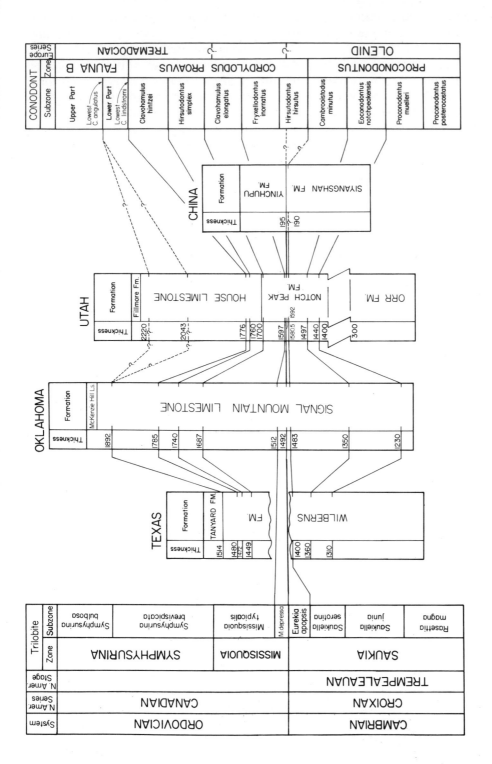

Fig. 16.3 — Correlation of strata in Texas, Oklahoma, Utah and China with biostratigraphic and chronostratigraphic units. Scale in feet as indicated for each section. Texas and Utah are composite sections. Correlation horizons are based on a combination of conodont and trilobite zones and subzones as recognized and used in the western USA. Each horizon is defined as the base of the respective zone. These same horizons are used as reference points in subsequent chemostratigraphic correlation charts.

depletion of cerium contents under suboxic to oxic conditions at the sediment–water interface or in the main water mass. Conversely, Ce_{anom} > −0.10 indicate enrichment of cerium under predominantly anoxic conditions (Wright, 1985; Wright *et al.*, 1985).

Fig. 16.4 shows the variations in Ce_{anom} in each section. The Cambrian–Ordovician boundary, as defined in each geographical region, is assumed to be synchronous, and the sections are 'hung' on this chronostratigraphic horizon. Additional biostratigraphic horizons are shown as full lines where known and broken lines where inferred. By using these reference horizons in USA sections, the shifts in Ce_{anom} are seen to correlate well across an interval from below the *Cordylodus proavus* Zone to above the *Missisquoia typicalis* Subzone. In the Oklahoma and Utah sections, Ce_{anom} values peak

above the *C. proavus* Zone, increase between the *M. depressa* and the *M. typicalis* subzones and decrease between the *C. proavus* Zone and the *M. depressa* Subzone and above the *M. typicalis* Subzone. A lowering of the sea level and resulting period of erosion during the Lange Ranch Eustatic Event is represented by the hiatus in the Texas section (Miller, 1984). In Fig. 16.4 a broken line connects samples below and above this unconformity surface. Equivalent samples are connected by broken lines in the Oklahoma and Utah sections. Despite the erosional event, the Texas section has variations in Ce_{anom} that are similar to those observed in Oklahoma and Utah, e.g. a peak at the *C. proavus* Zone and a decrease above the *M. typicalis* Subzone.

It is important to note that the basis of the Ce_{anom} correlations are three well-constrained

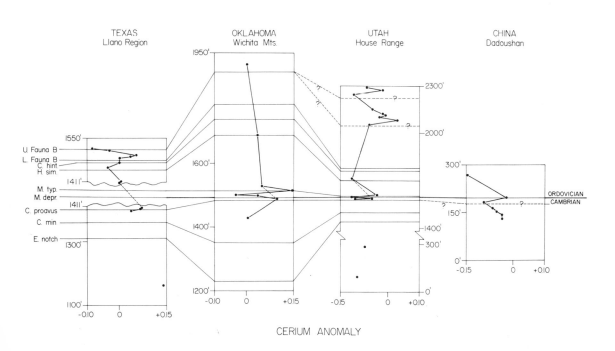

Fig. 16.4 — Correlation of Ce_{anom} in four sections. The folowing abbreviations for the correlation horizons are used in Figs. 16.4–16.8: E. notch., *Eoconodontus notchpeakensis*; C. min., *Cambrooistodus minutus*; C. proavus, *Cordylodus proavus*; M. depr., *Missisquoia depressa*; M. typ., *Missisquoia typicalis*; H. sim., *Hirsutodontus simplex*; C. hint., *Clavohamulus hintzei*; L. Fauna B, lower fauna B; U. Fauna B, upper fauna B. The broken line connects samples above and below unconformity surface in Texas and correlative samples in Oklahoma and Utah.

biostratigraphic zones that occur within a very short time interval. Using the biostratigraphic zones as tie lines between the sections, the highs and lows of Ce_{anom} values are observed to occur at the same stratigraphic position in all three western USA sections. This is even true in the Texas section where part of the interval was removed by erosion. The probability that five specific biostratigraphically constrained chemical correlations occurred by chance in three sections is extremely low.

The section from southeast China has Ce_{anom} values that are low at the inferred position of the *C. proavus* Zone and high at the *M. depressa* Subzone, which is the opposite of Ce_{anom} in the western USA. The lack of correlation of Ce_{anom} values from USA to China may be explained in several ways.

(1) The global correlation of biostratigraphic horizons *C. proavus* and *M. depressa* has not been established. When biostratigraphic studies are completed the question may be resolved by a precise placement of these horizons.

(2) If we assume that plate reconstructions in Fig. 16.1 are approximately correct, it is apparent that the USA and Chinese sections border on ocean basins that were probably not well 'connected' in terms of mixing times of surface waters (Parrish, 1982). Additionally, cerium and other rare earths have short residence times, e.g. 80 years for cerium (Schopf, 1980). Under these circumstances, heterogeneity would be reinforced so that short residence times coupled with the time necessary for mixing of surface waters between ocean basins could have prevented transmission of a chemical event.

(3) Cerium is very soluble in a water column dominated by anoxic conditions. This increased solubility could produce an overall increase in the sea-water concentration of cerium. Under dominantly anoxic conditions in the oceanic water column, the residence time of cerium would increase, so

that short-term local perturbations would not effect the overall balance of cerium in the world ocean (Wright, 1985).

The base of Upper Fauna B, thought to approximate the base of the Upper Tremadoc Series in the western USA, is well defined in the Texas and Oklahoma sections, but its exact position has only been inferred in Utah. In sections in Texas and Utah the sampling interval is close enough to show significant trends that might suggest the location of this horizon. There is a very obvious spike in Ce_{anom} (Fig. 16.4) between Lower Fauna B and Upper Fauna B in Texas. A similar spike occurs in the Utah section, which suggests that the base of Upper Fauna B probably occurs near 2100 ft in the measured Utah section.

16.6 RESULTS AND DISCUSSION: OTHER TRACE ELEMENTS

Variations in concentrations of nine additional trace elements (scandium, chromium, cobalt, nickel, zinc, arsenic, antimony, thorium and gold) are plotted for each section in Figs. 16.5–16.8. Correlation horizons *E. notchpeakensis* through Upper Fauna B (from Fig. 16.3), where known, are indicated on the right-hand side of each figure. Trace-element concentrations are in parts per million, except Au, which is in parts per billion. Trace-element fluctuations may be useful in determining the positions of correlation horizons *C. proavus* or *M. depressa* in China (Fig. 16.5). For reference the eight samples analysed for the China section are numbered 1–8 from oldest to youngest on the scandium profile in Fig. 16.5. All trace-element concentrations are low in samples 7 and 8. In sample 6, seven trace elements are enriched; only thorium and gold are not. Sample 5 has higher concentration of nickel, thorium and gold, but all other trace-element contents are lower than in the previous sample. In sample 4, concentrations of all trace elements are low. Sample 3 shows an increase in trace-element concentrations and thorium peaks here. In sample 2, only thorium concentrations remain high; all other trace elements are low. The oldest sample has high concentrations of scandium, chromium, antimony, nickel and gold. Gold concentrations reach a peak here, and scandium and antimony are at the second-highest concentration levels in the section. The greatest concentrations of trace elements, in decreasing order of enrichment, occur in samples 6, 1 and 5.

Trace-element concentrations for samples in the Texas section vary as shown in Fig. 16.6. For the correlative interval examined in the China section, only the base of the *C. proavus* Zone is present in Texas. The bases of both the *M. depressa* and the *M. typicalis* subzones are not present in the record because of a hiatus at this interval. For this portion of the section in Texas, concentrations of seven trace elements are highest in the sample just above the base of the *C. proavus* Zone; scandium and chromium are highest in the sample just below the base of the *C. proavus* Zone. Arsenic occurs in only two samples at detectable levels, just above the base of the *C. proavus* Zone and just above the erosional unconformity.

In the correlative part of the section from Oklahoma (Fig. 16.7) the greatest trace-element enrichments occur in the sample just above the base of the *C. proavus* Zone and in

the sample at the base of the *M. typicalis* Sub-zone. The base of the *C. proavus* Zone is en-riched in scandium, chromium, nickel, thorium, gold, cobalt and zinc. The first five are the highest concentrations of these elements in this part of the section. The sample below the base of the *M. typicalis* Subzone is enriched in cobalt, nickel and antimony, and is the only sample in which arsenic occurs. The sample at the base of the *M. typicalis* Subzone shows high concent-rations of chromium, thorium and gold. The horizon with the greatest enrichment of trace elements is at the base of the *C. proavus* Zone.

The section from Utah (Fig. 16.8) also shows a very enriched zone of all trace elements just above the base of the *C. proavus* zone. Western USA sections consistently have the same enriched zone of trace elements just above the base of the *C. proavus* Zone. In samples from Texas and Utah, this zone also coincides with a zone of high arsenic concentration.

Tables 16.3 and 16.4 summarize the distri-bution of the nine trace elements in the four sections. The section from China has the grea-test enrichment of trace elements at sample 6 (Table 16.3). Western USA sections (Table 16.4) show a similar enrichment at the base of *Cordylodus proavus* conodont zone which is equivalent to the *Eurekia aposis* trilobite sub-zone (Westrop and Ludvigsen, in press). The enrichment of trace elements at the same hori-

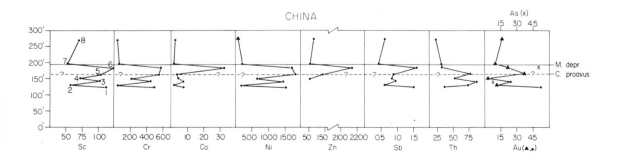

Fig. 16.5 — Variations in trace-element signatures in the China section: For explanation of abbreviations, see Fig. 16.4. In Figs. 16.5–16.8, squares are used to indicate a value with low precision (see Table 16.2 and text).

zon in the USA sections suggests that sample 6 in the section from China is the correlative

horizon. Biostratigraphic study of the section from China by Miller has shown that the base of

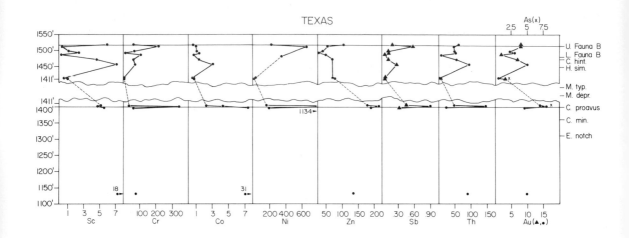

Fig. 16.6 — Variations in trace-element signatures in the Texas section. For explanation of abbreviations, see Fig. 16.4

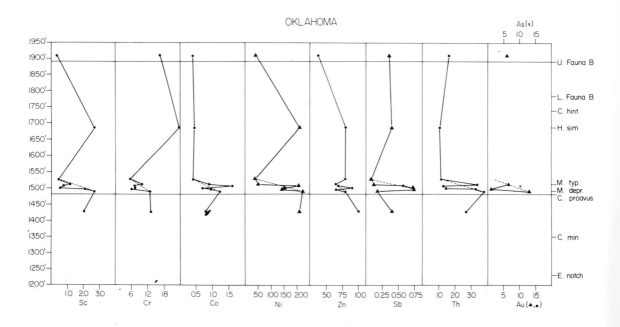

Fig. 16.7 — Variations in trace-element signatures in the Oklahoma section. For explanation of abbreviations, see Fig. 16.4

the *C. proavus* Zone in China occurs at sample 6. In this example the chemical signature predicted the location of the biostratigraphic horizon.

16.7 CONCLUSIONS

Detailed study indicates that sections at the Cambrian–Ordovician boundary interval display correlative variations in the trace-element contents of apatite of conodonts and inarticulate brachiopods. This research provides a new method of stratigraphic correlation, a method we refer to as chemostratigraphy. The use of a *group* of trace elements apparently is a reliable indicator of synchroneity in the stratigraphic record. Several lines of future research are indicated in order to establish fully the reliability of this method. First, sampling must be extended to global coverage of coeval strata to determine whether the chemical events recorded across the northern ocean also were significant in the southern hemisphere and the

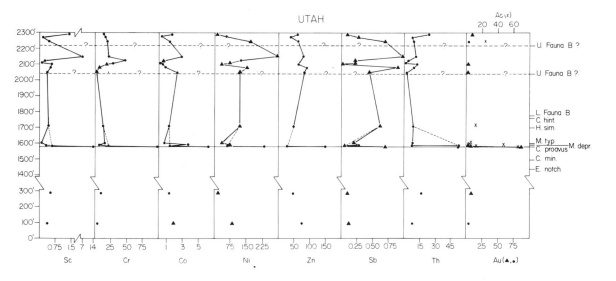

Fig. 16.8 — Variations in trace-element signatures in the Utah section. For explanation of abbreviations, see Fig. 16.4

Table 16.3 – Distribution of trace elements: China.

Sample	Sc	Cr	Co	Ni	Zn	As	Sb	Th	Au
8									
7									
6	X	X	X	X	X	X	X		
5	X	X	X					X	X
4									
3	X	X		X				X	
2								X	
1	X	X		X			X		X

X, peak in concentration level for this element.

Table 16.4 – Distribution of trace elements: western USA.

Sample	Sc	Cr	Co	Ni	Zn	As	Sb	Th	Au
Base of the M. typicalis Zone	O							O	O
Between			O	O	O	O	O		
Base of the M. depressa Zone									
Between									
Base of the C. proavus Zone	OU	TOU	OU	TOU	TOU	TU	TU	TOU	TOU
Below the C. proavus Zone	T		T						

T, peak in concentration for this element in the Texas section; O, peak in concentration for this element in the Oklahoma section; U, peak in concentration for this element in the Utah section.

world ocean. Such research may resolve some of the mysteries of oceanic circulation and mixing during the Palaeozoic Era. Results of further study also may confirm that certain trace-element variations are useful only for regional correlation whereas others are valid worldwide. Second, this method must be tested at additional stratigraphic intervals where there are also sufficient biostratigraphic data to provide an independent basis for correlation of samples. These broader studies will provide a sufficient data base to model the relationships that control the chemical signals. Trace-element chemostratigraphy will eventually become a reliable tool for correlating poorly fossiliferous strata which cannot be correlated by biostratigraphic methods.

16.8 ACKNOWLEDGEMENTS

Professor Lu Yan-hao and others at the Nanjing Institute of Geology and Palaeontology, China, assisted in the field work and arranged logistical support for collecting in China. Michael E. Taylor (US Geological Survey) assisted with shipping samples from China. Tim Carr, Dell Potter (ARCO Research, Dallas) and Steve Aronoff assisted with field work in the western USA. Richard S. Seymour (EG&G Ortec, Oakridge, Tennessee), Marilyn Lindstrom and Randy L. Korotev (Washington University, St. Louis, Missouri) were instrumental in the development and adaptation of instrumental neutron activation analysis procedures for the analysis of low-mass fossil material. We thank Richard J. Williams and Douglas P. Blanchard for providing laboratory facilities and irradiation costs at the Johnson Space Center, National Aeronautics and Space Administration. Ann Wright patiently typed and edited different versions of the manuscript. This research was supported by the National Science Foundation (Grants EAR 83-19429, EAR 81-15985 and EAR 78-21819 (Holser and Wright) and EAR 81-08621 (Miller)). This research represents part of the senior author's Ph.D. dissertation at the University of Oregon, Eugene.

16.9 REFERENCES

Budge, D. R. and Sheehan, P. M. 1980. The Upper Ordovician through Middle Silurian of the eastern Great Basin: Part 1, Introduction, historical perspective and stratigraphic synthesis. *Milwaukee Public Museum Contributions in Biology and Geology*, **28**, 1–26.
Deer, W. A., Howie, R. A. and Zussman, J. 1966. *An*

Introduction to the Rock Forming Minerals, Longman, London, 504–509.

Elderfield, H. and Greaves, M. J. 1982. The rare earth elements in sea water. *Nature,* **296,** 214–219.

Elderfield, H. and Pagett, R. In press. REE in ichthyoliths: variations with redox conditions and depositional environment. In J. P. Riley (Ed.), *The Science of the Total Environment.*

Haskin, L. A., Haskin, M. A., Frey, F. A. and Wildeman, T. R. 1968. Relative and absolute terrestrial abundance of the rare earths. In L. H. Ahrens (ed.), *Origin and Distribution of the Elements, Vol. 30,* Pergamon Press, London, 889–912.

Hintze, L. F. 1982. Ibexian Series (Lower Ordovician) type section, western Utah, USA. In R. J. Ross, Jr. and S. M. Bergström (Eds.), *The Ordovician System in the United States, Correlation Chart and Explanatory Notes, International Union of Geological Sciences,* **12,** 7–10.

Jacobs, J. W., Korotev, R. L., Blanchard, D. P. and Haskin, L. A. 1977. A well-tested procedure for instrumental activation analysis of silicate rocks and minerals. *Journal of Radioanalytical Chemistry,* **40,** 93–114.

Keto, L. S. and Jacobson, S. B. 1985. Implications for paleo-oceanography from Nd and Sr isotope ratios of Lower Paleozoic brachiopods. *EOS Transactions, American Geophysical Union,* **66,** (18), 424.

Korotov, R. L. In press. National Bureau of Standards coal fly ash (SRM-1633A) as a multielement standard for instrumental neutron activation analysis. *Journal of Radioanalytical and Nuclear Chemistry.*

Laul, J. C. 1979. Neutron activation analysis of geological materials. *Atomic Energy Reviews,* **17,** 603–695.

LeMone, D. V. 1975. Correlation aspects of the Ordovician of the southwestern United States. In J. M. Hills (Ed.), *Exploration from the Mountains to the Basin,* El Paso Geological Society, 1975 Guidebook, El Paso, Texas, 169–190.

Lindstrom, D. J. and Korotev, R. L. 1982. TEABAGS: computer programs for instrumental neutron activation analysis. *Journal of Radioanalytical Chemistry,* **70,** 439–458.

Lu, Y.-H. and Lin, H.-L. 1983. Uppermost Cambrian and lowermost Ordovician trilobites of Jiangshan–Changshan area. In *Papers for the Symposium on the Cambrian–Ordovician and Ordovician–Silurian Boundaries,* Nanjing Institute of Geology and Palaeontology, Academia Sinica, 6–11.

Miller, J. F. 1984. Cambrian and earliest Ordovician conodont evolution, biofacies and provincialism. In D. L. Clark (Ed.), *Conodont Biofacies and Provincialism, Geological Society of America Special Paper,* **196,** 43–68.

Miller, J. F., Taylor, M. E., Stitt, J. H., Ethington, R. L.

Hintze, L. F. and Taylor, J. F. 1982. Potential Cambrian–Ordovician boundary stratotype sections in the western United States. In M. G. Bassett and W. T. Dean (Eds.), *The Cambrian-Ordovician Boundary: Sections, Fossil Distributions and Correlations, Geological Series 3,* National Museum of Wales, Cardiff, 155–180.

Parrish, J. T. 1982. Upwelling and petroleum source beds, with reference to Paleozoic. *American Association of Petroleum Geologists Bulletin,* **66,** 750–774.

Pietzner, H., Vahl, J., Werner, H. and Ziegler, W. 1968. Zur chemischen Zusammensetzung und Mikromorphologie der Conodonten. *Palaeontographica,* **128,** 115–152.

Schopf, T. J. M. 1980. *Paleoceanography,* Harvard University Press, Cambridge, Massachusetts, 151.

Scotese, C. R., Bambach, R. K., Barton, C., Van Der Voo, R. and Ziegler, A. M. 1979. Paleozoic base maps. *Journal of Geology,* **87,** 240, Fig. 6.

Shaw, H. F. 1984. Sm–Nd and Rb–Sr systematics of tektites and other impactites, Appalachian mafic rocks, and marine carbonates and phosphates. *Ph.D. Dissertation,* California Institute of Technology, Pasadena, California, 1–282.

Shaw, H. F. and Wasserburg, G. J. 1985. Sm–Nd in marine carbonates and phosphates: implications for Nd isotopes in seawater and crustal ages. *Geochimica et Cosmochimica Acta,* **49,** 503–518.

Staudigel, H., Doyle, P. and Zindler, A. In press. Sr and Nd isotope systematics in fish teeth. *Earth and Planetary Science Letters.*

Westrop, S. R. and Ludvigsen, R. In press. Type species of the basal Ibexian trilobite *Corbinia* Walcott. *Journal of Paleontology.*

Wright-Clark, J. 1982. The geochemistry of conodont apatite: secular variations inferred by comparison with the cerium–iron system in the modern ocean. *Unpublished M.Sc Thesis,* University of Oregon, Eugene, Oregon, 1–67.

Wright, J. 1985. Rare earth element distributions in Recent and fossil apatite: implications for paleoceanography and stratigraphy. *Ph.D. Dissertation* University of Oregon, Eugene, Oregon, 1–259.

Wright, J. and Holser, W. T. 1981. Rare earth elements in conodont apatite as a measure of redox conditions in ancient seas. *Geological Society of America Abstracts,* **13,** 586.

Wright, J., Seymour, R. S. and Shaw, H. F. 1984. REE and Nd isotopes in conodont apatite: variations with geological age and depositional environment. In D. L. Clark (Ed.), *Conodont Biofacies and Provincialism, Geological Society of America Special Paper,* **196,** 325–340.

Wright, J., Schrader, H. and Holser, W. T. 1985. Paleoredox of seawater from cerium variations in fossil apatite. *Geological Society of America Abstracts,* **17,** 755.

17

The possibility of a Lower Devonian equal-increment time scale based on lineages in Lower Devonian conodonts

M. A. Murphy

The parsimonious explanation for a straight-line plot of three or more conodont lineage events when graphic correlation is used to compare two sections is a steady accumulation rate for the particular sections. It follows that either of the sections may then be subdivided into equal-thickness increments that represent equal intervals of time. This permits the development of an equal-increment time scale for the applicable interval.

The spacing of events in the Lower Devonian *delta* Zone conodont lineages in the genera *Ozarkodina*, *Ancyrodelloides*, *Icriodus* and *Amydrotaxis* suggests the possibility of establishing such a scale for the *delta* Zone.

17.1 INTRODUCTION

One of the major problems preventing palaeontologists from contributing more significantly to evolutionary biology is the inability to deal with rate problems. This is so because we do not have a method that permits us to measure time in equal increments. If we could devise a way of constructing such a time scale, it is possible that problems of relative rates of change in fossil organisms could be attacked (as well as other types of rate problem). At present we are unable to assess whether or not our data are adequate to evaluate any short-term rate problems.

In an important paper, Churkin *et al.* (1977) suggested that such a scale might be developed in basin sections where sedimentation was steady over relatively long periods of time. They applied the idea to black shale sequences of the Phi Kappa Formation in the Cordilleran region of North America, the Pete's Summit Formation in Nevada and the Descon Formation in Alaska, for the Ordovician and Early Silurian Periods.

The idea that the study of stratal thickness could furnish data for estimating geological time is not new (Hudson, 1964). Several classical studies, however, have indicated that gaps are present even in sequences where the sediments are almost uniform (Barrell, 1917; Brinkmann, 1929). Nevertheless, in some instances the rate of sedimentation can be tied to an annual absolute time scale (Bradley, 1929). Unfortunately, these instances almost always involve non-marine rocks or others to which correlation of the standard marine sections is impossible.

Even though most students of sedimentation will acknowledge that there is no such thing as uniform sedimentation rates and the recent attempt by Churkin *et al.* (1977) may suffer under the same types of analysis that were performed by Barrell (1917) and Brinkmann (1929), this is not the important point. If the method gives better accuracy or better precision than other methods, it will enable us to gain a better perspective on these problems than we now have. That is what is important.

Essentially, Churkin *et al.* (1977) have

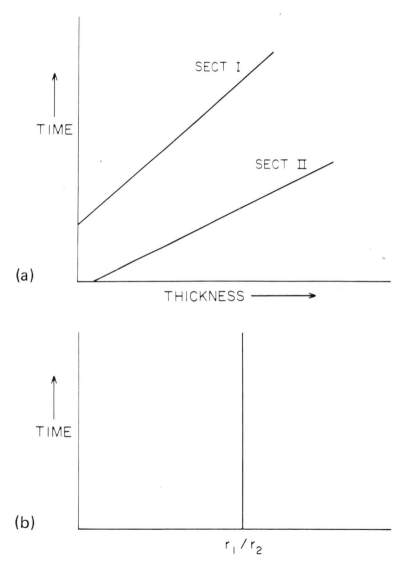

Fig. 17.1 — Assumptions of Shaw's method of graphic correlation: (a) steady rates of accumulation in two sections; (b) proportional rates.

assumed that virtually uniform lithology in a basin or quiet-water setting means a uniform rate of sediment accumulation. They point out that their assumption is supported, although not rigorously tested, by the fact that the most-often-reported zones in the literature correspond to the thickest zones in the Phi Kappa Formation (Churkin *et al.*, 1977, p. 454).

Another indication that the inference of Churkin *et al.* (1977) is valid is found in Shaw's method of graphic correlation. Shaw's (1964) method assumes steady rates of sediment accumulation or that the rates vary proportionally (Fig. 17.1). The results of a number of workers who have employed the method indicate that it works (Miller, 1977; Murphy and Edwards, 1977; Edwards, 1984; Murphy and Berry, 1983; Sweet, 1984). The underlying reason may have been given to us by Barrell (1917). It is simply that the rate of subsidence controls the rate of sedimentation. In any case, we need a test of the assumption that some sequences show steady sedimentation rates over geologically significant periods of time.

Steadiness, or lack thereof, of sedimentation has been defined as the measure of the variation in sedimentation rate as a function of time. Perturbations of a steady state are caused principally by climatic and tectonic fluctuations, which control the dynamics of the depositional processes (Sadler, 1981).

If the idea that uniform lithology in basinal sections suggests a lack of major perturbation of the sedimentation rate, these are the sections where a test of the idea should be attempted. The test should be as follows. If sedimentation rates in two sections are steady and the same, and if zonal boundaries are inferred to be the boundaries of time stratigraphic units, then chronozonal boundaries and other events should have the same spacing. If sedimentation rates are steady in each of the areas but not the same, then chronozonal boundaries and other events should be proportionally spaced (Fig. 17.2). In terms of the graptolite succession found in the Phi Kappa Formation this means that, if the graptolite zones in other sections of

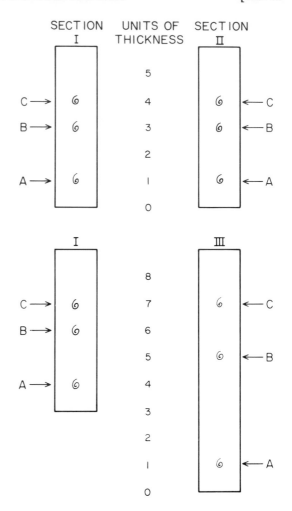

Fig. 17.2 — Test of steadiness of sedimentation rate: (a) steady and the same rate in the pertinent sections; (b) steady but not the same rate in the pertinent sections.

uniform lithology distant from the Phi Kappa Formation can be shown to have the same relative spacing as in the Phi Kappa Formation (e.g. that the *G. persculptus* Zone is just half as thick as the *A. acuminatus* Zone and the *C. vesiculosus* Zone is half again as thick as the *A. acuminatus* Zone in the Phi Kappa Formation), then Churkin *et al.* (1977) have given us an absolute time scale capable of resolving zonal scale problems.

Churkin *et al.* (1977) estimate that the graptolite zones in the three formations range in age from 0.2 ma to 7.7 ma. This is far better resolution for Early Palaeozoic rocks than any other method has been able to suggest. They tied their estimates to the radiometric scale, but this is not necessary for the solution of rate problems. The radiometric dates for Palaeozoic and Mesozoic rocks are few and the error margin is great in comparison with the resolution that can be obtained with fossils. For those parts of the column with which conodont workers deal, radiometric dates are also of questionable correlation with the biostratigraphic record. A non-annual equal-increment time scale could better serve the purpose of addressing rate problems in this interval.

I propose, using a slight modification of the above method, to assess the potential for the construction of a non-annual equal-increment time scale for a short interval within the Early Devonian Period. This can be done by comparing events within lineages that occur in different stratigraphic successions that are reasonably separated geographically, i.e. the sections are in different tectonic settings. The geographic separation is necessary to avoid the possibility that the rates of sedimentation could be controlled by the tectonics and, because of this, could have varied synchronously. If these events fall on a straight line (i.e. if they are proportionally spaced in the two sections) when the sections are compared with one another graphically (Shaw, 1964) (Fig. 17.3), the parsimonious explanation is that the sediments in the sections accumulated at a steady rate during the time indicated. We may then subdivide the stratigraphic section into equal thicknesses of sediment which will represent equal increments of time. Some measure of the actual duration of the total package may be estimated by a comparison with the radiometric scale. This gives additional perspective, but it is not necessary to tie the equal-increment scale to an absolute scale in order to compare relative rates. If this is done, however, estimates of completeness are possible using the method proposed by Sadler (1981) and these estimates may then be used to ensure that the scale of subdivision is not less than that expected for a complete section.

The above scenario depends upon our being

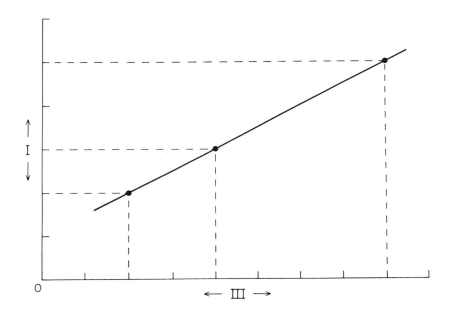

Fig. 17.3 — Graph showing two sections with steady but different sedimentation rates.

able to recognize time-specific biologic events. Murphy and Edwards (1977, p. 183) distinguished biologic and biostratigraphic events as follows: 'A biologic event is an evolutionary first occurrence, or an immigration into the study area (first occurrence), or an extinction, local extermination, or emigration from the study area (last occurrence). A biostratigraphic event is the lowest or the highest observed occurrence of a taxon in a given section in the study area. A biostratigraphic event may represent a corresponding biologic event (the unusual case) or may not represent a biologic event because of sampling, preservation, or identification problems.' The identification of a biologic event as opposed to a biostratigraphic event generally requires a large data base for each section considered and it requires that the taxa are defined or characterized.

The second requirement is a taxonomic problem. It deals with the limits of taxa. Most conodont workers have gone to great effort and expense to aid in setting up the criteria for stratigraphic boundary definition and in actually setting up the boundaries of stratigraphic units and have been involved in lengthy discussion about which species will define those boundaries by their first occurrences. Of course, each recognizes that the precision of the correlation depends on our ability to identify the defining and characterizing taxa and, yet, we have no comprehensively agreed method for establishing the boundaries of the taxonomic units that are used. When what we regard as the central tendency of a taxonomic unit is not being dealt with, a variety of hedging manouvres such as aff., cf., ex. gr., morph, etc., appear in faunal lists. Since stratigraphic ranges depend on firm identifications, such hedging identifications make ranges imprecise and, thus, inhibit precision in correlation. This, and the constant revision of taxonomic concepts, is what caused Shaw (1969) to raise the battle cry, 'Help stamp out species'. However, not many have joined Shaw's crusade. Since Shaw is not likely to win the battle to stamp out species, I propose the alternative of defining species, i.e.

setting taxonomic limits to our species and lesser-rank categories.

I submit that this can be done through the statistical study of lineages. It is only after you know the history of a group that you can classify it intelligently. In order to know the history of the group, you must know the extent of the variation in its populations both geographically and through time. The pattern will suggest the method of classification to be employed. This is because, when we look at the extreme modes of evolution, gradualism and punctuated equilibrium, it follows that there are two methods of classification that will lead to a more stable taxonomy and somewhat ameliorate the reaons for Shaw's battle cry.

In the case of gradualism, an arbitrary point in a continuum may be chosen to subdivide the lineage. This usually entails the selection of the first appearance of some morphologic character (a measurement, a count, character presence or absence). Much of our biostratigraphic work is done this way.

The other extreme case, that of punctuated equilibrium (Eldredge and Gould, 1972), is more difficult to handle satisfactorily from a theoretical viewpoint because it requires an act of faith. We must believe in rapid allopatric speciation in small short-lived populations since we can never hope to see them. In this case, we cannot set limits to our taxonomic categories because we do not know the evolutionary pathways. The only thing that we can do is to characterize the taxon, i.e. to point out its unique characteristics, those present during the stasis or equilibrium phase.

In either case, we have biologic events, which can be used for accurate correlation between measured sections and which can be tested for synchroneity by checking to see that they maintain their relative positions with respect to other events.

Once lineages are worked out, taxonomy and its attendant nomenclature should be stable. With defined and/or characterized taxa, precision in identification is possible. This is a necessary prerequisite to the objective description of the distributions of fossil taxa in the rocks. Success in this endeavour, of course, depends upon a rigorous and consistent application of an agreed-upon taxonomic method.

17.2 NEVADA DATA

For the Lochkovian (Devonian) of central Nevada a large data base exists for a few sections but it was not assembled for the specific purpose of establishing an equal-increment time scale and so requires some additional filling-in. However, it does provide an opportunity to assess the potential of the above-outlined method.

Conodont lineages based on studies of a number of sections in central Nevada have been inferred from a qualitative assessment of their morphologies. These studies suggested that the best and most complete sections in the region in terms of being continuously fossiliferous were the COP V–IV (Murphy *et al.*, 1981) and SP VII sections (Berry and Murphy, 1975). These two sections were trenched in order to measure them as accurately as possible and to expose all potentially fossil-bearing horizons. A reconnaissance sample of every limy bed capable of digestion in formic acid was processed for conodonts and every centimetre of shale was searched for graptolites.

The sections are located about 40 miles apart and were accumulated in somewhat different depositional environments. Both are mostly mudstones, siltstones and shales with interspersed, commonly graded beds of coarse pelloidal and bioclastic material. Secondary chert is common. Both are interpreted as basin facies (Matti *et al.*, 1975; Matti and McKee, 1977). The COP V–IV section at Copenhagen Canyon in the Monitor Range has much more oxidized iron, as shown by its yellow–orange weathering and replacement of its abundant sponge-spicule fauna by hydrated iron oxides. The mudstones are, in part, laminated and sparsely graptolitic but mostly massive and commonly preserve abundant examples of the trace fossil *Chrondrites*. The rocks of the SP

VII section at Coal Canyon in the Simpson Park Range, in contrast, are dark gray in colour and have few sponges. The iron is in the form of pyrite. The fine-grained rocks are laminated and more commonly graptolitic, but without *Chondrites*.

According to Shaw (1964), when steady sedimentation in two sections is assumed and the two sections are plotted against each other on a two-axis graph, the regression (correlation line) will be a straight line. Fig. 17.4 shows the data from the Lochkovian of Nevada, which suggest that the correlation line between COP V–IV and SP VII–V sections is straight or nearly so and that, for the interval considered, stratal thickness is proportional to accumulation time.

Evolutionary events in conodont lineages which have been established in the region with reasonable certainty (Murphy *et al.*, 1981; Murphy and Matti, 1983; Murphy and Cebecioglu, 1984) are shown in Table 17.1.

The *Amydrotaxis* lineage was suggested by Klapper and Murphy (1980, p. 497) and elaboratd by Murphy and Matti (1983, p. 29). *Amydrotaxis sexidentata* is present in the *eurekaensis* Zone in most sections of Nevada and in Alaska, but always in low numbers. Its last occurrence is always separated by a significant interval of strata from the first occurrence of *A. johnsoni* α, which it closely resembles and to which it is connected by a few specimens of intermediate morphology (Lane and Ormiston, 1979, e.g. Plate 1, Figs. 9 and 11). In contrast, *A. johnsoni* α occurs in significant numbers at five levels in SP VII, enabling statistical study of the lineage, which shows changes associated with the denticle size and number and the shape and proportions of the basal cavity with time (Murphy and Johnson, unpublished information). Although *A. johnsoni* α is represented at six levels in the COP V–IV section, the preservation is poor and the numbers too low to permit detailed comparison of evolutionary stages between COP V–IV and SP VII. The transition from *A. johnsoni* α to *A. johnsoni* ß apparently takes place at the top of the *delta* Zone, but so few specimens of

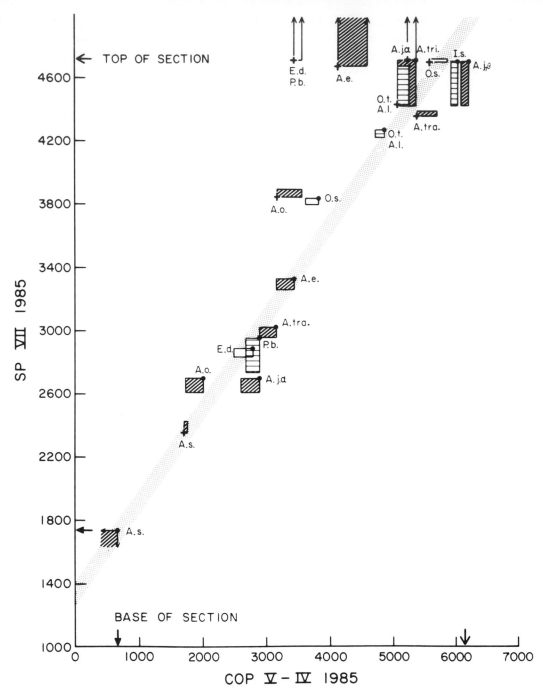

Fig. 17.4 — Graph comparing the COP I–IV and SP VII sections: *Amydrotaxis sexidentata;* A.j.α, *A. johnsoni* α; A.j.ß, *A. johnsoni* ß; A.o., *Ancyrodelloides omus;* A.tra., *A. transitans;* A.tri., *A. trigonicus;* A.e., *A. eleanorae;* A.l., *A. limbacarinatus;* E.d., *Erika divarica;* I.s., *Icriodus steinachensis;* O.s., *O. stygia;* O.t., *O. tuma.* Dots (●) represent first occurrences; pluses (+) represent last occurrences. The boxes show the limits of accuracy as the result of sample spacing: the cross-hatched boxes represent lineage events; the open boxes represent the range limits of cryptogenetic taxa. The scale is in inches.

johnsoni ß are known from these strata that its range needs additional verification.

The *Ancyrodelloides* plexus of species has been interpreted as having two branches by Murphy and Matti (1983, p. 13). The genus originates with *A. omus* from the lowest *delta* Zone. Forms intermediate between *A. omus* and *Ozarkodina remscheidensis* are known from subjacent beds. *A. omus* is not abundant in any of the samples encountered but ranges widely as evidenced by its occurrence in Czechoslovakia and Nevada.

A. omus is morphologically the ideal ancestor of both branches of the genus and intermediate morphologies between *A. omus* and *A. transitans* and between *A. omus* and *A. eleanorae* are known from the appropriate stratigraphic levels. However, again the numbers are low and these inferred transitions need better documentation. The difference between *A. transitans* and *A. trigonicus* is slight (Murphy and Matti, 1983, p. 20) but distinct; however, again the numbers generally are low. The two taxa are known to occur together in a number of places (Bischoff and Sannemann, 1958; Lane and Ormiston, 1979, p. 51; Murphy and Berry, 1983, Fig. 2), and so the lack of overlap in Fig. 17.4 indicates that one or both of these taxa is not showing its maximum range in one or both sections. This is indicated also by the fact that the last occurrence of *transitans* and the first occurrence of *trigonicus* plot off the correlation line.

In the other branch of the genus, the evolution of *A. limbacarinatus* from *A. delta* is easy to recognize because of their distinctive morphologies, but only *limbarcarinatus* has been recovered in the COP V–IV section and it is not possible to fix the event in that section.

Ozarkodina tuma has a very short range during which it apparently displaces the long-ranging *O. excavata* from which it is derived and clearly closely related. The two are not usually found together, but several samples with abundant *O. tuma* have also yielded a few *O. excavata*. *O. tuma* shows a punctuational mode of evolution (Murphy and Cebecioglu, in press) and is abundant in most of the samples in which it has been identified. It is also strongly associated with *A. limbacarinatus*. Its great abundance and short range have made it especially useful for correlation within Nevada.

Other taxa, such as *Icriodus steinachensis, Ozarkodina stygia, Pedavis breviramus* and *Erika divarica,* appear cryptogenetically. Their appearance may result from immigration or a punctuational event (Eldredge and Gould, 1972), but these records cannot be used as primary evidence until their stratigraphic ranges are established firmly. At present, the lower limits of the ranges of all four are reasonably documented, but the latter three are always in low numbers and minor adjustments should be expected. Only *I. steinachensis* has a well-established upper limit. Only the lower occurrence of *I. steinachensis* is within the part of the column being considered here.

E. divarica is such a distinctive taxon that it

Table 17.1

1.	*O. remscheidensis* → *omus*
2.	*Ancyrodelloides omus* → *transitans*
3.	*A. transitans* → *trigonicus*
4.	*A. delta* → *limbacarinatus*
5.	*Ozarkodina excavata* → *tuma*
6.	*O. remscheidensis* → *pandora* morphs
7.	*O. pandora* morphs → *Eognathodus sulcatus* morphs
8.	*Amydrotaxis sexidentata* → *A. johnsoni* α
9.	*A. johnsoni* α → *A. johnsoni* ß

can be identified by small fragments of any part of the apparatus. It has a considerable geographic range (Nevada and Australia); however, it is apparently restricted to particular habitats and was not abundant in its community and so is not found in large numbers in any sample. The two factors offset one another, but its cryptogenetic appearances in the record appear to be contemporaneous in the two sections.

Another cryptogenetic taxon in Nevada, *Ozarkodina stygia*, is also widespread (Nevada, Alaska and Austria). In the Carnic Alps, Schönlaub *et al.* (1980, p. 39) have suggested a sequence of morphotypes within *O. stygia* beginning with a straight-bladed form above the first occurrence of *Ancyrodelloides transitans* and, thus, well up in the *delta* Zone. This early form (α morph) has not been found in Nevada, but the next morph in the sequence, the ß morph, is apparently present (Murphy and Matti, 1983, Fig. 2). This may be an immigration event in the Cordilleran region, but not enough data are available at present to support this interpretation. Its first occurrences in the COP V–IV and SP VII sections plot somewhat above the correlation line in Fig. 17.4.

Pedavis breviramus is apparently on the main line of evolution within the genus, but its immediate ancestor is not known. The platform and M_{2b} elements are distinctive, but they are not abundant. For the present, it can be used as an indicator of the *delta* Zone on the basis of its characterization (Murphy and Matti, 1983), but no events have been identified in this lineage.

17.3 CONCLUSIONS

Conodont taxonomy can be stabilized through the statistical study of the geographic and chronologic variation of population samples. A stable taxonomy is a necessary prerequisite for an event chronostratigraphy and this in turn for the demonstration of steady sedimentation in particular sections.

The conodont lineages found in the *delta* Zone in central Nevada indicate that steady sedimentation in the COP V–IV and SP VII sections is not an unreasonable explanation for the present biostratigraphic data in the two sections. This in turn suggests that an equal-increment time scale that will open the door to the solution of some rate problems can be constructed if proportional event spacing can be demonstrated in another basin.

17.4 ACKNOWLEDGEMENTS

This work derived financial support from the National Science Foundation and the University of California, Riverside Intramural Research Fund.

17.5 REFERENCES

Barrell, J. 1917. Rhythms and the measurements of geologic time. *Geological Society of American Bulletin,* **28,** 745–904.

Berry, W. B. N. and Murphy, M. A. 1975 (imprint 1974). Silurian and Devonian graptolites of central Nevada. *University of California Publications in Geological Sciences,* **110,** 1–109, 15 plates.

Bischoff, G. and Sannemann, D. 1958. Unterdevonisch Conodonten aus dem Frankenwald. *Notizblatt hessische Landesamt für Bodenforschung,* **86,** 87–110.

Bradley, W. A. 1929. The varves and climate of the Green River epoch. *US Geological Survey Professional Paper,* **80–110,** 5 plates.

Brinkmann, R. 1929. Statistisch-Biostratigraphische Untersuchungen an mittlejurassischen Ammoniten über Artbegriff und Stammesentwicklung. *Abhandlunen Gesellschaft der Wissenschaften, Göttingen, N.F.,* **13** (3), 1–249.

Churkin, M. C. Jr., Carter, C. and Johnson, B. 1977. Subdivision of Ordovician and Silurian time scale using accumulation rates of graptolitic shale. *Geology,* **5,** 452–456.

Edwards, L. E. 1984. Insights on why Graphic correlation (Shaw's method) works. *Jounal of Geology,* **92,** 583–597.

Eldredge, N. and Gould, S. J. 1972. Punctuated equilibria: an alternative to phyletic gradualism. In T. J. M. Schopf (Ed.), *Models in Paleobiology,* Freeman–Cooper, San Fransciso, California, 82–115.

Hudson, J. D. 1964. Sedimentation rates in relation to the Phanerozoic time scale. In *The Phanerozoic Time Scale—A Symposium, Geological Society of London Quarterly Journal, Supplement,* **120s,** 37–42.

Klapper, G. and Murphy, M. A. 1980. Conodont zonal species from the *delta* and *pesavis* Zones (Lower Devonian) in central Nevada. *Neues Jahrbuch Geologie-Paleontologie Monatsheft,* **H8,** 490–504.

Lane, H. R. and Ormiston, A. R. 1979. Siluro-Devonian biostratigraphy of the Salmontrout River area, east-central Alaska. *Geologica et Palaeontologica*, **13**, 39–96.

Matti, J. C. and McKee, E. H. 1977. Silurian and Lower Devonian paleogeography of the outer continental shelf of the Cordilleran Miogeocline, central Nevada. In J. H. Stewart, C. H. Stevens and A. E. Fritsche (Eds.), *Paleozoic Paleogeography of the Western United States, Pacific Coast Paleogeography Symposium 1*, Pacific Section, Society of Economic Paleontologists and Mineralogists, Los Angeles, California, 181–216.

Matti, J. C., Murphy, M. A. and Finney, S. C. 1975. Silurian and Lower Devonian Basin and Basin Slope Limestones Copenhagen Canyon, Nevada. *Geological Society of American Special Paper*, **159**, 1–48.

Miller, F. X. 1977. The graphic correlation method in biostratigraphy. In E. G. Kauffman and J. E. Hazel (Eds.), *Concepts and Methods of Biostratigraphy*, Dowden, Hutchinson, and Ross, Stroudsburg, Pennsylvania, 165–186.

Murphy, M. A. and Berry, W. B. N. 1983. Early Devonian conodont-graptolite collation and correlations with brachiopod and coral zones, central Nevada. *American Association Petroleum Geologists Bulletin*, **67**, 371–379.

Murphy, M. A. and Cebecioglu, M. K. 1984. The *Icriodus steinachensis* and *I. claudiae* lineages (Devonian conodonts). *Journal of Paleontology*, **58**, 1399–1411.

Murphy, M. A. and Cebecioglu, M. K. 1986. Statistical study of *Ozarkodina excavata* (Branson and Mehl) and *O. tuma* Murphy and Matti (Lower Devonian,

delta Zone, conodonts, Nevada). *Journal of Paleontology*, **60**, 865–869

Murphy, M. A. and Edwards, L. E. 1977. The Silurian–Devonian boundary in central Nevada. In M. A. Murphy, C. A. Sandberg and W. B. N. Berry (Eds.), *Western North American: Devonian, University of California Campus Museum Contributions*, **4**, 183–189.

Murphy, M. A. Matti, J. C. 1983 (imprint 1982). Lower Devonian conodonts (*hesperius-kindlei* Zones), central Nevada. *University of California Publications in Geological Sciences*, **123**, 1–82, 8 plates.

Murphy, M. A., Matti, J. C. and Walliser, O. H. 1981. Biostratigraphy and evolution of the *Ozarkodina remscheidensis–Eognathodus sulcatus* lineage (Lower Devonian) in central Nevada. *Journal of Paleontology*, **55**, 747–772.

Sadler, P. M. 1981. Sediment accumulation rates and the completeness of stratigraphic sections. *Journal of Geology*, **89**, 569–584.

Schönlaub, H. P. *et al.* 1980. Field trip A: Carnic Alps. In H. P. Schönlaub (Ed.), *Second European Conodont Symposium (ECOS II), Abhandlungen Geologischen Bundesanstalt*, **35**, 1–213, 25 plates.

Shaw, A. B. 1964. *Time in Stratigraphy*, McGraw-Hill, New York, 1–365.

Shaw, A. B. 1969. Adam and Eve, paleontology and the non-objective arts. *Journal of Paleontology*, **43**, 1085–1098.

Sweet, W. C. 1984. Graphic correlation of upper Middle and Upper Ordovician rocks, North American Midcontinent Province, USA. In D. L. Bruton (Ed.), *Aspects of the Ordovician System, Palaeontological Contributions from the University of Oslo*, **295**, 23–35.

18

Conodont biostratigraphy of the Llanvirn–Llandeilo and Llandeilo–Caradoc Series boundaries in the Ordovician System of Wales and the Welsh Borderland

S. M. Bergström, F. H. T. Rhodes and M. Lindström

Samples from critical intervals in the type Llandeilo Series and other Llandeilo sections in south Wales, and from the lower part of the type Caradoc Series in the Welsh Borderland, have produced representative collections of relatively low-diversity but stratigraphically significant conodont faunas. The basal part of the Llandeilo Flags at Llandeilo contains *Amorphognathus inaequalis* and *Baltoniodus prevariabilis*, and the top part of the subjacent Ffairfach Group yields *Eoplacognathus lindstroemi*. This indicates that the base of the type Llandeilo Series is in the lowermost part of the *A. inaequalis* Subzone, which corresponds to a level in the uppermost *Glyptograptus teretiusculus* Zone. A species of *Amorphognathus* quite similar to *A. tvaerensis* occurs, together with *A. inaequalis* and *Icriodella* sp. cf. *I. praecox*, in the Upper Llandeilo Stage at Llandeilo,

and the same taxa, together with *Baltoniodus variabilis* and *Eoplacognathus elongatus*, are present in Llandeilo–Caradoc transition strata at Narberth, 50 km west south west of Llandeilo. Specimens similar to *A. tvaerensis* and *I.* sp. cf. *I. praecox* have been found also in the basal Caradoc (Costonian Stage) in the Caradoc Series type area in the Welsh Borderland. The type Llandeilo Series is directly overlain by shales of the *Nemagraptus gracilis* Zone and graptolites of that zone are known also from the Costonian Stage of the Welsh Borderland. The combined conodont and graptolite evidence indicates that the base of the Caradoc Series corresponds to a level in the lower *B. variabilis* Subzone, which is coeval with the upper *N. gracilis* Zone. Hence, only the basal portion of the type Llandeilo Series is of pre-*N. gracilis* Zone, i.e. *G. teretiusculus* Zone, age, leaving the main portion of the latter zone without series assignment if the common practice is followed to define the top of the Llanvirn Series as the top of the *Didymograptus murchisoni* Zone.

18.1 INTRODUCTION

Unlike some other systems such as the Silurian System and the Jurassic System, the Ordovician System has currently no internationally accepted series classification. In North America (Ross *et al.*, 1982) the Ordovician System is divided into the Ibexian, Whiterockian, Mohawkian and Cincinnatian Series, in Baltoscandia (Jaanusson, 1960) it is divided into the Oelandian, Viruan and Harjuan Series, and in other major regions such as Australia (Webby *et al.*, 1981) and China (Sheng, 1980) there are other provincial series schemes. However, in recent years the standard British series units (the Tremadoc, Arenig, Llanvirn, Llandeilo, Caradoc and Ashgill) have become increasingly widely used internationally for regional comparisons of faunas and rock successions. Unfortunately, in many cases the use of British units has been uncritical and based on very tenuous evidence. For instance, it has been common practice outside the British Isles to interpret the series units in terms of graptolite zones despite the fact that virtually all these units are based on shelly fossils and can be correlated directly with the graptolite zone succession only imprecisely. Furthermore, until recently is has not been widely recognized that the precise biostratigraphic scope of most of these series has not been established even in their respective type areas and none of these six series has formally designated stratotypes or boundary reference sections (Whittington *et al.*, 1984). Clearly, before these series units are used regionally, they should have a clearly defined biostratigraphic scope. A first step towards that goal should include a reassessment of each unit in its type area. Such studies are currently under way in Great Britain (see, for instance, Zalasiewicz, 1984) and the present chapter may be considered a contribution to that effort.

The Llandeilo Series is a good illustration of problems associated with biostatigraphic definition and correlation of a series unit. The term Llandeilo was introduced by Murchison (1835, 1839) for Ordovician rocks exposed at Llandeilo in south Wales (Fig. 18.1). During the late

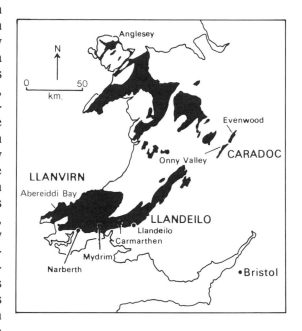

Fig. 18.1 — Sketch map of Wales and the Welsh Borderland showing the distribution of Ordovician rocks (black) and the location of type areas of series and collecting localities mentioned in the text.

nineteenth century and early twentieth century, the Llandeilo became a standard unit in the classification of the British Ordovician System, and, since the 1950s, the term has become widely used also internationally. For informative reviews of the history of the term and its faunal and stratigraphic basis, see Williams (1953), Williams *et al.* (1972) and Whittington *et al.* (1984). On the basis of the assumption that strata immediately below the typical Llandeilo Flags are coeval with the *Didymograptus murchisoni* Zone and the fact that shales with *Nemagraptus gracilis* Zone graptolites directly overlie the unit at Llandeilo (Williams, 1953), it was generally assumed for a long time that the Llandeilo Series corresponds precisely to the *Glyptograptus teretiusculus* Zone (Fig. 18.2),

STANDARD PRIOR TO 1972		GRAPTOLITE ZONES	WILLIAMS ET AL. 1972	
DICRANOGRAPTUS SHALES		*NEMAGRAPTUS GRACILIS*	N. GRACILIS SHALES	
			LLANDEILO FLAGS	U.
				M.
LLANDEILO FLAGS	U.	*GLYPTOGRAPTUS TERETIUSCULUS*		
	M.			L.
	L.			
FFAIRFACH GROUP		*DIDYMOGRAPTUS MURCHISONI*	FFAIRFACH GROUP	

Fig. 18.2 — Two alternative previous correlations between standard graptolite zones and the type Llandeilo Series and adjacent strata at Llandeilo. Note that in both these interpretations, the Ffairfach Group is correlated with an interval of the *Didymograptus murchisoni* Zone and considered to be of Llanvirn age. The pre-1972 correlation refers to that widely adopted during the 1950s and 1960s; as shown by Jaanusson (1960, Table II), prior to that period there was little agreement regarding the graptolite correlation of the Llandeilo Series.

and the unit was interpreted to have that scope in terms of the graptolite succession also outside Britain, despite the fact that stratigraphically diagnostic graptolites are virtually unknown in the Llandeilo Series at Llandeilo. Skevington's (1969) proposal to include the *G. teretiusculus* Zone in the *N. gracilis* Zone, and hence to refer the Llandeilo Series to the latter zone, has not been adopted in Britain or elsewhere and, for reasons given in Finney and Bergström (1986), we retain the *G. teretiusculus* Zone as defined by Finney and Bergström (1986).

During the early 1970s, independently obtained evidence from shelly fossils (Addison, in Williams *et al.*, 1972; Addison, 1974), graptolites (Toghill, 1970) and conodonts (Bergström, 1971a, 1971b) suggested that the correlation of the Llandeilo Series with the *G. teretiusculus* Zone was in error and that a substantial part of this series is equivalent to the *N. gracilis* Zone. Yet, up to now, the faunal evidence for this revised correlation has not been published in full detail.

The purpose of the present study is to assess the significance of conodonts to define the biostratigraphic scope of the typical Llandeilo Series as well as to clarify its relations to overlying and underlying series units. We shall first discuss the conodont succession through the Llandeilo Flags and subjacent beds of the pre-Llandeilo Ffairfach Group at Llandeilo; then we shall review conodont biostratigraphic evidence from some coeval successions elsewhere in south Wales which has bearing on the interpretation of the typical Llandeilo succession; finally, we shall make some comparisons with the basal Caradoc Series in its type area in the Welsh Borderland. The thrust of the present chapter is biostratigraphic, and we shall attempt to show how conodonts can be used for detailed correlations. It should be stressed that our biostratigraphic conclusions are based mainly on evidence from conodont taxa representing distinct morphological stages in four, or five, relatively rapidly evolving and stratigraphically overlapping lineages (Fig. 18.5), but we also consider megafossil evidence where appropriate. Because of page limitations, we shall publish species descriptions and taxonomic discussions in a companion paper elsewhere, but some taxonomic notes are given in the appendix.

18.2 PAST AND PRESENT CONODONT WORK

Conodonts were first described from the Llandeilo Flags by Rhodes (1953) and his descriptive work is still the most extensive published on conodonts from that unit. Some of Rhodes' taxa were discussed by Bergström (1964), who also recorded conodonts of Llandeilo age from the Castell Limestone in the Llanvirn type area in southwestern Wales. Other records of Llandeilo conodont occurrences in south Wales have been published by Bergström (1971a, 1971b, 1983), Bergström et al. (1974, 1984), Orchard (in Bergström and Orchard, 1985) and Addison (in Williams et al., 1972).

The present research, which has been carried out intermittently for a number of years, is based mainly on collections assembled by Bergström and on a large collection from the Ffairfach Group at Llandeilo made by Rhodes. We have also had the opportunity to examine many smaller collections made by others, including those of Dr. R. Addison (see Addison, in Williams et al., 1972; Addison, 1974).

Small numbers of conodonts have been obtained from many levels in the sections studied and some stratigraphically important samples yielded a rich fauna. Nevertheless, samples producing relatively abundant and taxonomically varied specimens are not very common, especially in the Llandeilo area, even when using a sample size of 3 kg or more. The state of preservation varies considerably from mediocre to good; as shown by Bergström (1980) and by Savage and Bassett (1986), Middle and Upper Ordovician specimens from south Wales exhibit a colour alteration index value of about 5, suggesting a heating of some 300 °C.

Most of our collections are currently housed at the Department of Geology and Mineralogy of The Ohio State University.

18.3 THE STANDARD SECTION AT LLANDEILO

Llandeilo-age rocks cover much of the Llandeilo district (Fig. 18.3) but most outcrops are small, and many faults disrupt the succession. A stratigraphically relatively complete and moderately well-exposed succession of the

Llandeilo Flags and subjacent Ffairfach Group can be pieced together from outcrops extending for a distance of almost 2 km along the River Cennen and in nearby railway cuts (Fig. 18.3). This has become known as the Llandeilo standard, or type section. For details of this section, see Williams (1953), Bassett (1982), Wilcox and Lockley (1981), Lockley and Williams (1981) and Williams et al. (1981). In this exposure, the Llandeilo Flags rest directly on the Ffairfach Group but the basal contact is covered, and the contact interval is poorly fossiliferous. Until recently, it was common belief that Upper Llandeilo strata were overlain unconformably by Silurian (Wenlock) strata in this section but recent work (Wilcox and Lockley, 1981, p. 287) suggests that the Dicranograptus Shales, which overlie the Llandeilo Flags locally in the Llandeilo district, are present on the top of the Llandeilo Flags also in the Cennen section. The total thickness of the Llandeilo Flags here is estimated at 716 m, about 63% of which are exposed (Wilcox and Lockley, 1981). Since the basic work by Williams (1953), trilobites have been used to subdivide the type Llandeilo into Lower, Middle and Upper Llandeilo Stages, each of which has been further subdivided into trilobite zones (Fig. 18.4). For a dicussion of these units and their faunal basis, see Wilcox and Lockley (1981).

In Fig. 18.4 we illustrate the known ranges of conodont species through the Llandeilo Flags and immediately subjacent strata at Llandeilo. As is clear from that figure, the conodont faunal diversity is relatively low and there is no conspicuous faunal change through the succession although a few species appear and others disappear. Nevertheless, as will be shown below, the conodonts present provide useful biostratigraphic information. The inclusion of conodont data from the upper part of the Ffairfach Group is both appropriate and significant because information from that unit is helpful to establish a maximum age of the base of the superjacent Llandeilo Flags in the standard section.

The Ffairfach Group has been informally

subdivided into five formations referred to in descriptive lithic terms (Williams, 1953; Williams *et al.*, 1981). Virtually all the conodont collections to be discussed here come from calcarenite lenses in the Flags and Grits Formation, although a single sample was collected from some limestone lenses in the lower part of the Rhyolitic Conglomerates Formation, the topmost unit of the Ffairfach Group (Fig. 18.3). The conodont fauna of the former unit is dominated by representatives of *Baltoniodus prevariabilis*, *Eoplacognathus lindstroemi*, *Panderodus* spp. and *Plectodina flexa*. Among these, *E. lindstroemi* and *B. prevariabilis* are of particu-

lar biostratigraphic significance because both belong to stratigraphically closely controlled evolutionary lineages and their ancestors and descendents are well established (Bergström, 1971a, 1973, 1983; Dzik, 1976, 1978) (Fig. 18.5). The evolution of *E. lindstroemi* from *E. robustus* is displayed in the Baltoscandic succession in the lower Uhaku Stage in strata coeval with the middle part of the *Glyptograptus teretiusculus* Zone (Bergström, 1973, Fig. 2); hence it is suggested that the occurrence of *E. lindstroemi* in the Flags and Grits Formation indicates an age not older than that graptolite zone interval. In fact, direct comparison with Balto-

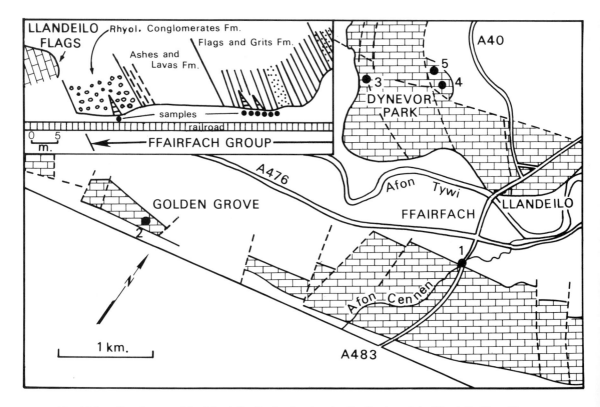

Fig. 18.3 — Sketch map of the Llandeilo district showing the distribution of the Llandeilo Flags (limestone pattern; based on Williams (1953)) and the location of our five most important collecting sites. Site 1 is the well-known type section of the Ffairfach Group along the railway near Ffairfach (SN 628212; see inset map for details and collecting levels); site 2 is the disused quarry in the Lower Llandeilo Flags at Llanfihangel Aberythych (SN 592198; loc. 1 of Rhodes (1953)); site 3 is the section along the dirt road 0.5 km westsouthwest of Dynevor Castle (SN 608223; Stop II-10 of Bassett *et al.* (1974)); site 4 is the disused quarry in the Upper Llandeilo Flags 0.25 km northeast of Dynevor Castle (SN 616227; Stop II-8 of Bassett *et al.* (1974)); site 5 is the disused quarry in the Upper Llandeilo Flags 0.3 km northnortheast of Dynevor Castle (SN 615229). The Llandeilo Series reference section is at outcrops along the Afon Cennen from just south of loc. 1 to just south of A in Afon.

scandian specimens shows that the Ffairfach representatives of *E. lindstroemi* compare favourably with specimens from the lower part of the *Pygodus anserinus* Zone, the *Amorphognathus kielcensis* Subzone, where also *B. prevariabilis* is common in the Baltoscandian faunas. This is an interval high in the *G. teretiusculus* Zone (Bergström, 1973, Fig. 2). Interestingly, according to Jaanusson (pers. commun. 1984), the Ffairfach Group brachiopod fauna (Williams *et al.*, 1981) is most similar

to that of the Uhaku Stage among those of the Middle Ordovician Stages in Baltoscandia.

The sample from the Rhyolitic Conglomerates Formation contains *Amorphognathus inaequalis*, an index of the *A. inaequalis* Subzone, suggesting that the *A. kielcensis–A. inaequalis* subzonal boundary corresponds to a level in the upper part of the Ffairfach Group. It is quite possible, and perhaps even likely, that the appearance of representatives of *A. inaequalis*

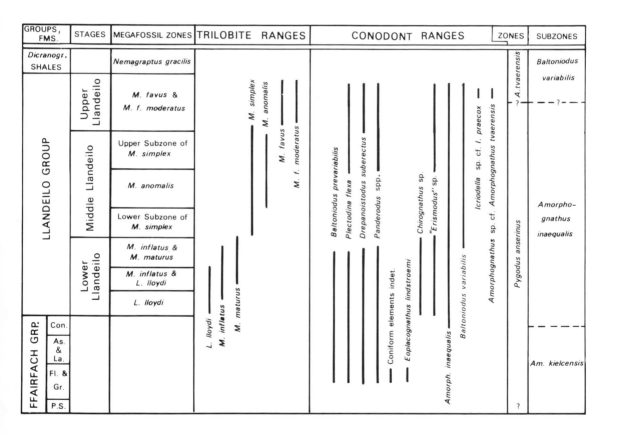

Fig. 18.4 — Biostratigraphic classification of the Llandeilo Group and adjacent strata at Llandeilo showing ranges of selected trilobites and conodonts and trilobite and conodont zones. The trilobite data are from Wilcox and Lockley (1981). Note that no conodonts are known from the uppermost 40–45 m of the Upper Llandeilo sequence which consists of poorly exposed thin-bedded flags grading into the *Dicranograptus* Shales. The latter contain graptolites of the upper *N. gracilis* Zone (Williams, 1953). The base of the Llandeilo Series, taken at the base of the Llandeilo Flags, is interpreted to be coeval with a level low in the *A. inaequalis* Subzone and the top of the series, taken at the base of the *Dicranograptus* Shales, with a level low in the *B. variabilis* Subzone. For explanation of abbreviations of Ffairfach Group formations and generic names of trilobites, see Fig. 18.3 and Wilcox and Lockley (1981), respectively.

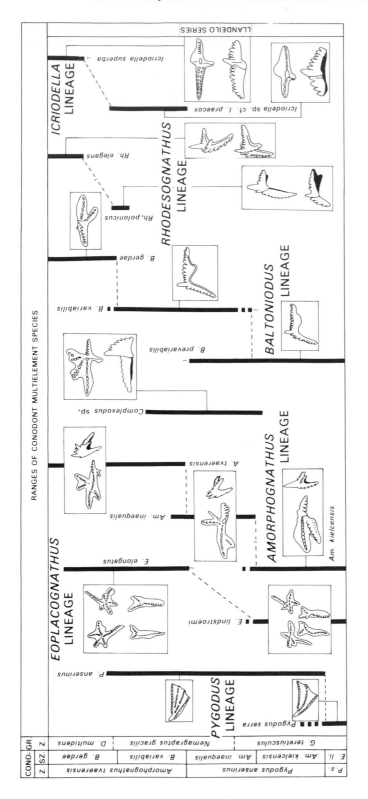

Fig. 18.5 — Known vertical ranges, in terms of conodont and graptolite zonal units, of species in six important Middle Ordovician evolutionary lineages, all of which except that of *Pygodus*, are represented in the Llandeilo and adjacent strata in Wales and the Welsh Borderland. *P.s.*, *Pygodus serra*; *E. li.*, *Eoplacognathus lindstroemi*; Cono., conodont; Z., Zone; SZ, Subzone; GR., graptolite. Illustrations of *Am. kielcensis*, *Complexodus* sp. and *Rh. polonicus* are based on Dzik (1978). The overlapping ranges of morphotypes in these rapidly evolving lineages provide a high-resolution biostratigraphic scheme. Note that the ranges shown are those established internationally and that the local Welsh ranges of several taxa, as now known, are more restricted than shown in the diagram.

is ecologically controlled at Ffairfach and that the species may be present in somewhat older strata elsewhere. However, if Dzik (1978) is correct in his interpretation of *A. inaequalis* as a direct descendent of *A. kielcensis* appearing in the *P. anserinus* Zone, then it is clear that the uppermost portion of the Ffairfach Group cannot be older than that zone. Representatives of *B. prevariabilis* are, as just noted, common in the *P. anserinus* Zone and the evolutionary successor of that species, *B. variabilis*, appears near the *A. kielcensis–A. inaequalis* subzonal boundary in Baltoscandia (Bergström, 1983). Our *Baltoniodus* specimens from the Ffairfach Group, like most of those of the Llandeilo Flags, are small and apparently immature, and such specimens are difficult to separate from the closely related *B. variabilis*. Pending the discovery of typical adult specimens, we here tentatively identify this form as *B. prevariabilis* (cf. Bergström, 1964). However, one of our samples of the Lower Llandeilo at Dynevor Park contains specimens identifiable as *B. variabilis*. An overlap in range between these species is known from Baltoscandia (Fig. 18.5), and the evidence at hand, also including the known relations between conodont and graptolite zones elsewhere (Bergström, 1986), indicates that the base of the typical Llandeilo Flags is no older than the upper part of the *Glyptograptus teretiusculus* Zone in the graptolite succession, and the *A. inaequalis* Subzone in the conodont succession. As will be discussed below (p. 308) this conclusion has significant implications not only for correlation of the Llandeilo Series regionally but also for the standard classification of British Series.

It should be noted that palaeoecologically the common occurrence of *E. lindstroemi* and *B. prevariabilis* in the Ffairfach Group is of interest because they occur in strata interpreted as having been deposited in quite shallow water with, at times, considerable current action (Williams *et al.*, 1981) during a regressive depositional phase. Also in Baltoscandia and eastern North America (Bergström, 1971a; Bergström

and Carnes, 1976), species of *Eoplacognathus* (including *E. lindstroemi*) and *Baltoniodus* are known from strata deposited in relatively shallow water. However, representatives of *E. elongatus* and *B. gerdae* occur also in relatively deep-water sediments (Repetski and Ethington, 1977; Burrett *et al.*, 1983), and the ecological range of representatives of these genera is clearly considerably greater than the 'shallow sublittoral environment' indicated by Barnes and Fåhraeus (1975).

Regrettably, the impure limestones, sandstones and flags of the Llandeilo standard section have proved to be poor in conodonts, and the siliceous nature of the limestones has prevented reasonably complete dissolution of many of our samples. To gain an idea of the conodont succession through the Llandeilo Flags, it has therefore been necessary to piece together a composite section based on several other localities with stratigraphically more limited successions. The disadvantage of not being able to tie our productive conodont samples to precise levels in the standard section is, however, not a very serious one for three reasons.

(1) Williams (1953) mapped in detail the horizontal distribution of the trilobite-based stages in the Llandeilo district and recent reassessment of his trilobite biostratigraphy (Wilcox and Lockley, 1981) has shown it to be reliable and in need of only minor modification. Hence, using the available megafossil information it is possible to date sections to stage, or even zone.

(2) Although the type Llandeilo sequence has a considerable thickness (>700 m), it evidently represents only a very limited biostratigraphic interval. For general age assessments it is therefore less necessary to have the sample levels located precisely stratigraphically than if we were dealing with a succession deposited during a substantial time period.

(3) The Llandeilo conodont succession, as now

known, is not one of great vertical and horizontal complexity; the assemblage of species is rather uniform throughout the unit, with few appearances and disappearances of significant taxa, and virtually the same assemblage of species can be traced in coeval strata throughout south Wales.

Among the localities examined by us in the Llandeilo district, the most productive ones are at Golden Grove (of Lower Llandeilo age; loc. 1 of Rhodes (1953)) and in Dynevor Park (of Lower and Upper Llandeilo age). For the location of these collecting sites, see Fig. 18.3. The known ranges of species are shown in Fig. 18.4.

The conodonts from the Lower Llandeilo Stage are closely similar to those of the uppermost Ffairfach Group, indicating that, if there is a hiatus at the base of the Llandeilo succession (as has been suggested by some researchers), it is of minor biostratigraphic magnitude. It is notable that not a single specimen of *Eoplacognathus lindstroemi* or its evolutionary descendent *E. elongatus* has been found in typical Llandeilo strata. Compared with the Ffairfach species assemblage, another difference is the presence in the Lower Llandeilo succession of two distinctive species, namely *Chirognathus* sp. and '*Erismodus*' sp. (Bergström, 1964). The former species is at present recorded only from the Llandeilo district but the latter is known also from Narberth (Fig. 18.7). A form similar to '*Erismodus*' sp. occurs in the Costonian Stage in the Welsh Borderland (Bergström, 1971a) but its affinities are still unclear. The British occurrences are unique because they represent the only known examples of these genera in Europe. Representatives of *Chirognathus* and *Erismodus* are widely distributed in the Middle Ordovician succession of the North American Midcontinent Region, where they are typically present in shallow-water strata. Wilcox and Lockley (1981), although acknowledging the difficulties involved in interpreting the depositional environments of the Llandeilo Flags on

the basis of the available lithic evidence, suggested that the unit was deposited in shallow subtidal, or even intertidal, to deep subtidal environments with a general trend towards deeper water, particularly during Upper Llandeilo time. The more shallow of these depositional environments, which are characteristic of the Lower Llandeilo succession, are similar to those preferred by representatives of *Chirognathus* and *Erismodus* in North America.

The small conodont collections obtained thus far from the Middle Llandeilo Stage are not significant biostratigraphically, but samples from the Upper Llandeilo Stage in Dynevor Park (Fig. 18.3) contain a more diagnostic species association, including specimens of a species very close to, if not identical with, the zonal index *Amorphognathus tvaerensis* (Plate 18.1, Fig. 14). There are also representatives of a species of *Icriodella*, in all probability the form identified as *I*. sp. cf. *I. praecox* by Bergström (1983), but the diagnostic element of the apparatus has not yet been found in the Dynevor Park samples. The presence of probable *A. tvaerensis*, which is the evolutionary descendent of *A. inaequalis*, indicates the *A. tvaerensis* Zone. The lower part of the latter zone is coeval with the upper portion of the *Nemagraptus gracilis* Zone (Berström, 1971b, 1986) which is in good agreement with the fact that the Upper Llandeilo strata are overlain, with gradational contact (Williams, 1953, p. 194), by the graptolitic *Dicranograptus* Shales that yield *N. gracilis* Zone graptolites (Williams, 1953). On the assumption that Williams' identifications are correct and that the taxa listed by him have the same range at Llandeilo as elsewhere, the presence of *Orthograptus calcaratus vulgatus* and common *Leptograptus* spp. suggests that the *Dicranograptus* Shales represent an interval in the upper *N. gracilis* Zone (Finney and Bergström, 1986). Indirect support for this conclusion is also provided by the reported occurrence (Wilcox and Lockley, 1981, p. 312) of *Corynoides* sp. at an unspecified level in the Llandeilo Flags, because representatives of that genus are not known from strata older than the *N. gracilis* Zone (Finney and Bergström, 1986, Fig. 1). We interpret the conodont and graptolite evidence at hand as indicating that the top of the Llandeilo Series at Llandeilo is coeval with a level in the *B. variabilis* Subzone of the *A. tvaerensis* Zone, although quite typical specimens of *A. tvaerensis* have not yet been found in the Upper Llandeilo strata of the type area. However, it should be noted that no conodonts are known from the uppermost 40–45 m of the Upper Llandeilo succession which consists of poorly exposed thin-bedded flags. In this interpretation, the type Llandeilo sequence corresponds to an interval from the upper(most) *Glyptograptus teretiusculus* Zone to the upper *N. gracilis* Zone (both zones taken in the sense of Finney and Bergström (1986)) in the graptolite succession (Fig. 18.8). In the following sections of the present chapter, we present additional evidence supporting this conclusion from other successions of Llandeilo age in south Wales.

18.4   THE CARMARTHEN AND MYDRIM SUCCESSIONS

Thanks to the courtesy of Dr. R. Addison, we have had the the opportunity of examining several conodont collections from Llandeilo rocks in the region just east of Carmarthen, 13–16 km west of Llandeilo. The best collections are from the Nantgaredig area, 7 km east of Carmarthen, among which a sample, from a level near the base of the Llandeilo Flags at an exposure 0.5 km northnortheast of Nantgaredig (SN 496222), contains excellent specimens of *Amorphognathus inaequalis* (Plate 18.1, Figs. 8–10) in association with what looks like a relatively primitive morphotype of *Baltoniodus variabilis*. This is in agreement with the ranges in Baltoscandia and Poland where the *A. inaequalis* Subzone overlaps the lower part of the range of *B. variabilis* (cf. Dzik, 1978; Bergström, 1983).

The lithology of the Llandeilo rocks near the town of Mydrim, about 35 km west of Llandeilo, differs conspicuously from that at Llandeilo, reflecting a westward facies change. That is, apart from the argillaceous Mydrim Limestone, which has yielded *Nemagraptus gracilis* and other graptolites of that zone, carbonate rocks are virtually absent in the Mydrim succession which is developed in shaly-facies bearing graptolites and some shelly fossils. Conodonts are as yet unknown from the area but it is appropriate to consider this sequence briefly, because it provides some biostratigraphic evidence useful for the interpretation of the sequence at Llandeilo. Although we have seen the Mydrim succession, we have no new biostratigraphic data and the present account is based on Addison's restudy of the area (Addison, in Williams *et al.*, 1972; Addison, 1974; Bassett *et al.*, 1974) and previously published data, especially Toghill (1970). For a review of the Mydrim succession with some biostratigraphic data, see Fig. 18.6.

In terms of regional correlation, the Mydrim succession is of particular interest in that it contains co-occurrences of stratigraphically diagnostic graptolites and trilobites used as zonal indices in the Llandeilo strata of Llandeilo (Fig. 18.4). Two such occurrences merit mention here.

(1) In the lowermost portion of the Hendre Shales, specimens of *Marrolithus inflatus* are associated with graptolites of the *Glyptograptus teretiusculus* Zone (Addison, 1974). This trilobite is a zonal index in, and restricted to, the Lower Llandeilo Stage at Llandeilo (Wilcox and Lockley, 1981) (Fig. 18.4). Assuming that the range of this trilobite is closely similar in the two areas, one may conclude that the lowermost Llandeilo Stage is coeval with a portion of the *G. teretiusculus* Zone.

(2) Within a stratigraphically slightly higher interval in the Hendre Shales 'well below the level of the Mydrim Limestone'

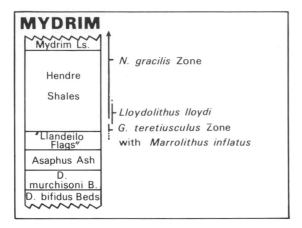

MYDRIM

Fig. 18.6 — Schematic columnar section of the succession in the Mydrim (Meidrim) area showing known ranges of graptolite zones and stratigraphically important trilobites. Based on the works of Addison (1974), Williams *et al.* (1972) and Toghill (1970). Fossil ranges are based on data from the road B4299 section and adjacent quarry 0.2–0.4 km south of Mydrim (SN 289206), and from two sections near Tynewydd (SN 35562197 and SN 36182208), 7 km east of Mydrim, which are described in detail by Addison (1974).

(Addison, in Williams *et al.*, 1972; Addison, 1974), specimens of *Lloydolithus lloydi* occur with *Nemagraptus gracilis*. At Llandeilo, *L. lloydi* is a zonal index in, and restricted to, the Lower Llandeilo Stage (Wilcox and Lockley, 1981) (Fig. 18.4). The co-occurrence of the latter two species in the Mydrim succession is an indication that a portion of the Lower Llandeilo Stage is coeval with the *N. gracilis* Zone, presumably its lowest part. When applied to the standard Llandeilo sequence at Llandeilo, the biostratigraphic information provided by these occurrences is in good agreement with the evidence of the conodonts in the latter succession as outlined above. Significantly, it strongly supports the idea that only a minor portion of the typical Llandeilo Series is older than the *N. gracilis* Zone.

18.5   THE NARBERTH SUCCESSION

The biostratigraphy of the lower Middle Ordovician outcrop area between Lampeter Velfrey and Crinow, some 15 km westsouthwest of Mydrim and 3–4 km east of Narberth (Fig. 18.1) has attracted considerable attention after a recent restudy (Addison, in Williams *et al.*, 1972; Addison, 1974; Bassett *et al.*, 1974) showed that the succession here may represent an unbroken Llandeilo–Caradoc transition developed in shelly facies. The scattered nature of the exposures and the absence of long continuous sections make it necessary to piece together a composite section (Fig. 18.7). Our conodont data are based on three sets of samples collected by ourselves and several samples from critical levels kindly provided by Dr. Addison. As shown in Fig. 18.7, the faunal diversity is not great and, by and large, the fauna is quite similar to that of the Upper Llandeilo strata at Llandeilo, although it contains at least two species not yet known in the latter succession.

Our samples from the lowermost carbonate unit, about 40 m thick, in the succession (the 'Bryn-glas Limestone Member' of Addison (1974)) come from exposures in the stream (SN 156143) and in the field just south of Lampeter Velfrey, and from the lowermost part of the succession in the old quarry southwest of Henllan (SN 131160). This unit contains *Marrolithus favus*, an index species of the Upper Llandeilo Stage at Llandeilo (Wilcox and Lockley, 1981; Fig. 18.4), associated with a brachiopod–trilobite assemblage showing affinities with both Llandeilo and Caradoc faunas (Addison, 1974). Our samples have yielded a conodont species association including *Eoplacognathus elongatus* and *Icriodella* sp. cf. *I. praecox* (Fig. 18.7) which may belong to the *Baltoniodus variabilis* Subzone or the next older subzone; in the absence of key species, the subzonal assignment remains uncertain. Samples through the slightly younger 70 m succession exposed in the Bryn-banc quarry 0.2 km northeast of Llan-mill (SN 141144) (Strahan *et* al., 1914, p. 32; Addison, 1974) have produced moderately abundant conodonts including specimens of *Amorphognathus inaequalis*, *A. tvaerensis*, *Baltoniodus variabilis* and *Eoplacognathus elongatus*, a species association best compared with that of the lower *B. variabilis* Subzone (Figs. 18.5 and 18.7). Some of the Bryn-banc samples have also yielded well-pre

served elements of *Icriodella* sp. cf. *I. praecox* (Bergström, 1983) (Plate 18.1, Figs. 1 and 3). A closely similar, although taxonomically less varied, fauna has been isolated from near the top of the carbonate sequence, about 25 m below the *Dicranograptus* Shales, at the Henllan quarry (Strahan *et al.*, 1914, p. 34). This conodont occurrence, in all likelihood the youngest

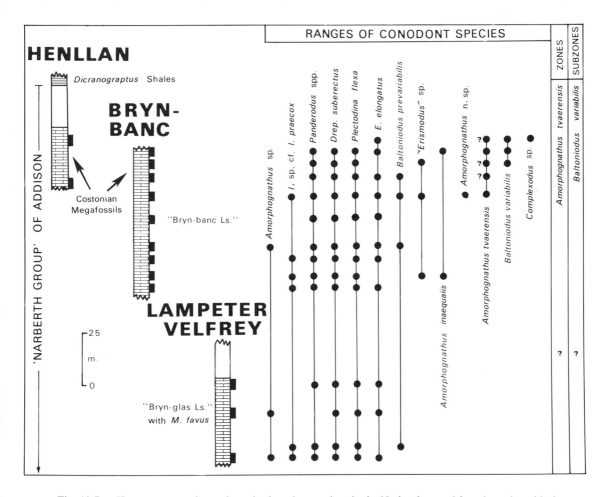

Fig. 18.7 — Known ranges of conodonts in three key sections in the Narberth area. Mutual stratigraphical relations of these sections are after Addison (1974). In accordance with Addison's (1974) interpretation that his 'Bryn-glas' and 'Bryn-banc' limestones are both represented in the thin limestone succession at Henllan, the conodonts from the lowermost Henllan sample are shown as being from a level near the base of the 'Bryn-glas Limestone' in the range diagram. Based on shelly fossils, Addison (1974, p. 61) tentatively correlated the 'Bryn-banc Limestone' and an undetermined part of the 'Bryn-glas Limestone' with the Costonian Stage (lowermost Caradoc Series) of the Welsh Borderland. The conodonts indicate that at least the upper part of the 'Bryn-banc Limestone' can be referred to the *Baltoniodus variabilis* Subzone but the exact zonal classification of the 'Bryn-glas Limestone' remains uncertain although the presence of *Eoplacognathus elongatus* throughout the unit suggests the *Baltoniodus variabilis* Subzone or, possibly, the upper part of the *Amorphognathus inaequalis* Subzone (Bergström, 1983, Fig. 2).

known in the Narberth area, is in rocks (the 'Bryn-banc Limestone Member' of Addison (1974)) that have Costonian (basal Caradoc) megafossils (Addison, in Williams *et al.*, 1972, p. 36; Addison, 1974) and a similar megafossil fauna is known from the succession in the Bryn-banc quarry (Addison, 1974). A notable conodont species found at Henllan is *Complexodus* sp. (Plate 18.1, Fig. 4). It is only the second record of this rare genus in the British Isles, the other being from the Garn Formation of Anglesey (Bergström, in Bergström and Orchard, 1985), where several specimens have been found in rocks of about the same age as those at Henllan.

In assessing the age of the topmost part of the Henllan carbonate succession, as well as that of the sequence in the Bryn-banc quarry, we have paid particular attention to the evolutionary stage of the specimens of *Eoplacognathus elongatus* present in our collections. As shown in Fig. 18.5, this species belongs to a relatively rapidly evolving evolutionary lineage. The morphological type present in the Henllan top sample compares closely with that in the *Baltoniodus variabilis* Subzone in Baltoscandia and is definitely less advanced than that present in the *B. gerdae* Subzone as illustrated by Repetski and Ethington (1977) and Burrett *et al.* (1983). Accordingly, we are confident that the top of the Narberth carbonate succession is not younger than the *B. variabilis* Subzone, which is coeval with an interval in the *Nemagraptus gracilis* Zone. This conclusion is in good agreement with the fact that the overlying *Dicranograptus* Shales contain *N. gracilis* and other graptolites of that zone (Strahan *et al.*, 1914, p. 41). Indeed, these conodont and graptolite data, combined with the known relations between conodont and graptolite zones elsewhere (Bergström, 1986), suggest that the top of the Henllan carbonate succession is not coeval with the very top of, but represents a level well down in, the *B. variabilis* Subzone.

The samples investigated by us suggest that there is no distinct conodont faunal turnover in

the Llandeilo–Caradoc Series boundary interval at Narberth. Addison's work (Addison, in Williams *et al.*, 1972; Addison, 1974) shows that the upper part of the Narberth carbonate succession contains Caradocian megafossils and the lower part Llandeilian megafossils, but currently it is not possible to fix the series boundary at a specific level. However, if the base of the Caradoc Series is taken at a level near the middle of the Narberth carbonate succession, it would evidently be rather close to the base of the *A. tvaerensis* Zone. In general, our Narberth conodont data appear to confirm the correctness of our suggestion, based on the Llandeilo succession, that the top of the Llandeilo Series corresponds to a level within the *B. variabilis* Subzone.

18.6   THE ABEREIDDI BAY SUCCESSION

In an attempt to clarify the scope of the Llandeilo Series and its relation to underlying and overlying series units, it is obviously appropriate to consider also the successions of the Llanvirn and Caradoc Series in their type areas. The type area of the Llanvirn Series is at Abereiddi Bay on the southwestern coast of Wales (Fig. 18.1). The term was introduced by Hicks (1881), and the Llanvirn Series has usually been taken to correspond to the *Didymograptus bifidus* and *D. murchisoni* zones in the graptolite succession. However, the detailed faunal succession in the Llanvirn–Llandeilo Series boundary interval at Abereiddi Bay is not very clear, and the data at hand do not permit a precise correlation with the succession at Llandeilo.

Virtually the entire Middle Ordovician succession at Abereiddi Bay is developed in shaly facies but graptolites are by no means present throughout the sequence which is, furthermore, disrupted by faults (Bassett *et al.*, 1974; Hughes *et al.*, 1982). The only notable limestone occurrence is an interval of interbedded carbonate beds and shaly strata referred to as the Castell Limestone. This unit is directly overlain by the *Dicranograptus* Shales which contain trilobites suggesting correlation with the *Nemagraptus gracilis* Zone (Black *et al.*, 1972; Bassett, in Williams *et al.*, 1972; Bassett *et al.*, 1974; Hughes *et al.*, 1982).

Samples from the basal Castell Limestone in an outcrop south of the quarry on the north side of Abereiddi Bay (SM 795301) have produced a low-diversity conodont fauna (Bergström, 1964) which in addition to a few coniform species of uncertain affinities, includes representatives of *Amorphognathus tvaerensis* (Plate 18.1, Figs. 15 and 16), *Drepanoistodus suberectus*, *Phragmodus* sp. (cf. *P. polonicus* Dzik; see Bergström and Orchard, 1985) and *Plectodina flexa*. In our collection is also a single very fragmentary specimen resembling an *Icriodella* (with a double denticle row) but additional specimens are required to confirm the presence of this genus in the Castell fauna. The presence of several typical Sc elements of *Phragmodus* is interesting because it is the only record of that genus in rocks of this age in the British Isles. Sc elements of *P. polonicus*, described by Dzik (1978) from Poland, as well as those of a presumably conspecific form in the Kukruse of Estonia (Bergström, 1971a, p. 106), look similar to those from the Castell Limestone and are from rocks of about the same age, but additional elements are needed to confirm the identity of the Castell Limestone form with that species. As a whole, the Castell Limestone conodont species association, with typical *A. tvaerensis*, compares closely with that of the Upper Llandeilo Stage at Llandeilo (Bergström, 1964) and is evidently of early *A. tvaerensis* Zone age; indeed, some of the M (holodontiform) elements of *A. tvaerensis* from the Castell Limestone appear to represent a morphologically slightly more advanced type than that in our collections from the Upper, but not uppermost, Llandeilo Flags at Llandeilo, possibly indicating a slightly younger age. However, the age of this fauna is in agreement with the fact that the *Dicranograptus* Shales which directly overlie

the Castell Limestone contain graptolites of the *N. gracilis* Zone (Hughes *et al.*, 1982).

18.7 THE TYPE CARADOC SUCCESSION

In the Caradoc type area in South Shropshire (Fig. 18.1), the basal part of the series rests unconformably on Precambrian, Cambrian or Tremadoc rocks, is faulted out or is poorly exposed. In addition, there is evidence that the base of the Caradoc Series is diachronous in the type area, and the significance of its dominantly shelly fossils for definition of a regionally recognizable base of the series is currently unclear (Whittington *et al.*, 1984). Accordingly, as noted by Dean (in Whittington *et al.*, 1984), there is little, if any, prospect of finding a suitable stratotype section for the base of the Caradoc Series in its type area.

Conodonts have been known from the Costonian Stage, the basal stage of the Caradoc Series, since the early 1970s (Bergström, 1971a, 1971b) but the faunas have not been described until recently (Savage and Bassett, 1986). Our Costonian collections are relatively rich in specimens but contain only a few species. The productive samples were collected from an old quarry (SO 41188624) in the Hoar Edge Grit just southwest of the River Onny in the Onny Valley standard section (Stop VIII-16 of Bassett *et al.* (1974)), from exposures of coeval rocks near the highway 0.2 km west of Glenburrell (SO 41198623), and from the disused quarry (SJ 551013) at Evenwood about 2 km eastsoutheast of Acton Burnell (Bergström, 1971a). The latter locality is of special biostratigraphic interest because it has yielded graptolites of the *N. gracilis* Zone (Bergström, 1971b, 1986).

The Costonian conodont collections are all essentially the same and include representatives of *Amorphognathus* sp. cf. *A. tvaerensis* (see Bergström, 1971a, p. 110), *Dapsilodus mutatus*, *Drepanoistodus suberectus*, *Icriodella* sp. cf. *I. praecox*, *Panderodus* spp., *Plectodina flexa*, *Erismodus?* sp. and *Rhodesognathus?* sp. aff. *R.? polonicus*. Not a single element of

Eoplacognathus has been found, and *Amorphognathus* and *Icriodella* are represented by just a few specimens among thousands of elements. Accordingly, there is a distinct difference in the faunal aspect compared with that of presumably coeval faunas from Narberth even if the species assemblages are similar in the two areas.

Attempts to classify the Costonian fauna in terms of the North Atlantic conodont zonal scheme are hampered by the fact that virtually no index species are present, the only exception being the form here identified as *A.* sp. cf. *A. tvaerensis* that suggests the *A. tvaerensis* Zone. It should be noted, however, that no M (holodontiform) element of the latter species has yet been found in the Costonian strata and the precise identity of our specimens remains somewhat uncertain, although they compare favourably with *A. tvaerensis* in observable morphological features. In Poland, *Rhodesognathus? polonicus (=R. elegans polonicus* in Dzik (1976)) was described from the upper *Baltoniodus variabilis*–lowermost *B. gerdae* subzones (Fig. 18.5); if the Costonian species is indeed the same, its occurrence suggests an age no younger than the *B. gerdae* Subzone, and it could be older. In view of the fact that the Evenwood conodonts are associated with *N. gracilis* Zone graptolites, a *B. variabilis* Zone age seems likely. Such an age would also agree with that of the Costonian strata at Narberth discussed above. Although it cannot be ruled out that additional conodont species of stratigraphic importance will be discovered in the Costonian rocks of the Caradoc Series type area, the fact that the collections of Bergström, and of Savage and Bassett (1986), together total several thousand specimens indicates that such species are likely to be quite rare.

18.8　CONCLUSIONS

The main results from our study may be summarized as follows.

(1) Among the conodonts recorded from the Llandeilo rocks at Llandeilo, Carmarthen, Narberth and Abereiddi Bay, and from the basal Caradoc rocks at Narberth and several localities in the Caradoc type area in the Welsh Borderland, there are several biostratigraphically diagnostic species that permit a classification of the rock successions in terms of the North Atlantic conodont zonal scheme.

(2) It is shown that the base and the top of the Llandeilo Series at Llandeilo are coeval with levels in the lowermost *Amorphognathus inaequalis* Subzone and the lower part of the *Baltoniodus variabilis* Subzone, respectively, but it should be noted that no conodonts are yet known from the topmost part of the Upper Llandeilo sequence at Llandeilo. Neither of these series boundaries appears to coincide precisely with a boundary in the conodont zonal scheme, although the top of the Llandeilo Series is likely to be rather close to the base of the *Amorphognathus tvaerensis* Zone judging from the successions at Narberth. However, it should be stressed that currently there are no formally selected boundary stratotypes for these series boundaries, which are still not very precisely defined biostratigraphically.

(3) There is excellent agreement between the conodont and graptolite biostratigraphy in those sections where graptolites are present. Also, there is no obvious conflict between conodont, brachiopod and trilobite data, suggesting that our conclusions regarding the biostratigraphic scope of the Llandeilo Series are in accord with the total faunal evidence currently available.

(4) In our interpretation, the base of the *Nemagraptus gracilis* Zone is coeval with a level somewhere in the middle of the Lower Llandeilo sequence at Llandeilo and the base of the Lower Llandeilo Stage with a level in the upper part of the *Glyptograptus teretiusculus* Zone (Fig. 18.8). Accordingly, if one follows current practice and defines the top of the Llanvirn Series as the

THIS PAPER	GRAPHITE ZONES	CONODONT ZONES ∟SUBZONES
CARADOC		*AMORPHOGNATHUS TVAERENSIS* va
LLANDEILO U.	*NEMAGRAPTUS GRACILIS*	
LLANDEILO M.		*PYGODUS ANSERINUS* in
L.		
UNCLASSIFIED	*GLYPTOGRAPTUS TERETIUSCULUS*	ki.
		li
	PYGODUS SERRA	ro
		re
		fo
LLANVIRN	*DIDYMOGRAPTUS MURCHISONI*	*EOPLACOGNATHUS SUECICUS*

Fig. 18.8 — Summary diagram showing the correlation between the top and the base of the Llandeilo Series, and the top of the Llanvirn Series, and the standard graptolite and North Atlantic conodont zonal schemes. For a review of the correlation between conodont and graptolite zones, see Bergström (1986). The abbreviations of conodont subzones follow Bergström (1983). If the base of the Llandeilo Series is taken to be the level of the base of the Llandeilo Flags at Llandeilo and if the top of the Llanvirn Series, following traditional practice, is located at the top of the *Didymograptus murchisoni* Zone, then the British succession includes a post-Llanvirn pre-Llandeilo interval (shaded) corresponding to a major part of the *Glyptograptus teretiusculus* Zone and at least three condont subzones. Hence, it is by no means an insignificant interval as is implied by Whittington *et al.* (1984).

top of the *Didymograptus murchisoni* Zone, then there is a post-Llanvirn pre-Llandeilo interval corresponding to a substantial part of the *G. teretiusculus* Zone which cannot be classified to series without changing the stratigraphic scope of one or both of these series (Fig. 18.8). This is not an insignificant interval; it includes at least three conodont subzones in the Baltoscandian succesion. Clearly, this classification problem will have to be addressed in cur-

rent reassessments of the biostratigraphic scope of the British Series and in the formal selection and definition of stratotypes. The easiest solution would probably be to extend the Llandeilo Series downwards to the top of the *D. murchisoni* Zone. A more drastic solution would be to extend the Caradoc Series downwards to the top of the Llanvirn Series and to include the Llandeilo as the basal stage of the Caradoc Series. In its present scope, the Llandeilo Series is small, with a range corresponding to the range of a graptolite and conodont zone. More significantly, this alternative would tie the base of the Caradoc Series to a widely recognized zonal boundary in the graptolite succession, a definite advantage compared with the current practice of having this series boundary based on shelly fossils that are virtually endemic to Wales and the Welsh Borderland. However, adopting one of these alternatives would make it difficult in south Wales at least to follow the preferable practice of defining a biostratigraphic boundary at the level of appearance, rather than disappearance, of a distinct assemblage of fossils. Alternatively, a conodont-based series boundary could be considered; as shown by the present study, it is possible to date both the top and the base of the Llandeilo Series in terms of conodont subzones of the standard North Atlantic conodont zonal scheme, which represents a considerable refinement compared with the correlation with the graptolite zonal scheme, which is, furthermore, only indirect. For instance, defining the base of the Caradoc Series as the base of the *Amorphognathus tvaerensis* Zone would probably not change the stratigraphic scope of this series significantly, and it would be tied to a level in an evolutionary lineage of a widely distributed conodont genus.

The present study illustrates how conodont lineages can be used, in conjunction with megafossils, to clarify a biostratigraphic problem that has been the subject of discussion for decades. Our results have direct bearing not only on the British Series classification but also on the use of the Llandeilo Series as a unit in local and global correlations. As indicated earlier in the present chapter, the growing precision and resolution in Ordovician biostratigraphy make it both necessary and timely to reassess the status of standard units such as the British Series. As shown recently (Whittington *et al.*, 1984), the Llandeilo Series is by no means the only series in need of such reassessment, and we hope that this contribution may stimulate further work of this type in Britain and elsewhere.

18.9 ACKNOWLEDGEMENTS

We are indebted to R. L. Ethington and M. G. Bassett for useful reviews of our manuscript, and to W. C. Sweet for informative discussions. Particular thanks are due to H. Jones, K. Tyler, T. Leonardi and S. Osborne for prompt and valuable technical assistance. Funds for much of the laboratory and scanning electron microscope work were made made available by the Department of Geology and Mineralogy, The Ohio State University, and the University College of Swansea, and for the senior author's fieldwork by Lund University, Sweden, the Royal Physiographic Society, Lund, Sweden, the Swedish Natural Science Research Council and The Ohio State University.

18.10 APPENDIX

The taxonomy of the conodont species from Wales and the Welsh Borderland listed in the present chapter will be dealt with in a separate publication. However, to clarify our multielement concept of some species, especially the stratigraphically important ones, it is appropriate to include some comments.

Amorphognathus inaequalis Rhodes 1953.

Includes *A. inaequalis* Rhodes 1953 (Pa), *Ambolodus* sp. Rhodes 1953 (Pb), *Ligonodina valma* Rhodes 1953 (Sc) and *Trichonodella inclinata* Rhodes 1953 (=*Hibbardella*? *inclinata* in Bergström 1964) (Sa).

Amorphognathus tvaerensis Bergström 1962. For multielement reconstruction, see Bergström (1971a). Separation from *A. inaequalis* is based primarily on the appearance of the M element (cf. Dzik, 1976).

Baltoniodus prevariabilis (Fåhraeus 1966). For multielement reconstruction, see Bergström (1971a). Includes *Dichognathus* cf. *D. typicus* Branson and Mehl in Rhodes (1953) (=*Prioniodus* n. sp. aff. *P.*? *variabilis* Bergström 1972 in Bergström (1964)) (Pa), *Paracordylodus lindstroemi* Bergström 1962 in Bergström (1964) (Sc) and *Tetraprioniodus asymmetricus* Bergström 1962? in Bergström (1964) (Sd).

Baltoniodus variabilis (Bergström 1962). For multielement reconstruction, see Bergström (1971a). Separation from *B. prevariabilis* follows Bergström (1971a).

Complexodus sp. The few specimens in our collections are morphologically less complex than those illustrated from the *A. inaequalis* Subzone of Poland by Dzik (1978) but they are typical representatives of the genus.

Eoplacognathus elongatus (Bergström 1972). For multielement reconstruction, see Bergström (1971a).

Eoplacognathus lindstroemi (Hamar 1964). For multielement reconstruction, see Bergström (1971a). Separation from *E. elongatus* follows Bergström (1971a).

Icriodella sp. cf. *I. praecox* Lindström, Racheboef and Henry 1974. Open nomenclature is used until we have had the opportunity to make direct comparisons between the French types and the Welsh specimens.

Plectodina flexa (Rhodes 1953). Includes Rhodes' (1953) *Ozarkodina tenuis* Branson and Mehl 1933 (Pa), *Cordylodus reclineatus*

Stauffer 1935 (=*Cordylodus delicatus* Branson and Mehl 1933 in Bergström (1964)) (Sc), *Cyrtoniodus complicatus* Stauffer 1935 (=*Cordylodus flexuosus* (Branson and Mehl 1933) in Bergström (1964)) (M), *Trichonodella flexa* Rhodes 1953 (Sa), and *Gyrognathus elongatus* Rhodes 1953 (=*Zygognathus deformis* (Stauffer 1935) in Bergström (1964)) (Sb). The characteristic pastinate pectiniform Pb element is present in topotype collections from Rhodes' loc. 1. This species includes *Plectodina* n. sp. of Bergström (1971a) and is a likely senior synonym of Savage and Bassett's (1986) *Plectodina bullhillensis*.

PLATE 18.1

The illustrated specimens (OSU) are in the type collection of the Orton Geological Museum, The Ohio State University.

Icriodella sp. cf. *I. praecox* Lindström *et al.* 1974
Plate 18.1, Figs. 1, 3. Fig. 1, lateral view, pastinate pectiniform Pa element, specimen OSU 37180, ×55. 'Narberth Group', Llandeilo–Caradoc transition beds, Bryn-banc quarries 3 km east of Narbeth, Wales, sample 79B40-1. Fig. 3, lateral view, tertiopedate element, specimen OSU 37182, ×55. 'Narbeth Group', Llandeilo–Caradoc transition beds, Bryn-banc quarries 3 km east of Narbeth, Wales, sample 79B40-1.

Baltoniodus variabilis (Bergström 1962)
Plate 18.1, Fig. 2. Posterior view, Pb element (note ledges along processes), specimen OSU 39066, ×85. 'Narberth Group', Costonian beds, Henllan 4 km east of Narbeth, Wales, sample 70/155.

Complexodus sp.
Plate 18.1, Fig. 4, Lateral view, P element, specimen OSU 39067, ×78. 'Narberth Group', Costonian beds, Henllan 4 km east of Narberth, Wales, sample 70/155.

Eoplacognathus elongatus (Bergström, 1962)
Plate 18.1, Figs. 5–7. Fig. 5, upper view, Pa element, specimen OSU 39068, ×60. 'Narberth Group', Costonian beds, Henllan 4 km east of Narbeth, Wales, sample 70/155. Fig. 6, upper view, one type of Pb element, specimen OSU 39069, ×42. 'Narbeth Group', Costonian beds, Henllan 4 km east of Narbeth, Wales, sample 70/155. Fig. 7, upper view, another type of Pb element, specimen OSU 39070, ×45. 'Narberth Group', Costonian beds, Henllan 4 km east of Narbeth, Wales, sample 70/155.

Amorphognathus inaequalis Rhodes 1953
Plate 18.1, Figs. 8–10. Fig. 8, upper view, one type of Pa element, specimen OSU 39071, ×30. Lower Llandeilo Flags, Nantgaredig, 7 km east of Carmarthen, Wales. Fig. 9, upper view, another type of Pa element (note that the posterior process is split), specimen OSU 39072, ×30. Lower Llandeilo Flags, Nantgaredig, 7 km east of Carmarthen, Wales. Fig. 10, posterior view, M element (note the straight apical denticles), specimen OSU 39073, ×120. Lower Llandeilo Flags, Nantgaredig, 7 km east of Carmarthen, Wales.

Eoplacognathus lindstroemi (Hamar 1964)
Plate 18.1, Figs. 11–13. Fig. 11, upper view, Pa element, specimen OSU 39074, ×40. Ffairfach beds, Ffairfach railway section, Wales, sample W62-2. Fig. 12, upper view, one type of Pb element, specimen OSU 39075, ×60. Ffairfach railway section, Wales, sample W62–2. Fig. 13, upper view, another type of Pb element, specimen OSU 39076, ×40. Ffairfach beds, Ffairfach railway section, Wales, sample W62–2.

Amorphognathus sp. cf. *A. tvaerensis* Bergström 1962
Plate 18.1, Fig. 14. Upper view, Pa element, specimen OSU 39077, ×42. Upper Llandeilo Flags, old quarry 0.25 km northeast of Dynevor Castle, Llandeilo, Wales, sample 74B70-3.

Amorphognathus tvaerensis Bergström 1962
Plate 18.1, Figs. 15, 16. Fig. 15, postero-lateral view, M element (note the reclined apical denticles), specimen OSU 39078, ×155. (Castell Limestone, Abereiddi Bay, Wales, sample W62-17b. Fig. 16, upper view, Pa element, specimen OSU 39079, ×65. Castell Limestone, Abereiddi Bay, Wales, sample W62-17b.

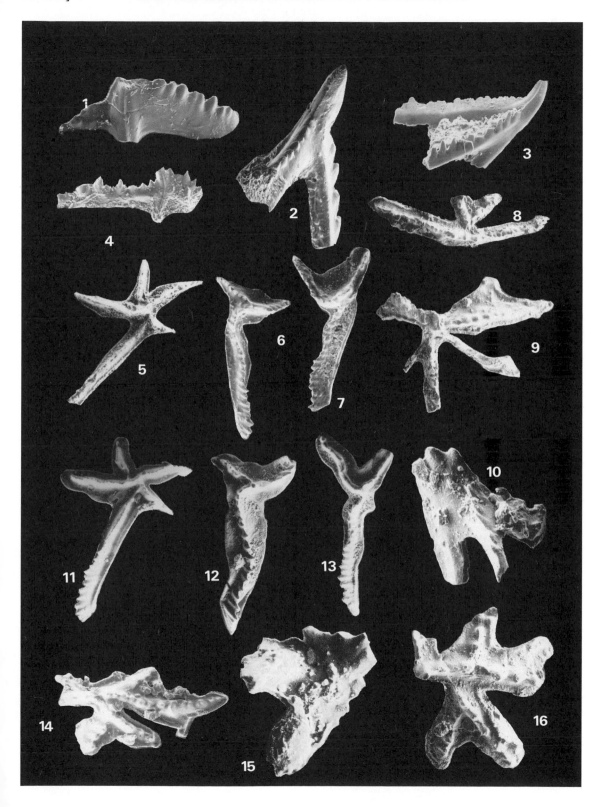

18.11 REFERENCES

Addison, R. 1974. The biostratigraphy of the Llandeilo facies of south Wales. *Unpublished Ph.D. dissertation*, Queen's University, Belfast.

Barnes, C. R. and Fåhraeus, L. E. 1975. Provinces, communities, and the proposed nectobenthic habit of Ordovician conodontophorids. *Lethaia*, **8**, 133–149.

Bassett, D. A., Ingham, J. K. and Wright, A. D. (Eds.) 1974. Field excursion guide to type and classical sections in Britain. *The Palaeontological Association Ordovician System Symposium, Birmingham, 1974*, 1–66.

Bassett, M. G. 1982. Ordovician and Silurian sections in the Llangadog–Llandilo area. In M. G. Bassett (Ed.), *Geological excursions in Dyfed, southwest Wales*, National Museum of Wales, Cardiff, 271–287.

Bergström, S. M. 1962. Conodonts from the Ludibundus Limestone (Middle Ordovician) of the Tvären area (S.E. Sweden). *Arkiv för Mineralogi och Geologi*, **3** (1), 1–61.

Bergström, S. M. 1964. Remarks on some Ordovician conodont faunas from Wales. *Acta Universitatis Lundensis Sectio II*, **3**, 1–66.

Bergström, S. M. 1971a. Conodont biostratigraphy of the Middle and Upper Ordovician of Europe and eastern North America. *Geological Socity of America Memoir*, **127**, 83–157.

Bergström, S. M. 1971b. Correlation of the North Atlantic Middle and Upper Ordovician conodont zonation with the graptolite succession. *Colloque Ordovicien–Silurien, Brest, Septembre 1971, Mémoirs du Bureau de Recherches géologiques et minières*, **73**, 177–187.

Bergström, S. M. 1973. Biostratigraphy and facies relations in the Lower Middle Ordovician of easternmost Tennessee. *American Journal of Science*, **273-A**, 261–293.

Bergström, S. M. 1980. Conodonts as paleotemperature tools in Ordovician rocks of the Caledonides and adjacent areas in Scandinavia and the British Isles. *Geologiska Föreningens i Stockholm Förhandlingar*, **102**, 377–392.

Bergström, S. M. 1983. Biogeography, evolutionary relationships, and biostratigraphic significance of Ordovician platform conodonts. *Fossils and Strata*, **15**, 35–58.

Bergström, S. M. 1986. Biostratigraphic integration of Ordovician graptolite and conodont zones — a regional review. In C. P. Hughes and R. B. Rickards (Eds.), *Palaeoecology and Biostratigraphy of graptolites*, Blackwell Scientific, Oxford, 61–78.

Bergström, S. M. and Carnes, J. B. 1976. Conodont biostratigraphy and paleoecology of the Holston Formation (Middle Ordovician) and associated strata in eastern Tennessee. *Geological Association of Canada Special Paper*, **15**, 27–57.

Bergström, S. M. and Orchard, M. J. 1985. Conodonts of the Cambrian and Ordovician Systems from the British Isles. In A. C. Higgins and R. L. Austin (Eds.), *A Stratigraphical Index of Conodonts*, Ellis Horwood, Chichester, West Sussex, 32–67.

Bergström, S. M., Rhodes, F. H. T. and Lindström, M. 1984. Conodont biostratigraphy of the type Llandeilo and•associated strata in the Ordovician of Wales. *Geo-

logical Society of America Abstracts with Programs*, **16**, 125.

Bergström, S. M., Riva, J. and Kay, M. 1974. Significance of conodonts, graptolites, and shelly faunas from the Ordovician of western and north–central Newfoundland. *Canadian Journal of Earth Sciences*, **11**, 1625–1660.

Black, W. W., Bulman, O. M. B., Hey, R. W. and Hughes, C. P. 1972. Ordovician stratigraphy of Abereiddy Bay, Pembrokeshire. *Geological Magazine*, **108**, 546–548.

Burrett, C., Stait, B. and Laurie, J. 1983. Trilobites and microfossils from the Middle Ordovician of Surprise Bay, southern Tasmania, Australia. *Memoir Association of Australian Palaeontologists*, **1**, 177–193.

Dzik, J. 1976. Remarks on the evolution of Ordovician conodonts. *Acta Palaeontologica Polonica*, **21** (4), 395–455.

Dzik, J. 1978. Conodont biostratigraphy and palaeogeographical relations of the Ordovician Mojcza Limestone (Holy Cross Mts., Poland). *Acta Palaeontologica Polonica*, **23**, 51–72.

Fåhraeus, L. E. 1966. Lower Viruan (Middle Ordovician) conodonts from the Gullhögen Quarry, southern central Sweden, *Sveriges Geologiska Undersökning*, **60**, 1–40, 4 plates.

Finney, S. C. and Bergström, S. M. 1986. Biostratigraphy of the Ordovician *Nemagraptus gracilis* Zone. In C. P. Hughes and R. B. Rickards (Eds.), *Palaeoecology and Biostratigraphy of Graptolites*, Blackwell Scientific, Oxford, 47–59.

Hamar, G. 1964. The Middle Ordovician of the Oslo Region, Norway. 17. Conodonts from the lower Middle Ordovician of Ringerike. *Norsk Geologisk Tidsskrift*, **44**, 243–292.

Hicks, H. 1881. The classification of the Eozoic and Lower Palaeozoic rocks of the British Isles. *Popular Science Review*, **5**, 289–308.

Hughes, C. P., Jenkins, C. J. and Rickards, R. B. 1982. Abereiddi Bay and the adjacent coast. In M. G. Bassett (Ed.), *Geological Excursions in Dyfed, South-west Wales*, National Museum of Wales, Cardiff, 51–63.

Jaanusson, V. 1960. On the series of the Ordovician System. *Proceedings of the 21st International Geological Congress*, **7**, 70–81.

Lindström, M., Racheboef, P. R. and Henry, J.-L. 1974. Ordovician conodonts from the Postolonnec Formation (Crozon Peninsula, Massif Armoricain) and their stratigraphic significance. *Geologica et Palaeontologica*, **8**, 15–28.

Lockley, M. G. and Williams, A. 1981. Lower Ordovician Brachiopoda from mid- and southwest Wales. *Bulletin British Museum (Natural History), Geology*, **33** (3), 1–75.

Murchison, R. I. 1835. On the Silurian System of rocks. *London and Edinburgh Philosophical Magazine*, **3** (7), 46–52.

Murchison, R. I. 1839. *The Silurian System founded on geological researches in the Counties of Salop, Hereford, Radnor, Montgomery, Caermarthen, Brecon, Pembroke, Monmouth, Gloucester, Worcester and Stafford: with Descriptions of the Coal-fields and Overlying Formations*, Murray, London, 1–768.

Repetski, J. E. and Ethington, R. L. 1977. Conodonts from graptolite facies in the Ouachita Mountains, Arkansas and Oklahoma. *Proceedings of the Symposium on the Geology of the Ouachita Mountains*, 1, Arkansas Geological Commission, 92–106.

Rhodes, F. H. T. 1953. Some British Lower Palaeozoic conodont faunas. *Philosophical Transactions of the Royal Society of London B*, **237**, 261–334.

Ross, R. J., Jr. *et al.* (27 co-authors) 1982. The Ordovician System in the United States. Correlation chart and explanatory notes. *International Union of Geological Sciences Publication*, **12**, 1–73.

Savage, N. M. and Bassett, M. G. 1986. Caradoc-Ashgill conodont faunas from Wales and the Welsh Borderland. *Palaeontology*. **28**, 679–713.

Sheng, S.-F. 1980. The Ordovician System in China. correlation chart and explanatory notes. *International Union of Geological Sciences Publication*, **1**, 1–7.

Skevington, D. 1969. The classification of the Ordovician System in Wales. In A. Wood (Ed.), *The Pre-Cambrian and Lower Palaeozoic Rocks of Wales*, University of Wales Press, Cardiff, 161–179.

Strahan, A., Cantrill, T. C., Dixon, E. E., and Thomas, H. H. and Jones, O. T. 1914. The geology of the South Wales Coalfield. Part XI. The county around Haverfordwest. *Memoir of the Geological Survey, England and Wales,*, 1–261.

Toghill, P. 1970. A fauna from the Hendre Shales (Landeilo) of the Mydrim area, Carmarthenshire. *Proceedings of the Geological Society of London*, **1663**, 121–129.

Webby, B. D. *et al.* (11 co-authors) 1981. The Ordovician System in Australia, New Zealand and Antarctica. Correlation chart and explanatory notes. *International Union of Geological Sciences Publication*, **6**, 1–64.

Whittington, H. B., Dean, W. T., Fortey, R. A., Rickards, R. B., Rushton, A. W. A., and Wright, A. D. 1984. Definition of the Tremadoc Series and the series of the Ordovician System in Britain. *Geological Magazine*, **121**, 17–33.

Wilcox, C. J. and Lockley, M. G. 1981. A reassessment of facies and faunas in the type Llandeilo (Ordovician), Wales. *Palaeogeography, Palaeoclimatology, Palaeoecology*, **34**, 285–314.

Williams, A. 1953. The geology of the Llandeilo district, Carmarthenshire. *Quarterly Journal of the Geological Society of London*, **108**, 177–208.

Williams, A., Lockley, M. G., and Hurst, J. M. 1981. Benthic palaeocommunities represented in the Ffairfach Group and coeval Ordovician successions of Wales. *Palaeontology*, **24**, 661–694.

Williams, A., Strachan, I., Bassett, D. A., Dean, W. T., Ingham, J. K., Wright, A. D. and Whittington, H. B. 1972. A correlation of Ordovician rocks in the British Isles. *Geological Society London Special Report*, **3**, 1–74.

Zalasiewicz, J. 1984. A re-examination of the type Arenig Series. *Geological Journal*, **19**, 105–124.

19

Conodont biostratigraphy of the Upper Carboniferous–Lower Permian rocks of Bolivia

M. Suárez Riglos, M. A. Hünicken and **D. Merino**

Limestones of the Copacabana Formation (250 m thick), cropping out at the classic region of Lake Titicaca and in other eastern and central Bolivian areas, were considered to be of Lower Permian age on the basis of the occurrence of fusulinids (Dunbar and Newell, 1946). The discovery of index conodonts within this formation (Hünicken, 1975; Suárez Riglos, 1984) at several Bolivian localities suggests that the limestones probably represent the Upper Carboniferous (Upper Pennsylvanian or Virgilian) and the Lower Permian ('Wolfcampian' and 'Leonardian') time intervals. The nine main localities where the Copacabana Formation crops out are in La Paz Department, Cochabamba Department and Santa Cruz Department. All localities yielded well-preserved conodonts of the *Streptognathodus elongatus* Assemblage Zone and the *Idiognathodus ellisoni* Assemblage Zone. They characterize the Upper Virgilian–Lower Wolfcampian time intervals in Bolivia. The Wolfcampian Stage is represented in Bolivia by the *Neogondo-*lella bisselli–Sweetognathus whitei* Assemblage Zone. The *Neostreptognathodus pequopensis–Sweetognathus behnkeni* Assemblage Zone is characteristic of the Leonardian Stage.

19.1 INTRODUCTION

The Upper Palaeozoic marine sequence (Copacabana Formation) in Bolivia occurs at several localities within a wide Andean belt that is oriented from westnorthwest to eastsoutheast and extends from Lake Titicaca and the Peruvian–Bolivian boundary in the west to the Santa Cruz and Chuquisaca Departments in the east. In the northnortheast the belt is bounded by the eastern border of the Andes (Subandean Belt) and in the southwest by the Desaguadero River and Pilcomayo River headwaters (Fig. 19.1).

Nine Bolivian localities with outcrops of the Copacabana Formation are considered in this study. Limestone samples yielded index conodont elements which permit a biostratigraphic zonation of Upper Carboniferous (Upper Pennsylvanian)–Lower Permian sequences to be established.

19.2 PREVIOUS WORK

The abundant and well-preserved megafossils of the Copacabana Formation (principally from the Yaurichambi and Apillapampa localities) which are very well known through the studies of d'Orbigny (1842), Steinmann (in Meyer, 1914) and Kozlowski (1914), are considered to be of Upper Carboniferous age. Dunbar and Newell (1946) studied the fusulinid fauna collected from the Lake Titicaca area and the Central Andes and assigned an Early Permian age to the Copacabana limestones. Hünicken (1975) studied the Permian conodonts from

Fig. 19.1 — Map showing the localities sampled for conodonts.

Torotoro (Potosí Department) and Suárez Riglos (1984) has presented a preliminary report of the Carboniferous–Permian conodont biostratigraphy.

There are several references to Bolivian Upper Palaeozoic stratigraphy. Cabrera La Rosa and Petersen (1936) worked in the Titicaca area and proposed the name 'Copacabana Group' for the Copacabana Limestone and the overlying Palaeozoic sequences. Newell *et al.* (1953) restricted the term to the limestone sequence. Ahlfeld and Branisa (1960) and Branisa (1965) used the term 'Copacabana Group' and accepted an Early Permian age. Chamot (1965) used the term 'Titicaca Group' and Calvo (1981) included the Copacabana, Collasuyo, San Pablo and Tiquina Formations in this group. Calvo (1981) and Suárez Riglos (1984) presented a generalized table for the Upper Palaeozoic stratigraphy, placing the Carboniferous–Permian boundary in the Lower Copacabana Limestone. This idea had previously been suggested by Reyes (1972) and Urdininea (1975). The Japanese Research Group (Sakagami, 1984) referred the Copacabana Formation (Jacha Khatawi Hill, near Yaurichambi) to the Lower Permian Series.

19.3 COPACABANA FORMATION: CONODONT LOCALITIES

Sample localities have been numbered as in Figs. 19.1 and 19.2 and Table 19.1. The main localities where the Copacabana Formation crops out are in La Paz Department (El Pelado (loc. 1), Cañada de Beu (loc. 2), Patapatani (loc. 3), Belén-Copacabana (loc. 4), Yaurichambi (loc. 5) and Colquencha-Collana (loc. 6)), in Cochabamba Department (Apillapampa (loc. 7)), in Potosí Department (Torotoro (loc. 8)) and in Santa Cruz Department (El Tunal (loc. 9)). All localities yielded conodonts.

(a) Locality 1: El Pelado
In the region of localities 1 and 2 (Díaz, 1959; Ahlfeld and Branisa, 1960), there is a Gondwanid sequence (140 m) with till containing fossil plants overlying Devonian rocks. The Copacabana Formation (250 m) follows conformably. Limestones were sampled by YPFB (Bolivian Petroleum Company) geologists and conodonts were obtained by Suárez Riglos. The El Pelado Mo-4941 sample yielded 29 elements of *Streptognathodus elongatus* Gunnell (Plate 19.1, Figs. 4–7, 12 and 13), and 16 m above, from the sample Mo-4951, the present authors obtained 14 specimens of *Idiognathodus ellisoni* Clark and Behnken.

(b) Locality 2: Cañada de Beu (Alto Rio Beni)
One sample from Cañada de Beu, Pal-1235, yielded two specimens of *I. ellisoni* (Plate 19.1, Fig. 20).

(c) Locality 3: Patapatani
The Abra of Patapatani is located northeast of Lake Titicaca (Achacachi–Chululaya Road). A thin development of the Copacabana Limestone overlies the Devonian sequence (Ahlfeld and Branisa, 1960). One sample from the middle of the limestone sequence, collected and processed by Suárez Riglos, yielded three specimens of *Sweetognathus whitei* (Rhodes) (Plate 19.2, Fig. 12), eight specimens of *Neostreptognathodus pequopensis* Behnken (Plate 19.3, Fig. 6) and ten elements of *Hindeodus* sp.

(d) Locality 4: Belén-Copacabana
A detailed map and generalized stratigraphic section of the Copacabana Peninsula near Lake Titicaca was given by Ahlfeld and Branisa (1960). There, the Copacabana Formation is 250 m thick. Two samples from the top, collected and processed by Suárez Riglos, provided several elements of *Hindeodus* sp. and fragments of *Sweetognathus behnkeni* Kozur.

(e) Locality 5: Yaurichambi
The Jacha Khatawi Hill is situated 50 km west-northwest of La Paz in the Yaurichambi area, near Lake Titicaca. This is the most important area of our study, because abundant conodonts

were obtained through the 215 m limestone succession (Sakagami, 1984) and enabled us to establish an Upper Carboniferous–Lower Permian conodont biostratigraphy for Bolivia, thus permitting correlation of other sampled localities. The first samples were collected by Calvo (1981) and Suárez Riglos (1984). The most recent sampling following the three parallel sections Ya, Yb and Yc studied by the Japanese Research Group (Sakagami, 1984) was made by

one of the present authors, Merino. 72 samples (51 with conodonts), taken approximately every 10 m, yielded conodont elements as indicated in Table 19.1 and Plates 19.2 and 19.3.

(f) Locality 6: Colquencha-Collana

This locality is 18 km west of La Paz–Oruro railway at Vilaque station. In the faulted syncline of Colquencha-Collana, overlying the Devonian and Gondwanid soft sandstone, there

Table 19.1 — Conodont localities.

Locality	Sample	species	Strepo-gnathodus elongatus Gunnell	Idio-gnathodus ellisoni Clark and Behnken	Neo-gondolella bisselli (Clark and Behnken)	Sweeto-gnathus whitei (Rhodes)	Neostrepto-gnathodus pequopensis Behnken	Sweeto-gnathus behnkeni Kozur	Hindeodus elements
La Paz Department									
1, El Pelado	4941		29	—	—	—	—	—	—
	4951		—	14	—	—	—	—	—
2, Cañada de Beu	Pal 1235		—	2	—	—	—	—	—
3, Patapatani	s/n		—	—	—	3	8	—	10
4, Belén	6–7		—	—	—	—	—	4	4
5, Yaurichambi	Ya 1		56	—	—	—	—	—	—
	Ya 3		—	—	—	1	6	—	22
	M 3		—	—	—	—	2	—	2
	Ya 5		—	—	—	—	2	1	3
	Ya 6		—	—	—	—	1	1	6
	Ya 103–105		34	—	—	—	—	—	—
	106		—	—	2	1	—	—	1
	107		—	—	1	—	—	—	—
	108–111		—	—	—	2	—	—	2
	116		—	—	—	—	1	—	3
	117–118		—	—	—	—	15	6	40
	119–122		—	—	—	—	—	5	17
	Yb 203,204		12	—	—	—	—	—	—
	205		—	3	—	—	—	—	—
	206–Yb11		—	—	24	—	—	—	5
	212–214		—	—	—	—	4	12	12
	215–217		—	—	—	—	—	8	27
	Yc 302,303		21	—	—	—	—	—	—
	304		—	—	3	—	—	—	—
	305		—	—	4	1	—	—	3
	307–309		—	—	—	2	—	—	2
	310–313		—	—	—	12	13	—	17
	314–316		—	—	—	—	48	20	97
	317–320		—	—	—	—	—	16	24
6, Colquencha-Collana	C 70		—	6	—	—	—	—	—
Cochabamba Department									
7, Apillapampa	as/n		—	—	—	2	2	—	13
Potosí Department	9-1		—	—	—	10	—	—	12
8, Torotoro	19		—	—	—	—	—	30	35
Sta. Cruz Department	5		—	44	—	—	—	—	—
9, El Tunal	11		—	18	—	—	—	—	—

is a cherty limestone sequence 292 m thick (Ahlfeld and Branisa, 1960) which consists mainly of limestone and mudstone accompanied by nodular chert. A limestone sample (sample C70, Bolivian Gulf Company) from the lower part provided six elements of *I.ellisoni* (Plate 19.2, Fig. 7).

(g) Locality 7: Apillapampa

This is situated in Cochabamba Department, 10 km south of the Caine River and Capinota Village, located 35 km south of Cochabamba City. The Copacabana Limestone is 175 m thick (Ahlfeld and Branisa, 1960; Chamot, 1965). A sample collected by Lobo (YPFB, Centro de Tecnología Petrolera) from the middle part of the sequence gave two specimens of *N. pequopensis* (Plate 19.3, Fig. 8), two specimens of *S. whitei* (Rhodes) and 13 elements of *Hindeodus* spp.

(h) Locality 8: Torotoro

This is in Charcas Province, Potosí Department, 10 km south of the Caine River. Fossiliferous limestone samples of the Copacabana Formation were collected by Dávila (YPFB geologist) and processed by Hünicken. Sample 9-1, from the middle part of the sequence, provided ten specimens of *S. whitei* (Plate 19.3, Figs. 12, 13 and 16) and 12 specimens of *Hindeodus* sp. Sample 19, 65 m above, yielded 30 specimens of *S. behnkeni* (Plate 19.3, Figs. 19, 20 and 23) and 35 elements of *Hindeodus* sp. nov.?

(i) Locality 9: El Tunal

This is located in Santa Cruz Department, along the El Tunal River, 11 km eastnortheast of Comarapa. This section (80 m thick) was sampled by Suárez-Riglos and Hünicken in 1978. Only the samples from the middle third of the sequence (samples 5 and 11, with a stratigraphic separation of 25 m) provided well preserved conodonts of the index species *I. ellisoni* (62 specimens, some of them illustrated in Plate

19.1, Figs. 14–19) and fragments, not recorded in Table 19.1, of *Streptognathodus* sp.

19.4 CONODONT BIOSTRATIGRAPHY

See Figs. 19.2 and 19.3.

(a) Streptognathodus elongatus Assemblage Zone

S. elongatus defines a faunal zone in the Virgilian Conemaugh Group of the Central Appalachian Pennsylvanian sequence (Merrill, in Lane *et al.*, 1971, p. 409).

At El Pelado (loc. 1) and in three sections of Yaurichambi (loc. 5), *S. elongatus* is restricted to the lower part of the Copacabana Formation and is the only species in the conodont fauna. Elements of the overlying *Idiognathodus ellisoni* Assemblage Zone are not associated with *S. elongatus* in the above-mentioned localities. Occurrences of *S. elongatus* in North America have been reported in Kansas by Gunnell (1933) and Ellison (1941) for the Wabaunsee Group (Upper Pennsylvanian or Virgilian age) and by Ellison (1941) for the Big Blue Group (Early Permian or Wolfcampian age), in Illinois, by Rhodes (1952) (who recorded *S.* cf. *elongatus* of Missourian age) and, in Wyoming, by Rhodes (1963) for the Tensleep Sandstone of the Eastern Big Horn Mountain, which was considered to be Wolfcampian in age, 'Although it would not by itself exclude the possibility of a highest Virgilian (Wabaunsee) age' (Rhodes, 1963, p. 405).

In South America, Rabe (1977, p. 198) reported *S. elongatus* from limestones north of Bucaramanga (Colombia) as 'Ober-Pennsylvanian bis Unter Perm'.

Fusulinids from the *Streptognathodus elongatus* Assemblage Zone in Bolivia (Yaurichambi, loc. 5, samples Ya 103, Ya 105, Yc 302 and Yc 303) were identified as *Triticites patulus* Dunbar and Newell and *Triticites* sp. by the Japanese Research Group (Sakagami, 1984). Urdininea and Yamagiwa (1980) established there the *Triticites* cf. *T. nitens* Subzone, which includes the following fusulinids: *Triticites patulus, Triticites* cf. *T. nitens, Triticites* sp. A., *Triticites* spp. and *Oketaella?* sp. indet. The *Triticites* cf. *nitens* Subzone is about 25 m thick in Yaurichambi and includes the *Streptognathodus elongatus* and the *Idiognathodus ellisoni* assemblage zones. An analysis of the sedimentary relationships (Sempere, pers. commun., 1985) shows that the first megasequence established at Yaurichambi finishes at the top with a sequence 25 m thick, which starts with thin mudstones and ends with thick grainstones, before a major subsidence of the basin occurs. This unit coincides with the *Triticites* cf. *nitens* Subzone.

On the basis of the wall structure and phylogenetic relationships of some of the major microgranular benthic foraminiferal families of the Late Palaeozoic, Boersma (1978, pp. 61–63) pointed out that, in the family Schwagerinidae, the genus *Triticites* is of Late Carboniferous age and does not extend into rocks of the Permian Period.

In conclusion, the present authors believe that the *Streptognathodus elongatus* Assemblage Zone reported herein could be referred to the Upper Pennsylvanian (Virgilian).

(b) Idiognathodus ellisoni Assemblage Zone

In the Moorman Ranch section in Nevada, *I. ellisoni* defines a faunal zone in a sequence that extends from the Upper Pennsylvanian, Hogan Formation, to the Lower Wolfcampian, Riepe

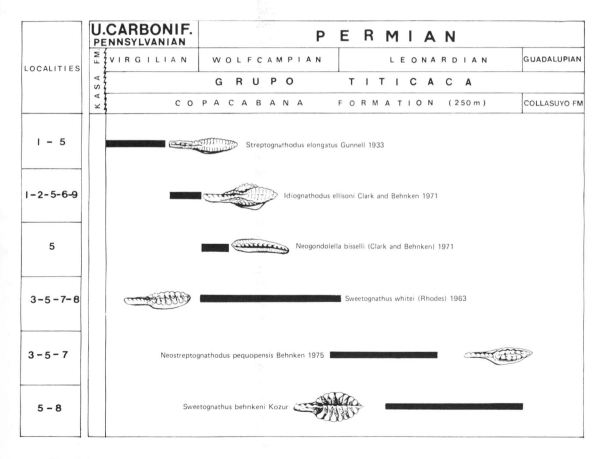

Fig. 19.2 — Stratigraphic distribution and location of Upper Carboniferous–Lower Permian conodont species in the Copacabana Formation, Bolivia.

Springs Formation (Clark and Behnken, 1971, Fig. 2, p. 420, pp. 423–424). Clark (1974, pp. 711–712) established 'that the *Idiognathodus ellisoni* Assemblage Zone ranges from the Pennsylvanian into the lower part of the Wolfcampian', in age. Clark *et al.* (1979) noted that the 'name-giving species characterizes an assemblage of predominantly Upper Pennsylvanian and Lower Permian species that range from Virgilian into basal Asselian and Tastubian rocks'. Kozur (1975, p. 27) suggested that the upper limit of the *Idiognathodus ellisoni* Zone is below the Carboniferous–Permian (Virgilian–Wolfcampian) boundary. The *Idiognathodus ellisoni* Assemblage Zone in the Bolivian samples (locs. 1,2,5,6 and 9) is, in general, represented by 10–12 m in Yaurichambi (loc. 5) to 25–30 m in El Tunal (loc. 9), where there are abundant elements of *I. ellisoni,* which dominate the well-preserved conodont fauna. Some specimens of *T. patulus* and *Triticites* sp. were reported by Sakagami (1984, p. 28 level Yc-7a) from the *Idiognathodus ellisoni* Zone in Yaurichambi (loc. 5). The *Streptognathodus elongatus* Assemblage Zone and the *Idiognathodus ellisoni* Assemblage Zone form part of a sedimentary sequence at the top of the Carboniferous megasequence. The megasequence coincides with the *Triticites* cf. *nitens* Subzone defined by Urdininea and Yamagiwa (1980) and occurs below a sedimentary discontinuity which marks a major subsidence of the basin. The discontinuity represents for the present authors the position of the Carboniferous–Permian boundary (Fig. 19.3) in Bolivia.

(c) Neogondolella bisselli–
Sweetognathus whitei Assemblage Zone

Elements of this fauna have been found at Moorman Ranch, Nevada, 'above the base of the Riepetown Sandstone, and the fauna ranges through strata that have been considered Middle and Upper Wolfcampian' (Clark and Behnken, 1971, p. 425). Merrill (1973, pp. 294, 310) reported *S. whitei* from· Permian (Wolfcam-

pian) rocks of the Council Grove and Chase Groups of Kansas. Clark (1974, p. 712) established that the *Neogondolella bisselli–Sweetognathus whitei* Assemblage Zone extends from the lower part of the Wolfcampian 'throughout the Upper Wolfcampian rocks, to just a few feet below the oldest Leonardian fusulinids'. Behnken (1975, p. 292) pointed out that the upper limit of this zonal fauna 'approximately coincides with the Wolfcampian–Leonardian boundary as defined in the Pequop Mountains'. Kozur (1975, p. 1) stated that 'the Carboniferous–Permian boundary is drawn at the base of the *whitei* Zone. It coincides (in Kansas) with the first appearance of the fusulinid genus *Pseudoschwagerina. . .*'. Also, 'in the Ural region of the USSR the first occurrence of *Pseudoschwagerina* is generally used to recognize the base of the type Permian System and this criterion was followed in the type Wolfcampian region in the Glass Mountains, Texas' (Sabins and Ross, 1963, p. 330). On the basis of previous work, Ziegler (1977, p. 547) defined the age and range of the species of the *bisselli–whitei* Assemblage Zone as Permian (Wolfcampian Series).

In South America, Rabe (1977, p. 175) has reported '*Gnathodus*' *whitei* in Wolfcampian limestone samples from Bucaramanga (Colombia). In Bolivia (Yaurichambi (loc. 5)) the *Neogondolella bisselli–Sweetognathus whitei* Assemblage Zone, developed throughout 60 m of limestone, is approximately recognized in the upper part of the lower half of the Copacabana Formation. It coincides with the first occurrence of the fusulinid genus *Pseudoschwagerina* in rocks which mark the beginning of the Permian System.

Fusulinids from the *Pseudoschwagerina texana* Subzone (Urdininea and Yamagiwa, 1980; Sakagami, 1984) were identified as *Pseudoschwagerina texana, Pseudoschwagerina patens, Pseudoschwagerina d'orbigny, Pseudoschwagerina* spp, *Schwagerina* sp. indet, *Chusenella* sp. indet. and *Pseudofusulina?* sp. They occur throughout the zone and coincide with the first sedimentary sequence of the second megasequence recognized in Yaurichambi (Sempere, pers. commun., 1985) which begins after the major subsidence of the basin.

N. bisselli (Plate 19.2, Figs. 14–20, and Plate 19.3, Figs. 1–5) is only present in the basal part (10 m) of this assemblage zone and *S. whitei* (Plate 19.2, Figs. 8–13, and Plate 19.3, Figs. 12–16 and 25) characterizes the upper 50 m. Only one sample (Yc-310), from the upper limit of this zonal fauna in Yaurichambi, has yielded

S. whitei together with *N.pequopensis* (see Table 19.1) and *Eoparafusulina* sp. (Sakagami, 1984, Sample Yc-21b). This point marks the beginning of the overlying Leonardian assemblage zone, partially modified by the present authors. Elements of *S. whitei* are also present in Belén (loc. 4) and in Torotoro (loc. 8) (see Table 19.1).

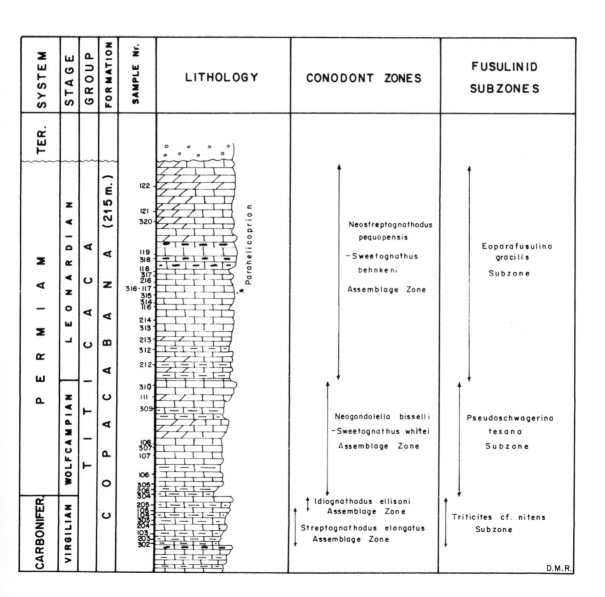

Fig. 19.3 — Generalized stratigraphic section in Yaurichambi.

(d) Neostreptognathodus pequopensis– Sweetognathus behnkeni Assemblage Zone

N. pequopensis defines a faunal zone in the Lower–Middle Leonardian Pequop Formation of Nevada (Clark, 1974; Behnken, 1975). As Behnken (1975, p. 310) has pointed out. *N. pequopensis* was 'found with *S. whitei* in the late Wolfcampian Ripetown Formation, Moorman Ranch section, Nevada'. In the Bolivian Copacabana Formation (Yaurichambi section, loc. 5, samples Ya-3 and Yc-310; see Table 19.1) we have observed the same association, corresponding to the lower limit of the *Neostreptognathodus pequopensis–Streptognathodus behnkeni* Assemblage Zone, as previously mentioned. The same association occurs also at Patapatani (loc. 3) and Apillapampa (loc. 7). *N. pequopensis*, the oldest species of the genus according to Behnken (1975, p. 296), 'has its lowest occurrence approximately coincident with the Wolfcampian–Leonardian boundary in Nevada'. In the Bolivian sequence of the Copacabana Formation, the first occurrence of *N. pequopensis* (Plate 19. 3, Figs. 6–11) is associated with the fusulinid identified as *Eoparafusulina* sp. (*Eoparafusulina gracilis* Subzone) (Urdininea and Yamagiwa, 1980; Sakagami, 1984) considered to be of Leonardian age. However, Kozur (1975, p. 28) has established the *Sweetognathus behnkeni* Zone above the *Sweetognathus whitei* Zone, characterized by *S. behnkeni* Kozur (Upper Wolfcampain Riepetown Formation, Moorman Ranch, Nevada) (Kozur, 1975, pp. 3–4).

In the Bolivian Copacabana Formation, we have obtained well-preserved specimens of *S. behnkeni* (Plate 19.3, Figs. 17, 18, 21 and 22–24) from 18 limestone samples collected from the upper 50–60 m of the sequence outcropping in Yaurichambi (loc. 5). This interval corresponds to Kozur's '*behnkeni* Zone', but it overlaps the upper 20 m of the *Neostreptognathodus pequopensis* Assemblage Zone. For this reason, we prefer to consider that both species together definine the *Neostrepto-gnathodus pequopensis–Sweetognathus behnkeni* Assemblage Zone in the upper part (90 m)

of the Copacabana Formation (Leonardian Stage) at Yaurichambi. In the lower third (locs. 3, 5 and 7) of this assemblage zone, *N. pequopensis* (Plate 19.3, Figs. 6 and 8) is associated with several elements of *Hindeodus* sp. and *Hindeodus* sp. nov.?, but *S. behnkeni* is not present. Ten samples from the middle third at Yaurichambi (loc. 5) yielded *N. pequopensis* (70 specimens, some of them shown in Plate 19.3, Figs. 7 and 9–11) associated with *S. behnkeni* (40 specimens), (Plate 19.3, Figs. 17, 21 and 22) and 155 elements of *Hindeodus* aff. *H. excavatus* (Behnken) and *Hindeodus* sp. nov.?. The upper third of this assemblage zone (locs. 4, 5 and 8) is characterized by *S. behnkeni* (29 specimens) (Plate 19.3, Figs. 19, 20 and 23) associated with 68 elements of *Hindeodus* aff. *H. excavatus* (Behnken) and *Hindeodus* sp. nov.?. In the lower part of this upper third in Yaurichambi (loc. 5), we found a beautiful specimen of an early edestid selachian of the genus *Parahelicoprion* Karpinsky (Janvier and Merino, unpublished). Clark and Ethington (1962) reported *Neostreptognathodus sulcoplicatus* Youngquist, Hawley and Miller associated with *Helicoprion* from the Meade Peak Member of the Phosphoria Formation, Idaho, 'regarded as uppermost Leonardian' (Behnken, 1975, p. 285). The *Neostreptognathodus pequopensis–Sweetognathus behnkeni* Assemblage Zone coincides, in Yaurichambi, with the second sedimentary sequence of the second megasequence defined by Sempere (pers. commun., 1985), here consisting of alternating wackestones and packstones. (Fig. 19.3).

19.5 CONCLUSIONS

Upper Pennsylvanian–Lower Permian conodonts of the Copacabana Formation (Bolivia), which crops out principally in La Paz Department (locs. 1–6) and in Cochabamba (loc. 7), Potosí (loc. 8) and Santa Cruz (loc. 9) Departments, provide a good basis for a biostratigaphic zonation of the Bolivian Virgilian, Wolfcampian and Leonardian sequences. Four assemblage zones have been distinguished

(1) *Streptognathodus elongatus* Assemblage Zone.
(2) *Idiognathodus ellisoni* Assemblage Zone.
(3) *Neogondolella bisselli–Sweetognathus whitei* Assemblage Zone.
(4) *Neostreptognathodus pequopensis–Sweetognathus behnkeni* Assemblage Zone.

19.6 ACKNOWLEDGEMENTS

The work was undertaken jointly at the Museo y Cátedra de Paleontología, Departamento de Geología, Facultad de Ciencias Exactas, Físicas y Naturales, Universidad Nacional de Córdoba, Argentina and at the Centro de Tecnología Petrolera, YPFB (Bolivian Petroleum Company) in Santa Cruz, Bolivia.

The field work in Bolivia was financially supported by YPFB (Centro de Tecnología Petrolera). Laboratory facilities were made available at the Universidad Nacional de Córdoba, the Consejo Nacional de Investigaciones Científicas y Técnicas (CONICET) in Argentina, and the Centro de Tecnología Petrolera, YPFB, in Bolivia.

The authors sincerely thank Alejandra Mazzoni (CONICET and Universidad Nacional de Córdoba) for reading, correcting and typing the manuscript, Dora Cortese (CONICET, Córdoba) and Martha Romero (Centro de Tecnología Petrolera, YPFB, Bolivia) for laboratory preparations, Professor Catalina Forte (CONICET, Córdoba) for providing the figures and plates and the geology student Norbeto Vaccari (Universidad Nacional de Córdoba) for assistance with the illustrations. The conodont photographs were taken with a scanning electron microscope, Laboratorio de Microscopía Electrónica de Barrido, Facultad de Odontología, Universidad de Buenos Aires, Argentina, with the help of Dante Giménez and were processed by Ricardo Munch of CONICET, Córdoba, and Museo Botánico, Universidad Nacional de Córdoba.

PLATE 19.1

The specimens denoted CTP-MP are deposited in the Centro de Tecnología Petrolera, YPFB, Santa Cruz, Bolivia. All these are pectiniform elements.

Streptognathadous elongatus Gunnell 1933
Plate 19.1, Figs. 1–13. Fig. 1, upper view, specimen CTP-MP 1-3,×67. Copacabana Formation. Yaurichambi, sample Ya 1. Fig. 2, lateral view, specimen CTP-MP 1-17,×67. Copacabana Formation, Yaurichambi, sample Ya 1. Fig. 3, upper lateral view, specimen CTP-MP 77-1,×100. Copacabana Formation, Yaurichambi, sample Yc 303. Fig. 4, upper lateral view, specimen CTP-MP 10-8,×67. Copacabana Formation, El Pelado, sample 4941. Fig. 5, upper lateral view, specimen CTP-MP 10-5,×100. Copacabana Formation, El Pelado, sample 4941. Fig. 6, lower view, specimen CTP-MP 10-2,×67. Copacabana Formation, El Pelado, sample 4941. Fig. 7, upper lateral view, specimen CTP-MP 10-10,×100. Copacabana Formation, El Pelado, sample 4941. Fig. 8, lower view, specimen CTP-MP 1-2,×67. Copacabana Formation, Yaurichambi, sample Ya 1. Fig. 9, upper lateral view specimen CTP-MP 1-8,×67. Copacabana Formation, Yaurichambi, sample Ya 1. Fig. 10, upper view, platform, specimen CTP-MP 1-1,×67. Copacabana Formation, Yaurichambi, sample Ya 1. Fig. 11, lateral view, platform and free blade, specimen CTP-MP 1-6,×67. Copacabana Formation, Yaurichambi, sample Ya 1. Fig. 12, upper view, specimen CTP-MP 10-3,×67. Copacabana Formation, El Pelado, sample 4941. Fig. 13, upper lateral view, specimen CTP-MP 10-4,×67. Copacabana Formation, El Pelado, sample 4941.

Idiognathodus ellisoni Clark and Behnken 1971.
Plate 19.1, Figs. 14–20. Fig. 14, lateral view, specimen CTP-MP 15-16,×67. Copacabana Formation, El Tunal, sample 11. Fig. 15, upper lateral view, specimen CTP-MP 15-13,×100. Copacabana Formation, El Tunal, sample 11. Fig. 16, lateral view, free blade, specimen CTP-MP 15-3,×67. Copacabana Formation, El Tunal, sample 11. Fig. 17, upper lateral view, specimen CTP-MP 14-1,×67. Copacabana Formation, El Tunal, sample 5. Fig. 18, upper lateral view, specimen CTP-MP 14-9,×67. Copacabana Formation, El Tunal, sample 5. Fig. 19, upper lateral view, specimen CTP-MP 14-12,×100. Copacabana Formation, El Tunal, sample 5. Fig. 20, upper lateral view, specimen CTP-MP 12-1,×48. Copacabana Formation, Cañada de Beu, sample Pal 1235.

Pl. 19.1]   Upper Carboniferous—Lower Permian rocks of Bolivia   327

PLATE 19.2

The specimens denoted CTP-MP are deposited in the Centro de Tecnología Petrolera, YPF-B, Santa Cruz, Bolivia. All these are pectiniform elements.

Idiognathodus ellisoni Clark and Behnken 1971
Plate 19.2, Figs. 1–7. Fig. 1, upper view, free blade and anterior part of platform fragment, specimen CTP-MP 11-2,×67. Copacabana Formation, El Pelado, sample 4951. Fig. 2, oblique upper lateral view, free blade and part of platform, specimen CTP-MP 11-1,×100. Copacabana Formation, El Pelado, sample 4951. Fig. 3, upper view, specimen CTP-MP 15-8,×67. Copacabana Formation, El Tunal, sample 11. Fig. 4, upper lateral view, specimen CTP-MP 15-1,×67. Copacabana Formation, El Tunal, sample 11. Fig. 5, upper view, specimen CTP-MP 14-4,×67. Copacabana Formation, El Tunal, sample 5. Fig. 6, lower view, specimen CTP-MP 14-3,×67. Copacabana Formation, El Tunal, sample 5. Fig. 7, upper view, platform and free blade fragment, specimen CTP-MP 13-2,×67. Copacabana Formation, Colquencha-Collana, sample C70.

Sweetognathus whitei (Rhodes 1963)
Plate 19.2, Figs. 8–13. Fig. 8, upper lateral view, early growth stage, specimen CTP-MP 85-2,×100. Copacabana Formation, Yaurichambi, sample Yc 310. Fig. 9, upper view, specimen CTP-MP 85-1,×100. Copacabana Formation, Yaurichambi, sample Yc 310. Fig. 10, upper lateral view, juvenile specimen, specimen CTP-MP 80-1,×140. Copacabana Formation, Yaurichambi, sample Yc 305. Fig. 11, upper lateral view, juvenile specimen, specimen CTP-MP 80-1, ×700. Copacabana Formation, Yaurichambi, sample Yc 305. Fig. 12, upper lateral view, juvenile specimen, specimen CTP-MP 16-1,×140. Copacabana Formation, Patapatni, sample s/n 12. Fig. 13, upper view, mature specimen, specimen CTP-MP 82-1,×140. Copacabana Formation, Yaurichambi, sample Yc 307.

Neogondolella bisselli (Clark and Behnken 1971)
Plate 19.2, Figs. 14–20. Fig. 14, upper view, posterior part, specimen CTP-MP 56–5,×67. Copacabana Formation, Yaurichambi, sample Yb 206. Fig. 15. oblique view, anterior part of upper surface, specimen CTP-MP 56-3,×67. Copacabana Formation, Yaurichambi, sample Yb 206. Fig. 16, upper view, surface, specimen CTP-MP 56-7,×67. Copacabana Formation, Yaurichambi, sample Yb 206. Fig. 17, upper view, surface, specimen CTP-MP 56-6,×67. Copacabana Formation, Yaurichambi, sample Yb 206. Fig. 18, upper view, surface, specimen CTP-MP 56-2,×67. Copacabana Formation, Yaurichambi, sample Yb 206. Fig. 19, upper view, surface, specimen CTP-MP 78-1,×100. Copacabana Formation, Yaurichambi, sample Yc 304. Fig. 20, upper view, surface, specimen CTP-MP 79-2,×100. Copacabana Formation, Yaurichambi, sample Yc 305.

Pl. 19.2]   **Upper Carboniferous—Lower Permian rocks of Bolivia**   329

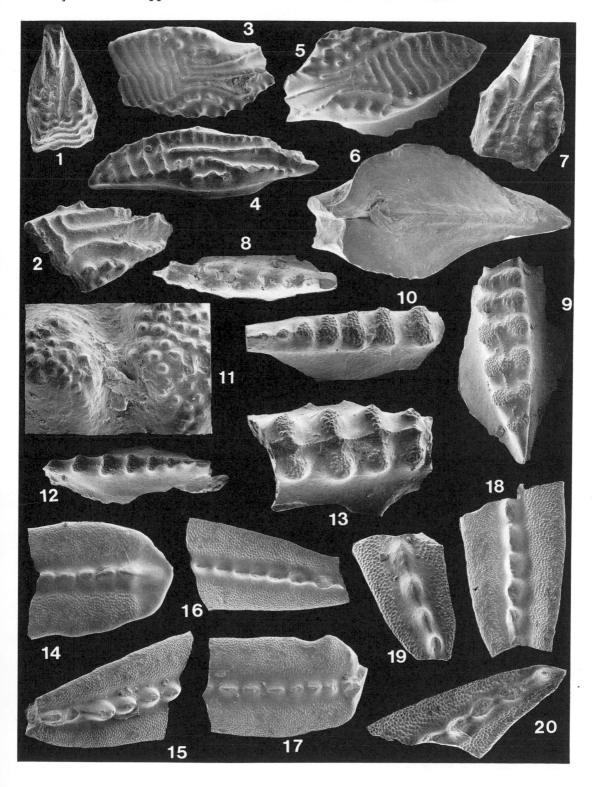

PLATE 19.3

The specimens denoted CTP-MP are deposited in the Centro de Tecnología Petrolera, YPFB, Santa Cruz, Bolivia, and the specimens denoted CORD-MP in the Museo y Cátedra de Paleontología, Universidad Nacional de Córdoba, Argentina. All these are pectiniform elements.

Neogondolella bisselli (Clark and Behnken 1971)
Plate 19.3, Figs. 1–5. Fig. 1, upper view, surface, specimen CTP-MP 79-1,×67. Copacabana Formation, Yaurichambi, sample Yc 305. Fig. 2, upper view, surface, specimen CTP-MP 78-3,×100. Copacabana Formation, Yaurichambi, sample Yc 304. Fig. 3, upper view, surface, specimen CTP-MP 56-1,×67. Copacabana Formation, Yaurichambi, sample Yb-206. Fig. 3, upper view, surface, specimen CTP-MP 56-1,×67. Copacabana Formation, Yaurichambi, sample Yb 206. Fig. 4, upper view, surface, specimen CTP-MP 56-4,×100. Copacabana Formation, Yaurichambi, sample Yb 206. Fig. 5, upper view, surface, specimen CTP-MP 56-4,×700. Copacabana Formation, Yaurichambi, sample Yb 206.

Neostreptognathodus pequopensis Behnken 1975
Plate 19.3, Figs. 6–11. Fig. 6, upper view, specimen CTP-MP 17-4,×100. Copacabana Formation, Patapatani, sample s/n 6. Fig. 7, upper view, specimen CTP-MP 98-1,×67. Copacabana Formation, Yaurichambi, sample Yc 3314. Fig. 8, upper lateral view, specimen CTP-MP 21-1,×100. Copacabana Formation, Apillapampa, sample as/n 8. Fig. 9, upper lateral view, platform and free blade, specimen CTP-MP 115-13, ×67. Copacabana Formation, Yaurichambi, sample Yc 315. Fig. 10, upper view, specimen CTP-MP 115-1,×67. Copacabana Formation, Yaurichambi, sample Yc 315. Fig. 11, upper view, specimen CTP-MP 100-11,×67. Copacabana Formation, Yaurichambi, sample Yc 315.

Sweetognathus whitei (Rhodes 1963)
Plate 19.3, Figs. 12–16, 25. Fig. 12, upper view, specimen CORD-MP 1-6,×100. Copacabana Formation, Torotoro, sample 9-1. Fig. 13, upper view, mature specimen, specimen CORD-MP 1-5,×100. Copacabana Formation, Torotoro, sample 9-1. Fig. 14, upper view, specimen CTP-MP 92-3,×100. Copacabana Formation, Yaurichambi, sample Yc 313. Fig. 15, upper view, specimen CTP-MP 92-2,×100. Copacabana Formation, Yaurichambi, sample Yc 313. Fig. 16, upper view, mature specimen, specimen CORD-MP 1-7,×100. Copacabana Formation, Torotoro, sample 9-1. Fig. 25, upper view, juvenile specimen, specimen CTP-MP 88-2,×210. Copacabana Formation, Yaurichambi, sample Yc 313.

Sweetognathus behnkeni Kozur 1975.
Plate 19.3, Figs. 17–24. Fig. 17, upper view, specimen CTP-MP 97-2,×67. Copacabana Formation, Yaurichambi, sample Yc 314. Fig. 18, upper, view, mature specimen, specimen CTP-MP 107-2,×67. Copacabana Formation, Yaurichambi, sample Yc 317. Fig. 19, upper lateral view, specimen CORD-MP 1-2,×100. Copacabana Formation, Torotoro, sample 19. Fig. 20, upper view, specimen CORD-MP 1-3,×67. Copacabana Formation, Torotoro, sample 19. Fig. 21, upper view, specimen CTP-MP 92-2,×100. Copacabana Formation, Yaurichambi, sample Yc 316. Fig. 22, upper view, specimen CTP-MP 60-1,×67. Copacabana Formation, Yaurichambi, sample Yb 213. Fig. 23, lower view, specimen CORD-MP 1-4,×67. Copacabana Formation, Torotoro, sample 19. Fig. 24, upper lateral view, specimen CTP-MP 107-1,×67. Copacabana Formation, Yaurichambi, sample Yc 317.

Pl. 19.3]　　　**Upper Carboniferous–Lower Permian conodonts of Bolivia**　　　331

19.7 REFERENCES

Ahlfeld, F. and Branisa, L. 1960. In D. Bosco (Ed.), *Geología de Bolivia*, Instituto Boliviano del Petróleo, 1–245, 12 plates.

Behnken, F. H. 1975. Leonardian and Guadalupian (Permian) conodont biostratigraphy in western and southwestern United States. *Journal of Paleontology*, **49** (2), 284–315, 2 plates.

Boersma, A. 1978. Foraminifera. In B. Haq and A. Boersma (Eds.), *Introduction to Marine Micropaleontology*, 2, Elsevier, Amsterdam, 19–77.

Branisa, L. 1965 Los Fósiles guías de Bolivia. I: Paleozoico. *Boletín del Servicio Geológico de Bolivia*, 6, 1–282, 80 plates.

Cabrera La Rosa, A. and Petersen, G. 1936. Reconocimiento geológico de los yacimientos petrolíferos del Departamento de Puno. *Boletín del Cuerpo de Ingenieros de Minas y Petróleo del Perú*, **115**, 1–100.

Calvo, J. C. 1981. Estudio estratigráfico y sedimentológico de las unidades litoestratigráficas del Paleozoico Superior en el área comprendida entre las poblaciones de Tiquina, Cumaná y Yaurichambi (Departamento de La Paz). (Unpublished) *Tésis de la Universidad Mayor de San Andrés*, La Paz, Bolivia, 1–122.

Chamot, G. A. 1965. Permian section at Apillapampa, Bolivia and its fossil content. *Journal of Paleontology*, **39** (6) 1112–1124, 3 plates.

Clark, D. L. 1974. Factors of Early Permian conodont paleoecology in Nevada. *Journal of Paleontology*, **48** (4) 710–720, 2 plates.

Clark, D. L. and Behnken, F. H. 1971. Conodonts and biostratigraphy of the Permian. In W. C. Sweet and S. M. Bergström (Eds.), *Proceedings of the Symposium on Conodont Biostratigraphy, Geological Society of América Memoir*, **127**, 415–433, 1 plate.

Clark, D. L., Carr, T. R., Behnken, F. H., Wardlaw, B. R. and Collinson, J. W. 1979. Permian Conodont Biostratigraphy in the Great Basin. In C. A. Sandberg and D. L. Clark (Eds.), *Conodont Biostratigraphy of the Great Basin and Rocky Mountains, Brigham Young University Geology Studies*, **26**, 143–147.

Clark, D. L. and Ethington, R. L. 1962. Survey of Permian conodonts in western North America. *Brigham Young University Geology Studies*, 9 (2), 102–114, 1 plate.

Díaz, H. 1959. Comunicación acerca de las condiciones geológicas presentes en el curso superior del Río Beni. *Boletín Técnico de Yacimientos Petrolíferos Fiscales Bolivianos*, 1 (2), 21–32.

Dunbar, C. O. and Newell, N. D. 1946. Marine Early Permian of the Central Andes and its fusuline faunas. *American Journal of Science*, **244** (6), 377–402; **244** (7), 457–491.

Ellison, S. P., Jr. 1941. Revision of Pennsylvanian conodonts. *Journal of Paleontology*, **15**, 107–143, 3 plates.

Gunnell, F. G. 1933. Conodonts and fish remains from Cherokee, Kansas City, and Wabaunse Groups of Missouri and Kansas. *Journal of Paleontology*, **7**, 261–297, 3 plates.

Hünicken, M. A. 1975. Sobre los hallazgos de conodontos en el Pérmico Inferior (Formación Copacabana) del área de Torotoro, Departamento de Potosí, Bolivia. *Actas del Primer Congreso Argentino de Paleontología y Bioestratigrafía*, **1**, 319–328, 2 plates.

Kozlowski, R. 1914. Les Brachiopodes du Carbonifere Supérieur de Bolivie. *Annales de Paléontologie*, **9**, 1–100, 11 plates.

Kozur, H. 1975. Beiträge zur Conodontenfauna des Perm. *Geologische Paläontologische Mitteilungen Innsbruck*, **5** (4), 1–44, 4 plates.

Lane, H. R., Merrill, G. K., Straka II, J. J. and Webster, C. D. 1971. North American Pennsylvanian conodont biostratigraphy. In W. C. Sweet and S. M. Bergström (Eds.), *Proceedings of the Symposium on Conodont Biostratigraphy, Geological Society of America Memoir*, **127** 395–414, 1 plate.

Meyer, L. F. 1914. Carbonfaunas aus Bolivien und Peru. *Neues Jahrbuch Mineralogie, Geologie und Paläontologie*, 37, 590–651.

Merrill, G. K. 1973. Pennsylvanian nonplatform conodont genera, I: *Spathognathodus*. *Journal of Paleontology*, **47** (2), 289–314, 3 plates.

Newell, N. D., Chronic, B. J. and Roberts, T. G. 1953. Upper Paleozoic of Peru. *Geological Society of America Memoir*, **58**, 1–276, 40 figs., 43 plates.

d'Orbigny, A. D. 1842. *Voyages dans l'Amérique Méridionale 3 (4:Paléontologie)*, 1–188, 22 plates.

Rabe, E. H. 1977. Zur Stratigraphie des ostandinen Raumes von Kolumbien. II. Conodonten des jungeren Paläozoikum der Ostkordillere, Sierra Nevada de Santa Marta und Serranía de Perijá. *Giessener Geologische Schriften*, **11**, 101–223, 4 plates.

Reyes, F. C. 1972. On the Carboniferous and Permian of Bolivia and Northwestern Argentina. *Anais da Academia Brasileira de Ciencias (Supplement)*, **4**, 261–277.

Rhodes, F. H. T. 1952. A Classification of Pennsylvanian conodont assemblages. *Journal of Paleontology*, **26**, 866–901, 4 plates.

Rhodes, F. H. T. 1963. Conodonts form the topmost Tensleep Sandstone of the eastern Big Horn Mountains, Wyoming. *Journal of Paleontology*, **37** (2), 401–408, 1 plate.

Sabins, F. F., Jr. and Ross, C. A. 1963. Late Pennsylvanian–Early Permian Fusulinids from southeast Arizona. *Journal of Paleontology*, **37** (2), 323–365, 6 plates.

Sakagami, S. (Ed.) 1984. Biostratigraphic study of Paleozoic and Mesozoic Groups in Central Andes (an interim report of the research group). 3.Geology of Cerro Jacha Khatawi, near Yaurichambi. *Rep.*, Department of Earth Science, Faculty of Science, Chiba University, Japan 8–35.

Suárez Riglos, M. 1984. Introducción a los conodontes del Permocarbónico de Bolivia. *Memoria del Tercer Congreso Latinoamericano de Paleontología, México*, 125–129.

Urdininea, M. 1975. Necesidad de un estudio bioestratigráfico en el Grupo Copacabana. *Revista Técnica de Yacimientos Petrolíferos Fiscales Bolivianos*, **4** (3), 81–88.

Urdininea, M. and Yamagiwa, N. 1980. Paleontological Study on the Copacabana Group at the Hill of Jacha Khatawi in the Yaurichambi Area, Bolivia, South America. Part 1: Fusulinids, Professor Saburo Kanno Memorial Volume, Japan, 277–289, 3 plates.

Ziegler, W. (Ed.) 1977. *Catalogue of Conodonts, III*, Schweizerbart'sche, Stuttgart, 1–574, 39 plates.

20

Lithic and faunistic ratios of conodont sample data as facies indicators

K. Weddige and W. Ziegler

Each conodont sample in addition to a stratigraphic value contains several facies indicators. Within the pelagic upper Lower Devonian of the Cerveny lom section (Barrandian, CSSR) a characteristic vertical shift of facies is indicated essentially by the frequency ratios of the following.

(1) Tentaculites and crinoidal stem fragments as seen in rock thin sections.
(2) Pyrite and oxidized ironstones retained in the insoluble residues.
(3) Clay and carbonate contents identified by the insoluble residues, ignition losses and X-ray fluorescence analyses.
(4) The conodont genera *Latericriodus* and *Polygnathus*.
(5) The *Polygnathus* taxa of the '*linguiformis*' and '*costatus* branch'.

The recognized lateral differentiation of conodonts obviously points to a correspondence of ecological and evolutionary patterns.

20.1 INTRODUCTION

The present palaeoecological investigation refers frequency data of lithic and faunistic characters to gradations of facies. In the neritic facies area of the Eifel (Rhenish Slate Mountains), changes in facies are marked by changes in carbonate content and in conodont fauna. The carbonate content, which is essentially reef detritus, increases towards the shallow-water reef and thus increasing carbonate content indicates a higher-energy facies. The associated conodont faunas also change their composition reefwards; an increase in abundance of the genus *Icriodus* coincides with a decrease in abundance of the genus *Polygnathus* (Weddige and Ziegler, 1976). A simple comparison of frequencies of both characters, i.e. carbonate content and the ratios of the two conodont genera, thus indicates a clear facies pattern.

20.2 THE CERVENY LOM SECTION

As a follow-up to the first study (Weddige and Ziegler, 1976), conodont samples from a pelagic facies area have now been investigated. The section at Cerveny lom (Fig. 20.1) is located in the Barrandian near Suchomasty and exposes the Suchomasty Limestone and the lowest part of the overlying Acanthopyge Limestone in a thickness of 24 m. The full thickness was sampled continuously, yielding 82 samples (Fig. 20.2). The age of the limestones ranges from the

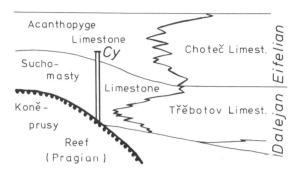

Fig. 20.1 — General biostratigraphic solution of the Cerveny lom section (Cy) after Chlupac *et al.* (1979).

laticostatus to the *partitus* conodont zones (Fig. 20.2), i.e. through the Dalejan Stage into the base of the Eifelian Stage (Chlupac *et al.*, 1979, pp. 134–5, Figs. 1–3 and 14).

The basal Suchomasty Limestone overlaps the Koneprusy reef of Pragian age and these basal deposits therefore represent a higher-energy facies. Upwards the facies energy decreases to a minimum at sample levels Cy30–Cy40, above which there is a gradual return to the higher-energy facies at the top of the section (Fig. 20.3, column a). This succession of facies with its characteristic A–B–A repetition has been interpreted by summarizing all lithic and faunistic characters of the 82 samples obtained from thin sections, insoluble residues and geochemical analyses. The most significant components are six in number, forming three pairs. In each pair the two components are negatively correlated and can thus be displayed graphically as ratios (Fig. 20.3, columns b–e).

(a) Tentaculites: crinoid fragments

These two components were identified in thin sections (Fig. 20.3, column b). The negative correlation between them is very distinct, as is the A–B–A shift. Since these two components are known to have preferred contrasting facies, tentaculites (the low-energy basinal facies) and

crinoids (the shallow high-energy facies), their relative ratio could be used to predict the conodont distribution.

(b) Pyrite: iron oxides

In the heavy-mineral fraction of the insoluble residues, pyrite and iron oxides show a negative correlation (Fig. 20.3, column c). As is well known, pyrite forms in anaerobic low-energy conditions, for instance in the deep-water facies of basins. Once formed, pyrite has a high stability through diagenesis and metamorphism, becoming unstable only in weathering conditions. Its presence is therefore indicative of reducing conditions. In the Cerveny lom sections it is not confined to the low-energy facies, but in the higher-energy facies it is accompanied by iron oxide in the form of red and brown lithified flakes, grains or crystals. Although some of these iron oxide grains may be the products of recent weathering, their distribution is so closely associated with the higher-energy facies as recorded by crinoid content that it seems clear that most are of primary origin and therefore record better bottom aeration during deposition. Persistence of pyrite into the higher-energy facies probably records local reducing conditions.

(c) Clay: carbonate ratio

The carbonate content, predominantly debris from crinoid gardens growing in the higher energy facies, is estimated from the loss on acid digestion (Figs. 20.3, columns d, e). The acid-insoluble residue consists partly of the iron minerals referred to above but is dominated by detrital clays, which were preferentially deposited in the low-energy facies. Determination of these two components is fraught with difficulties, particularly in the case of the clays. There is no practicable rapid method of separating the clay after acid digestion; sieving the residues reveals lithified clay flakes in the sand fraction as well as in the finer fraction. The absence of quartz and felspar from the sand fraction permits us to use the total light fraction of the

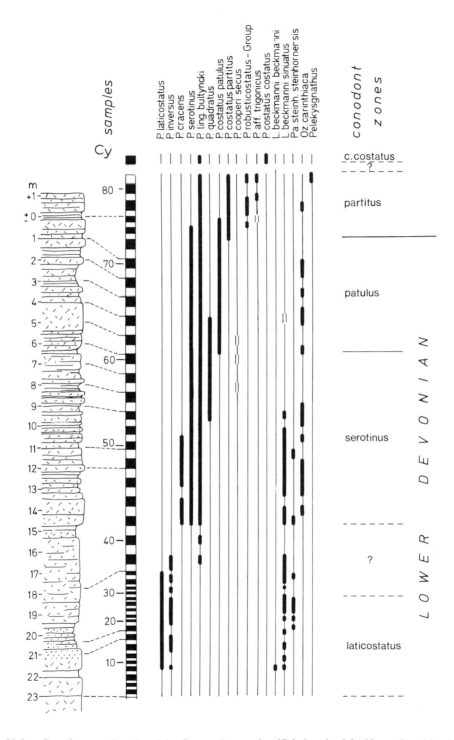

Fig. 20.2 — Conodont stratigraphy of the Cerveny lom section (Cy). Levels of the 82 samples with reference
to the profile (columnar section on the left-hand side) of Chlupac *et al.* (1979).

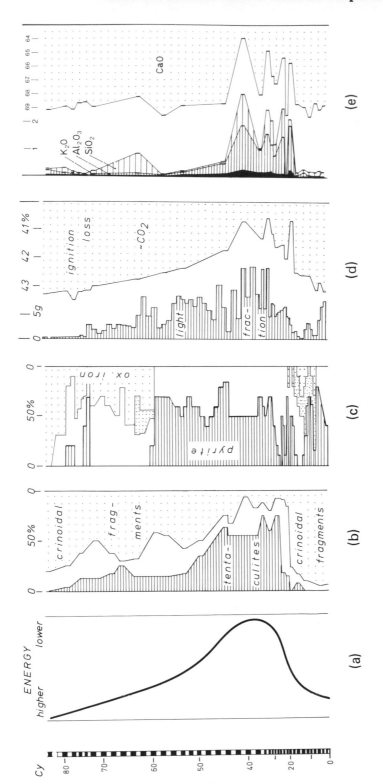

Fig. 20.3 — (a) General vertical facies development of the Cerveny lom section (Cy); (b) proportions of tentaculites and crinoidal fragments of the megafaunas in the thin sections of the Cerveny lom samples Cy1–Cy82; (c) proportions of pyrite and oxidized iron in the insoluble residues of the Cerveny lom samples Cy1–Cy82 (brown claystone portions are finely dotted; portions of red crusts and grains are more heavily dotted); (d) light fractions of the insoluble residues and ignition losses (in weight per kilogram of limestone) of the Cerveny lom samples Cy1–Cy–82 (diagram of ignition losses is inverse!); (e) quantities of main elements in Cerveny lom samples Cy1–Cy82 determined by X-ray fluorescence analyses (diagram of CaO quantities is inverse!).

insoluble residue as an approximation of the clay content.

An alternative estimate of the carbonate content is by loss on ignition, a necessary preliminary to geochemical analysis. The bulk of the loss on ignition is carbon dioxide from the carbonate minerals. It will also include combined water from any clay minerals but the Cerveny lom limestones are so pure that this is insignificant and the ignition-loss curve in Fig. 20.3, column d, corresponds very closely to that of the general facies change (Fig. 20.3, column a).

As a final check on the clay and carbonate contents, half of the Cerveny lom samples were subjected to X-ray analysis for potassium, aluminium and silicon, the characteristic elements of clay minerals, and calcium as the representative of the carbonate minerals. The results are plotted as oxides in Fig. 20.3, column e, which demonstrates the characteristic A–B–A curve.

20.3 CONODONTS AS FACIES INDICATORS

Comparison of the absolute frequencies of the conodonts from Cerveny lom, using histogram of numbers of individuals per kilogram (Fig. 20.4) reveals the existence of two statistical communities. Representatives of one community are more abundant in the higher-energy facies, whereas members of the other community predominate in the lower-energy facies.

The lower-energy facies shows enrichment in simple cones, mainly *Bellodella* sp. and *Coelocerodontus* sp. with lesser numbers of *Neopanderodus* sp., perhaps as a result of resedimentation of these light elements. Another simple conodont of this association, *Acodina* sp., has been counted separately, as it is thought to be part of an icriodontid multielement association (Klapper and Philip, 1971, pp. 438–9, type 4), and specifically of a *Latericriodus* multielement (Weddige, 1977, p. 279) assemblage. Indeed, the abundances of *Acodina* and *Latericriodus beckmanni sinuatus* Klapper *et al.* (1978) show direct covariation between individual samples (Fig. 20.4). Both, together with

Pandorinellina steinhornensis steinhornensis Ziegler (1956), are restricted to the lower-energy facies interval of the Cerveny lom section, although their overall stratigraphic ranges extend from much earlier (e.g. Klapper and Ziegler, 1979, Fig. 20.2) into the lowest part of the Eifelian Stage, in the case of *L.b. sinuatus* (Weddige and Requadt, 1985).

The conodont community of the higher-energy facies is particularly characterized by the genus *Polygnathus*, whose Pa element abundances correlate highly with those of their ramiform multielement partners. In addition, *Ozarkodina carinthiaca* (Schulze, 1968), and probably *Pelekysgnathus* sp., are associated. This community is best developed in the upper higher-energy facies; the earlier higher-energy facies contains much sparser faunas, perhaps reflecting higher sedimentation rates. In this part of the section, relative conodont frequencies are more diagnostic than absolute frequencies.

Latericriodus and *Polygnathus* are the diagnostic representatives of the two communities. Their antithetic variation is clearly displayed in Fig. 20.5, which plots the relative abundances of the conodonts, excluding all the simple cones. In particular, it picks out the earlier higher-energy facies at the base of the Cerveny lom section in spite of the generally sparse faunas at this level. It is particularly noteworthy that *Latericriodus* is confined to the lower half of the Cerveny lom section.

Several taxa are present in the *Polygnathus* community. Their distribution is recorded in Fig. 20.6, which, by comparison with Fig. 20.3, column a, suggests an ecological influence. The higher-energy facies is associated with *Polygnathus laticostatus* Klapper and Johnson (1975), *Polygnathus costatus patulus* Klapper (1971), *Polygnathus costatus partitus* Klapper *et al.* (1978) and *Polygnathus costatus costatus* Klapper (1971) whereas *Polygnathus inversus* Klapper and Johnson (1975), *Polygnathus linguiformis bultyncki* Weddige (1977) and *Polygnathus cracens* Klapper *et al.* (1978) occur in the lower-energy facies. *Polygnathus serotinus*

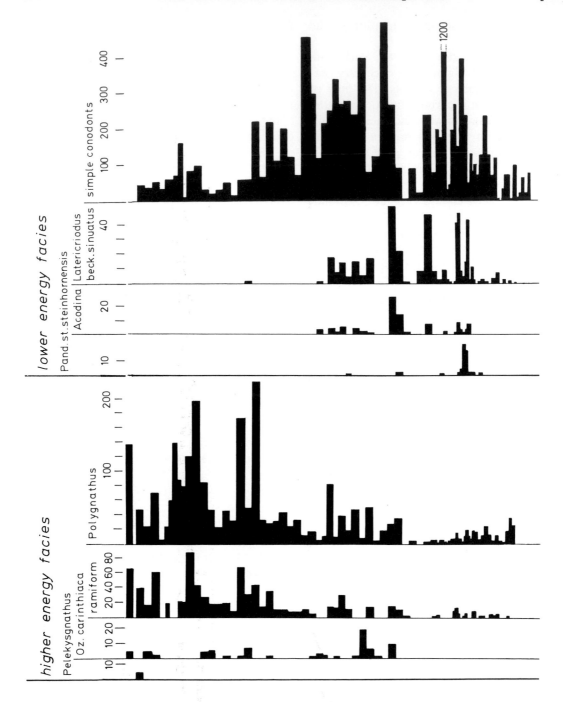

Fig. 20.4 — Absolute frequencies of the conodonts (individuals per kilogram) from the Cerveny lom samples Cy1–Cy82.

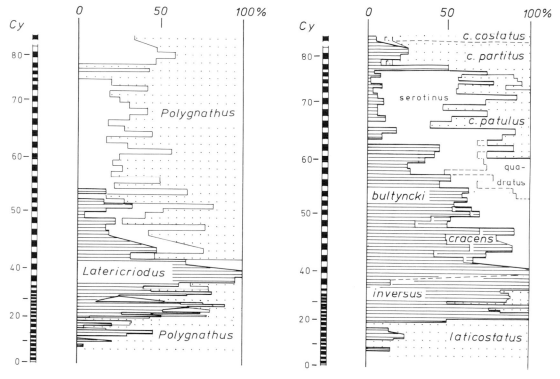

Fig. 20.5 — Relative frequencies of the genera *Latericriodus* and *Polygnathus* according to the total conodont fauna less simple cones from the Cerveny lom samples Cy1–Cy82.

Fig. 20.6 — Relative frequencies of the *Polygnathus* taxa from the Cerveny lom samples Cy1–Cy82. Taxa of the *'linguiformis* branch' are shown shaded with horizontal lines, those of the *'costatus* branch' are shown dotted.(r, *robusticostatus* group).

Telford (1975) and *Polygnathus quadratus* Klapper *et al.* (1978) occur in both facies associations, although *P. serotinus* sometimes dominates the higher-energy faunas. An identical grouping of the genus *Polygnathus* has already been made on phylogenetic grounds (Weddige, 1977, pp. 304–5, Fig. 4). The *P. costatus* group of taxa, here assigned to a higher-energy facies, were assigned to the *'costatus branch'*, the lower-energy facies group to the *'linguiformis* branch' and the two ecologically uncertain taxa to the blind-ended *'serotinus* branch'.

20.4   CONCLUSIONS

Our ecological investigation is aimed ultimately at the production of a facies-range chart of the conodonts, analogous to the vertical-range chart of already proven use in conodont biostratigraphy. Our facies-range chart is defined as a series of opposing parameters thus.

Bathymetric: lower energy . . . .higher energy
               deeper water . . shallower water
Lithological: clay rich . . . . . carbonate rich
               pyrite . . . . iron oxides
Macrofaunal: tentaculites . . . . crinoids
Conodonts: *Latericriodus* . . . . .*Polygnathus*
               'linguiformis   *'costatus*
               branch'   branch'

The palaeoecological differentiation of the *Polygnathus* taxa, which has recently been recognized in early Middle Devonian faunas (Sparling, 1984, pp. 115–7, Fig. 2), obviously corresponds with the phylogenetic branching. Weddige and Ziegler (1979) had previously identified such a connection in the neritic distributions of *Icriodus* taxa, and Weddige and Ziegler (1976) reported similar ecological associations in the Middle Devonian *Icriodus–Polygnathus* faunas. Putting all these associations together produces an overall *Icriodus–Polygnathus–Latericriodus* facies range from shallow-water neritic to deep-water pelagic. This separation of the branches of the icriodontids by a *Polygnathus* facies appears to support the still-disputed taxonomic differentiation of the icriodontids as a consequence of the principle relationship that ecological and evolutionary patterns are in strict correspondence because of adaption mechanisms.

20.5 REFERENCES

Chlupac, I., Lukes, P. and Zikmundova, J. 1979. The Lower–Middle Devonian boundary beds in the Barrandian area, Czechoslovakia. *Geologica et Palaeontologica*, **13**, 125–156, 3 plates.

Klapper, G. 1971. Sequence within the conodont genus *Polygnathus* in the New York lower Middle Devonian. *Geologica et Palaeontologica*, **5**, 59–79, 1 fig., 3 plates.

Klapper, G. and Johnson, D. B. 1975. Sequence in conodont genus *Polygnathus* in Lower Devonian at Lone Mountain, Nevada. *Geologica et Palaeontologica*, **9**, 65–77, 3 plates.

Klapper, G. and Philip, G. M. 1971. Devonian conodont apparatuses and their vicarious skeletal elements. *Lethaia*, **4**, 429–452.

Klapper, G. and Ziegler, W. 1979. Devonian conodont biostratigraphy. In M. R. House, C. T. Scrutton and M. G. Bassett (Eds.), *The Devonian System, Special Papers in Palaeontology*, **23**, 199–224.

Klapper, G., Ziegler, W. and Mashkova, T. V. 1978. Conodonts and correlation of Lower–Middle Devonian boundary beds in the Barrandian area of Czachoslovakia. *Geologica et Palaeontologica*, **12**, 103–116. 2 plates.

Schulze, R. 1968. Die Conodonten aus dem Paläozoikum der mittleren Karawanken (Seebergebeit). *Neues Jahrbuch für Geologie und Paläontologie*, **130**, 133–245, 5 plates.

Sparling, D. R. 1984. Paleoecologic and paleogeographic factors in the distribution of lower Middle Devonian conodonts from north-central Ohio. *Geological Society of America Special Papers*, **196**, 113–125.

Telford, P. G. 1975. Lower and Middle Devonian conodonts from the Broken River Embayment North Queensland, Australia, *Special Papers in Palaeontology*, **15**, 1–96, 16 plates.

Weddige, K. 1977. Die Conodonten der Eifel-Stufe im Typusgebiet und in benbenachbarten Faziesgebieten. *Senckenbergiana Lethaea*, **58**, 271–419, 6 plates.

Weddige, K. and Requadt, H. 1985. Conodonten des Ober-Emsium aus dem Gebiet der Unteren Lahn (Rheinisches Schiefergebirge). *Senckenbergiana Lethaea*, **66**, 1–20, 4 plates.

Weddige, K. and Ziegler, W. 1976. The significance of *Icriodus: Polygnathus* ratios in limestones from the type Eifelian. In C. R. Barnes (Ed.), *Conodont Paleoecology, Geological Association of Canada Special Paper*, **15**, 187–199.

Weddige, K. and Ziegler, W. 1979. Evolutionary patterns in Middle Devonian conodont genera *Polygnathus* and *Icriodus. Geologica et Palaeontologica*, **13**, 157–164.

Ziegler, W. 1956. Unterdevonische Conodonten, insbesondere aus dem Schönauer und dem Zorgensis-Kalk. *Notizblatt des hessischen Landesamtes für Bodenforschung*, **84**, 93–106, 2 plates.

21

Reconstructing a lost faunal realm: conodonts from mega-conglomerates of the Ordovician Cow Head Group, western Newfoundland

S. L. Pohler, C. R. Barnes and N. P. James

Ancient continental margins, particularly the shelfbreak and upper slope environments, are especially difficult to study and yet they yield critical information in terms of sedimentology, stratigraphy, palaeoecology and biostratigraphy. This study attempts a reconstruction of the inner margin of the northwestern edge of the Iapetus Ocean during Arenig (Early Ordovician) time. It employs data from the lithostratigraphy, carbonate sedimentology and conodont biofacies and biostratigraphy of the large clasts in the mega-conglomerate beds 10, 12, and 14 of the Cow Head Group, western Newfoundland. The clasts provide the record of the varied shelfbreak and slope facies which accumulated in apron debris flows at the toe of the slope.

From over 300 samples from the coarse proximal facies, the conodont data (1) give precise ages to many of the clasts in the three beds, (2) reveal the various conodont biofacies assemblages across the inner margin, (3) demonstrate the development of local endemic faunas on marginal platforms and build-ups with their subsequent interior migration during transgressive events, (4) provide biostratigraphic ties between the North Atlantic and Midcontinent provincial faunas and between the slope and platform facies and (5) support developing models concerning the evolution of this Iapetus margin from a passive gentle ramp in the early Arenig to an oversteepened and collapsing margin by late Arenig time.

The complexity of the margin facies is emphasized. This type of study is only possible in areas of good exposure and stratigraphic continuity and where adequate base studies in sedimentology and palaeontology have been completed. The particular sampling methods and the data on conodont biofacies, biostratigraphy and carbonate sedimentology are applied to resolve a broad range of important problems associated with ancient continental margin evolution and palaeoceanographic and palaeobiogeographic events.

21.1 INTRODUCTION

One of the most difficult depositional environments to examine and understand in the ancient stratigraphic record is the shelfbreak–upper slope setting. Through recent studies on modern examples, the sedimentary processes and faunal distributions have become better known, encouraging comparison with ancient analogues. This work has been documented within recent symposium volumes (e.g. Doyle and Pilkey, 1979; Cook *et al.* 1983; Stanley and Moore, 1983).

The morphology of ancient and modern margins is complex and is governed by factors such as their tectonic origin, whether they are divergent, convergent or transform, and their level of maturity with modification by progradation or carbonate build-ups. Ancient margins commonly were subjected to destructive processes of deformation associated with continental or island-arc collision. Consequently, they became complexly deformed, metamorphosed, covered by nappes or obducted oceanic crust and generally obscured in the geological record. An actual margin represents a narrow depositional belt which may be preserved and exposed only rarely along the length of an orogenic belt. Thus, in those few places were the ancient shelfbreak–upper slope can be examined, detailed study is essential. Such a place is in western Newfoundland where evidence of shelfbreak and slope environments are preserved in conglomerates of the Middle Cambrian–Middle Ordovician Cow Head Group.

Cambro-Ordovician continental margins are still poorly known, with some examples reviewed by Cook (1979, 1983) and James and Mountjoy (1983). James and Mountjoy (1983, Fig. 1) illustrated the global distribution of early Palaeozoic carbonate platform margins. In some of these, the Cambro-Ordovician margins reflect a transition from a carbonate platform to a deeper shale basin (e.g. Hazen Trough, Canadian Arctic Archipelago; Selwyn Basin, Northwest Territories–Yukon; Spitsbergen). In these areas, the shelfbreak–upper slope facies is not well preserved nor well represented in clasts since extensive slope conglomerates are rare, possibly because of the gentle gradient of the margin. In eastern Australia, slope conglomerates are more closely related to small carbonate

rims on volcanic island arcs (Webby, 1976) and the true continental margin is not evident except in restricted areas of Tasmania (Burrett *et al.*, 1984). In the Appalachian Orogen, the Cambro-Ordovician continental margin is likewise preserved in few places. The thin-skinned tectonics characteristic of the southern Appalachians (e.g. Cook *et al.*, 1979) has obscured the ancient margin. In the northern Appalachians, slope deposits and prominent conglomerates are known from Quebec (Mystic Formation (Barnes and Poplawski, 1973) and Lévis Formation (Uyeno and Barnes, 1970)) and western Newfoundland (Cow Head Group (Kindle and Whittington, 1958; James and Stevens, 1982, in press)). By far the most spectacular are those of the Cow Head Group with a range of proximal to distal facies, with bedded slope deposits intercalated with debris-flow conglomerates and with clasts up to 200 m in length. Their exposure in imbricate thrust sheets extends over a 50 km × 15 km area (Fig. 21.1).

In these sequences the actual shelfbreak is not exposed directly but is represented by clasts carried downslope to accumulate as a thick wide apron at the toe of the slope (James and Stevens, in press) (Fig. 21.2). In the Cow Head Group, and especially in the proximal facies, the size, abundance and diversity of clasts permits a partial reconstruction of the shelfbreak and upper slope facies. This chapter provides an example of how such rare facies can be reconstructed. The shelfbreak–upper slope region is commonly a critical region where major oceanic water masses interact. The position of the thermocline may intersect the depositional surface in this belt, resulting in the close proximity of two distinct provincial faunas. This has been well demonstrated for Late Cambrian trilobites, with comparisons with modern isopod analogues, by Cook and Taylor (1976) and Taylor (1977). The shelly faunas that characterized the early Middle Ordovician margin of the ancient North American craton were referred to as the Toquima–Table Head faunal realm by Ross and Ingham (1970). The faunal community approach, using both trilo-

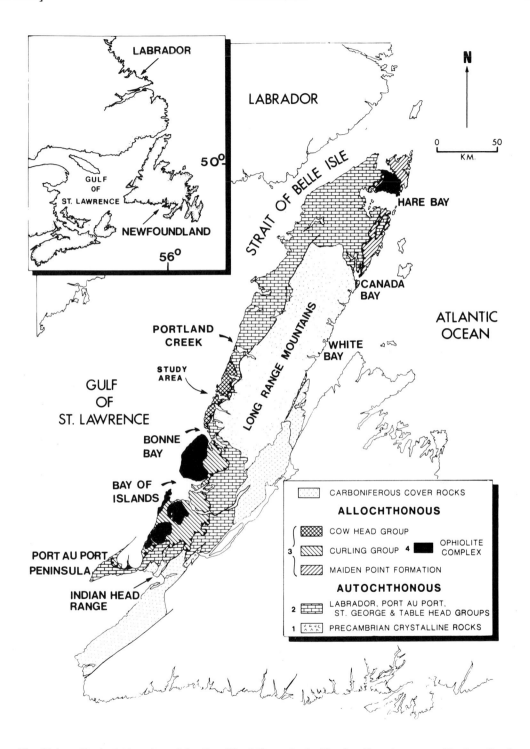

Fig. 21.1 — Geological setting of the Cow Head Group in the Humber Zone in western Newfoundland.
(From James and Stevens (in press)).

bites and conodonts (Barnes and Fåhraeus, 1975; Fortey, 1975; Fortey and Barnes, 1977) provided a reassessment of structure and extent of communities and provinces along the Early Ordovician Iapetus Ocean margin exposed in eastern North America and Spitsbergen.

In terms of the analysis of the ancient shelf-break–upper slope deposits, the main problems lie with an initially narrow facies belt compounded by later tectonic fragmentation and poor exposure. The facies can be reconstructed from several imbricated thrust belts (e.g. Taylor and Cook, 1976) or from the clasts in slope conglomerates if they are sufficiently large, stratigraphically extensive and well exposed (e.g. James and Stevens, in press; this chapter). In both cases, correlation of strata is critical to effect the reconstruction. Detailed lithofacies and biofacies studies must be available together

with precise biostratigraphic work. Fortunately, this work has now attained sufficient detail to allow a reconstruction of the western shelf-break–upper slope for the Iapetus Ocean during the early Ordovician Period in western Newfoundland. Analysis of lithofacies and regional stratigraphic work have been completed by Schuchert and Dunbar (1934), Kindle and Whittington (1958), Baird (1960), James and Stevens (1982, in press) and Williams *et al.* (1985). Key publications on the biostratigraphy and biofacies of the Cow Head Group include those of Johnson (1941), Ruedemann (1947), Kindle and Whittington (1958, 1959), Whittington (1963), Kindle (1982), Fåhraeus and Nowlan (1978) and Fortey *et al.* (1982). Studies concerned with the sedimentology and tectonic setting of the Cow Head Group include those of Oxley (1953), Rodgers and Neale (1963),

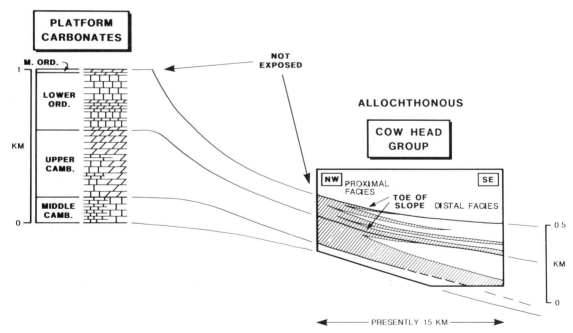

Fig. 21.2 — Schematic diagram showing the depositional relationship between the autochthonous platform carbonates and the allochthonous Cow Head Group. Note the difference in thickness between both sequences and the gradual steepening of the platform rim from Middle Cambrian to Middle Ordovician times. The striped lines in the Cow Head diagram indicate the debris-flow facies which accumulated at the toe of the slope. This facies yields clasts derived from the part of the slope marked 'not exposed', i.e. shelfbreak, upper slope and partly lower slope. (From James and Stevens (in press)).

Rodgers (1968), Stevens (1970), Callahan (1974), Jansa (1974), Fåhraeus *et al.* (1979), Hiscott and James (1985) and Coniglio (1985, in press). The general palaeogeography and eustatic changes were discussed by Barnes (1984). Previous work is thoroughly reviewed by James and Stevens (in press) but the studies noted above reflect the prolonged and increasingly intensive investigations of the Cow Head Group in western Newfoundland which have provided the necessary framework for this particular study.

21.2 GEOLOGICAL SETTING

The Cow Head Group (Kindle and Whittington, 1958) is located in the Humber Zone (Williams, 1975) of the northern Appalachians. The geological setting of the Humber Zone in western Newfoundland is generally accepted to represent a well-preserved, although tectonically complicated, fragment of the northwestern continental shelf, margin and adjacent basin floor of the Early Palaeozoic Iapetus Ocean. The generation, evolution and destruction of this continental margin lasted from late Pre-Cambrian to at least early Middle Ordovician time and is recorded in the rocks composing the Humber Zone (Williams, 1979). Rocks of a shelf sequence unconformably overlie Grenvillian basement and begin with Cambrian rift clastics followed by Cambro-Ordovician carbonates of the Port au Port Group, St. George Group and Table Head Group (Fig. 21.1). The narrow clastic shelf, which was established following the opening of the Iapetus Ocean, developed into a wide carbonate shelf platform over the course of the Cambrian-Middle Ordovician time interval (Barnes, 1984). The nature of the sediments indicates deposition in low latitudes. The foundering of the platform associated with orogenic processes is recorded in the Table Head Group which passes upwards from shelf carbonates through to deep-water mudstones and turbidites and finally to easterly derived flysch deposits (James and Stevens, 1982).

The platform sequences are overlain structurally by coeval slope and ocean-floor deposits of the Humber Arm and Hare Bay allochthons, which were thrust onto the platform from the east during Middle Ordovician time (Rodgers and Neale, 1963). The allochthonous rocks are thought to represent a slope segment of the western continental margin of the Iapetus Ocean, originally situated oceanwards of the carbonate platform noted above. The sedimentary rocks of the Humber Arm Allochthon are referred to as Humber Arm Supergroup and subdivided into Cow Head Group and Curling Group. The sediments of the Curling Group (Stevens, 1970) are interpreted as slope and rise deposits of the Early Palaeozoic continental margin (James and Stevens, 1982, in press). They are mainly equivalent to the more proximal slope sediments of the Cow Head Group at the type locality at Cow Head.

The Cow Head Group extends in age from late Middle Cambrian to early Middle Ordovician (Kindle and Whittington, 1958). The sequence is 300–500 m thick and consists of alternating successions of shales, limestones, conglomerates and marls with minor cherts, dolostones, sandstones and phosphorites. Kindle and Whittington (1958) informally numbered the strata as beds 1–14, based on distinct lithostratigraphic and biostratigraphic characteristics of each unit. Bed 15 was subsequently recognized in later studies (James and Stevens, 1982, in press) as forming an additional horizon completing the sequence. The abundant conglomerate horizons and slide scars interrupting thinly bedded limestone and shale successions indicate deposition on a sloping bottom in front of a carbonate platform. The most spectacular units within the Cow Head Group are several mega-conglomerate horizons containing boulders tens of metres in size.

21.3 SCOPE OF STUDY

The clasts and boulders within the three Arenig mega-conglomerate horizons (beds 10, 12 and 14) are the focus of this study. These deposits are interpreted as debris flows (Hiscott and James, 1985) which involved unstable coeval shelf and slope sediments and eroded locally into stratigraphically older strata. The flows came to rest on the lower slope or the toe of the slope (James and Stevens, in press), having assembled boulder– to pebble–sized clasts from the various facies belts across the shelfbreak down through the upper slope (Fig. 21.2). The setting of the Cow Head Group on the lower slope or deeper is indicated by the nature of the stratified sequences between the debris sheets (i.e. shales and thin-bedded limestones). Furthermore, large rafts of bedded fine-grained peri-platform sediments within the conglomerates are different from underlying lithologies and must have been acquired upslope (Fig. 21.3).

cies and conodont biofacies analysis in order to determine their provenance and previous stratigraphic position. This approach should enable the reconstruction of the segment of the margin preserved in the mega-conglomerate horizons and detail its evolution through the Arenig Epoch. This is of particular interest because the inner marginal environment is a rarely preserved facies of continental margins, certainly of the North American margin as it existed during Early–Middle Ordovician time. The present chapter attempts to outline the practical approach to this problem and gives a few preliminary results to exemplify the applicability of conodont data to this type of reconstructive study.

Fig. 21.3 — Clast of soft-deformed bedded peri-platform carbonate derived from the upper slope, observed at Lower Head, bed 14.

Other 'exotic' boulders are of shallow-water origin (James, 1981). The clasts in the debris flows are fragments of a carbonate platform margin facies that was obliterated by later tectonism; all that is left are juxtaposed pieces of a monster jig-saw puzzle and, as in a puzzle, it may be possible to reconstruct these fragments.

The purpose of this study is to recognize and interpret the different clast lithologies through the combined application of carbonate lithofa-

21.4 COW HEAD STRATIGRAPHY AND DEPOSITIONAL SETTING DURING EARLY ORDOVICIAN TIME

The rocks of the Cow Head Group are arranged into several northeast–southwest trending outcrop belts, reflecting structural complexity which is probably the result of imbrication (Fig. 21.4). In spite of the problematical tectonic history of the area the good lithostratigraphic and biostratigraphic control allows the tracing of individual beds from outcrop to outcrop over a distance of more than 20 km (James and Stevens, in press) (Fig. 21.5). Facies changes recognizable from belt to belt indicate a facies transition from proximal sediments in the northwest to distal sediments in the south and east, suggesting a southeast-dipping palaeoslope (unless major rotation occurred).

Proximal facies are characterized by successions of carbonate-dominated limestone–shale rhythmites and green–grey–black shales punctuated by massive polymict conglomerate horizons. Distally, the conglomerates decrease in thickness, coarseness and abundance and the red shales dominate over the limestones and grey–green shales. Conglomerates of the proximal facies have been investigated at Lower Head, Cow Head and on Stearing Island (Figs.

21.1 and 21.6); those of the distal facies were examined at St. Pauls Inlet, Western Brook Pond and Martin Point (Figs. 21.1 and 21.7).

The character of the Cow Head conglomerates also changes through time; the bed thickness and boulder size increase from bed 10 upwards to bed 14. Increase in boulder size is attributed to progressive steepening of the carbonate platform margin through time (James and Stevens, in press). Coeval platform deposits (St. George Group) are three times as thick as the adjacent slope deposits because of higher sedimentation rates on the shallow platform. This progradation resulted in an increase in platform relief above the basin floor and led to progressive instability of the sediments accumulating near the shelfbreak. This in turn produced an increase in competence of debris flows (Hiscott and James, 1985).

21.5 INVESTIGATIVE TECHNIQUES

In studying the clasts in the debris flows it is necessary to consider their possible origin. Depending on where the flow originated on the

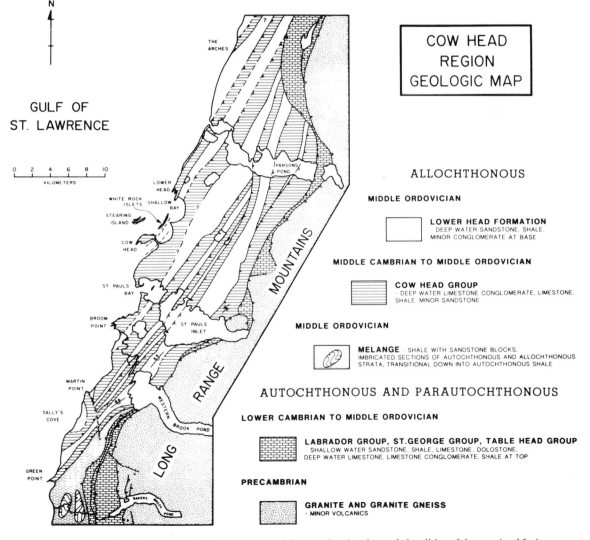

Fig. 21.4 — Map of outcrop belts of the Cow Head Group, showing the study localities of the proximal facies at Lower Head, on Stearing Island and Cow Head Peninsula and of the distal facies at St. Pauls Inlet, Western Brook Pond and Martin Point. (From James and Stevens (in press)).

COW HEAD STRATIGRAPHY

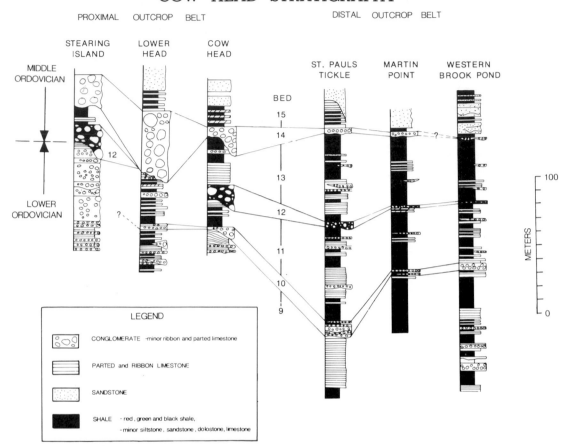

PROXIMAL   OUTCROP   BELT   DISTAL   OUTCROP   BELT

STEARING ISLAND   LOWER HEAD   COW HEAD   ST. PAULS TICKLE   MARTIN POINT   WESTERN BROOK POND

MIDDLE ORDOVICIAN

LOWER ORDOVICIAN

BED
15
14
13
12
11
10
9

100

METERS

0

LEGEND

CONGLOMERATE - minor ribbon and parted limestone

PARTED and RIBBON LIMESTONE

SANDSTONE

SHALE - red, green and black shale,
- minor siltstone, sandstone, dolostone, limestone

Fig. 21.5 — Lithostratigraphic sections examined in this study, showing the correlation of mega-conglomerate beds 10, 12 and 14 from proximal to distal facies belts.

Fig. 21.6 — Conglomerate of the proximal facies at Lower Head, bed 14, with clasts up to 200 m in length.

Fig. 21.7 — Conglomerate of the distal facies at Martin Point, bed 14, with clasts up to only 0.5 m in length.

slope the clasts incorporated in the debris sheets may have been derived from the shelf or the shelf edge, from the upper slope, from the middle-lower slope or from older underlying strata. To identify the nature and the abundance of clasts contributed from each of the environments cited above, the following procedures have been (or will be) applied.

(a) Determination of all different clast lithologies in the field

This includes classification of the boulders and clasts contained in the conglomerate beds and description of their lithologies. To understand the lateral and vertical facies relationships, detailed study focussed mainly on the large boulders (30 cm in diameter or length and larger) because they display the entire spectrum of variability within each lithofacies. A reasonably large clast size is also necessary to take conodont samples which must have to weigh at least 1 kg (an average of 2 kg in this study) in order to attain a sufficiently abundant fauna and to avoid contamination with the surrounding matrix. For these reasons the field work concentrated on outcrops of the proximal conglomerates which yield the largest clasts.

(b) Determination of abundances of the different clasts

To recognize all the lithologies present in selected outcrops, 5 m^2 areas were chosen and demarcated, in which all clasts larger than 10 cm maximum diameter were classified according to their lithofacies type and then point-counted using a grid of spacing 10 cm (Fig. 21.8). The results of the count were converted to percentages. Knowledge of the abundance of the clast facies solves the following problems. The most abundant clast lithologies should theoretically be the penecontemporaneous clasts which are expected to be the main contributors to the flows. It has to be taken into consideration, however, that unlithified sediments may not be preserved in the form of clasts but may supply the bulk of the matrix. Thus, early lithified

Fig. 21.8 — A 5 m^2 square area in which all clasts larger than 10 cm in diameter were point-counted, bed 14 at Lower Head.

sediments from marginal build-ups and the upper slope will be overrepresented. Less abundant clast lithologies will be (1) old clasts which have been eroded from underlying strata, (2) weakly lithified sediments which disintegrated during transport and (3) clasts derived from far outside the depositional environment (the inner shelf, for example). The study of the change of the clast lithologies from proximal to distal beds, i.e. the change in the ratio of endemic to exotic boulders, will help to recognize upper and lower slope lithologies more clearly and may also add to the knowledge of the mode of transport of the clasts in the flow. In this context, it is noteworthy that clasts from the lower slope become progressively more common in increasingly distal sections.

(c) Conodont sampling

To study the conodont biofacies, over 300 conodont samples have been collected from the conglomerate beds. An additional 30 were taken from the different lithofacies of the intercalated bedded sequences, especially from strata immediately over and underlying the conglomerates. Depending on the clast size, the sample weight ranged from 1 kg to 4 kg and averaged 2 kg. The largest number of samples was taken from proximal beds containing the largest boulders and the highest portion of exotic clasts. Within these beds, bed 14, with the largest variety of different clast lithologies, was sampled most extensively and collecting concentrated on boulders at Lower Head and on Cow Head Peninsula. For each lithology recognized, at least two samples were taken from two different clasts to check on the identification. The most abundant 'exotic' clasts were sampled most thoroughly as they probably represented fragments of the upper slope and the marginal build-up facies which is the main concern of this study. In more distal beds, some additional lithofacies types were recognized and samples collected where clast size permitted.

At Lower Head, bed 14 yields clasts eroded from strata as old as Late Cambrian (James and Stevens, in press) and the amount of different lithologies is so large that, despite collecting 150 samples, probably not all lithofacies have been recognized. Commonly a certain lithofacies type is represented by only one clast which could not be found elsewhere in the entire outcrop. The sampling of these sparse lithologies was therefore random.

21.6 APPLICATIONS OF CONODONT DATA

(a) Biostratigraphic use of conodonts
The dominating species in samples from bed 10 (Fig. 21.9) is *Prioniodus elegans* Pander, indicating the affinity of the fauna to the North Atlantic Province (Barnes *et al.*, 1973; Sweet and Bergström, 1974; Lindström, 1976) and allowing correlation with the *P. elegans* Zone of the standard Baltoscandian sequence. The species occurs in boulders which are most probably derived from the upper slope as well as in clasts of bedded lithologies from the lower slope. *Prioniodus elegans* is accompanied by faunal elements which have been recognized previously from Australia (McTavish, 1973), Argentina (Serpagli, 1974), Scandinavia (Lindström, 1955; Van Wamel, 1974; Löfgren, 1978) and North America (Bergström *et al.* 1972; Ethington and Clark, 1981; Repetski, 1982; and

others). Typical North Atlantic forms encountered are listed in decreasing abundance: *Paroistodus parallelus* (Pander), *Scolopodus peselephantis* Lindström, *Paracordylodus gracilis* Lindström. Forms known from Argentina and Australia (most of them with representatives in North America) are *Oistodus* n. sp. 1 Serpagli, *Acodus sweeti* (Serpagli), *Walliserodus australis* Serpagli, *Bergstroemognathus extensus* Serpagli, *Microzarkodina*? sp. aff. *M.*? *adentata* McTavish and *Protoprioniodus simplicissimus* McTavish. Conodont species known mainly from North America include *Drepanoistodus gracilis* (Branson and Mehl), *Glyptoconus quadraplicatus* (Branson and Mehl) and *Scolopodus gracilis* Ethington and Clark.

Three different species dominate the faunas in samples from the clasts of bed 12: *Periodon flabellum* Lindström, *Periodon aculeatus* Hadding and *Prioniodus (O.) evae* Lindström. The occurrence of the latter species assigns the fauna to the *P. evae* Zone of the Baltoscandian sequence. As in bed 10, the accompanying faunas are of mixed affinity and are known from North America, Australia, China and Argentina. They include *Jumodontus gananda* Cooper, *Protoprioniodus aranda* Cooper, *Acodus sweeti* (Serpagli), *Walliserodus australis*, *Oistodus*? *striolatus* Serpagli, *Microzarkodina*? *marathonensis* (Bradshaw), *Prioniodus (O.) communis* (Ethington and Clark) and *Strachanognathus parva* Rhodes. A few endemic forms have also been found, e.g. a new species of *Microzarkodina*?.

In samples from bed 14 a large variety of different taxa occurs because of the large number of reworked 'old' clasts in the flow (see discussion below). Boulders which have previously been recognized by their trilobite faunas as being equivalent in age to bed 13 (Kindle and Whittington, 1958) yield mainly *Periodon aculeatus*. The fauna generally is sparse in boulders from the upper slope and the marginal build-up facies. The accompanying elements are of North Atlantic and Midcontinent affinity or endemic to the adjacent carbonate platform sequence of equivalent age. The components of the fauna (Figs. 21.12 and 21.14) make precise stratigraphic assignment difficult, because Mid-continent as well as North Atlantic zone fossils are lacking. Kindle and Whittington (1958) correlated the rocks of the middle Table Head Group with the 'young' boulders of bed 14 at Lower Head based on the occurrence of similar trilobite faunas from the two localities and assigned bed 14 to the Llanvirn Series. Recent study of the macrofauna, however, suggests an

Arenig age for the clasts penecontemporaneous with bed 13 which occur in bed 14 (Fortey, 1980; James and Stevens, in press).

In order to recognize the youngest clasts in the flows, samples have been collected from the matrix and the immediately overlying bed. The conodont faunas retrieved from the matrix samples are generally sparse and of mixed ages. This is because much of the matrix is demonstrably injected just before the debris flow

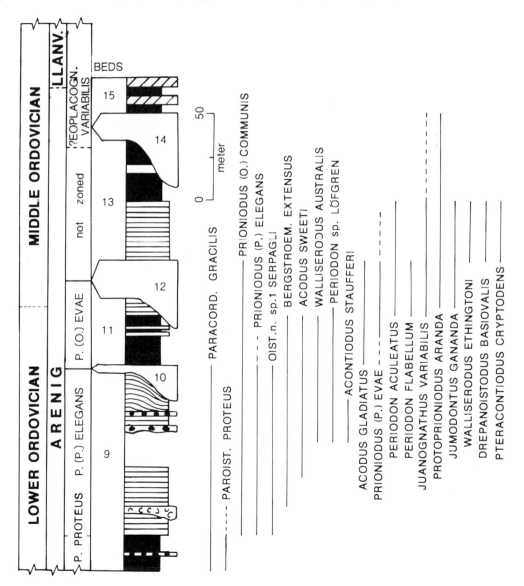

Fig. 21.9 — Stratigraphic ranges of selected species in the bedded sequence (beds 9, 11 and 13) of the Cow Head Group. (Data provided by D. I. Johnston).

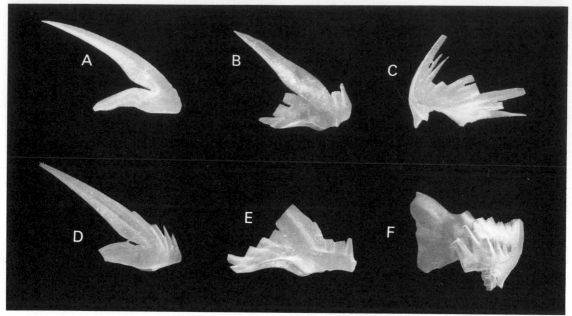

Fig. 21.10 — (A)–(C) *Periodon aculeatus* Hadding from older clast of marginal build-up facies (lateral views of hypotypes; GSC 81354–81356; ×100): (A) oistodontiform element; (B) ozarkodiniform element; (C) ramiform element. (D)–(F) *Periodon aculeatus* Hadding, later advanced form (lateral views of hypotypes; GSC 81357–81359): (D) oistodontiform element, ×110; (E) ozarkodiniform element, ×110; (F) Ramiform element, ×120. Note the more strongly developed denticulation in the more advanced form.

comes to rest and is composed of both directly underlying lithologies and contemporaneous fine grained sediments (Hiscott and James, 1985). The samples from the overlying beds proved to be more useful for dating the youngest clasts. In bed 14, boulders composed of lithoclasts yield an advanced species of *Periodon aculeatus* Hadding comparable with the forms reported from the Womble Shale (Repetski and Ethington, 1977). Similar forms have been found in samples from bed 15 at St. Pauls Inlet, suggesting that the fauna from the clast is equivalent in age to the bed 15 fauna. Advanced species differ from presumed older forms in processing a significantly higher number of denticles (Fig. 21.10). Löfgren (1978) demonstrated that the increase in denticles is an evolutionary trend in *Periodon*. Similar relationships may help to determine the youngest clast lithologies in other beds which are important for the reconstruction of the margin.

(b) Recognition of the oldest clast lithologies
The determination of the oldest clast lithologies and their abundance helps to evaluate the erosional strength of the flow and more importantly to distinguish penecontemporaneous clasts which are time-equivalent to the immediately underlying sequences from older clasts. In bed 10 a new species of *Prioniodus* has been retrieved from a clast at Lower Head. The form is probably the ancestor of *Prioniodus elegans*. This relationship is suggested by the elemental composition of the apparatus of *P*. n. sp., which is similar to *P. elegans* but in which the denticles are minute. The evolutionary trend towards higher denticulation in *P. elegans* has been suggested by Bagnoli and Stouge (1985; pers. commun., 1985) and the new species is therefore considered to be older than *P. elegans*. *P.* n. sp. has not been recorded previously in the adjacent shelf nor in the lower slope sediments. This observation, together with the obvious shallow-water origin of the boulder (a grain-

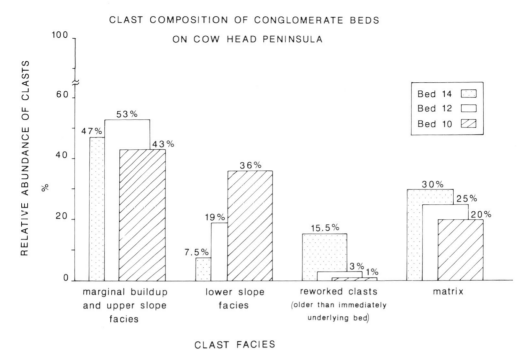

Fig. 21.11 — Diagram showing clast-composition of conglomerate belt on Cow Head Peninsula. Diagram shows increase in abundance of reworked clasts older than immediately underlying bed from bed 10 to bed 14, indicating an increase in erosional depth of the debris flows. Data were derived by applying results of the point-count analysis to the biostratigraphic data.

stone with shell debris), suggests that it was derived from high on the slope, probably eroded directly at the slide scar. The low number of these older clasts in the flow indicates that these conglomerates yield mainly penecontemporaneous clasts. The erosion of the flow was restricted therefore to bed 9 and equivalent sediments on the upper slope (Fig. 21.11).

The oldest fauna retrieved from a clast in bed 12 at Cow Head yielded *Paroistodus proteus* (Lindström) and *Paracordylodus gracilis*, together with *Teridontus nakamurai* (Nogami), *Cordylodus intermedius* Furnish and *Proconodontus muelleri* Miller. The fauna is of mixed age, ranging from latest Cambrian to early Arenig. Another boulder contained a fauna dominated by *P. elegans* and *Paracordylodus gracilis*, suggesting reworking from bed 9 or 10. The application of the biostratigraphic data to the counts of the clast lithologies indicates that the

amount of eroded 'old' lithologies is low and only slightly higher than in bed 10. In bed 14, some bedded limestone clasts yielded trilobites (R. K. Stevens, pers commun., 1984) as well as *Cordylodus proavus* Müller and *Eoconodontus notchpeakensis* Miller, indicating an early Tremadoc age. Another boulder of shallow-water facies contained conodonts indicative of fauna C of the North American Province (Ethington and Clark, 1971), yielding *Loxodus bransoni* Furnish and *Rossodus manitouensis* Repetski and Ethington. Fortey (in James and Stevens, in press) found an Upper Cambrian trilobite fauna in a boulder at Lower Head. These results indicate that substantial downcutting of the flow into the underlying strata occurred, although the immediately underlying beds at Cow Head for example record only minor erosion. The occurrence of old clasts such as the boulder with Midcontinent fauna C suggests that deep erosion must have occurred high on the upper slope. This deep erosion produced a wide range of facies and stratigraphic age for the reworked clasts in bed 14 and explains the large variety of different clast lithologies. Kindle and Whittington (1958) suggested that the clasts in bed 14 are approximately contemporaneous with the flow and can be used for correlation with the carbonate platform facies. The above evidence, however, confirms, as indicated by James and Stevens (in press), that the composition of bed 14 clasts is much more complex than originally anticipated and must be studied carefully to avoid confusion between young and old lithologies. Fig. 21.11 shows that the highest amount of clasts reworked from strata older than the immediately underlying bed is present in bed 14, and the lowest amount in bed 10. This indicates that there was a gradual increase in the erosional strength from the oldest to the youngest debris flow.

(c) Distinction of penecontemporaneous clasts
Within the clasts equivalent in age to the beds immediately underlying the conglomerates, the good biostratigraphic control achieved by other workers (Fåhraeus and Nowlan, 1978; Johnston *et al.*, 1985) allows a distinction between older and younger clasts incorporated from the underlying bed. For example, bed 10 clasts yield two distinct faunal assemblages: a sparse *Paroistodus proteus* fauna and a dominating *Prioniodus elegans* fauna. The ranges of conodonts from the underlying bed 9 sequence (Fig. 21.9) indicate that the lower part of the sequence belongs to the *P. proteus* Zone, while the upper part belongs to the *P. elegans* Zone. This same stratigraphic range can be recognized in the different clasts and enables the recognition of older and younger lithologies. Beginning with bed 11, a significant faunal change occurs, expressed in the species accompanying the zone fossil *Prioniodus (O.) evae* which is present throughout beds 11–13 in decreasing abundance (Fig. 21.9). The lower part of bed 11 includes an assemblage dominated by *P. evae* together with

Acodus sweeti, Bergstroemognathus extensus and *Walliserodus australis* (Fig. 21.9). This assemblage changes to one dominated by *Periodon flabellum* and *Periodon aculeatus* accompanied by *Juanognathus variabilis* Serpagli and *Protoprioniodus aranda* (Fig. 21.9). This assemblage, together with *Strachanognathus parva, Protoprioniodus papiliosus* Van Wamel and *Microzarkodina*? n. sp., is present in bed 12 clasts (Fig. 21.12). Although not all clasts yielded a fauna sufficiently abundant and diverse to recognize a particular assemblage, some of the upper slope lithologies can be assigned, thus preserving a record of the facies change occurring on the upper slope during bed 11 deposition. The lack of diagnostic assemblages within bed 13 does not yet allow a similar subdivision of those clasts as in bed 14. Further investigation of additional samples from bed 13 may however establish an internal biostratigraphy.

21.7 APPLICATIONS OF PALAEOECOLOGICAL DATA

The major problem in correlating North American conodont zones with the Baltoscandian zonal scheme is the lack of short-ranging species which occur in both Lower Ordovician faunal provinces (the North Atlantic and Midcontinent provinces). In the Cow Head area, 'mixed' faunas of Midcontinent and North Atlantic affinity are present. The setting offers an opportunity to tie both provinces together with a higher precision than previously possible. The continental margin should preserve the interface region between the two rather distinct and separate faunal provinces. Rather than recording time equivalence, many faunal shifts may reflect event equivalence in response to slow environmental changes. This relationship can be demonstrated by comparing faunas from clasts representing a marginal build-up facies in bed 14 (Fig. 21.12) with time-equivalent conodonts from shelf carbonates of the St. George Group and Table Head Group (Fig. 21.13).

In order to correlate the shelf sequences with the time-equivalent strata of the St. George Group, it is important to understand that slope sedimentation is strongly controlled by the events on the shelf. The reason for this is that no significant calcareous planktonic microfossils existed in Early Palaeozoic times and the main sediment contribution was derived from the shelf, particularly the shelf margin. the slope sediments therefore reflect the character of the margin (McIlreath and James, 1984). This relationship requires an evaluation of the factors which influence slope sedimentation on the platform and at the platform edge. Generally an open carbonate platform with sand shoals will deliver large amounts of sand to the slope, whereas a platform margin rimmed by organic build-ups or mounds will have a slope consisting of alternating successions of fine-grained sediments and conglomerates. Relative sea level is another main controlling factor of platform sedimentation and consequently of slope deposition. In the case of shallow platform flooding, sediment accretion on the shelf will be either vertical, if the sea-level rise is in equilibrium with the rate of carbonate sedimen-

tation, or horizontal, if the sea-level rise is slower than the carbonate production. In the latter case, seaward progradation of the platform will result. If sea level rises quickly, the outer shelf may fall below the depth of rapid carbonate sediment production (about 50–70 m) amd slope sedimentation would be reduced, hardgrounds would form and non-carbonates (shales, cherts and phosphates) would prevail (James and Stevens, in press). In the case of platform exposure duc to sea-level fall the source of carbonate slope sediments is also cut off and starved sedimentation would prevail. If the shelf edge is subaerially exposed and the exposure is prolonged, karstification of the platform rim may result in its collapse. Another possible response to the lowered sea level could be the development of a new narrow shelf rooted on upper slope sediments. This also may eventually cause overloading and failure of weakly lithified slope sediments and extensive debris flows could result (James et al., 1979).

The conglomerates of beds 12 and 14 are probably the deep-water expression of a regression (Barnes, 1984, Fig. 4) and possibly prolonged exposure of shelf carbonates together

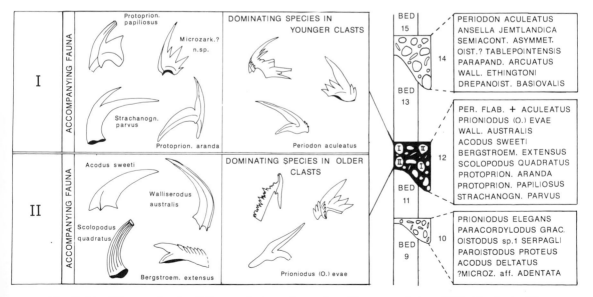

Fig. 21.12 — Diagram showing differences between faunas from (I) younger clasts and (II) older clasts of the upper slope facies in bed 12. These differences allow a distinction between older and younger lithofacies within the range of the *Prioniodus evae* Zone.

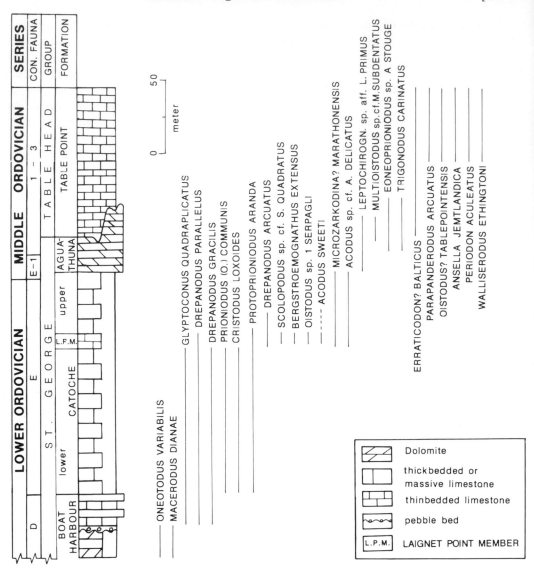

Fig. 21.13 — Stratigraphic ranges of selected species in the platformal sequence of the St. George Group and lower part of the Table Head Group. (Data from Stouge (1982, 1984)).

with synsedimentary faulting on the platform, represented by the Aguathuna Formation. The Aguathuna Formation consists of a series of peritidal carbonates, now dolostones, at the top of the St. George Group. These strata were deposited during a time of synsedimentary folding or faulting as indicated by the varying thicknesses (6–84 m) of the unit (Lane, 1984; Knight and James, in press). The top of the formation is erosional in many places. Stratiform breccia horizons imply several breaks in sedimentation and possibly subaerial exposure.

The overlying Table Head Group begins with dark-grey subtidal limestones of the Table Point Formation followed by ribbon limestones and shales of the Table Cove Formation. The three sections of the platform sequence record initial shallow-water deposition, possibly interrupted by phases of prolonged subaerial exposure, followed by 'drowning', probably as a result of foundering of the platform margin. The consequences of these events should be visible in the slope deposits as discussed above. Most probably the occurrence of massive debris flows such as beds 12 and 14 are the expression of regressions on the platform recorded in the Aguathuna Formation. This event correlation is supported by graptolites from the basal Aguathuna Formation which suggest an age of Bendigonian 4 of the Australian zonation (James and Stevens, in press; S. H. Williams, pers. commun., 1985). This level correlates with the lower part of bed 11. The upper beds of the Aguathuna Formation are basal Middle Ordovician in age, yielding trilobites of zone L and the lower parts of zone M of Utah and Nevada. The overlying basal Table Point Formation belongs to the upper part of zone M.

In the slope sequence the base of the North American Whiterock Stage lies probably within the upper part of bed 11 (Fig. 21.9). Boulders in bed 12 yielded the Whiterock brachiopod *Orthidiella* (R. J. Ross, Jr., and N. P. James, pers. commun., 1985). The youngest clasts in bed 14 contain faunas (trilobites (James and Stevens, in press) and brachiopods (R. J. Ross, Jr., and N. P. James, pers. commun., 1985)), suggesting an age equivalant to zone L in the North American trilobite zonation. Macrofossils from the Table Point Formation belong to the *Anomalorthis* Zone of the Whiterock Stage. The position of the Arenig–Llanvirn boundary is suggested by the occurrence of *Paraglossograptus tentaculatus* and of *Glyptograptus austrodentatus* in bed 15 (James and Stevens, in press). This age probably corresponds to the middle part of the Table Head Group. The results of event stratigraphy and biostratigraphy now indicate that the youngest conglomerates (i.e. bed 14) of the Cow Head Group correlate to the upper Aguathuna Formation (Fortey, 1980; James and Stevens, in press).

Fortey (1984) considered the biological effects of transgressive–regressive cycles on shelf and slope areas. For transgressions, he predicted the occurrence of high speciation rates in shallow epeiric seas because of the extension of environments unsuitable for shallow-water faunas. Furthermore, generation of endemic faunas may result from isolation of the epicontinental area (i.e. physical or climatic barriers). With proceeding transgression, extracratonic faunas are expected to migrate onto the shelf. This may be diachronous, depending on the rates of transgression and adaptation. In tropical areas such faunas will populate marginal platforms and build-ups.

During regressive phases, sharp taxonomic changes (e.g. extinction) are expected to characterize cratonic faunas whereas off-shelf biofacies will remain relatively unaltered. Mound faunas will develop near the edge of the shelf in tropical areas and debris slides will probably form during maximum regression. Mounds and islands may provide refuges for the retreating shelf faunas.

The mega-conglomerates of the Cow Head Group together with the platformal rocks of the Table Head Group and St. George Group provide an excellent site to test these predictions. Bed 14, for example, yields huge boulders of shallow-water limestones which are apparently the remains of extensive algal-sponge build-ups (James, 1981). The character and faunas of the equivalent Aguathuna Formation, with its peritidal carbonates, imply that the build-up formed in slightly deeper water probably seawards from the shelfbreak. These boulders yield a conodont fauna of mainly North Atlantic affinity such as *Periodon aculeatus*, *Walliserodus ethingtoni* Fåhraeus, *Ansella jemtlandica* (Löfgren), *Drepanoistodus basiovalis* Sergeeva (Fig. 21.12) and *Drepanoistodus venustus* Stauffer. This fauna is accompanied by forms which are of Midcontinent affinity, e.g. *Semiacontiodus asymmetricus* (Barnes and Poplawski) or new immigrants such as *Oistodus? tablepointensis* Stouge, *Parapanderodus arcuatus* Stouge and *Erraticodon* sp. The coeval shelf sediments of the Aguathuna Formation consist of a completely different fauna, with *Multioistodus* sp. cf. *M. subdentatus* Cullison, *Trigonodus carinatus* Stouge, *Leptochirognathus* sp. aff. *L. primus* Branson and Mehl and *Eoneoprioniodus* sp. A

Stouge (Stouge, 1982) (Figs. 21.13 and 21.14). Over the course of the regressive phase the faunas of the interior shelf disappear, to be replaced by Midcontinent and endemic faunas which develop in response to the subsequent transgression recorded in the Table Point Formation. Midcontinent forms are represented by *Ansella sinuosa* (Stouge), *Parapaltodus angulatus* (Bradshaw) and others (Stouge, 1984) (Fig. 21.14); a rich endemic fauna consists of *Histio-*

della tableheadensis Stouge, *Juanognathus serpaglii* Stouge, *Polonodus tablepointensis* Stouge and many others (Fig. 21.14). With ongoing platform deepening, the North Atlantic and endemic forms first recovered from the marginal build-up facies preserved in bed 14 clasts migrated onto the shelf and mixed with the Midcontinent and endemic fauna of the lower and middle Table Head Group (Fig. 21.14). This pattern of faunal interaction

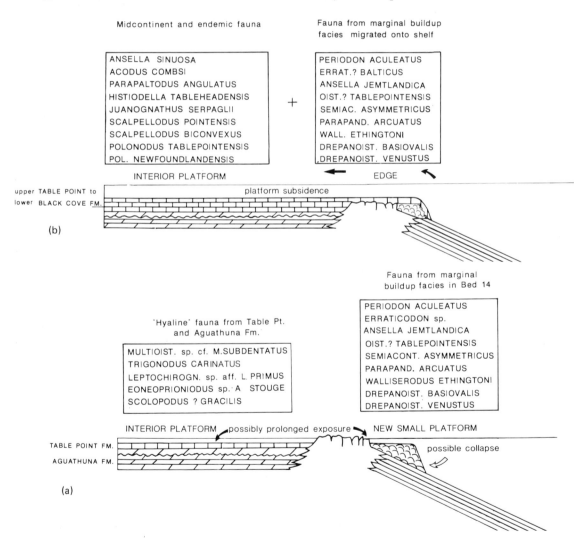

Fig. 21.14 — Diagram showing the interaction of platform and marginal faunas during the late Arenig–early Llanvirn interval: (a) development of different shallow platform and marginal faunas during Aguathuna––lower Table Point time; (b) the subsidence of platform during upper Table Point–lower Black Cove time caused replacement of the shallow platform taxa by a new Midcontinent and endemic fauna mixed with the fauna migrating onto platform from drowned margin.

between shelf and slope faunas matches the predicted effects discussed by Fortey (1984). Further study may possibly allow the same precise biofacies control exemplified for bed 14 to be revealed for earlier stratigraphic intervals.

21.8 SUMMARY

The shelfbreak and upper slope is a rarely preserved facies of ancient continental margins. This facies is critical for correlations between coeval slope and platform facies and for understanding the interactions between different provincial faunas commonly separated by the thermocline close to the shelfbreak area. A fragmented record of the Early Ordovician North American continental margin exists as clasts and boulders in debris-flow deposits of the Cow Head Group in western Newfoundland. Beds 10, 12 and 14 in the coarse proximal facies, where clasts reach up to 200 m in size, were the focus of this study. These beds can be traced into a more distal facies to enable a three-dimensional reconstruction of the flows which accumulated at the toe of the slope and which preserve a wide range of shelfbreak and lower slope facies in the clast lithologies. To reconstruct this segment of the inner marginal environment, carbonate lithofacies and conodont biofacies analyses are being employed. Both the theoretical and the practical approaches to the problem are addressed. This type of study hinges on a detailed base knowledge of stratigraphy, sedimentology and palaeontology.

Field work involved the following.

(1) Determination of different clast lithologies in the flows.
(2) Point-counting of clasts to determine abundances of lithologies.
(3) Collection of over 300 conodont samples to provide both a biostratigraphic and biofacies framework for the diverse clast lithologies.
(4) Consideration of the possible origin of the clasts 'frozen' in the conglomerates.

Lithofacies analysis from petrographic studies will be described in a later paper.

Study of the conodont faunas retrieved from the clasts allows correlation with the *Prioniodus elegans* and *P. (O.) evae* zones of the Baltoscandian sequence. Younger faunas are less readily correlated with that sequence. The age of the youngest clasts in each of the three mega-conglomerate horizons can be determined by comparison with faunas from beds immediately overlying each of the debris-flow deposits. This is exemplified for the uppermost mega-conglomerate (bed 14) in which the youngest clasts yield an advanced stage of *Periodon aculeatus*, identical with the forms found in the lower part of the overlying bed 15. The oldest conodont faunas found in clasts of each of the mega-conglomerate beds determine the depth of scouring of the flow into underlying strata. Furthermore, the character of the assemblages, together with the study of the clast lithologies, indicates the original provenance of the clast. Application of the conodont data to the results of the clast point-count proves that the depth of scouring increased from older to younger mega-conglomerate beds and that erosion occurred higher on the slope. Good stratigraphic control, achieved by other conodont studies (Fåhraeus and Nowlan, 1978; Johnson *et al.*, 1985) for the bedded strata intercalated with the mega-conglomerates, provides a refined biostratigraphy which allows the distinction of younger and older faunas within the range of the *P. (.O.) evae* Zone. This in turn enables the reconstruction of the development of the sedimentary environments on the upper slope and near the shelf break.

Shelf faunas of Midcontinent affinity from the adjacent carbonate platform (St. George Group and Table Head Group) can be shown to interact with faunas from the marginal build-up and upper slope facies preserved in boulders from bed 14. The changing faunal assemblages from the shelf and slope reflect regressions and transgressions by their pattern of migration. They also add further evidence to show that the northwest margin of the Iapetus Ocean evolved from a passive gentle ramp to an oversteepened collapsing margin between early and late Arenig time.

21.9 ACKNOWLEDGEMENTS

We greatly appreciate the ideas, discussions and field knowledge shared by R. K. Stevens. Similarly, D. I. Johnson provided valuable information on his study of Arenig conodonts from the bedded Cow Head strata. Much technical support was given by W. Marsh, M. Moore, L. Nowlan, F. O'Brien and A. Pye. Barnes and James gratefully acknowledge continued research funding from the Natural Sciences and Engineering Research Council of Canada. The cooperation of the administration and staff of Gros Morne National Park (Parks Canada) is also acknowledged.

21.10 REFERENCES

Bagnoli, G. and Stouge, S. 1985. Species of *Prioniodus* Pander, 1856 from Bed 9 (Lower Ordovician), Cow Head Group, Western Newfoundland. In R. J. Aldridge, R. L. Austin and M. P. Smith (Eds.), *Fourth European Conodont Symposium (ECOS IV), Abstracts*, Private publication, University of Southampton, 3–4.

Baird, D. M. 1960. Observations on the nature and origin of the Cow Head Breccias of Newfoundland. *Geological Survey of Canada Paper*, **60–3**, 1–26.

Barnes, C. R. 1984. Early Ordovician eustatic events in Canada. In D. L. Bruton (Ed.), *Aspects of the Ordovician System, Palaeontological Contributions from the University of Oslo*, **295**, 51–63.

Barnes, C. R. and Fåhraeus, L. E. 1975. Provinces communities, and the proposed nektobenthic habit of Ordovician conodontophorids. *Lethaia*, **8**, 133–149.

Barnes, C. R. and Poplawski, M. L. S. 1973. Lower and Middle Ordovician conodonts from the Mystic Formation, Québec, Canada. *Journal of Paleontology*, **47**, 760–790, 5 plates.

Barnes, C. R., Rexroad, C. B. and Miller, J. F. 1973. Lower Paleozoic conodont provincialism. In F. H. T. Rhodes (Ed.), *Conodont Paleozoology, Geological Society of America Special Paper*, **141**, 157–190.

Bergström, S. M., Epstein, A. G. and Epstein, J. B. 1972. Early Ordovician North Atlantic Province conodonts in eastern Pennsylvania. *US Geological Survey Professional Paper*, **800D**, D37–D44.

Burrett, C., Stait, B., Sharples, C. and Laurie, J. 1984. Middle-Upper Ordovician shallow platform to deep basin transect Tasmania, Australia. In D. L. Bruton (Ed.), *Aspects of the Ordovician System, Palaeontological Contributions, University of Oslo*, **295**, 149–158.

Callahan, R. K. M. 1974. Statigraphy, sedimentation and petrology of the Cambro-Ordovician Cow Head Group at Broom Point and Martin Point, western Newfoundland. *Unpublished M.Sc. Thesis*, University of Massachusetts, Amherst, Massachusetts, 1–119.

Coniglio, M. 1985. Origin and diagenesis of fine-grained slope sediments: Cow Head Group (Cambro-Ordovician), western Newfoundland. *Unpublished Ph.D. Thesis*, Memorial University of Newfoundland, 1–684.

Coniglio, M. In Press. Submarine slope failure and the origin of intrafolial deformation in deep-water slope carbonates, Cow Head Group, western Newfoundland. *Canadian Journal of Earth Sciences*.

Cook, H. E. 1979. Ancient continental slope sequences and their value in understanding modern slope development. In L. S. Doyle and O. H. Pilkey (Eds.), *Geology of Continental Slopes, Society of Economic Paleontologists and Mineralogists Special Publication*, **27**, 287–305.

Cook, H. E. 1983. Introductory perspectives, basic carbonate principles, and stratigraphic and depositional models. In H. E. Cook, A. C. Hine and H. T. Mullins (Eds.), *Platform Margin and Deepwater Carbonates, Society of Economic Paleontologists and Mineralogists Short Course*, **12**, 1-1-1-89.

Cook, F. A., Albaugh, D. S., Brown, L. D., Kaufman, S., Oliver, J. E. and Hatcher, R. D., Jr., 1979. Thin-skinned tectonics in the crystalline southern Appalachians; COCORP seismic reflection profiling of the Blue Ridge and Piedmont. *Geology*, **7**, 563–567.

Cook, H. E., Hine, A. C. and Mullins, H. T. (Eds.) 1983. Platform margin and deep water carbonate environments. *Society of Economic Paleontologists and Mineralogists Short Course*, **12**, 1–56.

Cook, H. E. and Taylor, M. E. 1976. Early Paleozoic continental margin sedimentation, trilobite biofacies, and the thermocline, western United States. *Geology*, **3**, 559–562.

Doyle, L. and Pilkey, O. H. (Eds.) 1979. Geology of continental slopes. *Society of Economic Paleontologists and Mineralogists Special Publication*, **27**, 1–374.

Ethington, R. L. and Clark, D. L., 1971. Lower Ordovician conodonts in North America. In W. C. Sweet and S. M. Bergström (Eds.), *Symposium on Conodont Biostratigraphy, Geological Society of America Memoir*, **127**, 63–82, 2 plates.

Ethington, R. L. and Clark, D. L. 1981. Lower and Middle Ordovician conodonts from the Ibex area, Western Milland County, Utah. *Brigham Young University Geology Studies*, **28** (2), 1–159 p, 14 plates.

Fåhraeus, L. E. and Nowlan, G. S. 1978. Franconian (Late Cambrian) to Early Champlainian (Middle Ordovician) conodonts from the Cow Head Group, western Newfoundland. *Journal of Paleontology*, **52**, 444–471, 3 plates.

Fåhraeus, L. E., Slatt, R. M. and Nowlan, G. S. 1974. Origin of carbonate pseudopellets. *Journal of Sedimentary Petrology*, **44**, 27–29.

Fortey, R. A. 1975. Early Ordovician trilobite communities. *Fossils and Strata*, **4**, 331–352.

Fortey, R. A. 1980. The Ordovician of Spitsbergen, and its relevance to the base of the Middle Ordovician in North America. *Virginia Polytechnic Institute and State University, Department of Geological Sciences, Memoir*, **2**, 33–40.

Fortey, R. A. 1984. Global early Ordovician transgression and regressions and their biological implications. In D. L. Bruton (Ed.), *Aspects of the Ordovician System, Palaeontological Contributions from the University of Oslo*, **295**, 37–50.

Fortey, R. A. and Barnes, C. R. 1977. Early Ordovician conodont and trilobite communities of Spitsbergen: influence on biogeography. *Alcheringa*, **1**, 297–309.

Fortey, R. A., Landing, E. and Skevington, D. 1982. Cambrian–Ordovician boundary sections in the Cow Head Group, western Newfoundland. In M. G. Bassett and W. T. Dean (Eds.), *The Cambrian-Ordovician Boundary: Sections, Fossil Distributions, and Correlations, National Museum of Wales Geological Series*, **3**, 95–129.

Hiscott, R. N. and James, N. P. 1985. Carbonate debris flows, Cow Head Group, western Newfoundland. *Journal of Sedimentary Petrology*, **55**, 735–745.

Hubert, J. F., Suchecki, R. K. and Callahan, R. K. M. 1977. The Cow Head Breccia: Sedimentology of the Cambro-Ordovician continental margin, Newfoundland. In H. E. Cook and P. Enos (Eds.), Deep Water Carbonate Environments, *Society of Economic Paleontologists and Mineralogists Special Publication*, **25**, 125–124.

James, N. P. 1981. Megablocks of calcified algae in the Cow Head Breccia, western Newfoundland: vestiges of a Cambro-Ordovician platform margin. *Geological Society of America Bulletin*, **92**, 799–811.

James, N. P. and Mountjoy, E. W. 1983. Shelf-slope break in fossil carbonate platforms: an overview. In D. J. Stanley and G. T. Moore (Eds.), *The Shelfbreak: Critical Interface on Continental Margins, Society of Economic Paleontologists and Mineralogists Special Publication*, **33**, 189–206.

James, N. P. and Stevens, R. K. 1982. Anatomy and evolution of a Lower Paleozoic continental margin. *International Association of Sedimentologists Guidebook*, **2B**, 1–75.

James, N. P. and Stevens, R. K. In press. Stratigraphy and correlation of the Cow Head Group (Cambro-Ordovician), western Newfoundland. *Geological Survey of Canada Bulletin*.

James, N. P., Stevens, R. K. and Fortey, R. A. 1979. Correlation and timing of platform margin megabreccia deposition, Cow Head and related groups, western Newfoundland (Abstract). *American Association of Petroleum Geologists Bulletin*, **63**, 508.

Jansa, L. F. 1974. Trace fossils from the Cambro-Ordovician Cow Head Group, Newfoundland, and their paleobathymetric implications. *Palaeogeography, Palaeoclimatology, Palaeoecology*, **15**, 233–234.

Johnson, H. 1941. Paleozoic lowlands of northwestern Newfoundland. *New York Academy of Sciences Transactions Series II*, **3**, 141–145.

Johnston, D. I., Barnes, C. R. and Fåhraeus, L. E. 1985.

Early Ordovician (Arenig) conodonts from St. Pauls Inlet and Martin Point, Cow Head Group, western Newfoundland, Canada. In R. J. Aldridge, R. L. Austin, and M. P. Smith (Eds.), *Fourth European Conodont Symposium (ECOS IV)*, Abstracts, Private publication, University of Southampton, 15.

Kindle, C. H. 1982. The C. H. Kindle Collection: Middle Cambrian to Lower Ordovician trilobites from the Cow Head Group, western Newfoundland. *Geological Survey of Canada Paper*, **82–1C**, 1–17.

Kindle, C. H. and Whittington, H. B. 1958. Stratigraphy of the Cow Head Region, western Newfoundland. *Geological Society of America Bulletin*, **69**, 315–342.

Kindle, C. H. and Whittington, H. B. 1959. Some stratigraphic problems of the Cow Head area in western Newfoundland. *New York Academy of Sciences Transactions*, **22**, 7–18.

Knight, I. and James, N. P. In press. Stratigraphy of the St. George Group, Lower Ordovician, western Newfoundland. *Canadian Journal of Earth Sciences*.

Lane, T. E. 1984. Preliminary classification of carbonate breccias, Newfoundland Zinc Mines, Daniels Harbour, Newfoundland, *Geological Survey of Canada Paper*, **84–1A**, 505–512.

Lindström, M. 1955. Conodonts from the lowermost Ordovician strata of south-central Sweden. *Geologiska Föreningens i Stockholm Förhandlingar*, **76**, 517–604, 7 plates.

Lindström, M. 1976. Conodont palaeogeography of the Ordovician. In M. G. Bassett (Ed.), *The Ordovician System, Proceedings of a Palaeontological Association Symposium, Birmingham, September 1974*, University of Wales Press, Cardiff, and National Museum of Wales, Cardiff, 502–522.

Löfgren, A. 1978. Argenigian and Llanvirnian conodonts from Jämtland, northern Sweden. *Fossils and Strata*, **13**, 1–129, 16 plates.

McIlreath, I. A. and James, N. P. 1984. Carbonate slopes. In R. G. Walker (Ed.), *Facies Models* (2nd edition), *Geoscience Canada Reprint Series*, **1**, 245–257.

McTavish, R. A. 1973. Prioniodontacean conodonts from the Emanuel Formation (Lower Ordovician) of Western Australia. *Geologica et Palaeontologica*, **7**, 27–58, 3 plates.

Oxley, P. 1953. Geology of Parsons Pond–St. Pauls area, west coast, Newfoundland. *Newfoundland Geological Survey Report*, **5**, 1–53.

Repetski, J. E. 1982. Conodonts from El Paso Group (Lower Ordovician) of westernmost Texas and southern New Mexico. *New Mexico Bureau of Mines and Mineral Resources Memoir*, **40**, 1–121, 28 plates.

Repetski, J. E. and Ethington, R. L. 1977. Conodonts from graptolitic facies in the Ouachita Mountains, Arkansas and Oklahoma. In *Proceedings of a Symposium on the Geology of the Ouachita Mountains*, **1**, 92–106, 2 plates.

Rodgers, J. 1968. The eastern edge of North American continent during the Cambrian and early Ordovician. In E. A. Zen, W. S. White, J. B. Hadley, and J. B. Thompson, Jr. (Eds.), *Studies of Appalachian Geology: Northern and Maritime*, Interscience, New York, 141–149.

Rodgers, J. and Neale, E. R. W. 1963. Possible 'Taconic' klippen in western Newfoundland. *American Journal of Science*, **261**, 713–730.

Ross, R. J., Jr. and Ingham, J. K. 1970. Distribution of the Toquima–Table Head (Middle Ordovician Whiterock) faunal realm in the northern hemisphere. *Geological Society of America Bulletin*, **81**, 393–408.

Ruedemann, R. 1947. Graptolites of North America. *Geological Society of America Memoir*, **19**, 1–652, 92 plates.

Schuchert, C. and Dunbar, C. O. 1934. Stratigraphy of western Newfoundland. Geological Soceity of America Memoir, **1**, 1–123.

Serpagli, E. 1974. Lower Ordovician conodonts from Precordilleran Argentina (Province of San Juan). *Bolletino della Società Paleontologica Italiana*, **13**, 17–98, 31 plates.

Stanley, D. J. and Moore, G. T. 1983. The shelfbreak: critical interface on continental margins. *Society of Economic Paleontologists and Mineralogists Special Publication*, **33**, 1–467.

Stevens, R. K. 1970. Cambro-Ordovician flysch sedimentation and tectonics in west Newfoundland and their possible bearing on a proto-Atlantic Ocean. In J. Lajoie (Ed.), *Flysch Sedimentology in North America, Geological Association of Canada Special Paper*, **7**, 165–177.

Stouge, S. 1982. Preliminary conodont biostratigraphy and correlation of Lower to Middle Ordovician carbonates of the St. George Group, Great Northern Peninsula, Newfoundland. *Newfoundland Department of Mines and Energy, Mineral Division, Report*, **80–3**, 1–59.

Stouge, S. 1984. Conodonts of the Middle Ordovician Table Head Formation, western Newfoundland. *Fossils and Strata*, **16**, 1–145, 18 plates.

Sweet, W. C. and Bergström, S. M. 1974. Provincialism exhibited by Ordovician conodont faunas. In C. A. Ross (Ed.), *Paleogeographic Provinces and Provinciality, Society of Economic Paleontologists and Mineralogists Special Publication*, **21**, 189–202.

Taylor, M. E. 1977. Late Cambrian of western North America: trilobite biofacies, environmental significance, and biostratigraphic implications. In E. G. Kauffman and J. E. Hazel (Eds.), *Concepts and Methods of Biostratigraphy*, Dowden, Hutchinson, and Ross, 397–426.

Taylor, M. E. and Cook, H. E., 1976. Shelf to slope facies transition in the early Paleozoic of Nevada. In R. A. Robinson and A. J. Rowell (Eds.), *Cambrian of Western North America, Brigham Young University Geology Studies*, **2** (2), 181–214.

Uyeno, T. T. and Barnes, C. R. 1970. Conodonts from the Levis Formation (Zone D1) (Middle Ordovician), Lévis, Québec. *Geological Survey of Canada Bulletin*, **187**, 99–123, 4 plates.

Van Wamel, W. A. 1974. Conodont biostratigraphy of the Upper Cambrian and Lower Ordovician of northwestern Öland, south-eastern Sweden. *Utrecht Micropalaeontological Bulletins*, **10**, 1–125, 8 plates.

Webby, B. D. 1976. The Ordovician System in southeastern Australia. In M. G. Bassett (Ed.), *The Ordovician System, Proceedings of a Palaeontological Association Symposium, Birmingham, September 1974*, University of Wales Press, Cardiff, 417–446.

Whittington, H. B. 1963. Middle Ordovician trilobites from Lower Head, western Newfoundland. *Bulletin of the Museum of Comparative Zoology, Harvard College*, **129** (1), 1–118.

Williams, H. 1975. Structural succession, nomenclature, and interpretation of transported rocks in western Newfoundland. *Canadian Journal of Earth Sciences*, **12**, 1874–1894.

Williams, H. 1979. Appalachian Orogen in Canada. *Canadian Journal of Earth Sciences*, **16**, 792–807.

Williams, H., James, N. P. and Stevens, R. K. 1985. Humber Arm Allochthon and nearby groups between Bonne Bay and Portland Creek, western Newfoundland. *Geological Survey of Canada Paper*, **85-1A**, 399–406.

22

Conodont biofacies in the Famennian Stage of Pomerania, northwestern Poland

H. Matya

Biofacies and lithofacies relationships are analysed within the sedimentary environments of the Late Devonian (Famennian Stage) from the *crepida* to the *praesulcata* zones in Pomerania, northwestern Poland. Six conodont biofacies (palmatolepid, palmatolepid–polygnathid, polygnathid–bispathodid, polygnathid, polygnathid–'icriodid' and polygnathid–pelekysgnathid) are recognized, based on the precentages of platform genera in settings ranging from off-shore pelagic to shallow near-shore. A nearshore peritidal clydagnathid, scaphignathid and pandorinellinid biofacies is also recognized which consists of washed elements from the shallowest biofacies mixed with those of the nearby polygnathid–pelekysgnathid biofaces.

The chronostratigraphic distribution of lithofacies and biofacies reflects a regressive tendency during most of the Famennian Stage and a general trend of gradual progradation of near-shore facies from northwest to southeast progressively with time. A relationship with worldwide transgressive and regressive events is substantiated.

22.1 INTRODUCTION

Upper Devonian rocks are widespread in Poland. They extend from the well-known outcrops of the Holy Cross Mountains and other regions in southern Poland northwestwards into the subsurface of Pomerania. There, Devonian rocks are concealed under a considerable thickness of younger sediments and have been reached only occasionally by boreholes.

Famennian conodont faunas (*crepida* to *praesulcata* zones) in Pomerania are analysed on the basis of several hundred conodont specimens obtained from 25 boreholes (Fig. 22.1). The conodont frequencies are not very high and range from about 20–30 specimens to only a few specimens per 1–2 kg of rock sample (core drill). These rather low conodont frequencies in Pomerania lead to uncertainty with regard to the identification of some biofacies. Most of the biofacies recognized by Sandberg (1976), Sandberg and Ziegler (1979) and Ziegler and Sandberg (1984) are represented in the Famennian rocks of Pomerania. Minor exceptions and changes are as follows. Biofacies I, distinguished in the Famennian Stage of Pomerania, is called simply, the palmatolepid biofacies as in the Belgian model (Dreesen and Thorez, 1980; Sandberg and Dreesen, 1984). This change is necessary since the genus *Bispathodus* did not evolve until the *marginifera* Zone; hence it was non-existent in the pelagic faunas of the *crepida* Zone. The polygnathid biofacies II recognized in this chapter does not exist in the model of Sandberg (1976). Recently it has been distinguished by Klapper and Lane (1985) and Dreesen *et al.* (1985).

22.2 FAMENNIAN LITHOFACIES AND BIOFACIES IN POMERANIA

A pelagic facies dominates most of the Frasnian succession in northern Poland. The basinal sediments are characterized by black bituminous shales with rare thin intercalations of marls and marly limestones which indicate a quiet environment well below the wave base. Cephalopods, tentaculitoids, thin-shelled bivalves (*Buchiola*), small thin-shelled brachiopods (*Lingula*), planktonic ostracods (entomozoids) and rare conodonts characterize the faunal assemblages. Comparable sedimentation continued up to the early Famennian. However, thin-bedded shales, marls and marly limestones of *crepida* Zone age are characterized by increasing numbers of benthic organisms, a progressive loss of fine lamination and a lighter

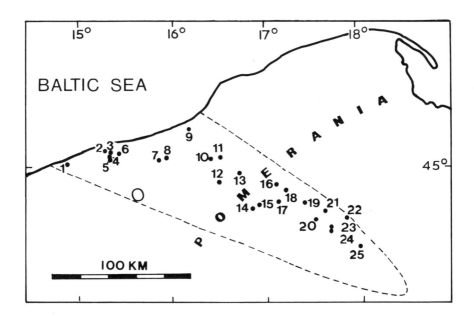

Fig. 22.1 — Localities of boreholes which were the source of information used for lithofacies and biofacies analysis. Boreholes 1–13 belong to the Northern Facies Belt; boreholes 14–25 belong to the Southern Facies Belt: 1, Strzezewo 1; 2, Gorzyslaw 8; 3, Gorzyslaw 9; Gorzyslaw 14; 5, Gorzyslaw 11; 6, Karcino 2; 7, Karlino 1; 8, Daszewo 3; 9, Jamno IG-1; 10, Klanino 1; 11, Karsina 1; 12, Chmielno 1; 13, Drzewiany 1; 14, Bielica 2; 15, Bielica 1; 16, Koczala 1; 17, Rzeczenica 1; 18, Brda 1; 19, Babilon, 1; 20, Chojnice 4; 21, Krojanty 1; 22, Chojnice 3; 23, Chojnice 1; 24, Chojnice 2; 25, Byslaw 2.

colouration of rocks up the section, all indicative of better oxygenation of the environment.

Since the Middle–Upper *crepida* zones, two facies belts may be recognized. These are a northern facies belt (Fig. 22.2, area 1) with a relatively shallow-shelf sedimentation, and a southern belt (Fig. 22.2, area 2) with a relatively deeper-shelf sedimentation. Their distributions were determined by the vicinity of the Old Red Continent (Fig. 22.2) during most of the Famennian Stage.

(a) The Northern Facies Belt

Shales, marls and marly limestones probably extend in the northern facies belt up to the Upper *crepida* Zone (Fig. 22.3). These rocks contain conodonts typical of the palmatolepid biofacies (I), consisting mainly of *Palmatolepis* species in the well-dated Middle–Upper *crepida* zones. Species of *Palmatolepis* comprise 82% of

all the platform (Pa) elements (Table 22.1) and include *Palmatolepis wolskajae*, *Palmatolepis circularis*, *Palmatolepis quadrantinodosalobata*, *Palmatolepis crepida* and *Palmatolepis tenuipunctata* (Plate 22.1). Other species constitute a minor component and are represented by *Polygnathus communis communis*, *Polygnathus procerus* and *Polygnathus brevilaminus* and comprise 18% of the platform (Pa) elements.

Nodular limestone near the top of the Upper *crepida* Zone are indicative of a regressive phase. This type of deposition continued until the Lower *marginifera* Zone. The associated biota is diverse and includes crinoids, bryozoans, brachiopods, benthic ostracods and conodonts. Wackestones dominate but mudstones are also common. The limestones contain conodont faunas typical of the palmatolepid–polygnathid biofacies (II) (Figs. 22.3 and 22.4 and

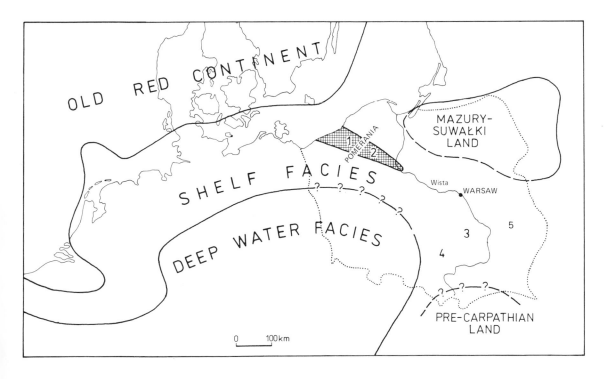

Fig. 22.2 — Late Devonian palaeogeography: cross-hatched area, present extension of the Devonian rocks in Pomerania; 1, Northern Facies Belt; 2, Southern facies Belt; 3, Holy Cross Mountains; 4, Olkusz–Zawiercie area; 5, Lublin region. (Mainly after Ziegler, 1982; Polish area after Czermiński and Pajchlowa, 1974; Pozaryski and Dembowski, 1984.)

Table 22.1). However, it should be noted that, in the Upper *rhomboidea* Zone, species of *Palmatolepis* (represented by *Palmatolepis klapperi*, *Palmatolepis glabra pectinata*, *Palmatolepis glabra acuta*, *Palmatolepis glabra prima* and *Palmatolepis rhomboidea*) still constitute about 71% of the platform (Pa) elements. The *Polygnathus nodocostatus* group (4% of the total platform fauna) is represented by *Polygnathus triphyllatus*. Thus, 75% of the platform (Pa) elements are of pelagic or off-shore origin

(Sandberg 1976; Sandberg and Ziegler, 1979; Sandberg and Dreesen, 1984). Representatives (about 25% of the platform (Pa) elements) of the near-shore, presumably nektobenthonic *Polygnathus semicostatus* group are *Polygnathus* aff. *procerus*, *Polygnathus webbi* and *Polygnathus semicostatus*. *Palmatolepis* conversely decreased to 17% in the Lower *marginifera* Zone, the *Polygnathus nodocostatus* group (*Polygnathus triphyllatus* only) to 8% and representatives of the *Polygnathus semicostatus*

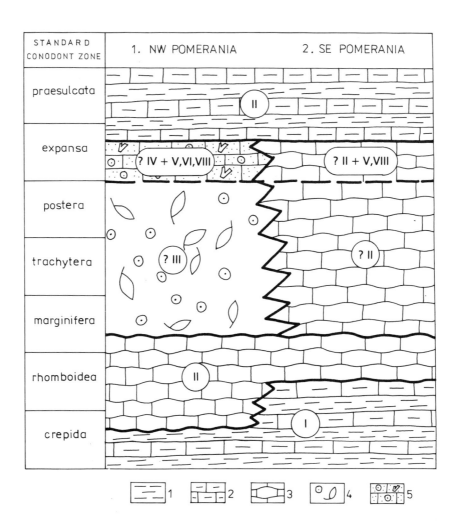

Fig. 22.3 — Lithofacies (1, marls; 2, marly limestones; 3, nodular limestones; 4, crinoid–brachiopod packstones; 5, grainstones) and biofacies (biofacies are indicated by roman numerals; for key to biofacies see text) distinguished in the Famennian Stage of Pomerania.

group (*Polygnathus semicostatus*, *Polygnathus szulczewskii* and *Polygnathus* aff. *procerus* increased to 75% of the platform (Pa) elements. Pelagic or off-shore forms decreased therefore from 75% to 25% and the shallow-water (near-shore) forms increased from 25% to 75% in a short stratigraphic interval.

Accumulations of crinoid debris and scattered bryozoans (crinoid–bryozoan wackestone) are observed in some sections of the Lower *marginifera* Zone. Fissure fillings (neptunian dykes?) also have been found which may have been fragments of reef-like structures. Conodonts mainly of the *Polygnathus semicostatus* group and *Palmatolepis* are sparse within this crinoid–bryozoan wackestone.

The richest conodont fauna comes from a thin bed of crinoid–brachiopod packstone, above the reef-like structures, which probably represents a condensed sequence. Most of the conodonts indicate the *marginifera* Zone, although it is possible that condensation extends up to the Upper *postera* Zone. The thin bed contains a conodont fauna representative of the polygnathid–'icriodid' biofacies (III) (Fig. 22.3, Table 22.1 and Plate 22.3). *Palmatolepis* was not found and about 64% of the platform (Pa) elements are of shallow-water origin. These include mainly representatives of the *Polygnathus semicostatus* group (44%), 'Icriodus' chojnicensis (18%) and rare *Pelekysgnathus inclinatus* (2%). The biofacies contains 28% ubiquitous forms which are interpreted as having lived in the euphotic zone since they occur in all biofacies (Sandberg, 1976; Sandberg and Zielgler, 1979). The forms are *Polygnathus communis communis*, *Mehlina strigosa* and also probably belonging to this group *Polygnathus glaber glaber*, *Polygnathus brevilaminus* and *Polygnathus glaber bilobatus*. Only 8% of the total platform fauna consists of pelagic or off-shore forms; the *Polygnathus nodocostatus* group is represented by *Polygnathus rhomboideus*, *Polygnathus perplexus* and *Polygnathus nodoundatus*.

A rather abrupt facies change is recognized during the Lower–Middle *expansa* zones.

Microfacies types (peloidal grainstone with ostracods, foraminifera and abraded crinoid debris; kamaenid algae grainstone with intraclasts of unfossiliferous micrite, or kamaenid algal wackestone) and sedimentary structures (often small-angle cross-bedding and lamination) are characteristic of high-energy shallow-water shoals with sedimentation of calcareous sands. These characteristic rocks yield conodonts which probably should be placed within the polygnathid–pelekysgnathid biofacies (IV) (Fig. 22.3, Table 22.1 and Plates 22.4 and 22.5). Analysis of the conodonts shows that 64% of the platform (Pa) elements are representatives of the near-shore shallow-water environments. These include *Polygnathus semicostatus* (54%), 'Icriodus' raymondi and 'Icriodus' darbyensis (6%) and *Pelekysgnathus inclinatus* (4%). Ubiquitus forms such as *Bispathodus stabilis*, *Bispathodus aculeatus aculeatus*, *Branmehla bohlenana* and *Mehlina strigosa* together with *Polygnathus communis communis* and *Polygnathus inornatus* (Sandberg, 1976; Sandberg and Gutschick, 1979; Sandberg and Dreesen, 1984; Ziegler and Sandberg, 1984) comprise 16% of the platform (Pa) elements. In addition to representatives of the polygnathid–pelekysgnathid biofacies mentioned above, *Clydagnathus ormistoni*, *Scaphignathus peterseni*, *Scaphignathus ziegleri* and *Pandorinellina* cf. *P. insita* occur also. These are typical of the clydagnathid (V), scaphignathid (VI) and pandorinellinid (VIII) biofacies respectively which are characteristic of more near-shore, restricted and peritidal environments (Sandberg, 1976; Sandberg and Ziegler, 1979; Ziegler and Sandberg, 1984). Shallow marine transport processes are invoked to have occasionally washed the conodonts out of these environments and mixed them with the faunas of the polygnathid–pelekysgnathid biofacies.

(b) The Southern Facies Belt

In the more seaward successions of southeastern Pomerania, lithological changes are less distinct and they have a rather gradual character. The deposition of shales, marls and marly

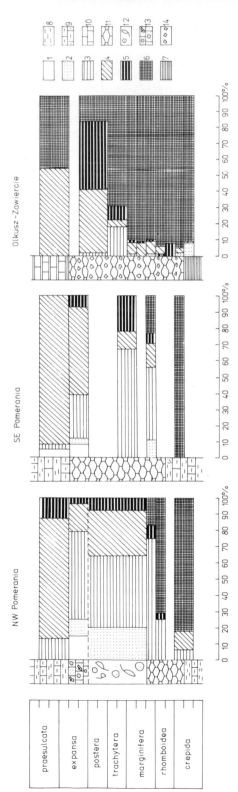

Fig. 22.4 — Summary of the distribution of selected platform conodont groups in relation to the lithofacies in the Famennian Stage of Pomerania (northwestern Poland) and the Olkusz-Zawiercie area (southwest Poland). For detailed information on percentages see Tables 22.1–22.3. Conodont groupings for 1–6 mainly after Sandberg (1976), Sandberg and Ziegler (1979) and Sandberg and Dreesen (1984): 1, asymmetric shallow-water genera *Clydagnathus*, *Scaphignathus* and *Pandorinellina*; 2, shallow-water icriodontids '*Icriodus*' and *Pelekysgnathus*; 3, shallow-water (near-shore) *Polygnathus semicostatus* Group (here including *Po. semicostatus*, *Po. szulczewskii*, *Po. aff. procerus* and *Po. delicatulus*); 4, uniquitous forms (all biofacies) *Polygnathus communis* and probably, *Po. glaber glaber*, *Po. glaber biolobatus*, *Po. brevilaminus*, *Po. inornatus*, and *Hemiistrona*, *Polygnathellus*; 5, off-shore or deep-water *Polygnathus nodocostatus* group (here including *Po. bouckaerti*, *Po. granulosus*, *Po. nodocostatus*, *Po. rhomboideus*, *Po. triphyllatus*, *Po. diversus*, *P. nodoundatus*, *Po. styriacus*, *Po. homoirregularis*, *Po. perplexus* and *Po. experplexus*), *Pseudopolygnathus*, *Alternognathus* and *Siphonodella*; 6, off-shore or pelagic *Palmatolepis*; 7, shale; 8, marl; 9, marly limestone; 10, micritic limestone; 11, nodular limestone (wackestone and packstone); 12, crinoid-brachiopod packstone; 13, grainstone; 14, detrital limestone.

limestones with conodonts typical of the palmatolepid biofacies continued in this facies belt probably until the Lower *rhomboidea* Zone, i.e. to a somewhat younger horizon than in the Northern Facies Belt (Fig. 22.3). Sedimentation of the overlying nodular limestone continued up to the Middle *expansa* Zone. Detailed studies, however, provide evidence for a continuous upward shallowing of the succession. In the Upper *rhomboidea* Zone, mudstones dominate and, in the Lower *marginifera* Zone, nodular wackestones with a brachiopod, crinoid and bryozoan fauna are common. The latter contain a conodont fauna representative of the palmatolepid–polygnathid biofacies (Fig. 22.3, Table 22.2, and Plate 22.2). This biofacies is composed of *Palmatolepis marginifera marginifera, Palmatolepis quadrantinodosa quadrantinodosa, Palmatolepis quadrantinodosa inflexa, Palmatolepis quadrantinodosa inflexoidea, Palmatolepis glabra pectinata* and *Palmatolepis stoppeli* which comprise 23% of the platform component. Only 3% of the total platform

fauna consists of members of the off-shore *Polygnathus nodocostatus* group (*Polygnathus bouckaerti* only). The little-known *Alternognathus pseudostrigosus* also forms 3%. Ubiquitous forms are *Mehlina strigosa, Polygnathus communis communis* and probably also *Polygnathus glaber glaber, Polygnathus fallax* and *Polygnathus brevilaminus*, which comprise 15%. The *Polygnathus semicostatus* group. (*Polygnathus semicostatus, Polygnathus szulczewskii* and *Polygnathus* aff. *P. procerus*) constitute 45% of the fanua. It should be noted that specimens of '*Icriodus*' *chojnicensis* (of a small size only) are also present in this polygnathid–palmatolepid biofacies and they constitute 11% of the platform component.

The overlying wackestones and packstones with abundant tubular problematical ?algae, and with foraminifera, rare crinoid debris, fragments of bryozoans and brachiopods, have yielded sparse and rather monotonous conodont faunas, indicative of the (?Upper) *marginifera* and Lower *trachytera* zones. Representa-

Table 22.1 — Distribution and percentage frequencies of Pa elements of condont species in the Famennian Stage of the Northern Facies Belt in Pomerania.

STANDARD CONODONT ZONE		Scaphignathus petersseni	S. ziegleri	Pandorinellina cf. insita	Clydagnathus ormistoni	Pelekysgnathus inclinatus	"Icriodus" chojnicensis	"I." costatus darbyensis	"I." raymondi	Polygnathus webbi	Po. aff. Po. procerus	Po. semicostatus	Po. szulczewskii	Po. delicatulus	Po. inornatus	Po. communis communis	Siphonodella praesulcata	Mehlina strigosa	Branmehla bohlenana	Bispathodus aculeatus	Bi. stabilis	Bi. costatus	Polygnathus glaber glaber	Po. glaber bilobatus	Po. brevilaminus	Po. triphyllatus	Po. rhomboideus	Po. nodoundatus	Po. perplexus	Po. experplexus	Palmatolepis	Total number of specimens
praesulcata	U																																
	M																																
	L	–	–	–	–	–	–	–	–	–	13	12	12	13	–	–	–	–	50	–	–	–	–	–	–	–	–	–	–	–	–	16
expansa	U																																
	M																																
	L	2	4	2	7	4	–	2	4	–	–	54	–	–	1	7	–	4	1	2	2	–	–	–	–	–	–	–	4	–	–	55
postera	U																																
	L																																
trachytera	U																																
	L	–	–	–	–	2	18	–	–	–	3	35	6	–	–	21	–	2	–	–	–	–	2	1	2	–	2	2	4	–	–	57
marginifera	Um																																
	U																																
	L	–	–	–	–	–	–	–	–	–	8	50	17	–	–	–	–	–	–	–	–	–	8	–	–	–	–	–	–	–	17	24
rhomboidea	U	–	–	–	–	–	–	–	–	2	2	21	–	–	–	–	–	–	–	–	–	–	4	–	–	–	–	–	–	–	71	56
	L																																
crepida	U																																
	M	–	–	–	–	~	–	–	–	–	6	–	–	–	3	–	–	–	–	–	–	–	9	–	–	–	–	–	–	–	82	33
	L																																

Table 22.2 — Distribution and percentage frequencies of Pa elements of conodont species in the Famennian Stage of the Southern Facies Belt in Pomerania.

Standard conodont zone	Pandorinellina cf. insita	Pa. plumula	Clydagnathus ormistoni	Pelekysgnathus inclinatus	"Icriodus" chojnicensis	Polygnathus aff. Po. procerus	Po. semicostatus	Po. szulczewskii	Po. delicatulus	Po. inornatus	Po. communis communis	Po. communis carina	Mehlina strigosa	Branmehla sp. nov.	Br. inornata	Br. suprema	Bispathodus aculeatus	Bi. stabilis	Bi. costatus	Bi. ultimus	Polygnathus brevilaminus	Po. fallax	Po. glaber glaber	Po. glaber bilobatus	Hemilistrona pulchra	Polygnathellus giganteus	Alternognathus pseudostrigosus	Al. regular s	Polygnathus bouckaerti	Po. granulosus	Po. nodocostatus	Po. perplexus	Palmatolepis	Total number of specimens
praesulcata U																																		
praesulcata M																																		
praesulcata L	-	5	-	-	-	-	-	3	1	32	10	3	-	4	1	4	7	1	29	-	-	-	-	-	-	-	-	-	-	-	-	-	-	73
expansa U																																		
expansa M																																		
expansa L	4	-	4	4	-	4	9	-	14	-	5	-	16	7	-	-	18	4	-	-	-	-	-	-	2	2	-	-	-	-	5	2	-	44
postera U																																		
postera L																																		
trachytera U																																		
trachytera L	-	-	-	-	56	11	-	-	-	-	-	-	-	-	-	-	-	-	-	-	-	-	-	-	-	11	-	-	11	-	11	-	-	18
marginifera Um																																		
marginifera U																																		
marginifera L	-	-	-	-	11	15	29	1	-	-	3	-	1	-	-	-	-	-	-	-	1	9	1	-	-	-	3	-	3	-	-	-	23	74
rhomboidea U																																		
rhomboidea L																																		
crepida U	-	-	-	-	-	-	-	-	-	-	-	-	-	-	-	-	-	-	-	-	-	-	-	-	-	-	-	-	-	-	-	-	100	9
crepida M																																		
crepida L																																		

Table 22.3 — Distribution and percentage frequencies of Pa elements of conodont species in the Famennian Stage of the Olkusz–Zawiercie area. (Based on data of Narkiewicz, 1978.)

Standard conodont zone	Pelekysgnathus planus	"Icriodus" cornutus	Polygnathus webbi	Po. semicostatus	Po. szulczewskii	Po. communis communis	Mehlina strigosa	Branmehla bohlenana	Br. inornata	Br. werneri	Bispathodus stabilis	Bi. bispathodus	Bi. aculeatus	Bi. costatus	Bi. ultimus	Polygnathus brevilaminus	Po. glaber glaber	Po. fallax	Po. glaber bilobatus	Po. lagowiensis	Polygnathellus sublaevis	Pol. postsublaevis	Polylophodonta confluens	Polylop. pergyrata	Alternognathus regularis	Al. beulensis	Pseud. marburgensis trigonicus	Polygnathus nodocostatus ovatus	Po. rhomboidea	Po. bouckaerti	Po. nodocostatus nodocostatus	Po. triphyllatus	Po. diversus	Po. nodoundatus	Po. granulosus	Po. perplexus	Po. znepalensis	Po. styriacus	Po. homoirregularis	Palmatolepis	Total number of specimens
praesulcata U																																									
praesulcata M																																									
praesulcata L	-	-	-	-	-	5	-	17	-	11	15	2	2	1	-	-	-	-	-	-	-	-	-	-	-	-	1	-	-	-	-	-	-	-	-	-	-	-	-	46	83
expansa U																																									
expansa M																																									
expansa L																																									
postera U	-	-	-	2	-	<1	8	2	2	2	26	-	-	-	-	-	-	-	-	-	-	-	-	-	-	-	-	-	-	-	-	-	-	-	8	23	9	3	16	474
postera L																																									
trachytera U																																									
trachytera L	-	-	7	11	-	-	4	<1	-	-	-	-	-	-	-	-	-	-	-	-	-	-	-	-	3	<1	-	-	-	-	-	-	-	-	-	-	2	3	-	69	239
marginifera Um	-	-	<1	<1	-	-	5	<1	-	-	-	-	-	-	-	-	-	-	-	-	-	-	-	-	<1	<1	-	-	-	<1	<1	-	-	-	<1	-	-	-	-	90	415
marginifera U	-	-	-	-	-	-	2	-	-	-	-	-	-	-	-	-	-	-	-	-	-	-	-	<1	3	2	-	-	-	-	-	-	-	-	<1	1	-	-	-	91	430
marginifera L	-	1	-	-	-	-	4	-	-	-	-	-	-	-	-	1	2	<1	-	1	<1	<1	-	-	-	-	-	-	-	-	-	-	-	<1	<1	-	-	-	-	90	762
rhomboidea U	-	-	-	-	<1	-	-	-	-	-	-	-	-	-	-	-	-	2	-	-	-	-	-	-	3	-	-	-	-	-	-	-	<1	<1	<1	-	-	-	-	94	270
rhomboidea L	-	-	-	-	-	-	-	-	-	-	-	-	-	-	-	-	-	-	-	-	-	-	-	-	-	8	-	-	-	-	-	-	-	-	-	-	-	-	-	92	12
crepida U	2	-	-	-	-	-	-	-	-	-	-	-	-	3	-	-	-	-	-	-	-	-	-	-	-	-	-	-	-	-	-	-	-	-	-	-	-	-	-	95	42
crepida M	8	-	-	-	-	-	-	-	-	-	-	-	-	1	-	-	-	-	-	-	-	-	-	-	-	-	-	-	-	-	-	-	-	-	-	-	-	-	-	91	190
crepida L																																									

tives of the off-shore fauna are *Alternognathus regularis* and members of the *Polygnathus nodocostatus* group (Sandberg, 1976; Sandberg and Ziegler, 1979; Ziegler and Sandberg, 1984) and they comprise about 22% of the platform component; 11% are probably ubiquitous (*Polygnathus glaber bilobatus*) and 67% are representatives of the near-shore (nectobenthonic) *Polygnathus semicostatus* group (Table 22.2). Thus, the biofacies is composed mainly of species of *Polygnathus* and contains neither true pelagic forms such as *Palmatolepis* nor shallow-water forms such as '*Icriodus*' and *Pelekysgnathus*. The fauna may be treated as representing a biofacies intermediate between the palmatolepid–polygnathid (II) and the polygnathid–'icriodid' (III) biofacies. A similar biofacies, dominated by species of *Polygnathus*, is known from the Frasnian sequence of Canada (Klapper and Lane, 1985) and also from Famennian reef-like structures in the Ardenne-Rhenish Massif (Dreesen *et al.*, 1985).

?Algal–foraminiferal wackestones and packstones of the Lower–Middle *expansa* zones are interbedded with bioclastic grainstones often with imbricate textures. They contain worn and abraded crinoid and brachiopod debris and small intraclasts. This sediment was deposited in an environment of periodic wave and/or current action. The limestones contain a conodont fauna representative of the polygnathid–bispathodid biofacies composed of species of *Polygnathus*, mainly of the *Polygnathus semicostatus* group (*Polygnathus delicatulus*, *Polygnathus semicostatus* and *Polygnathus* aff. *P. procerus*) with subordinate examples of the *Polygnathus nodocostatus* group (*Polygnathus nodocostatus* and *Polygnathus perplexus*) which together comprise 34% of the platform (Pa) elements; *Bispathodus* (*Bispathodus aculeatus aculeatus* and *Bispathodus stabilis*) together with the morphologically similar *Mehlina* and *Branmehla* comprise 45%. Other ubiquitous forms such as *Hemilistrona pulchra*, *Polygnathellus giganteus* and *Polygnathus communis communis* (Sandberg, 1976; Sandberg and Ziegler, 1979; Sandberg and Dreesen,

1984) constitute 9% of the platform (Pa) elements (Table 22.2). It is possible that the polygnathid-bispathodid biofacies occupied a similar palaeoecological niche to the palmatolepid–polygnathid biofacies. As well as the forementioned components of the polygnathid–bispathodid biofacies there occur also *Pandorinellina* cf. *insita*, *Clydagnathus ormistoni* and *Pelekysgnathus inclinatus*, derived representatives more near-shore environments.

In the Upper *expansa*–Lower *praesulcata* zones a change in facies is observed all over Pomerania (Fig. 22.3). Black marls and marly limestones represent a different type of pelagic limestone. In addition to cephalopods, entomozoid ostracods and conodonts, benthonic organisms such as trilobites (phacopids), pelecypods, gastropods, brachiopods and solitary corals occur (Korejwo, 1975, 1976), associated with benthonic ostracods and encrusting foraminifera. This lithofacies contains conodonts indicative of the polygnathid–bispathodid biofacies. It differs, however, from the polygnathid–bipathodid biofacies distinguished within the ?algal–foraminiferal nodular wackestones and packstones of the Lower–Middle *expansa* zones. The polygnathid–bispathodid biofacies of the Upper *expansa*–Lower *praesulcata* zones is composed mainly of ubiquitous forms such as *Bispathodus*, *Mehlina*, *Branmehla* and *Polygnathus communis communis* (Sandberg, 1976; Sandberg and Ziegler, 1979; Sandberg and Dreesen, 1984); other conodont species occur only as minor components (Tables 22.1 and 22.2 and Plate 22.6). It is worthy of note that specimens of *Siphonodella praesulcata* are also present in the polygnathid–bispathodid biofacies. A continuous Famennian–Tournaisian sequence with a siphonodellid succession is observed in the off-shore open marine environment.

22.3 COMPARISON WITH OTHER REGIONS

Comparison of the Famennian lithofacies and conodont biofacies from the north and south of Poland is difficult. The Famennian outer shelf facies of southern Poland is distinctly different, being mostly pelagic in character. In the Holy Cross Mountains the Famennian Stage (Fig. 22.2 area 3,) is known to occur in two facies types, namely in alternating marly limestones and shales and as a condensed reef cover in the form of cephalopod or crinoid–brachiopod lime-stones (Szulczewski, 1971, 1973, 1978). Between Olkusz and Zawiercie (Fig. 22.2, area 4) the Famennian Stage is represented by muddy limestone with a significant amount of coarse intraclasts (calcirudites and coarse-grained calcarenites) due to intraformational reworking (Narkiewicz, 1978). The lithostratigraphic units distinguished in the Olkusz–Zawiercie area are generally characterized by the co-occurrence of marly micritic beds; these yielded a poor benthonic and an abundant planktonic fauna with detrital intercalations abounding in intraclasts together with abraded and sorted remains of echinoderms, diverse calcareous algae, foraminifera, etc. These marly micritic beds are interpreted by Narkiewicz (1978) as autochthonous sediments of the deeper shelf. The detrital intercalations are interpreted (Narkiewicz, 1978) as allochtonous sediments redeposited from shallower areas (carbonate platform). In another paper, Narkiewicz and Racki (1985) argue that the appearance of these tempestite interbeds was associated with a regression.

Successions in Pomerania and southern Poland are quite distinct, particularly when the conodont assemblages are taken into account. According to data presented by Narkiewicz (1978) from the Olkusz–Zawiercie area, only the palmatolepid, palmatolepid–polygnathid and palmatolepid–bispathodid biofacies are present (Table 22.3). These represent more offshore environments according to Sandberg's model. According to data provided by Szulczewski (1971) it can be stated with certainty that the palmatolepid biofacies was dominant in the Holy Cross Mountains at least up to the *rhomboidea* Zone. The conodont genera and species present in younger strata (Wolska, 1967); Szulczewski and Zakowa, 1976; Zakowa

et al., 1983) also indicate a dominance of biofacies associated with deeper environments. Unfortunately, these papers contain no information on the numerical abundance of the conodont species.

The Famennian Stage in Pomerania can be compared with that in the Lublin region located to the east of the Holy Cross Mountains. Sedimentation commenced there with calcareous mudstones and muddy limestone which represent probably the *triangularis–crepida* zones with the overlying nodular limestone belonging to the *crepida–marginifera* zones (Matyja and Zbikowska, 1985). In some sections the topmost parts of this limestone include sandy limestone interbeds, marking incipient regression. The sediments of the Upper Famennian Stage are of lagoonal-continental character (Mitaczewski, 1981) and do not contain conodonts. Similarity of lithofacies in both the areas in question and a common tendency to regression, resulting from their location near land areas (Fig. 22.2), lead me to expect the occurrence of a similar conodont biofacies. Although the evidence is fragmentary (Matyja and Zbikowska, 1985), the conodont assemblages in both areas show distinct affinities up to the Upper *marginifera* Zone.

The regressive sequence seen in the Famennian sediments of Pomerania and the conodont biofacies encountered therein seem to correlate with the Famennian successions of the Ardennes (Dreesen and Thorez, 1980; Sandberg and Dreesen, 1984, Fig. 8) despite the different depositional regimes. The general coincidence of the Famennian transgressive deepening and regressive shallowing events (Johnson *et al.*, 1985, Fig. 2) in both these areas, exposed on the southern periphery of the Old Red Sandstone Continent, is noteworthy. In both these areas during the Famennian Stage, the regressive tendencies marked by changes in lithofacies and biofacies are to be noted near the Upper *crepida* Zone and both show a gradual regressive tendency up to the Middle *expansa* Zone (see Fig. 22.3 and Johnson *et al.* 1985, Fig. 2). A minor transgressive pulse is noted in Belgium in the Lower and Upper *marginifera* zones (the presence of Baelen mudmounds (see Dreesen *et al.*, 1985; Johnson *et al.*, 1985)). One may presume that reef-like structures occur also in Pomerania in the same stratigraphic horizons (see p. 370) but it is still an open question. A new transgressive episode began over the whole of Pomerania, as is the case in the Ardennes and the Rheinische Schiefergebirge, about the Upper *expansa* Zone. According to the nomenclature used by Johnson *et al.* (1985), the Pomeranian sediments correspond to the transgressive–regressive cycle defined as IIe Upper *triangularis*–Upper *postera* and to the inception of the next T-R-IIf cycle.

22.4 CONCLUSIONS

Following a study of conodont biofacies and lithofacies it is possible to demonstrate periods of relative deepening and shallowing within the Famennian sedimentary basin of Pomerania. The first sedimentary cycle commenced with a Frasnian transgression and terminated with a long and gradual regressive sequence. Culmination of this regressive sequence took place during the deposition of the grainstone beds which are assigned to the Lower–Middle *expansa* zones. A transgressive event is indicated by the change towards more argillaceous sedimentation and the incursion of a pelagic fauna. Such an onlap situation is reflected by the presence of muddy limestones and shales within the Upper *expansa*–Lower *praesulcata* zones.

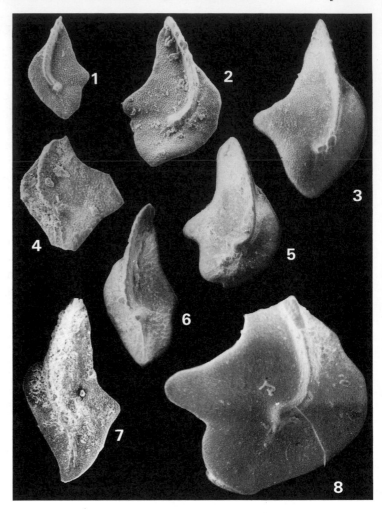

PLATE 22.1

The palmatolepid biofacies (I) of the Northern Facies Belt. Middle–Upper *crepida* zones. All are Pa elements.

Palmatolepis wolskajae Ovnatonva 1969
Plate 22.1, Figs. 1, 3. Fig. 1, upper view, specimen ING C 26, ×47. Borehole Karsina 1. Fig. 3, uupper view, specimen ING C 27, ×47. Borehole Karsina 1.

Palmatolepis circularis Szulczewski 1971
Plate 22.1, Figs. 2, 5 and 8. Fig. 2, upper view, specimen ING C 28, ×47. Borehole Karsina 1. Fig. 5, upper view, specimen ING C 29, ×30. Borehole Karsina 1. Fig. 8, upper view, specimen ING C 61, ×45. Borehole Gorzyslaw 8.

Palmatolepis quadrantinodosalobata Sannemann 1955
Plate 22.1, Fig. 4, upper view, specimen ING C 31, ×47. Borehole Karlino 1.

Palmatolepis crepida Sannemann 1955
Plate 22.1, Fig. 6, upper view, specimen ING C 32, ×30. Borehole Karcino 2.

Palmatolepis tenuipunctata Sannemann 1955
Plate 22.1, Fig. 7, upper view, specimen ING C 35, ×45. Borehole Koczala 1.

Pl. 22.2]   **Conodont biofacies in the Famennian Stage of Pomerania**   375

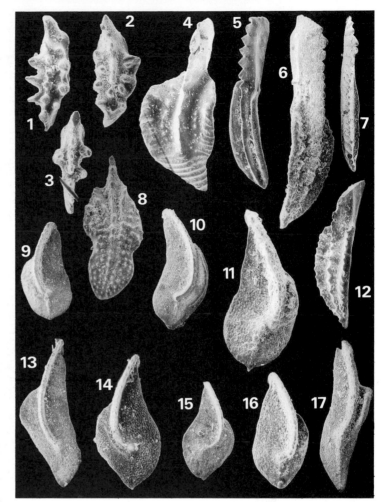

PLATE 22.2

The palmatholepid–polygnathid biofacies (II) of the Southern Facies Belt. Lower *marginifera* Zone. All are Pa elements.

'Icriodus' chojnicensis Matyja 1972
Plate 22.2, Figs. 1, 2, 3. Fig. 1, upper view, specimen ING C 1, ×47. Borehole Chojnice 2. Fig. 2, upper view, specimen ING C 2, ×47. Borehole Chojnice 2. Fig. 3, upper view, specimen ING C 3, ×47. Borehole Chojnice 2.

Polygnathus semicostatus Branson and Mehl 1934
Plate 22.2, Figs. 4, 12. Upper view, specimen ING C 17, ×20. Borehole Chojnice 4, Fig. 12. Upper view, morphotype 1, specimen ING C 4, ×30. Borehole Chojnice 2.

Polygnathus aff. *P. procerus* Sannemann 1955
Plate 22.2, Fig. 5. Upper view, specimen ING C 5, ×47. Borehole Chojnice 2.

Polygnathus fallax Helms and Wolska 1967
Plate 22.2, Fig. 6. Upper view, specimen ING C 6, ×47. Borehole Chojnice 2.

Alternognathus pseudostrigosus (Dreesen and Dusar 1974)
Plate 22.2, Fig. 7. Upper view, specimen ING C 7, ×30. Borehole Chojnice 2.

Polygnathus triphyllatus (Ziegler 1960)
Plate 22.2, Fig. 8. Upper view, specimen ING C 8, ×30. Borehole Chojnice 2.

Palmatolepis quadrantinodosa quadrantinodosa Branson and Mehl 1934
Plate 22.2, Figs, 9, 11. Fig. 9, upper view, specimen ING C 9, ×20. Borehole Chojnice 2. Fig. 11, upper view, specimen ING C 10, ×47. Borehole Chojnice 2.

Palmatolepis marginifera marginifera Helms 1959
Plate 22.2, Fig. 10. Upper view, specimen ING C 11, ×20.

Palmatolepis stoppeli Sandberg and Zeigler 1973
Plate 22.2, Figs. 14, 15. Fig. 14, upper view, specimen ING C 13, ×47. Borehole Chojnice 2. Fig. 15, upper view, specimen ING C 14, ×20. Borehole Chojnice 2.

Palmatolepis quadrantinodosa infelxa Müller 1956
Plate 22.2, Fig. 16, upper view, specimen ING C 15, ×47. Borehole Chojnice 2.

Palmatolepis glabra pectinata Ziegler 1962
Plate 22.2, Fig. 17, upper view, specimen ING C 16, ×47. Borehole Chojnice 2.

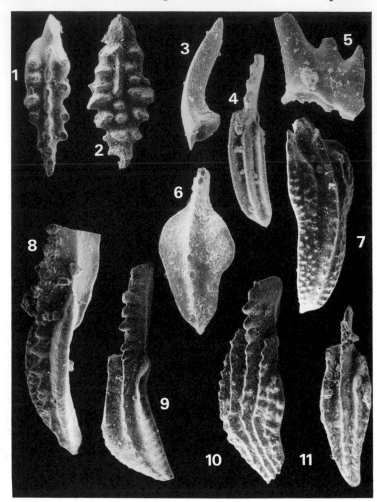

PLATE 22.3

The polygnathid–'icriodid' biofacies? (III) of the Northern Facies Belt. From *marginifera* to ?Upper *postera* zones.

'Icriodus' chojnicensis Matyja 1972
Plate 22.3, Figs. 1, 2. Fig. 1, upper view I element, specimen ING C 47, ×45. Borehole Gorzyslaw 8. Fig. 2, upper view I element, specimen ING C 48, ×45. Borehole Gorzyslaw 8.

Simple cone, undetermined
Plate 22.3, Fig. 3, Side view, specimen ING C 49, ×45. Borehole Gorzyslaw 8.

Polygnathus aff. *P. procerus* Sannemann 1955
Plate 22.3, Fig. 4. Upper view, specimen ING C 50, ×45. Borehole Gorzyslaw 8.

Pelekysgnathus inclinatus Thomas 1949
Plate 22.3, Fig. 5. Side view, I element, specimen ING C 51, ×30. Borehole Gorzyslaw 8.

Polygnathus glaber bilobatus Zeigler 1962
Plate 22.3, Fig. 6. Upper view, Pa element, specimen ING C 52, ×45. Borehole Gorzyslaw 8.

Polygnathus rhomboideus Ulrich and Bassler 1926
Plate 22.3, Fig. 7. Upper view, Pa element, specimen ING C 53, ×45. Borehole Gorzyslaw 8.

Polygnathus glaber glaber Ulrich and Bassler 1926
Plate 22.3, Fig. 8. Side view, Pa element, specimen ING C 54, ×45. Borehole Gorzyslaw 8.

Polygnathus szulczewskii Matyja 1974
         Plate 22.3, Fig. 9. Upper view, Pa element, specimen ING C 55, ×45. Borehole Gorzyslaw 8.

Polygnathus perplexus Thomas 1949
Plate 22.3, Fig. 10. Upper view, Pa element, specimen ING C 56, ×45. Borehole Gorzyslaw 8.

Polygnathus ex. gr. *nodocostatus* Branson and Mehl 1934
Plate 22.3, Fig. 11. Upper view, specimen ING C 57, ×45. Borehole Gorzyslaw 8.

Pl. 22.4] **Conodont biofacies in the Famennian Stage of Pomerania** 377

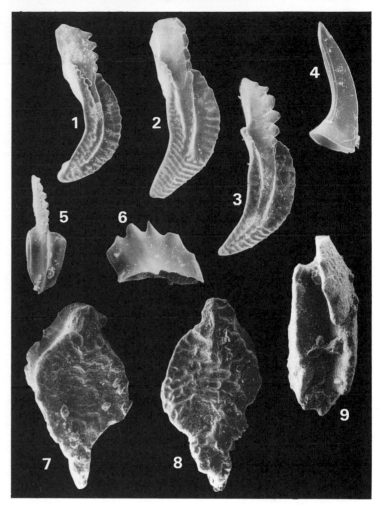

PLATE 22.4

The polygnathid–pelekygnathid biofaces ?(IV) and admixtures of biofacies V, VI and VII in the Northern Facies Belt. Lower–Middle *expansa* zones.

Polygnathus szulczewskii Matyja 1974
Plate 22.4, Figs. 1, 3. Fig. 1, upper view, Pa element, specimen ING C 36, ×30. Borehole Koczala 1. Fig. 3, upper view, Pa element, specimen ING C 37, ×30. Borehole Koczala 1.

Polygnathus semicostatus Branson and Mehl 1934
Plate 22.4, Fig. 2. Upper view, Pa element, specimen ING C 38, ×30. Borehole Koczala 1.

Simple cone, undetermined
Plate 22.4, Fig. 4. Side view, specimen ING C 38, ×30. Borehole Koczala 1.

Polygnathus communis communis Branson and Mehl 1934
Plate 22.4, Fig. 5. Upper view, Pa element, specimen ING C 33, ×47. Borehole Karcino 2.

Pelekysgnathus inclinatus Thomas 1949
Plate 22.4, Fig. 6. Side view, I element, specimen ING C 34, ×47. Borehole Karcino 2.

'*Icriodus*' *raymondi* Sandberg and Ziegler 1979
Plate 22.4, Figs. 7, 8. Fig. 7, upper view, I element, specimen ING C 58, ×45. Borehole Gorzyslaw 8. Fig. 8, upper view, I element, specimen ING C 59, ×45. Borehole Gorzyslaw 8.

Scaphignathus peterseni Sandberg and Zielger 1979
Plate 22.4, Fig. 9. Upper view, Pa element, specimen ING C 60, ×45. Borehole Gorzyslaw 8.

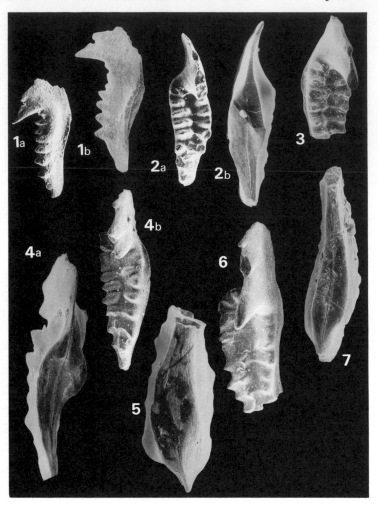

PLATE 22.5

The polygnathid–pelekysgnathid biofacies? (IV) and admixtures of biofacies V, VI and VIII in the Northern Facies Belt. Lower–Middle *expansa* zones. All Pa elements.

Clydagnathus ormistoni Beinert, Klapper, Sandberg and Zielger 1971
Plate 22.5, Figs. 1, 2, 3, 4. Fig. 1a, upper view, specimen ING C 40, ×45. Borehole Koczala 1. Fig. 1b, side view, specimen ING C 40, ×47. Borehole Koczala 1. Fig. 2a, upper view, specimen ING C 41, ×45. Borehole Koczala 1. Fig. 2b, lower view, specimen ING C 41, ×47. Borehole Koczala 1. Fig. 3, upper view, specimen ING C 42, ×47. Borehole Koczala 1. Fig. 4a, lower view, specimen ING C 43, ×47. Borehole Koczala 1. Fig. 4b, upper view, specimen ING C 43, ×45. Borehole Koczala 1.

Scaphignathus ziegleri Druce 1969
Plate 22.5, Figs. 5, 6, 7. Fig. 5, lower view, specimen ING C 44, ×47. Borehole Koczala 1. Fig. 6, upper view, specimen ING C 45, ×47. Borehole Koczala 1. Fig. 7, lower view, specimen ING C 47, ×47. Borehole Koczala 1.

Pl. 22.6] **Conodont biofacies in the Famennian Stage of Pomerania** 379

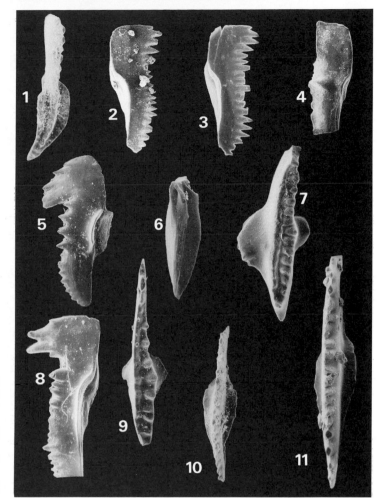

PLATE 22.6

The polygnathid–bispathodid biofacies (II) in the Northern and Southern Facies Belts. Upper *expansa*–Lower *praesulcata* zones. All are Pa elements.

Polygnathus delicatulus Ulrich and Bassler 1926
Plate 22.6, Fig. 1. Upper view, specimen ING C 18, ×47. Borehole Babilon 1.

Mehlina strigosa (Branson and Mehl 1934)
Plate 22.6, Figs. 2, 3. Fig. 2, side view, specimen ING C 62, ×30. Borehole Krojanty 1. Fig. 3, side view, specimen ING C 19, ×47. Borehole Babilon 1.

Bispathodus aculeatus anteposicornis (Scott 1961)
Plate 22.6, Fig. 4. Side view, specimen ING C 20, ×20. Borehole Babilon 1.

Pandorinellina plumula (Rhodes, Austin and Druce 1969)
Plate 22.6, Fig. 5. Side view, specimen ING C 21, ×30. Borehole Babilon 1.

Siphonodella praesulcata Sandberg 1972
Plate 22.6, Fig. 6. Lower view, specimen ING C 30, ×47. Borehole Karsina 1.

Bispathodus ultimus (Bischoff 1957)
Plate 22.6, Fig. 7. Upper view, specimen ING C 25, ×30. Borehole Rzeczenica 1.

Bispathodus aculeatus aculeatus (Branson and Mehl 1934)
Plate 22.6, Fig. 8. Side view, specimen ING C 22, ×30. Borehole Babilon 1.

Bispathodus costatus (Branson 1934)
Plate 22.6, Fig. 9. Upper view, morphotype I, specimen ING C 23, ×30. Borehole Babilon 1.

Bispathodus stabilis (Branson and Mehl 1934)
Plate 22.6, Figs. 10, 11. Fig. 10, upper view, specimen ING C 24, ×47. Borehole Babilon 1. Fig. 11, upper view, specimen ING C 63, ×47. Borehole Krojanty 1.

22.5 REFERENCES

Beinert, R. J., Klapper, G., Sandberg, C. A. and Ziegler, W. 1971. Revision of *Scaphignathus* and descriptions of *Clydagnathus? ormistoni* n. sp. (Conodonta, Upper Devonian). *Geologica et Palaeontologica*, **5**, 81–91, 2 plates.

Bischoff, G. 1957. Die Conodonten-Stratigraphie des rheno-herzynischen Unterkarbons mit Berücksichtigung der Woklumeria-Stufe und der Devon/Karbon-Grenze. *Hessisches Landesamt Bodenforschung Abhandlungen*, **19**, 1–64, 6 plates.

Branson, E. B. 1974. Conodonts of the Hannibal formation of Missouri. *University of Missouri Studies*, **8**, 301–334, 4 plates.

Branson, E. B. and Mehl, M. G. 1934. Conodont studies No. 3: Conodonts from the Grassy Creek Shale of Missouri. *University of Missouri Studies*, **8**, 171–259, 9 plates.

Dreesen, R., Bless, M. J. S., Conil, R., Flajs, G. and Laschet, Ch. 1985. Depositional environments paleoecology and diagenetic history of the 'Marbre rouge a crinoides de Baelan' (late Upper Devonian, Verviers synclinorium, eastern Belgium). *Annales de la Société Géologique de Belgique*, **108**, 311–359, 19 plates.

Dreesen R. and Dusar, M. 1974. Refinement of conodont biozonation in the Famennian type area: In J. Bouckaert and M. Streel (Eds.). *International Symposium on Belgian Micropaleontological Limits from Emsian to Visean*, Namur, September 1–10, 1974. **13**, 1–36, 7 plates.

Dreesen, R. and Thorez, J. 1980. Sedimentary environments, conodont biofacies and paleoecology of the Belgian Famennian (Upper Devonian) — an approach. *Annales de la Société Géologique de Belgique*, **103**, 97–110.

Druce, E. C. 1969. Devonian and Carboniferous conodonts from the Bonapart Gulf Basin, Northern Australia. *Bulletin, Australian Bureau Mineral Resources Geology and Geophysics*, **98**, 1–242, 43 plates.

Helms, J. and Wolska, Z. 1967. New Upper Devonian conodonts from Poland and West Germany. *Acta Palaeontologica Polonica*, **12**, 227–238.

Johnson, J. G., Klapper, G. and Sandberg, C. A. 1985. Devonian eustatic fluctuations in Euramerica. *Geological Society of America Bulletin*, **96**, 567–587.

Klapper, G. and Lane, H. R. 1985. Upper Devonian Frasnian conodonts of the *Polygnathus* biofacies, N. W. T., Canada. *Journal of Paleontology*, **59**, 904–952. 11 plates.

Korejwo, K. 1975. The Lowermost Dinantian from the Babilon 1 column Western Pomerania (in Polish). *Acta Geologica Polonica*, 25, 451–504, 22 plates.

Korejwo, K. 1976. The Carboniferous of the Chojnice area (Western Pomerania). *Acta Geologica Polonica*, **26**, 541–555, 8 plates.

Matyja, H. 1972. Biostratigraphy of the Upper Devonian from the borehole Chojnice 2 (western Pomerania). *Acta Geological Polonica*, **22**, 735–750, 4 plates.

Matyja, H. 1974. A new conodont species from the Famennian of Poland. *Bulletin de l'Academie Polonaise des Sciences*, **22**, 785–787, 1 plate.

Matyja, H. and Zbikowska, B. 1985. Stratigraphy of Devonian carbonate sequence in borehole sections in the Lublin area (in Polish). *Przeglad Geologiczny*, **5**, 259–263.

Milaczewski, L. 1981. The Devonian of the south-eastern part of the Radom–Lublin area, eastern Poland (in Polish). *Prace Instytutu Geologicznego*, **61**, 1–90.

Müller, K. J. 1956. Zue Kenntnis der Conodonten-Fauna des europäischen Devons, 1. Die Gattung *Palmatolepis Abhandlungen der Senckenbergischen Naturforschenden Gessellschaft*, **30**, 494, 1–70, 11 plates.

Narkiewicz, M. 1978. Stratigraphy and facies development of the Upper Devonian in the Olkusz–Zawiercie area, southern Poland (in Polish). *Acta Geologica Polonica*, **28**, 415–467, 30 plates.

Narkiewicz, M. and Racki, G. 1985. Major features of the Late Devonian paleogeography in the nearshore shelf area of southern Poland. *Przeglgd Geologiczny*, **5**, 271–274.

Ovnatonova, N. S. 1969. Novye verkhnedevonskie Konodonty Centralnykh rajonov Russkoj Platformy i Timane. *Trudy UNIGNI*, **93**, 139–141 and 292–293, 1 plate.

Pozaryski, W. and Dembowski, Z. 1984. *Geological Map of Poland and Adjoining Countries without Cenozoic, Mesozoic and Permian formations; Scale 1:1 000 000*, Wydawnictwa Geologiczne.

Rhodes, F. H. T., Austin, R. L. and Druce, E. C. 1969. British Avonian (Carboniferous) conodont faunas, and their value in local and intercontinental correlation. *British Museum (Natural History) Geology, Bulletin*, Supplement, **5**, 1–313, 31 plates.

Sandberg, C. A. 1972. In Sandberg, C. A., Streel, M. and Scott, R. A. Comparison between conodont zonation and spore assemblages at the Devonian–Carboniferous boundary in the western and central United States and in Europe. Compte Rendu 7th Congrès International. *Stratigraphie et Geologi du Carbonifère Krefeld*, **1**, 179–203, 4 plates.

Sandberg, C. A. 1976. Conodont biofacies of Late Devonian *Polygnathus styriacus* Zone in western United States. *Geological Association of Canada Special Paper*, **15**, 171–186.

Sandberg, C. A. and Dreesen, R. 1984. Late Devonian icriodontid biofacies models and alternate shallow-water conodont zonation. In D. L. Clark (Ed.), *Conodont Biofacies and Provincialism*, *Geological Society of America Special Paper*, **196**, 143–178, 4 plates.

Sandberg, C. A. and Gutschick, R. C. 1979. Guide to conodont biostratigraphy of Upper Devonian and Mississippian rocks along the Wasatch Front and Cordilleran Hingeline, Utah. In C. A. Sandberg, and D. L. Clark (Eds.), *Conodont Biostratigraphy of the Great Basin and Rocky Mountains, Brigham Young University Geology Studies*, **26**, 107–134.

Sandberg, C. A. and Ziegler, W. 1973. Refinement of standard Upper Devonian conodont zonation based on sections in Nevada and West Germany. *Geologica et Palaeontologica*, **7**, 97–122, 5 plates.

Sandberg, C. A., and Ziegler, W. 1979. Taxonomy and biofacies of important conodonts of Late Devonian *styriacus* Zone, United States and Germany. *Geologica and Palaeontologica*, **13**, 173–212. 3 plates.

Sannemann, D. 1955. Oberdevonische Conodonten (to IIα). *Senekenbergiana Lethaea*, **36**, 123–156, 6 plates.

Scott, A. J. 1961. Three new conodonts from the Louisiana

Limestone (Upper Devonian) of Western Illinois. *Journal of Paleontology, 35*, 1223—1227.

Szulczewski, M. 1971. Upper Devonian conodonts, stratigraphy and facial development in the Holy Cross Mts. *Acta Geologica Polonica*, 21, 1–129, 34 plates.

Szulczewski, M. 1973. Famennian-Tournaisian neptunian dykes and their conodont fauna from Dalnia in the Holy Cross Mts. *Acta Geologica Polonia*, 23, 15–59, 6 plates.

Szulczewski, M. 1978. The nature of unconformities in the Upper Devonian–Lower Carboniferous conodont sequence in the Holy Cross Mts. *Acta Geologica Polonica*, 28, 283–298.

Szulczewski, M. and Zakowa, H. 1976. New data on the Famennian of the Galezice Synclinc (in Polish), *Biuletyn Instytutu Geologicznego*, 296, 51–72.

Thomas, L. A. 1949. Devonian–Mississippian formations of southeast Iowa. *Geological Society of America Bulletin*, 60, 403-438, 4 plates.

Ulrich, E. O. and Bassler, R. S. 1926. A classification of the toothlike fossils conodonts, with descriptions of American Devonian and Mississippian species. *U.S. National Museum Proceedings*, 68, Article 12, 1–63, 11 plates.

Wolska, Z. 1967. Upper Devonian conodonts from the south-west region of the Holy Cross Mts., Poland (in Polish). *Acta Palaeontologica Polonica*, 12, 363–435, 19 plates.

Ziegler, P. A. 1982. *Geological Atlas of Western and Central Europe*, Shell International Petroleum Maotschappij, 1–130, 40 maps.

Ziegler, W. 1960. Conodonten aus dem Rheinischen Unterdevon (Gedinnium) des Remscheider Sattels (Rheinisches Schiefergebirge). *Palaontologische Zeitschrift*, 34, 169–201, 3 plates.

Ziegler, W. 1962. Taxonomie und Phylogenie Oberdevonischer Conodonten und ihre stratigraphische Bedeutung. *Hessisces Londesamt Bodenforschung Abhandlungen*, 38, 1–166, 14 plates.

Ziegler, W. and Sandberg, C. A. 1984. *Palmatolepis*-based revision of upper part of standard Late Devonian conodont zonation. In D. L. Clark (Ed.), *Conodont Biofacies and Provincialism, Geological Society of America Special Paper*, 196, 179–194, 2 plates.

Zakowa, H., Szulczewski, M. and Chlebowski, R. 1983. The Upper Devonian and Carboniferous of the Borkow Syncline (in Polish). *Biuletyn Instytutu Geologicznego*, 345, 5–134, 19 plates.

23

Biofacies-based refinement of Early Permian conodont biostratigraphy, in central and western USA

S. M. Ritter

The post-'Early Permian crisis' Wolfcampian interval (*Neogondolella bisselli–Sweetognathus whitei* Zone) was dominated by species of the novel *Sweetognathus* lineage and a new species of *Neogondolella*.

Late Wolfcampian peritidal to off-shore carbonates and shales in Kansas (Chase Group), Nevada, Utah (Riepetown Formation) and west Texas (upper Hueco Group) contain low diversity conodont faunas adapted to different lithofacies that reflect the inception of post-'crisis' biofacies. *Neogondolella bisselli*, *Sweetognathus behnkeni* and *Sweetognathus whitei* are most common in off-shore lithofacies and high percentages of these species define biofacies I. Conodont faunas from normal marine shelf and epicontinental lithofacies (biofacies II) are characterized by *Sweetognathus* n. sp. A, *S. whitei* and *Sweetognathus* n. sp. B in the lower part of the *N. bisselli–S. whitei* interval. In the upper part of this interval, *Rabeignathus bucaramangus* becomes a dominant component of shelf faunas of Utah and Nevada. New Genus sp. A replaces *Rabeignathus* in shelf limestones of west Texas. Although conodonts are largely excluded from restricted subtidal to supratidal lithofacies, a nearly monospecific fauna of

Sweetognathus n. sp. C (biofacies III) characterizes laminated dolomicrites of the upper part of the Barneston Limestone in Kansas.

Biofacies distinction allows for the subdivision of the *N. bisselli–S. whitei* interval into three sweetognathid-based subzones in the shallow shelf (biofacies II). Late Wolfcampian shelf and epicontinental sediments in Utah and Kansas are divided, in ascending order, into the *Sweetognathus whitei–Sweetognathus* n. sp. A and *Rabeignathus bucaramangus* subzones. The New Genus sp. A Subzone, based upon the locally restricted pre-*Neostreptognathodus* occurrence of the nominative species in west Texas, is coeval with the *Rabeignathus bucaramangus* Subzone.

23.1 INTRODUCTION

Recognition of palaeogeographical, palaeoecological and evolutionary controls on the distribution of *Neogondolella bisselli–Sweetognathus whitei* Zone conodonts allows establishment of biofacies and refinement of late Wolfcampian biostratigraphy. The current North American late Wolfcampian zonation was based upon the ranges of *Neogondolella bisselli* (Clark and Behnken) and *Sweetognath-*

us whitei (Rhodes) in the deep water stratotype section located at Moorman Ranch, Nevada (Clark and Behnken, 1971; Clark, et al., 1979). The absence or limited range of these species in shallow epicontinental and carbonate shelf deposits has obscured conodont-based interregional correlations (Clark, 1972; Wind, 1973; Rabe, 1977).

The purposes of this contribution are as follows: (1) to document the distribution and ranges of late Wolfcampian species in the central and western USA; (2) to establish late Wolfcampian conodont biofacies; (3) To present a facies-dependent subzonation of the *Neogondolella bisselli–Sweetognathus whitei* Zone.

The results reported in this chapter are based upon a study of conodont collections from four widely separated localities in the central and western USA (Fig. 23.1). New species are designated informally in this chapter. Formal species names appear in Ritter, S. M., 1986, *Geologica et Palaeontologica,* **20,** (Formal names also should be substituted into the biostratigraphic nomenclature proposed in this chapter.)

Conodonts experienced a worldwide faunal turnover in the late Wolfcampian. This event has been referred to as the Early Permian crisis (Clark, 1972). Herein the terms *turnover* and *crisis* are used interchangeably and refer to this faunal replacement. 'Turnover' is preferred to 'crisis' in that the latter may misleadingly connote sudden biological calamity as the cause of the observed change in Early Permian conodont faunas. The usage of the terms fauna A and fauna B essentially conforms to that of Clark (1974) in referring to the pre-turnover and post-turnover conodont faunas respectively.

23.2 STRATIGRAPHY AND LOCAL RANGES OF CONODONTS

(a) Cordilleran

Wolfcampian rocks of east–central Nevada and west–central Utah comprise, in ascending order, the Riepe Springs Limestone and Riepe-

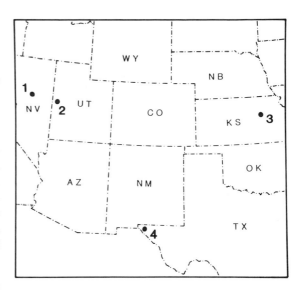

Fig. 23.1 — Index map showing location of sections in Kansas (KS), Nevada (NV), Texas (TX) and Utah (UT). The other states shown are as follows: WY, Wyoming; NB, Nebraska, CO, Colorado; AZ, Arizona; NM, New Mexico; OK, Oklahoma. Detailed location of the sections is provided in the locality register (appendix).

town Formation. These formations paraconformably overlie the Morrowan–Desmoinesian Ely Limestone. The 100 m thick Riepe Springs Limestone is a time-transgressive basal Permian unit throughout much of the central Cordillera. Sediments at present assigned to the overlying Riepetown Formation range from dominantly conglomerate and siltstone in east–central Nevada to sandy dolomite and limestone in western Utah. This lateral variation in lithofacies reflects a complex of Permian palaeotectonic features, the most important of which are the Dry Mountain Trough, Deep Creek–Tintic Uplift, West–Central Utah Highland, Oquirrh Basin and Antler Belt (Fig. 23.2). Detailed sections of the uppermost Riepe Springs Limestone and Reipetown Formation were studied at Moorman Ranch, Nevada, and in the Burbank Hills–Confusion Range area of western Utah. These sections represent end members on a basin to shelf (west to east) transect of the central Cordilleran miogeosyncline. No intervening shelf edge, characterized by abrupt vertical relief, has been recognized, which led some workers to visualize a ramp-depositional model for this area (e.g. Stevens, 1979).

(i) Nevada
At Moorman Ranch, Nevada (Fig. 23.2) the Riepetown Formation consists of more than 300 m of thin-bedded very fine sandstones and siltstones punctuated with infrequent thin (15 cm) to massive (6 m) limestone interbeds. Carbonates comprise less than 8% of this interval, which contrasts with the carbonate-dominated Riepetown Formation in western Utah. Fusulinid dating reported by Bissell (1964) indicates that the Riepetown Formation ranges from medial to late Wolfcampian at Moorman Ranch.

The three predominant rock types exposed in the Riepetown Formation at Moorman Ranch are as follows.

(1) Thin-bedded very fine sandstone to siltstone.
(2) Thin-bedded and graded skeletal carbonate wackestone to grainstone.
(3) Massive-bedded crinoidal packstone to grainstone.

The thin, rhythmically bedded very fine sandstone to siltstone units are interpreted to represent distal clastic turbidite deposits. These graded units are from 2 to 5 cm thick and possess sandy lower halves lacking internal structure and rippled upper halves. *Zoophycos*, *Phycosiphon*? and *Scolicia* characterize this clastic microfacies in the lower 30 m of the Riepetown Formation. *Phycosiphon*? is also present higher in the section. Clark (1974) suggested that this fodinichnial trace-fossil assemblage indicates deposition in at least 70 m of water.

Graded fine-grained skeletal grainstone to silty spiculiferous wackestone beds (5–15 cm) are interpreted as distal limestone turbidites. Broken and abraded fragments of crinoid columnals, coiled and detached encrusting foraminifera and *Tubiphytes* are commonly the

dominant constituents of this fine-grained microfacies.

Coarse-grained crinoidal packstones to grainstones comprise the third Riepetown microfacies at Moorman Ranch. Ledge-forming units of this microfacies are from 2 to 6 m thick and lack apparent internal structure.

Relative deepening of the Deep Creek Trough (with respect to its depth during deposition of the Riepe Springs Limestone) and influx of fine clastic material led to deposition of the Riepetown Formation on a ramp slope at Moorman Ranch. During the late Wolfcampian, more than 300 m of fine sand and silt were deposited by waning turbidity currents. Ichnofossils indicate low nutrient levels at various intervals. Periodic cessation of clastic influx allowed for development of stenohaline crinoid-dominated communities on the ramp slope. Bioclastic sediment was carried downslope at infrequent intervals, resulting in the emplacement of bioclastic limestone turbidites. Reworking of more proximally located crinoid communities resulted in the development of the massive crinoidal microfacies.

Studies pertaining to the biostratigraphy, evolution and palaeoecology of Early Permian conodonts from the Moorman Ranch section include those of Clark and Behnken (1971) and Clark (1972, 1974). This section is the stratotype for the *Neogondolella bisselli–Sweetonathus whitei* Zone as defined by Clark and Behnken (1971).

A total of 448 conodont elements were recovered from 21 1–3 kg samples recollected from the lower 294 m of the Riepetown Formation. Seven species representing six genera were identified. The individual species ranges are shown in Fig. 23.3.

Conodont occurrences are discontinuous and are restricted to the infrequent limestone units. Thin limestone turbidite samples (MR 1 2, 8, 9, 10 and 11) generally produced significantly higher yields than samples from the more massive ledge-forming crinoidal packstone to grainstone units.

The first occurrences of *S. whitei* 10 m above

the base of the Riepetown Formation, marks the base of the *Neogondolella bisselli–Sweetognathus whitei* Zone. *N. bisselli* appears 25 m above the base of this zone. The mutual disappearance of *N. bisselli* and *S. whitei* occurs 294 m above the base of the Riepetown Forma-

tion. Poorly preserved specimens of an indeterminate species of *Neostreptognathodus* were recovered 30 m below the last occurrence of *N. bisselli* and *S. whitei*. A similar overlap of *Neostreptognathodus* and elements of the *N. bisselli–S. whitei* fauna was reported by Behn-

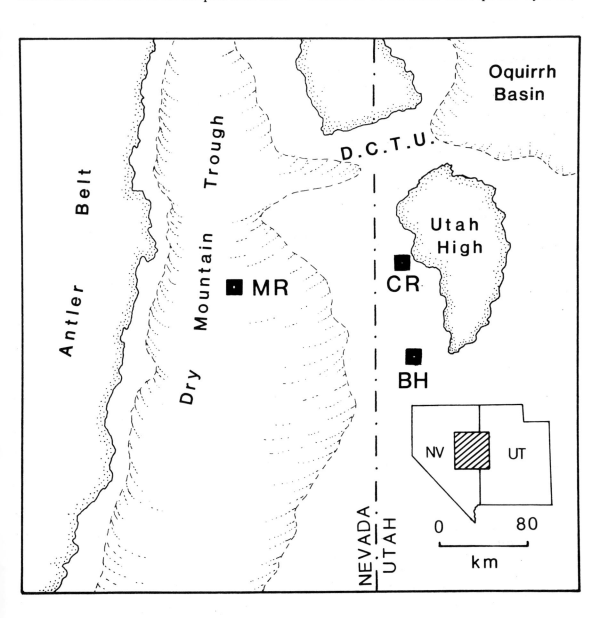

Fig. 23.2 — Palaeogeographical setting of central Cordilleran miogeosyncline showing palaeotectonic features that affected Early Permian sedimentation. DCTU, Deep Creek–Tintic Uplift: MR, Moorman Ranch; CR, Confusion Range; BH, Burbank Hills. (Modified from Stevens (1979).)

ken (1975) in the Pequop Mountains of north-eastern Nevada.

Sweetognathus behnkeni Kozur replaces S. whitei in a 15 m interval extending from 76.5 m to 91.5 m above the base of the Neogondolella bisselli–Sweetognathus whitei Zone. Within this interval, specimens of S. whitei are absent or outnumbered greatly by those of S. behnkeni. The inverse abundances between Sweetognathus and Neogondolella, described by Clark (1974), seem to be related in part to lithofacies differences. Specimens of Sweetognathus, especially S. behnkeni, are more abundant in limestone turbidites whereas N. bisselli is more common in the massive crinoidal microfacies.

The base of the Neogondolella bisselli–Sweetognathus whitei Zone also marks the initial occurrence of the post-turnover fauna B of Clark (1974). This initial occurrence of fauna

B is separated from the terminal occurrence of fauna A by a barren interval which corresponds to the upper 60–70 m of the Riepe Springs Limestone.

Comparison of the upper Wolfcampian sequence exposed at Moorman Ranch with that exposed in the Confusion Range–Burbank Hills region of western Utah (Fig. 23.2) reveals (1) a different suite of lithofacies and (2) a significantly different succession of conodonts.

(ii) Western Utah

In west–central Utah, Wolfcampian rocks are well exposed on the limbs of a north-plunging syncline 120 km long (Fig. 23.4). Sections of the uppermost Riepe Springs Limestone and Riepetown Formation were measured, sampled and described in the southern Burbank Hills (section BH) and in the northern end of the Confusion Range (section CR). At section BH the Riepetown Formation has an exposed thick-

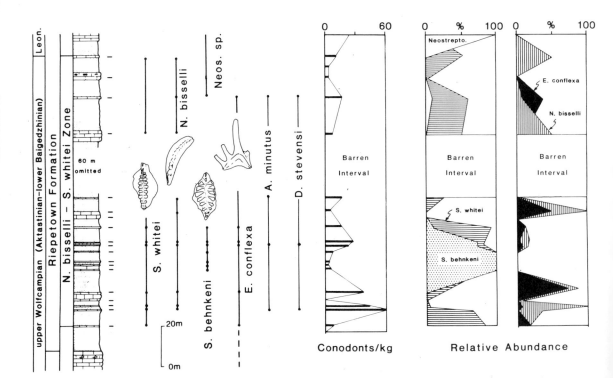

Fig. 23.3 — Distribution of conodonts in the Riepetown Formation, Moorman Ranch, Nevada. The relative abundances of individual species are shown on two curves for clarity. Consequently, the unpatterned area occupying the right-hand side of each curve corresponds to the patterned area on the opposing curve.

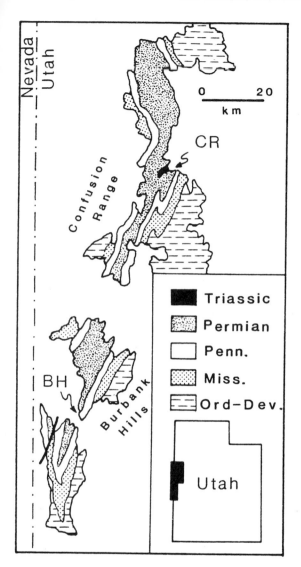

Fig. 24.4 — Geological map of the Burbank Hills–Confusion range area of western Utah showing outcrop of Permian rocks and location of measured sections BH and CR (Simplified from Hintze (1975).)

Springs and Riepetown strata crop out at section BH. By contrast, less than 100 m of equivalent strata are present at section CR in the northern end of the Confusion Range where the Riepe Springs Limestone and lower part of the Riepetown Formation are absent. The northward thinning of Lower Permian strata reflects the influence of the Permian West–Central Utah Highland (Fig. 23.2).

Limestone and dolomites in the Burbank Hills-Confusion Range region represent deposition in normal shelf to restricted marine conditions respectively. Limestones are compositionally and texturally heterogeneous but generally comprise a suite of normal shelf microfacies. Wackestone and packstone textures dominate but grainstones are not uncommon. Common skeletal components include brachiopods, bryozoans, palaeotextularid foraminifera, fusulinids and echinoids. Other constituents are fragments of *Tubiphytes*, ostracodes, encrusting foraminifera, corals and peloids. The biotic associations within the limestone microfacies are compositionally similar to those used to define the foraminiferal, coral and brachiopod communities recognized in the Permian rocks of the central Cordilleran by Stevens (1966). Brachiopod and coral communities were interpreted to have developed at depths of 4–10 m and 10–30 m respectively. Fusulinids were believed to have occupied slightly deeper water (20–50 m). Grainstone microfacies are interpreted to have developed in shoal-water conditions. Normal salinities are indicated by the abundance and diversity of invertebrates represented in these microfacies.

Dolomitic rocks are diagnosed as dolomicrites with crystal fabrics averaging 0.02–0.04 mm. Evaporite moulds and small chalcedony nodules with evaporite inclusions occur at some horizons and macrofossils are lacking throughout. Dolomites represent deposition in restricted subtidal near-shore environments. This interpretation is based upon the presence of evaporite minerals, abundant quartz silt, dolomitic composition and absence

ness of 220 m. The lower 120 m consists of sandy dolomite with infrequent limestone interbeds. The upper 100 m is characterized by wackestone to grainstone units 1–23 m thick which comprise 40% of the upper interval (Fig. 23.5).

Approximately 367 m of combined Riepe

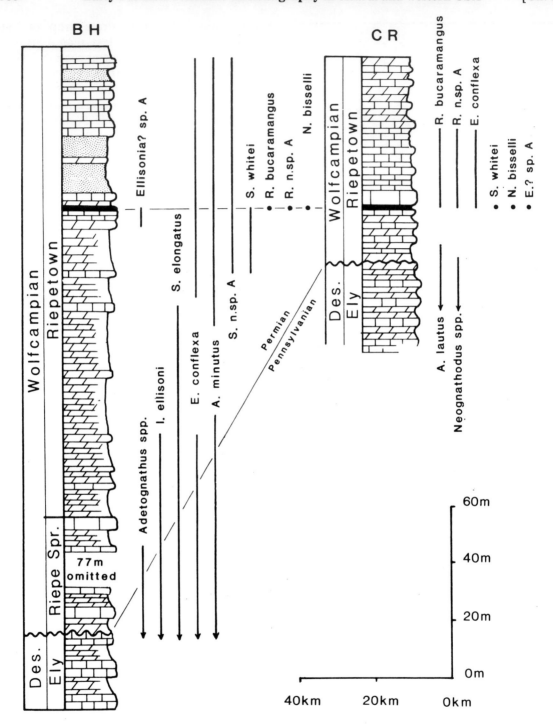

Fig. 23.5 — Lithological columns and ranges of conodonts in sections BH (Burbank Hills) and CR (Confusion Range), western Utah: An., *Anchignathodus*; S., *Sweetognathus* (except *Streptognathodus elongatus*); R., *Rabeignathus*; N., *Neogondolella*; E., *Ellisonia*; I., *Idiognathodus*; A., *Adetognathus*. The ranges of *Streptognathodus elongatus* and *Ellisonia conflexa* are after Clark (1972). The datum upon which correlations are based is the *Rabeignathus* bed.

of fossils. It is possible that some dolomites were deposited in intertidal to supratidal settings, although indicators of subaerial exposure such as desiccation cracks, intraclast horizons, fenestrae, sheetcracks and domal structures were not observed.

Samples from the Riepetown Formation in the Burbank Hills and Confusion Range produced discontinuous low- abundance conodont faunas. Limestone samples had an average yield of 12 specimens per kilogram. A total of nine late Wolfcampian species representing six genera were recovered in this area. The vertical ranges of these taxa are shown in Fig. 23.5.

Occurrences of the index species of the *Neogondolella bisselli–Sweetognathus whitei* Zone are limited in western Utah. Only 15 specimens of *S. whitei* were recovered from four samples at locality BH where the species ranges from 372 m to 397 m above the base of the Riepetown. *N. bisselli* was represented by a combined total of only six specimens in sections BH and CR.

The longest-ranging platform species in section BH is *Sweetognathus* n. sp. A. This species first appears in a brachiopod–bryozoan packstone 372 m above the base of the Riepetown Formation and continues to the top of the section. The last occurrence, 5 m below strata bearing *Parafusulina leonardensis*, is at or near the Wolfcampian–Leonardian boundary as defined in the central Cordillera by Stevens *et al.* (1979). *Sweetognathus* n. sp. A faunas have been recovered from fusulinid-foraminiferal packstone, brachiopod–bryozoan packstone, dolomitic mudstone and peloidal grainstone microfacies, indicating a broad environmental distribution. *Sweetognathus* n. sp. A was also recovered from a rounded algal-clast grainstone overlying a blue–green algal stromatolite.

Rabeignathus bucaramangus (Rabe) and *Rabeignathus* n. sp. A, neither of which were recovered at Moorman Ranch, first occur in a brachiopod-rich packstone 397 m above the base of the Riepetown Formation at locality BH. Samples from this horizon, collected over an 800 km^2 area, consistently yielded from 20 to 35 specimens of *Rabeignathus* per kilogram. The *Rabeignathus* horizon is overlain by 30 m of sandstone at locality BH and specimens of *Rabeignathus* spp. are restricted to this single horizon. At locality CR, the *Rabeignathus* bed is overlain by a normal marine limestone sequence characterized by the development of brachiopod–bryozoan and fusulinid microfacies. Here both species of *Rabeignathus* occur up to 32 m above the *Rabeignathus* bed. Younger beds have not been sampled, and so the local upper limits of *R. bucaramangus* and *Rabeignathus* n. sp. A are at present unresolved. Specimens were recovered only from samples assignable to a brachiopod–bryozoan microfacies. The uppermost *Rabeignathus*-bearing bed was also characterized by particularly robust specimens of Pb, Sb, Sc and Sa elements of *Ellisonia conflexa* (Ellison).

E. conflexa and *Anchignathodus minutus* sensu formo (Ellison) range throughout all sections and occur in a wide variety of normal shelf microfacies.

(b) Kansas

The Chase Group of eastern Kansas (19 members comprising seven formations) consists of approximately 100 m of cyclically disposed carbonates and shales, representing deposition in subaerial to normal marine settings. Detailed studies of the stratigraphy, cyclicity, lithofacies and depositional environments of the Chase Group include those of Elias (1937), Hattin (1957) and Wind (1973).

The Chase Group may be divided into three more or less complete and four incomplete cyclothems designated A through F in Fig. 23.6. The Speiser Shale (uppermost Council Grove Group), Wreford Limestone and lower part of the Matfield Shale comprise two recognizable cyclothems (A and B) that together constitute the Wreford Megacyclothem of Hattin (1957). Cycle C is incomplete and reflects a brief transgressive phase to the molluscan facies. The upper Matfield Shale–lower Doyle Shale constitutes the Barneston cycle (D). This asymmetric cycle, the single thickest cycle in the Chase

Group, is characterized by a relatively thin transgressive and thick regressive phase. Younger units in the Chase Group demonstrate continuation of cyclic sedimentation but reflect deterioration of cyclothem completeness and symmetry. Dolomitization obscures recognition of depositional fabrics in upper Chase Group carbonate units.

The vertical distribution of conodont species within the Chase Group, first studied by Wind (1973), is summarized in Fig. 23.6. Here, as in

Utah, the index species of the *Neogondolella bisselli–Sweetognathus whitei* Zone are absent or limited in their stratigraphic range.

For comparison with conodont distributional data, seven main marine microfacies, corresponding closely to the depositional phases of Hattin (1957), were differentiated on the basis of thin-section analysis. These microfacies (in addition to a generalized non-marine lithotope) are shown with respect to position along an offshore to onshore gradient in Fig. 23.7. Con-

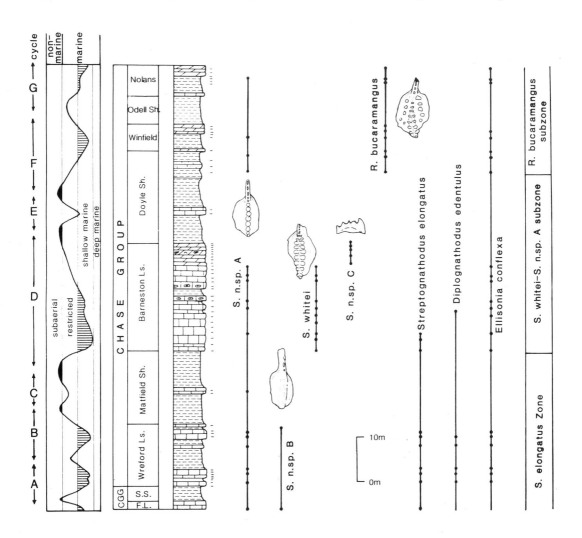

Fig. 23.6 — Lithological column, cycles and distribution of conodonts in the uppermost Council Grove (CGG) Speiser Shale, S.S.; Funston Limestone, F. L. and Chase Groups, eastern Kansas. Cycles modified from Elias (1937).

odonts are common in calcareous shales and normal marine limestones, sparse in mollusc-rich limestones and dolomicrites and absent in peritidal to non-marine shales. *Streptognathodus* is restricted to the chalky and cherty limestones which represent the deeper-water core of cyclothems A–C. The failure of *Streptognathodus* to recur in higher cycles is attributed to extinction of the genus, as it also disappears in Utah shortly after the appearance of *S. whitei*. The preference of *Streptognathodus* for the deep-water core lithofacies is similar to that in the Pennsylvanian cyclothems of Kansas (Heckel and Baesemann, 1975) and in offshore lithofacies elsewhere (Merrill and von Bitter, 1976; Driese *et al.*, 1984).

S. whitei is restricted to the Florence Limestone (lowest member of the Barneston Limestone) and lower 5 m of the Fort Riley Limestone (upper member of the Barneston Limestone). This species is most abundant in the bryozoan--brachiopod lithofacies of the lower Florence Limestone but continues into the mollusc-poor submicrofacies of the Fort Riley Limestone. *S.*

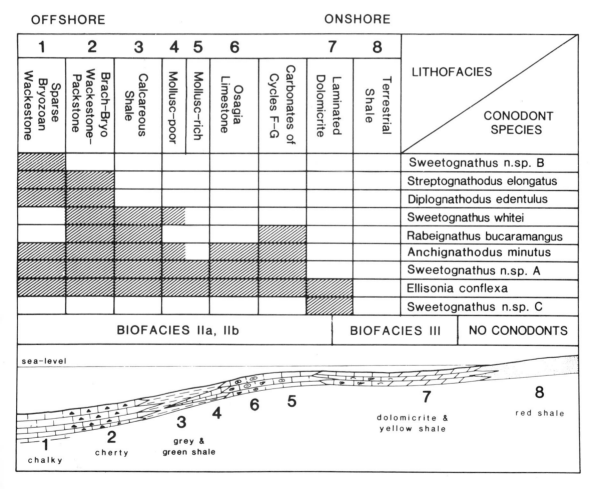

Fig. 23.7 — Environmental distribution of conodonts in the Chase Group, eastern Kansas. Lithofacies are based upon thin-section analysis and field relationships. Inferred depositional setting of lithofacies and range of conodont biofacies are shown in the lower part of the Figure.

whitei does not recur in younger Chase Group carbonates or marine shales where *R. bucaramangus* is common. The failure of *S. whitei* to recur may reflect intolerance of near-shore conditions represented by lithofacies comprising the upper part of the Chase Group. The absence of *S. whitei* is not attributed to extinction, as its range overlaps that of *R. bucaramangus* in the western USA. *R. bucaramangus* first appears in the upper part of the Doyle Shale and is present in calcareous shales and carbonates (except lower limestone member of the Nolans Limestone) of cycles F–G. As in Utah, this species is considered to have thrived in shallow normal marine shelf settings.

Sweetognathus n. sp. C is known only from restricted marine laminated dolomicrites comprising the upper 5 m of the Fort Riley Limestone. This species evolved from *Sweetognathus* n.sp. A in response to progressive restriction of the Kansas sea.

A. minutus sensu formo, *Sweetognathus* n.sp. A and *E. conflexa* are environmentally broad-ranging species that recur in cycles A–G. All these species occur in lithofacies ranging from chalky limestone to *Osagia* microfacies.

No specimens of the deeper-water species *N. bisselli* have been recovered from the Chase Group in eastern Kansas.

(c) Texas

Harbour (1972) described and mapped more than 760 m of Lower Permian Hueco Group limestones on the west side of the Franklin Mountains just north of El Paso, Texas. The Hueco Group is divided, in ascending order, into the Hueco Canyon, Cerro Alto and Alacran Mountain Formations (Williams, 1966). The Hueco Canyon Formation, as recognized herein, corresponds to units 1–90 of Harbour (1972, Plate 2) and is 300 m thick. The Hueco Canyon Formation consists chiefly of medium- to thick-bedded bioclastic wackestone and packstone with intervals of algal biostrome development. The overlying Cerro Alto Formation (130 m) corresponds to Harbour's units 95–125 and consists of medium-bedded dark-

grey wackestone interbedded with siltstone and shale. The Alacran Mountain Formation (115 m) consists of cherty, light-grey bioclastic wackestone, which is commonly medium to thick bedded. Platy algal biostromes are common in the lower and upper parts of the formation (Jordan, 1975).

Fusulinid dating reported by Williams (1966), Jordan and Wilson (1971) and Harbour (1972) indicates that most of the Hueco Group in the Franklin Mountains is Wolfcampian in age. Harbour (1972) suggested an early Leonardian age for the upper Alacran Mountain Formation on the basis of fusulinids and lithological similarity with Leonardian rocks in the nearby Hueco Mountains.

A total of 100 samples was collected from sections corresponding to sections 14a, 14b, 14c and 13 of Harbour (1972, Plates 1, 2). The continuous normal marine carbonate sequence of the Hueco Group was studied with the hope of recovering a relatively complete Early Permian conodont succession. Unfortunately, only 162 identifiable conodont elements were recovered from 44 productive samples of the upper Cerro Alto and Alacran Mountain Formations. No conodonts were recovered in 56 samples from the lower Cerro Alto and Hueco Canyon Formations.

Seven species of conodonts representing five genera were identified. The ranges of these species in the upper Cerro Alto and Alacran Mountain Formations are shown in Fig. 23.8. Most species are long-ranging forms with little stratigraphic value. Species with restricted ranges or biostratigraphic potential are *Sweetognathus* n. sp. D and New Genus sp. A.

Three specimens of *Sweetognathus* n. sp. D were recovered from a single horizon (unit 102 of Harbour, 1972) 76 m above the base of the Cerro Alto Formation. At present, this species is known only from this one locality.

New Genus sp. A first occurs 28 m below the top of the Cerro Alto Formation and ranges into the upper Alacran Mountain Formation. The range of this species defines the locally important zone of New Genus sp. A. Fusulinid-bear-

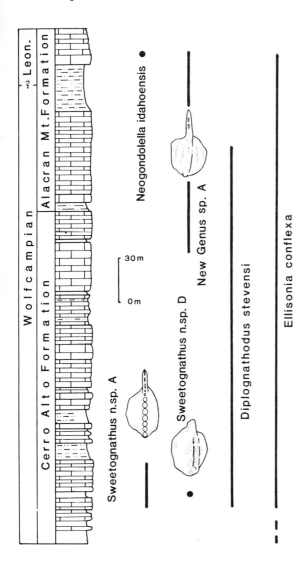

Fig. 23.8 — Lithological column and vertical distribution of conodonts in the upper Cerro Alto and Alacran Mountain Formations, Franklin Mountains, west Texas.

species is an index of the late Wolfcampian *Pseudoschwagerina convexa* fusulinid zone of Stevens *et al.* (1979).

The upper range of New Genus sp. A overlaps the range of *Neogondolella idahoensis* (Youngquist, Hawley and Miller) in the uppermost Alacran Mountain Formation. The association of these two species indicates that New Genus sp. A persisted into the early Leonardian and confirms Harbour's fusulinid-based early Leonardian age designation for the upper part of the Alacran Mountain Formation.

23.3 LATE WOLFCAMPIAN BIOFACIES

Lithofacies-related differences in *Neogondolella bisselli–Sweetognathus whitei* Zone conodont faunas from the central and western USA allow definition of five major biofacies. To a large extent, biofacies distinction reflects the successful radiation of species of *Sweetognathus* into a variety of marine environments. Biofacies are designated with Roman numerals (I, IIA, IIB, IIC and III) in order to avoid confusion with formal taxonomic and biostratigraphic nomenclature.

(a) Biofacies I

This biofacies is characterized by large percentages of *N. bisselli*, *S. whitei* and *S. behnkeni* (Tables 23.1 and 23.2). *E. conflexa*, *Diplognathodus stevensi* Clark and Carr, *A. minutus* sensu formo and *Sweetognathus* n. sp. A comprise a small percentage of this biofacies. The biofacies is best developed in the graded limestone turbidites of the Riepetown Formation at Moorman Ranch, Nevada, where the lithofacies, trace-fossil assemblage and inferred palaeogeographical setting suggest deposition in moderately deep open-marine conditions. The empirical percentages of this biofacies almost certainly reflect downslope transport of faunas and may be different in more autochthonous basinal lithofacies. The role of *N. bisselli* and *S. behnkeni* as deeper-water indicators is supported, however, by the virtual absence of these species in more shelf-

ing beds containing *Schwagerina eolata*, *Pseudoschwagerina*? *laxissima* and *Parafusulina linearis* occur approximately 10 m below the base of this zone. In addition, Williams (1966) reported *Schwagerina bellula*, *S. eolata* and *Pseudoschwagerina convexa* in the lower 50 m of the Alacran Mountain Formation. The latter ward lithofacies. The moderately common occurrence of *S. whitei* in shelf settings as well as in deep-water turbidites suggests that this species lives in, or tolerated, a shallower higher-energy zone than either *S. behnkeni* or *N. bisselli*.

Table 23.1 — Relative abundance of conodont species in biofacies of the upper part (*R. bucaramangus* Subzone) of the *Neogondolella bisselli–Sweetognathus whitei* Zone. For detailed discussion of subzonal nomenclature and definitions see section on biostratigraphy. Relative abundance counts of *Ellisonia conflexa* are based upon entire apparatus. All other species counts based upon Pa elements only.

Species	Amount (%) for the following biofacies and localities				
	I	IIb		IIc	Restricted marine
	Nevada	Utah	Kansas	Texas	Utah and Kansas
New Genus sp. A	—	—	—	28.5	No conodonts
Rabeignathus spp.	—	55	85	—	No conodonts
Sweetognathus n.sp. A	2.6	15.2	—	—	No conodonts
Sweetognathus n.sp. B	—	—	—	—	No conodonts
Sweetognathus whitei	26.5	13.2	—	—	No conodonts
Sweetognathus behnkeni	18.7	—	—	—	No conodonts
Neogondolella bisselli	40.5	2.0	—	—	No conodonts
Ellisonia conflexa	4.5	7.6	9.0	10.5	No conodonts
Anchignathodus minutus	5.2	3.6	6.0	22	No conodonts
Diplognathodus spp.	1.1	0.3	—	23.5	No conodonts

Table 23.2 — Relative abundance of conodont species in biofacies of the lower part (*S. whitei—S. inornatus* Subzone) of the *Neogondollela bisselli–Sweetognathus whitei* Zone. For detailed discussion of subzonal nomenclature and definitions see section on biostratigraphy. Relative abundance counts of *Ellisonia conflexa* are based upon entire apparatus. All other species counts based upon Pa elements only.

Species	Amount (%) for the following biofacies and localities				
	I	IIa			III
	Nevada	Utah	Kansas	Texas	Kansas
Sweetognathus n.sp. C	—	—	—	—	92
Sweetognathus n.sp. A	2.6	27	27	7.9	—
Sweetognathus n.sp. B	—	—	—	—	—
Sweetognathus whitei	26.5	21	22	2.6	—
Sweetognathus behnkeni	18.7	—	—	—	—
Neogondolella bisselli	40.5	—	—	—	—
Ellisonia conflexa	4.5	22	27	39.5	8.0
Anchignathodus minutus	5.2	29	16	29	—
Diplognathodus spp.	1.1	—	5.0	15.8	—

(b) Biofacies II

This biofacies developed in shallow epicontinental and carbonate shelf settings in Kansas, Utah and Texas. Evolution and provincialism within this palaeoenvironmental setting allows recognition of three compositionally distinct biofacies herein designated biofacies IIA–IIC.

(i) Biofacies IIA

This biofacies, characterizes shelf rocks in the lower *Neogondolella bisselli–Sweetognathus whitei* Zone and is comprised of nearly equal percentages of *S. whitei*, *Sweetognathus* n. sp. A, *A. minutus* sensu formo and *E. conflexa*. Stratigraphically, this biofacies occurs in the Wreford Limestone of Kansas and in the Riepetown Formation (100–125 m above the base) in western Utah. In the upper Cerro Alto Formation in the Franklin Mountains, *E. conflexa*, *A. minutus* sensu formo and *D. stevensi* dominate a meagre fauna and this biofacies is poorly developed.

In the upper half of the *Neogondolella bisselli–Sweetognathus whitei* Zone, sweetognathids with widely flared inflated basal cups and exotic ornamentation (*Rabeignathus* and New Genus) evolved in shallow marine settings.

(ii) Biofacies IIB

This biofacies is dominated by *R. bucaramangus* in association with *S. whitei*, *Sweetognathus* n. sp. A, *E. conflexa* and *A. minutus* sensu formo. In Kansas, *R. bucaramangus* comprises nearly 85% of faunas recovered from the upper third of the Chase Group. Skeletal wackestones and packstones in the upper 60 m of the Riepetown Formation in western Utah contain this biofacies. *Rabeignathus* spp., however, are most common in the lower 30 m of this interval, above which *Sweetognathus* n. sp. A and *S. whitei* are important components.

(iii) Biofacies IIC

This biofacies is characterized by New Genus sp. A in association with *D. stevensi*, *E. conflexa* and *A. minutus* sensu formo. This biofa-

cies is best developed in the upper 30 m of the Cerro Alto Formation and overlying Alacran Mountain Formation in the Franklin Mountains, west Texas. Species of *Rabeignathus* and *Sweetognathus*, which comprise a significant component of shelf biofacies in Kansas and Utah, are virtually absent from biofacies IIB. Conversely, New Genus sp. A is unknown in shelf deposits outside west Texas, suggesting provincialism in Permian conodonts. North to south changes in Permian invertebrate biotas of North America define a temperate Cordilleran Province and a tropical Grandian Province (Yancey, 1975). New-Genus-bearing faunas are known only from the latter province. Thus, *Rabeignathus* spp. and New Genus sp. A allow differentiation of two coeval, palaeogeographically distinguished biofacies, IIB and IIC.

New Genus sp. A occurs with *N. bisselli* and *S. whitei* (biofacies I) in the lower Hess Formation in the Glass Mountains, west Texas (Carr, 1977). The lower Hess Formation was deposited on the narrow 'Southern Shelf', which separated the Marathon Fold Belt and Delaware Basin. The presence of off-shore biofacies indicators in New-Genus-bearing shelf deposits may be explained by the close juxtaposition of basin and shelf conditions.

(c) Biofacies III

Conodonts are largely excluded from restricted subtidal to supratidal lithofacies. However, a nearly monospecific fauna of *Sweetognathus* n. sp. C characterizes laminated dolomicrites of the upper 5 m of the Barneston Limestone in Kansas. Juvenile specimens of *E. conflexa* comprise 8% of the fauna recovered from this interval.

23.4 BIOSTRATIGRAPHY

The foregoing discussion reveals that late Wolfcampian conodonts are not uniformly distributed in the central and western USA. Short-ranging species of the rapidly evolving *Sweetognathus* lineage (those with the greatest potential for biostratigraphic refinement) are strongly facies controlled and regionally restricted. More ubiquitous species are long ranging and have little stratigraphic application. Recognition of differences in the habitat and ranges of late Wolfcampian conodont species permits lateral and vertical subdivision of the *N. bisselli-S. whitei* interval (Fig. 23.9). This ecostratigraphic scheme, comprised of one off-shore zone and three shelf subzones, allows wider biostratigraphic applicability than that afforded by the previous single-zone scheme. The off-shore zone, defined by the ecological association of *N. bisselli* and *S. whitei*, is essentially the post-crisis Wolfcampian *Neogondolella bisselli–Sweetognathus whitei* Zone of Clark and Behnken (1971). On the shelf, the combined durations of the *Rabeignathus bucaramangus* and the *Sweetognathus whitei–Sweetognathus*

n. sp. A subzones cover roughly the same time interval, the chronostratigraphic classification of which is presently unresolved. Clark *et al.* (1979) stated that this zone represents an upper Sterlitimakian interval in the western USA. Kozur (1977) indicates that the concurrent ranges of *N. bisselli* and *S. whitei* are restricted to an Aktastinian–lowest Baigendzhinian interval in Eurasia.

Fusulinids and corals are used to date first and last occurrences of some conodont species and permit crude correlation of sections lacking similar sequences of conodonts.

(a) *Neogondolella bisselli–Sweetognathus whitei* Zone
This zone, as orginally defined by Clark and Behnken (1971), was based upon the mutual ranges of *N. bisselli* and *S. whitei* throughout an

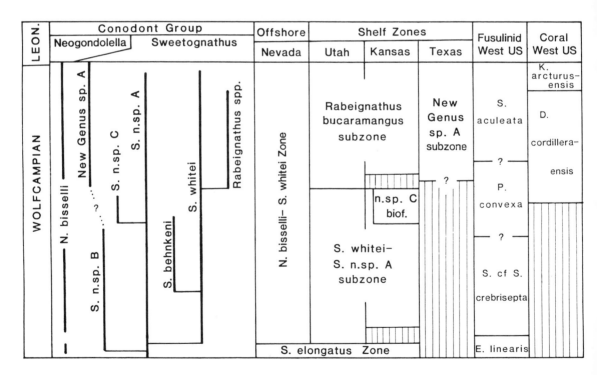

Fig. 23.9 — Phylogeny of selected late Wolfcampian conodonts and correlation of conodont, fusulinid and coral zones. Fusulinid and coral zonation after Stevens (1979) and Stevens *et al.* (1979). Conodonts: S., *Sweetognathus*; D., *Diplognathodus*; N., *Neogondolella*; *S. elongatus*, *Streptognathus elongatus*. Fusilinids: S., *Schwagerina*; P., *Pseudoschwagerina*; E., *Eoparafusulina*. Corals: K. *Kleopatrina*; D., *Diphyphyllum*.

interval extending from approximately 33 m to 100 m above the base of the Riepetown Formation at Moorman Ranch, Nevada. Subsequent studies suggest that *N. bisselli* evolved somewhat earlier than *S. whitei*. In the Ural Mountain region, *N. bisselli* and *Adetognathus lautus* (Gunnell) occur together in Sakmarian strata immediately below the appearance of *S. whitei* (Kozur, 1977). In British Columbia, joint occurrences of *N. bisselli* and *Adetognathus paralautus* Orchard precede the appearance of *S. whitei* (Orchard, 1984). The base of the *Neogondolella bisselli–Sweetognathus whitei* Zone is herein defined as the first appearance of *S. whitei* (which evolved shortly after the extinction of *Adetognathus*). *Streptognathodus elongatus* Gunnell occurs with *S. whitei* in the lower part of the zone.

In their original paper, Clark and Behnken (1971) stated that the upper stratigraphic limit of the zone was unknown. Later, Clark (1974) showed that ranges of the index species extended upwards to overlap early occurrences of *Neostreptognathodus* A similar overlap in the ranges of *N. bisselli* and *S. whitei* with that of *N. pequopensis* Behnken (referred to as *N. clarki* by Kozur and Mostler 1976) occurs in late Wolfcampian strata in the Pequop Mountains, Nevada (Behnken, 1975, Fig. 2), where the last occurrence of *S. whitei* is believed to coincide with the Wolfcampian–Leonardian boundary. In the Pequop Mountains, *N. bisselli* survives the apparent extinction of *S. whitei* and continues through more than 80 m of lower Leonardian strata (Behnken, 1975). In the Ural Mountain region the disappearance of *N. bisselli* reportedly coincides with that of *N. pequopensis* (Kozur, 1977), which gives it a significantly younger upper limit than in North America. Because timing of the extinction of *N. bisselli* is uncertain (or at least geographically variable), the top of the *N. bisselli–S. whitei* Zone is herein defined as the first occurrence of *N. pequopensis*.

These formal boundaries define a concurrent range zone essentially identical with that proposed by Clark and Behnken (1971) but allow for more precise recognition of the zone.

In the sections studied, stratigraphically continuous occurrences of *S. whitei* and *N. bisselli* occur only in the Moorman Ranch section. In epicontinental and shelf settings, the chronostatigraphic interval encompassed by the *Neogondolella bisselli–Sweetognathus whitei* Zone can be subdivided by the appearance of *Rabeignathus* and New Genus (Fig. 23.10 and Tables 23.1 and 23.2).

(i) Sweetognathus whitei–Sweetognathus n. sp. A Subzone

This subzone is defined by the pre-*Rabeignathus* occurrence of *S. whitei* and *Sweetognathus* n. sp. A. The base is defined by the appearance of *S. whitei* and the top by the appearance of *R. bucaramangus* or *Rabeignathus* n. sp. A. Conodonts which range into this zone are *Diplognathodus edentulus* (von Bitter) and *S. elongatus*. Other common conodonts are *E. conflexa* and *A. minutus* sensu formo. This subzone is recognized in the Riepetown Formation of western Utah and within the Barneston Limestone of Kansas. This subzone is the shallow-shelf equivalent of the lower *Neogondolella bisselli–Sweetognathus whitei* Zone.

(ii) Rabeignathus bucaramangus Subzone

The base of this subzone is defined by the first occurrence of *Rabeignathus* and the top is the first occurrence of *N. pequopensis*. Other species occurring in this subzone are *Rabeignathus* n. sp. A., *E. conflexa* and *A. minutus* sensu formo. This subzone is recognized in the Riepetown Formation of western Utah, in the upper Chase Group of Kansas and in Colombia, South America. This subzone supersedes the more limited New Genus A assemblage of Clark *et al.* (1979). As currently defined, *R. bucaramangus* is unknown in Nevada and reported Texas occurrences are now recognized as belonging to New Genus sp. A (non-New Genus A of Clark *et al.* (1979)).

Kozur (1978) designated *R. bucaramangus*

as an index for the much younger Leonardian Stage in the Cis–Ural region. In the southern Confusion Range of western Utah the initial appearance of *R. bucaramangus* is 3 m above a coral biostrome horizon containing Wolfcampian *Diphyphyllum connersensis* and 7 m below a fusulinid packstone bearing *Schwagerina wellsensis*, *Schubertella kingi* and an advanced species of *Triticites*. In Kansas and Utah, *R. bucaramangus* is associated with *Sweetognathus* n. sp. A. In Utah, the last occurrence of *Sweetognathus* n. sp. A is 5 m below a fusulinid horizon which bears specimens of earliest Leonardian *Parafusulina leonardensis*. In the USA the *Rabeignathus bucaramangus* Subzone is indisputably restricted to the uppermost Wolfcam-

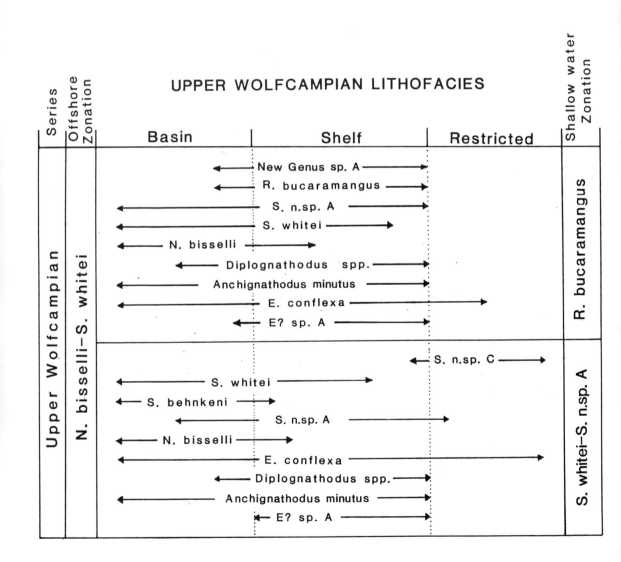

Fig. 23.10 — Summary chart showing environmental and temporal distribution of late Wolfcampian conodonts in central and western USA; R., *Rabeignathus*; S., *Sweetognathus*; N., *Neogondolella*; E., *Ellisonia*.

pian and correlates with the upper part of the *Neogondolella bisselli–Sweetognathus whitei* Zone.

(iii) New Genus sp. A Subzone

The base of this subzone is drawn at the first occurrence of New Genus sp. A and the top at the first occurrence of *N. pequopensis* Behnken. This subzone is currently recognized only in the Glass, Franklin and Hueco Mountains of west Texas. In the Franklin Mountains the first occurrence of New Genus sp. A is related to late Wolfcampian fusulinids but is difficult to correlate with ranges of other late Wolfcampian conodonts.

New Genus sp. A occurs with *N. bisselli* and *S. whitei* in the lower Hess Formation in the Glass Mountains (Carr, 1977) but disappears prior to the appearance of the Leonardian index *Neostreptognathodus*. The lower range of New Genus sp. A is truncated by a local unconformity. This species is also present in the Cerro Alto Formation in the Hueco Mountains (F. Behnken, pers. commun., 1984).

The association of New Genus sp. A with *N. bisselli* and *S. whitei* in the Glass Mountains and the appearance of New Genus sp. A following the initial appearance of *P. convexa* in the Franklin Mountains, west Texas, indicates that the New Genus sp. A Subzone is largely correlative with the *Rabeignathus bucaramangus* Subzone but may also correlate with the upper part of the *Sweetognathus whitei–Sweetognathus* n. sp. A Subzone.

23.5 CONCLUSIONS

Detailed successions of conodonts from coeval late Wolfcampian sections in Kansas, Nevada, Texas and Utah are presented. A study of the spatial distribution of conodonts in these sections indicates the following.

(1) Late Wolfcampian conodonts were not uniformly distributed with respect to either lithofacies (Fig. 23.10) or palaeogeography.

(2) *N. bisselli* and species of the *Sweetognathus* lineage define three major biofacies along a basin to bank transect. Specimens of *N. bisselli*, *S. behnkeni* and *S. whitei* dominate faunas in deep-water limestones at Moorman Ranch, Nevada. *Rabeignathus* spp. and New Genus sp. A characterize shallow shelf and epicontinental deposits from the upper part of the *N. bisselli–S. whitei* interval. The paucity of conodonts in dolomitic intervals suggests that peritidal to restricted marine environments were generally unsuitable for conodonts. *Sweetognathus* n. sp. C, however, adapted to restricted marine conditions represented by dolomites comprising the upper Barneston Limestone in Kansas.

(3) Those species showing little facies dependence, namely *E. conflexa* and *A. minutus* sensu, formo, are also long-ranging forms with little biostratigraphic value.

(4) In the absence of rapidly evolving, eurytopic conodonts, biostratigraphic refinement of the post-turnover Wolfcampian interval is best achieved through establishment of coeval facies-dependent subzones. The *Neogondolella bisselli–Sweetognathus whitei* Zone, as conceived at present, is best suited to correlation of off-shore lithotopes. Coeval shelf deposits are better correlated on the basis of *R. bucaramangus* or New Genus sp. A in the upper part of the *N. bisselli–S. whitei* interval, or *S. whitei* and *Sweetognathus* n. sp. A in the pre-*Rabeignathus* interval.

23.6 ACKNOWLEDGEMENTS

I wish to thank Dr. David L. Clark, University of Wisonsin, for his support of this project. The geographical scope of this study was made possible through the pioneering work of D. L. Clark (Utah and Nevada), T. R. Carr, R. L. Simpson, (west Texas) and F. H. Wind (Kansas). C. H. Stevens kindly identified fusulinid and coral specimens. G. L. Wilde provided additional fusulinid identifications. Kathie Ritter's assist-

PLATE 23.1

Specimens are deposited in the Department of Geology and Geophysics, University of Wisconsin-Madison (UW) and in the Paleontological Collections, University of Illinois, Urbana-Champaign (X).

Sweetognathus n. sp. A
Plate 23.1, Figs. 1, 2. Fig. 1, upper view, specimen UW 1776/61, ×62. Riepetown Formation, Burbank Hills, Utah, sample BH–33. Fig. 2, upper view, specimen UW 1776/98, ×66. Tensleep Sandstone, Wyoming, sample WYO–g..

Sweetognathus whitei (Rhodes 1963)
Plate 23.1, Figs. 3, 11. Fig. 3, upper view, specimen X/6644, figured specimen 1/19 of Rhodes (1963), ×62. Upper Tensleep Sandstone, Wyoming. Fig. 11, upper view, specimen UW 1776/141, ×60. Riepetown Formation, Moorman Ranch, Nevada, sample MR-1.

Rabeignathus n. sp. A
Plate 23.1, Figs. 4, 9. Fig. 4, upper view, specimen UW 1776/118, ×62. Riepetown Formation, Burbank Hills, Utah. sample BH-17. Fig. 9, upper view, specimen UW 1776/106, ×60. Riepetown Formation Confusion Range, Utah, sample CR-21.

Sweetognathus n. sp. D
Plate 23.1, Fig. 5. Upper view, specimen UW 1776/212, ×62. Cerro Alto Formation, Franklin Mountains, west Texas, sample CA–11.

Sweetognathus n. sp. C
Plate 23.1, Figs. 6, 8. Fig. 6, lateral view, specimen UW 1776/209, ×62. Fort Riley Member, Barneston Limestone, Chase Group. Kansas. Fig. 8, lateral view, specimen UW 1776/273, ×62. Fort Riley Member, Barneston Limestone, Chase Group, Kansas, samples FR–g and FR–10 respectively.

Neogondolella bisselli (Clark and Behnken 1971)
Plate 23.1, Fig. 7. Lateral view, specimen UW 1776/144, ×37. Riepetown Formation, Moorman Ranch, Nevada, sample MR-1.

Sweetognathus behnkeni Kozur 1975
Plate 23.1, Figs. 10, 12. Fig. 10, upper view, specimen UW 1776/134, ×62. Riepetown Formation, Moorman Ranch, Nevada, sample MR-9. Fig. 12, upper view, specimen UW 1776/135, ×62. Riepetown Formation, Moorman Ranch, Nevada, sample MR-9.

Rabeignathus bucaramangus Rabe 1977
Plate 23.1, Fig. 13. Upper view, specimen UW 1776/108, ×62. Riepetown Formation, Burbank Hills, Utah, sample BH-18.

New Genus sp. A
Plate 23.1, Figs. 14, 15. Fig. 14, upper view, specimen UW 1776/94, ×58. Alacran Mountain Formation, Franklin Mountains, west Texas, sample Am–4. Fig. 15, upper view, specimen UW 1776/86, ×54. Cerro Alto Formation, Franklin Mountains, west Texas, sample EL–22.

Sweetognathus n. sp. B
Plate 23.1, Fig. 16. Upper view, specimen UW 1776/4, ×70. Threemile Limestone Member, Wreford Limestone, Chase Group, Kansas, sample TM–1.

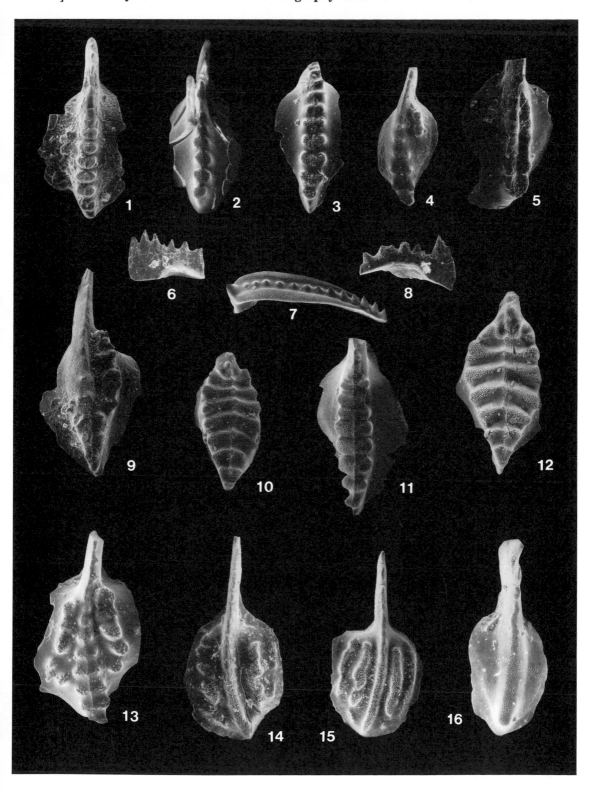

ance in the field is appreciated. This work was supported in part by the National Science Foundation under Grant EAR-8205675 to D. L. Clark.

23.7 LOCALITY REGISTER

(a) Nevada

(A) Riepetown Formation, Mormon Ridge––Butte Mountains. North of US 50, SW 1/4 section 7, T 17 N, R 58 E, White Pine Country, Nevada; Illipah 1:62 500 quadrangle. (Moorman Ranch)

(b) Utah

(A) Riepetown Formation (section BH), Burbank Hills. In centre of section 13, T 24 S, R 19 W, Millard County, Utah; Mormon Gap 1:24 000 quadrangle.

(B) Riepetown Formation (section CR), Confusion Range. Latitude, 39° 15'; longitude 113° 40', Millard County, Utah; Cowboy Pass. 1:62 500 quadrangle.

(c) Wyoming

(A) Tensleep Sandstone, approximately 5 miles west of Mayoworth on Mayoworth Road, Johnson County, Wyoming. Mayoworth 1:24 000 quadrangle.

(d) Kansas

(A) Threemile Limestone through Schroyer Limestone, Chase Group. South side of US 40 in NE 1/4 section, 34, T 11 S, R 6 E, Geary County, Kansas, Junction City 1:24 000 quadrangle.

(B) Kinney Limestone Member, Matfield Shale, Chase Group. South of junction of Kansas 177 and Kansas 113, NW 1/4 section 35, T 9 S, R 7 E, Riley County, Kansas; Manhattan 1:24 000 quadrangle.

(C) Florence Limestone Member, Barneston Limestone, Chase Group. North side of Kansas 177, 1 mile west of junction of Kansas 177 and Kansas 113, SW 1/4 section 22, T 9 S, R 7 E, Keats 1:24 000 quadrangle.

(D) Oketo–Towanda Limestone, Chase Group. East side of US 77 approximately 0.6 miles south of US 24, centre of section 33, T 11 S, R 5 E, Geary County, Kansas; Junction City 1:24 000 quadrangle.

(E) Gage Shale Member, Doyle Shale, Chase Group. South side of Kansas 244, 4.7 miles west of US 77, NW 1/4 section 35, T 11 S, R 4 E, Geary County, Kansas; Alida 1:24 000 quadrangle.

(F) Winfield Limestone, Chase Group. Adjacent to Kansas 244, 3.8 miles west of US 77, NW 1/4 section 36, T 11 S, R 4 E, Geary County, Kansas; Alida 1:24 000 quadrangle.

(G) Upper Odell Shale through Paddock Shale Member, Nolans Limestone, Chase Group. North side of small road 1 mile south of US 77–Kansas 177 junction, SE 1/4 section 34, T 7 S, R 6 E, Riley County, Kansas; Randolph 1:24 000 quadrangle.

(H) Herington Limestone Member, Nolans Limestone, Chase Group. North side of road on crest of hill, 5 miles west of Randolph, Kansas, SE 1/4 section 17, T 7 S, R 5 E, Riley County, Kansas; Randolph 1:24 000 quadrangle.

(e) Texas

(A) Upper Cerro Alto and Alacran Mountain Formations, Hueco Group, Franklin Mountains, El Paso County, Texas. Samples collected along line of section 13 of Harbour (1972).

(B) Hueco Canyon and lower Cerro Alto Formation, Hueco Group, Franklin Mountains, El Paso County, Texas. Samples collected west of and adjacent to Tom Mays Park along sections 14a, 14b and 14c Harbour (1972).

23.8 REFERENCES

Behnken, F. H. 1975. Leonardian and Guadalupian (Permian) conodont biostratigraphy in western and southwestern United States. *Journal of Paleontology*, **49**, 284–315.

Bissell, H. J. 1964. Ely, Arcturus and Park City Groups (Pennsylvanian–Permian) in eastern Nevada and western Utah. *American Association of Petroleum Geologists Bulletin*, **48**, 565–636.

Carr, T. R. 1977. Conodont biostratigraphy of the Skinner Ranch and Hess Formations (Permian), Glass Mountains, West Texas. *Unpublished M.S. Thesis*, Texas Tech University, 1–43.

Clark, D. L. 1972. Early Permian crisis and its bearing on Permo-Triassic conodont taxonomy. In M. Lindstrom and W. Ziegler (Eds.), *Symposium on Conodont Taxonomy, Geologica et Palaeontologica*, SB 1, 147–158.

Clark, D. L. 1974. Factors of early Permian conodont paleoecology in Nevada. *Journal of Paleontology*, **48**, 710–720.

Clark, D. L. and Behnken, F. H. 1971. Conodonts and biostratigraphy of the Permian. In W. C. Sweet and S. M. Bergström (Eds.), *Symposium on Conodont Biostratigraphy, Geological Society of America Memoir*, **127**, 147–158.

Clark, D. L., Carr, T. R., Behnken, F. H., Wardlaw, B. R. and Collinson, J. W. 1979. Permian conodont biostratigraphy in the Great Basin. In C. A. Sandberg and D. L. Clark (Eds.), *Conodont Biostratigraphy of the Great Basin and Rocky Mountains, Brigham Young University Geology Studies*, **26** (3), 143–150.

Driese, S. G., Carr, T. R. and Clark, D. L. 1984. Quantitative analysis of Pennsylvanian shallow-water conodont biofacies, Utah and Colorado. In D. L. Clark (Ed.), *Conodont Biofacies and Provincialism, Geological Society of America Special Paper*, **196**, 233–251.

Elias, M. K. 1937. Depth of deposition of the Big Blue (Late Paleozoic) sediments in Kansas. *Geological Society of America Bulletin*, **48**, 403–432.

Harbour, R. L. 1972. Geology of the northern Franklin Mountains, Texas and New Mexico. US Geological Survey *Bulletin*, **1298**, 1–129.

Hattin, D. E. 1957. Depositional environment of the Wreford Megacyclothem (Lower Permian) of Kansas. *Kansas Geological Survey Bulletin*, **124**, 1–50.

Heckel, P. H. and Baesemann, J. F. 1975. Environmental interpretation of conodont distribution in Upper Pennsylvanian (Missourian) megacyclothems in eastern Kansas. *American Association of Petroleum Geologists Bulletin*, **59**, 486–509.

Hintze, L. F. 1975. Geological highway map of Utah. *Brigham Young University Geology Studies Special Publication*, **3**.

Jordan, C. F., Jr. 1975. Lower Permian (Wolfcampian) sedimentation in the Oro Grande Basin, New Mexico, *26th Field Conference Guidebook*, New Mexico Geological Society, 109–117.

Jordan, C. F., Jr. and Wilson, J. L. 1971. The late Paleozoic section of the Franklin Mountains. *Field Conference Guidebook*, Society of Economic Paleontologists and Mineralogists, 77–86.

Kozur, H. 1975. Beitrage zur Conodontenfauna des Perm. *Geologisch-Palaontologische Mitteilungen Innsbruck*, **5**, 1–44.

Kozur, H. 1977. Beitrage zur Stratigraphie des Perms: teil II. Probleme der Abgrenzung und Gliederung des Perms. *Freidberger Forshungshefte*, **319**, 79–121.

Kozur, H. 1978. Beitrage zur Stratigraphie des Perms: teil II. Die Conodontenchronologie des Perms. *Freiberger Forschungshefte*, **334**, 85–161.

Kozur, H. and Mostler, H. 1976. Neue Conodonten aus dem Jungpalaozoikum und der Trias. *Geologisch-Palaontologische Mitteilungen Innsbruck*, **6**, 1–33.

Merrill, G. K. and von Bitter, P. H. 1976. Revision of conodont biofacies nomenclature and interpretations of environmental controls in Pennsylvanian rocks of eastern and central North America. *Life Sciences Contributions, Royal Ontario Museum*, **108**, 1–46.

Orchard, M. J. 1984. Early Permian conodonts from the Harper Ranch beds, Kamloops area, southern British Columbia. In *Current Research, part B, Geological Survey of Canada Paper*, **84–1B**, 207–215.

Rabe, E. H. 1977. Zur Stratigraphie des ostandinen Raumes von Kolombien. *Gliessener Geologishe Schriften*, 11, 1–223.

Rhodes, F. H. T. 1963. Conodonts from the topmost Tensleep Sandstone of the eastern Big Horn Mountains, Wyoming. *Journal of Paleontology*, **37**, 401–408.

Stevens, C. H. 1966. Paleoecologic implications of Early Permian fossil communities in eastern Nevada and western Utah. *Geological Society of America Bulletin*, **77**, 1121–1130.

Stevens, C. H. 1979. Lower Permian of the central Cordilleran Miogeosyncline. *Geological Society of America Bulletin, Part I*, **90**, 381–455.

Stevens, C. H., Wagner, D. B. and Sumsion, R. S. 1979. Permian fusulinid biostratigraphy, central Cordilleran Miogeosyncline. *Journal of Paleontology*, **53**, 29–36.

Williams, T. E. 1966. Permian Fusulinidae of the Franklin Mountains, New Mexico–Texas. *Journal of Paleontology*, **40**, 1142–1156.

Wind, F. H. 1973. Stratigraphic zonation and paleoecology of conodonts of the Chase Group, Lower Permian, Kansas. *Unpublished M.S. Thesis*, Florida State University, 1–223.

Yancey, T. E. 1975. Permian biotic provinces in North America. *Journal of Paleontology*, **49**, 758–766.

Taxonomic Index

General Index

Abereiddi Bay, 295, 306
absolute time scale, 286
absorption signals (ESR spectra), 230–237
acanthopyge Limestone, 333, 334
acetate peels, 149
acetic acid, 18–21, 23, 27, 35, 37, 39, 45, 46, 48, 49–51, 136, 205, 259
acetone, 27
acetylene tetrabromide (*see* tetrabromoethane)
acquisition of measurements, 169–172
Acre Limestone, 207
acritarch, 51
Acritarch Alteration Index (AAI), 189, 190
Adoyama Formation, 95
Aguathuna — lower Table Point time, 357
Aguathuna Formation, 356, 357
Aktastinian, 386, 396
Alacran Mountain Formation, 392, 393, 395, 401, 402
Alaska, 96, 174, 289, 292
Alberta, 96
albid (conodont structure), 190
alcohol, 27, 31, 259
algae, 51, 218, 357, 369, 372, 389
algal (clast) grainstone, 367, 389
Allegheny Frontal Zone, 189
allochthonous, 343, 344, 347, 372
allometry, 179, 181
allopatric speciation, 288
alluminium, 337
ammonium chloride, 31
ammonium hydroxide, 23
ammonium molybdate solution, 30
ammonium molybdophosphate, 30
Amorphognathus inaequalis Subzone, 294, 299–301, 303, 305, 308, 310
Amorphognathus kielcensis Subzone, 299–301
Amorphognathus lineage, 300
Amorphognathus tyaerensis Zone, 300, 302, 303, 305–309
amperage (settings), 26, 27, 67–69, 72
analysis of shape in conodonts, 168–186
anchizone, 209–215, 217, 224
ancient continental margins, 341, 342, 346, 359
Ancyrognathus triangularis Zone, 128

Andes, 317
Anglesey, 295
Anisian, 105, 106, 232
annual absolute time scale, 285
Anomalorthis Zone, 358
Antarctic, 257
anthracite, 217
antimony, 275, 279
Antler Belt, 385
Antler Formation, 97, 98, 100, 101, 106, 110, 112
apatite, 29, 55, 63, 66, 67, 95, 150, 231–234, 242, 243, 253, 256, 257
Appalachian (Basin), 188–190
Appalachian (s), 189, 192, 199, 247, 320, 342, 345
Appalachian orogen, 342
applications of quantitative morphology, 178–184
Aquathona Formation, 356
Arabia, 257
Arabic gum, 30
araldite, 148
Arctic Archipelago, 189, 342
Ardenne, 372, 373
Arenig, 136, 147, 232, 295, 342, 345, 346, 351, 353, 357, 358, 360
Argentina(e), 136, 138, 139, 350
Argentine Central Precordillera, 136–139
Arkansas, 120, 130
Arkansas Novaculite, 120, 130
Arrhenius diagram, 204
arsenic, 279, 275
Artinskian, 104
Asaphus Ash, 304
Aserian, 140
Ashgill, 295
Ashlock Formation, 244
Asselian, 322
Assymmetricus Zone, 128
Atlantic, 196–198
atomic number, 150
Australia, 141, 188, 189, 194, 195, 199, 249, 251, 292, 295, 350, 357
Austria, 292
Austrian Alps, 210
autochthonous, 343, 344, 347, 372, 393

surface waters, 278
Svalbard, 232
Sweden, 247
Sweetognathus behnkeni Zone, 323–324
Sweetognathus lineage, 395
Sweetognathus n. sp. A Subzone, 398, 399
Sweetognathus inornatus Subzone, 394
Sweetognathus whitei (Sub)zone, 324, 394
Sweetognathus whitei — *Sweetognathus* n. sp. A
    Subzone, 382, 390, 396–399
Sylvester Group, 97, 98, 105, 112, 114
systematic palaeontology (of conodonts), 139–141
systematics, 179

Tabberabberan Orogeny, 196
Table Cove Formation, 356
Table Head Formation, 140
Table Head Group, 343, 345, 354–359
Table Point Formation, 357, 358
Taconic Orogeny, 195
Tasmania, 188, 196, 200, 342
Tastubian, 322
taxonomic concepts, 288
taxonomic method, 289
taxonomic notes, 140–141, 296, 310
technical assistance, 36
technicians time, 37, 42
techniques for conodont concentration, 54–76
techniques for conodont extraction, 35–52
techniques of morphologic data analysis, 172–184
techniques of quantitative data acquisition and analysis,
    168
tectonic setting(s), 96, 257, 344
tectonic studies, 99
tectonic window(s), 192, 195, 199
temperature (of water), 241, 242, 245–254
temperature (across an intrusion), 204
temperature dependence of limestone dissolution, 39–42
Tennessee, 195
Tensleep Sandstone, 320, 401, 402
tentaculites, 333, 364
terbium, 275
Tesnus Formation, 80, 85–87, 122, 123, 129
tetrabromoethane (T.B.E.), 27, 28, 56, 62–67, 259
tetrachloroethylene, 29
terminal velocity, 63, 65
Texas, 77–92, 120, 121, 130, 256–265, 275–280, 282, 322,
    382, 383, 392–395, 397, 399, 401, 402
texture alteration, 212–215
texture(s), 213–215, 218, 220
thermal maturation, 189–192, 195, 196, 199
thermal maturity, 188
thin-section(s), 81, 82, 122, 334, 390
thioglycollic acid, 24
thorium, 275, 279
Threemile Limestone, 401, 402
time-temperature index (TTI), 193
Tiquina Formation, 318
'Titicaca Group', 318
toe of slope, 341, 344

Tournaisian, 98, 232, 371
tower apparatus, 37
tower method, 35, 37, 38
trace elements, 242, 253, 256, 257, 260–275, 278–282
trace fossils, 154, 289, 384
trachytera Zone, 366, 368–370
transgression, 358, 359, 363
transgressive (cycles), 372, 389
transmission electron microscopy (TEM), 31, 149, 162
Třebotov Limestone, 334
Tremadoc Series, 307, 353
Trempealeauan, 258
triangularis Zone, 372
Triassic, 18, 94–97, 99–108, 186, 191, 195, 198, 210–212,
    218, 220, 222, 230, 232, 234, 241–243, 245, 249–251,
    253, 387
Triassic conodont localities, 251
tribromomethane, 27
trichloroethane, 63
trilobites, 136, 275, 297, 301, 303, 307, 308, 342, 351, 353,
    357, 371
trilobites ranges, 299
trilobite zones, 299
trisodium phosphate, 22
Triticites cf. *nitens* Subzone, 320–323
'truss analysis', 171
Tsaybahe Group, 101, 105
turbidites, 77, 80, 94, 103, 345, 384, 386, 394
turbidity currents, 123, 126, 384
Turkey, 257
Tyler Formation, 161, 162, 164
type areas, 295, 307, 308
type section, 297

Uhaku, 299
ultra-structure, 31, 32, 146, 147, 149
ultraviolet radiation techniques, 162
unconformities, 76, 197, 198, 279, 399
uncrushed sample, 41
United States, 188, 200, 256, 257, 275, 277–280, 382, 392,
    393, 359, 396, 399
universal stage, 31
Upper Middle Limestone Group (Brigantian), 205
upper slope, 342, 350, 353, 355, 359
upper Table Point — lower Black Cove time, 358
Ural Mountains, 397, 398
uranium, 233, 234, 275
USSR, 141, 257, 322
Utah, 257, 258, 269–271, 275–279, 281, 357, 382–390, 392,
    394, 395, 396–400, 401, 402
U-Th-Ph. dating, 242

Vancouver, 96
Valley and Ridge Province, 189
varsol, 55
vertebrates, 154
video-disc, 172
video image, 171, 172